LABORATORY HANDBOOK OF
PETROGRAPHIC TECHNIQUES

Laboratory Handbook of Petrographic Techniques

CHARLES S. HUTCHISON

Associate Professor of Geology
University of Malaya, Kuala Lumpur, Malaysia

A WILEY-INTERSCIENCE PUBLICATION

JOHN WILEY & SONS, New York ● London ● Sydney ● Toronto

Library of Congress Cataloging in Publication Data:

Hutchison, Charles Strachan.
  Laboratory handbook of petrographic techniques.

  "A Wiley-Interscience publication."
  Bibliography: p.
  1. Petrology—Laboratory manuals. I. Title.

QE433.H87   552'.0028   73-17336

ISBN 0-471-42550-8

Printed in the United States of America

10 9 8 7 6 5 4 3 2 1

For ANN, HELENE, and TIMOTHY

"What sholde I tellen ech proporcioun
Of thinges whiche that we werche upon,
As on fyve or sixe ounces, may wel be,
Of silver or som other quantite,
And bisie me to telle yow the names
Of orpiment, brent bones, yren squames,
That into poudre grounden been ful smal?

Our fourneys eek of calcinacioun,
And of watres albificacioun,
Unslekked lym, chalk, and gleyre of an ey,
Poudres diverse, asshes, dong, pisse, and cley,"

Geoffrey Chaucer. *The Canterbury Tales:*

Chanouns Yemannes Tale.

PREFACE

The days when petrographic research consisted solely of a cursory hand specimen and a detailed thin-section description have come and gone. Although thin-section study under a polarizing microscope will and must continue to form the fundamental basis for igneous and metamorphic petrography, many additional petrographic techniques have come into everyday use. Many of them, however, are not conveniently available within the covers of a single book and it is the purpose of this volume to bring together in the form of a laboratory handbook many of the commonly used petrographic techniques that are generally applicable to igneous and metamorphic rocks. Of course, many of them will apply equally to sedimentary rocks, but it is the " hard rock " techniques with which this book is essentially concerned.

The techniques described or outlined herein should be performed only after the rocks have been thoroughly examined according to the conventional methods of field and thin-section study. This point cannot be too strongly emphasized. There is, for example, little value in knowing the structural state of the alkali feldspar in a granitic rock unless one first knows the field relationships of this rock and its texture and mineralogy as determined by thin-section analysis. The value of thin-section modal analysis by point-counting methods also cannot be overstressed in petrographic study.

The field relationships of any rock-suite must be the first step in petrographic research. Rocks form an integral part of the earth's crust and are not just hand specimens that can be found in a geological laboratory or museum. Geology must always continue to be essentially a field study and laboratory methods, such as those listed in this book, should be used only so that we can better understand the field relations of the rocks that have been studied in this way.

The methods of field geology are to be found, for example, in R. R. Compton, (1962), *Manual of Field Geology*, published by John Wiley and Sons, New York, 378 p. or in the well-known F. H. Lahee, (1961), *Field Geology*, 6th ed., published by McGraw-Hill Book Co., New York, 926 p.

The special problems encountered in mapping folded and metamorphic rocks are described in E. H. T. Whitten (1966), *Structural Geology of Folded Rocks*, published by Rand McNally and Co., Chicago, 663 p.

The specimens collected in the field are then sectioned for study under the petrographic microscope. The methods of study of crystals and an understanding of the optical properties of minerals as seen in grains and thin sections is well covered by E. E. Wahlstrom, (1969) in *Optical Crystallography*, 4th ed., published by John Wiley and Sons, New York, 489 p. This work has become a standard textbook for a course on the methods of the thin-section study of crystalline materials.

The next stage is the systematic identification of minerals in the rocks we have collected. Two well used books which systematically list the optical properties of all the common rock-forming minerals in thin section and in grains are P. F. Kerr (1959),

Optical Mineralogy, 3rd ed., published by McGraw-Hill Book Co., New York, 442 p., and E. Wm. Heinrich, (1965), *Microscopic Identification of Minerals*, published by McGraw-Hill Book Co., New York, 414 p.

These books are adequate for all undergraduate work but may be supplemented by research workers for additional details of certain mineral groups and for rarer minerals not included in the above texts, by W. E. Tröger, H. U. Bambauer, F. Taborszky, and H. D. Trochim, *Optische Bestimmung der gesteinsbildenden Minerale*, Vol. 1 (1971) *Bestimmungstabellen*, 4th ed., 188 p.; Vol. 2 (1969) *Textband*, 2nd ed., 882 p., both published by E. Schweizerbart'sche Verlagsbuchhandlung (Nägele u. Obermiller), Stuttgart, Germany, and both presently being translated into English, A. N. Winchell, and H. Winchell, (1951), *Elements of Optical Mineralogy*, Part II, 4th ed., published by John Wiley and Sons, New York, 551 p., and by W. A. Deer, R. A. Howie, and J. Zussman (1966), *An Introduction to the Rock-forming Minerals*, published by John Wiley and Sons, New York and by Longmans, Green and Co., London, 528 p., or by the 5 volume set by the same three authors: *Rock-forming Minerals:* Volume 1 (Ortho- and ring silicates, 333 p.; Volume 2 (Chain silicates), 379 p.; Volume 3 (Sheet silicates), 270 p.; Volume 4 (Framework silicates), 435 p.; and Volume 5 (Non-silicates), 371 p. All published by Longmans, Green and Co. in London or by John Wiley and Sons in New York, they provide a comprehensive reference to all the commonly occurring rock-forming minerals.

Following on a description of the minerals comes a petrographic description of the rock itself. Some standard books on petrography are H. Williams, F. J. Turner, and C. M. Gilbert, (1954), *Petrography*, published by W. H. Freeman and Co., San Francisco, 406 p., and W. W. Moorehouse (1959), *The Study of Rocks in Thin-Section*, published by Harper and Row, New York, 514 p.

Useful books on metamorphic petrography are A. Spry (1969), *Metamorphic Textures*, published by Pergamon Press, Oxford, London, and New York, 350 p., and A. Harker (1950), *Metamorphism*, 3rd ed., published by Methuen and Co., London, or by E. P. Dutton and Co., New York, 362 p.

Accordingly, it can be seen that the methods of thin-section study of minerals and rocks are more than adequately covered by a large number of books, the most outstanding of which have been listed here. However, there exists a large additional list covering identical ground to those already mentioned, many of which are less distinguished and accordingly less widely used.

No really comprehensive account of petrographic methods applied to the crystalline rocks has been published since the now classic A. Holmes (1921), *Petrographic Methods and Calculations*, published by Thomas Murby and Co., London, 515 p., and the recently reprinted A. Johannsen (1918), *Manual of Petrographic Methods*, published by McGraw-Hill Book Co., New York, 649 p.

A book that brings together the petrographic methods applied to sedimentary rock study is now available in English translation: G. Müller (translated from the German by H-U Schmincke (1967), *Methods in Sedimentary Petrology*, published by Hafner Publishing Co., New York, 283 p. However, H. B. Milner, A. M. Ward, and F. Higham, Eds., (1962), *Sedimentary Petrography*, Vol. 1, *Methods in Sedimentary Petrography* (643 p.), and Vol. 2, *Principles and Applications* (715 p.), 4th ed., published by George Allen & Unwin Ltd., London, still continues to be the foremost treatise on sedimentary petrography.

It is hoped to present within these covers, a similar handbook of the methods applied to crystalline rocks.

This book stresses the practical aspects of the petrographic methods—theoretical discussions are kept to a minimum—and, it is hoped, should prove useful for an undergraduate course on petrographic methods as well as to research workers in general.

The trend in the study of crystalline rocks is toward a more extensive use of the electron microprobe with or without a scanning electron miscroscope combination. The use of this instrument is excluded from this book for several reasons. First, several monographs on the instrument and methods are available. Second, inclusion in this book would involve an expansion beyond a reasonable size; and third, most geology departments throughout the world do not now possess and are not likely in the near future to acquire an electron microprobe. On the other hand, X-ray diffraction and X-ray fluorescence spectrometry are generally available and these techniques are included.

For those who wish to graduate to the more sophisticated electron microprobe, the following references will be useful: T. D. McKinley, K. F. J. Heinrich, and D. B. Wittry (1966), *The Electron Microprobe*, published by John Wiley and Sons, New York; J. V. P. Long (1967), *Electron Probe Microanalysis*, Chapter 5 (pp. 215-260) in *Physical Methods in Determinative Mineralogy*, J. Zussman, Ed., published by Academic Press, London, 514 p.; and T. R. Sweatman, and J. V. P. Long, (1967) *Quantitative electron-probe microanalysis of rock-forming minerals*, pp. 332-379 in *Journal of Petrology*, Vol. 10, No. 2, June 1969, published by the Clarendon Press, Oxford, England.

In a way there will be a part overlap between this book and *Physical Methods in Determinative Mineralogy*, edited by J. Zussman, 1967, published by Academic Press, London, 514 p. Although several papers in that book are practical, others unfortunately give descriptions of determinative techniques in nothing but very general terms and hence are of little value to the practical petrologist.

It is hoped that the present handbook will be a down-to-earth guide to the practical petrologist and will also provide a good framework for a laboratory course in petrographic methods for hard-rock geology students.

C. S. HUTCHISON

Kuala Lumpur, Malaysia
May 1973

ACKNOWLEDGMENTS

I am grateful to the following companies for their help, either in the form of information or photographs for inclusion in this book: Carl Zeiss, Postfach 35/36, 7082 Oberkochen, Germany; Columbia Scientific Industries, 3625 Bluestein Boulevard, P.O. Box 6190, Austin, Texas, 78762; Ernst Leitz GmbH., D.633 Wetzlar, Postfach 210/211, Germany; James Swift & Son Ltd., Joule Road, Houndmills Industrial Estate, Basingstoke, Hampshire, England; Joint Committee on Powder Diffraction Standards, 1601 Park Lane, Swarthmore, Pennsylvania, 19081; Microtec Development Laboratory, P. O. Box 1441, Grand Junction, Colorado, 81501; Varian Techtron Pty. Ltd., 679-687 Springvale Road, P. O. Box 222, North Springvale, Victoria, 3171 Australia; and Vickers Limited, Vickers Instruments, Haxby Road, York, YO3 7SD England.

My thanks are due to the following friends and colleagues for supplying information and advice which I have found so useful: Mr. R. F. Allbrook, Dr. B. P. Fabbi, Dr. F. E. H. Haser, Dr. P. R. Hooper, Dr. K. R. Chakraborthy, Dr. S. H. Chan, Dr. K. Norrish, Dr. G. R. Parslow, Mr. S. Sandrasagaram, Dr. H. P. Stauffer, and Dr. D. J. Swaine.

I especially wish to thank my dear friend Professor K. F. G. Hosking for his continual help and support in the form of suggestions and encouragement while I was engaged in writing the manuscript.

I must express my gratitude to Miss Anne Chong for typing the typescript, to Miss A. M. Chua for her typing assistance, to Mr. S. Sriniwass for drafting the figures, to Mr. Jaafar bin Haji Abdullah for photographic assistance, and to Tengku Ismail bin Tengku Mohammed for technical assistance.

Finally I wish to thank the editors and staff of John Wiley and Sons for their help in processing the book.

C.S.H.

CONTENTS

Chapter 1   Thin-Section Preparation   1

EXERCISE 1.1   Cutting the Hand Specimen, 1
EXERCISE 1.2   To Impregnate a Porous or Friable Rock, 2

Method A (for most friable rocks), 2
Method B (for extremely friable rocks), 2

EXERCISE 1.3   Mounting the Specimen, 3

Method A, 3
Method B, 4
Method C, 4

EXERCISE 1.4   Sawing and Grinding, 5
EXERCISE 1.5   Conventional Covering, 9
EXERCISE 1.6   Polished Thin Sections, 9

Method A, 10
Method B, 11
Method C, 12
Method D, 12

EXERCISE 1.7   Polished Thick Sections, 13
EXERCISE 1.8   Grain Thin Sections, 14
EXERCISE 1.9   Removal of Thin Section, 14

Chapter 2   Aids in Thin-Section Study   15

A. Staining Techniques, 15

FELDSPARS, 16

EXERCISE 2.1   Staining Rock Slabs and Thin Sections for K-Feldspar,
Plagioclase, and Quartz, 16

Method for Rock Slabs, 17
Method for Uncovered Thin Sections, 18

EXERCISE 2.2   Staining Rock Slabs for K-Feldspar, Plagioclase, and Quartz, 18

Method for Rock Slabs, 18

EXERCISE 2.3   Staining Rock Slabs and Thin Sections for K-Feldspar, Plagioclase, and Cordierite, 19

Method for Rock Slabs, 19
Method for Thin Sections, 19

EXERCISE 2.4   Staining for Anorthoclase, 20

CORDIERITE, 20

EXERCISE 2.5   For Distinguish K-Feldspar, Plagioclase, and Cordierite, 20
EXERCISE 2.6   For Distinguish K-Feldspar, Plagioclase, and Cordierite, 22

MICAS, 23

EXERCISE 2.7   Thin-Section Staining for Paragonite, 23

FELDSPATHOIDS, 24

CARBONATES, 24

EXERCISE 2.8   Thin-Section Carbonate Staining, 25
EXERCISE 2.9   Staining Thin Sections for Dolomite and Brucite, 25
EXERCISE 2.10   Staining Thin Sections for Dolomite, 26
EXERCISE 2.11   A Complete Staining Scheme for Carbonates, 26

OPAQUE MINERALS, 30

B.   Use of Ultraviolet Light, 30
C.   Opaque Mineral Identification, 32

EXERCISE 2.12   Determination of Reflectivity, 33
EXERCISE 2.13   Choice of Microindentation Hardness Method, 35
EXERCISE 2.14   Determination of Indentation Hardness, 37

SYSTEMATIC SCHEME OF OPAQUE MINERAL IDENTIFICATION, 40

Chapter 3   Grain Size, Modal Analysis, and Photomicrography   **44**

A.   Grain-Size Determination, 44

EXERCISE 3.1   To Calibrate the Eyepiece Micrometer, 44
EXERCISE 3.2   To Calibrate the Field of View, 45
EXERCISE 3.3   Grain Size of Rocks, 46

B.   Modal Analysis, 47

EXERCISE 3.4   Selection of Area for Modal Analysis, 48

VISUAL ESTIMATION OF AREAL MODE, 51

EXERCISE 3.5   Principle of Modal Analysis by Point Counting, 51
EXERCISE 3.6   Modal Analysis of Rock Slabs, 53

Method A (for porphyritic rocks), 53
Method B, 54
Method C, 55
Method D, 56

EXERCISE 3.7   Modal Analysis of Thin Sections, 56
EXERCISE 3.8   Modal Method for Porphyritic Rocks, 59
EXERCISE 3.9   Modal Analysis by X-ray Diffraction, 61
EXERCISE 3.10   Modal Analysis of Granite by X-ray Diffraction, 64
EXERCISE 3.11   Modal Analysis by Mineral Separation, 65

C.   Photomicrography, 65

EXERCISE 3.12   Choice of Objective and Eyepiece Combination, 66
EXERCISE 3.13   Defining the Scale of a Photomicrograph, 68
EXERCISE 3.14   The Simplest Photomicrography System, 69
EXERCISE 3.15   Photomicrography with a Photomicrographic Camera, 69
EXERCISE 3.16   Low-Power Photomicrography, 71
EXERCISE 3.17   Photography of Irradiated Specimens, 72

Chapter 4   Rotation Methods for the Polarizing Microscope   **73**

SOME PRELIMINARY EXERCISES, 73

EXERCISE 4.1   Calibration of the Micrometer Eyepiece for the Measurement of *BXa* Interference Figures, 74

Method A, 74
Method B, 75

EXERCISE 4.2   To Derive $2V$ from $N\alpha$, N_β, and $N\gamma$, 76

THE SPINDLE STAGE, 78

EXERCISE 4.3   Constructing and Setting up a Spindle Stage, 78

Design by Wilcox, 78
Design by Jones, 78
Design by Roy, 79
The Hartshorne Spindle Stage, 79

PROCEDURE FOR ATTACHING THE CRYSTAL, 80

EXERCISE 4.4   Measurement of $N\omega$ and $N\varepsilon$ or $N\alpha$, N_β, and $N\gamma$ on a Spindle Stage, 80
EXERCISE 4.5 $2V$ Determination on the Spindle Stage, 82

Method A, 82
   When the Two Optic Axes Are Accessible, 82
   When Only One Optic Axis Is Accessible, 85
Method B, 86
Method C, 88
Method D, 89

UNIVERSAL STAGE, 90
NOMENCLATURE OF AXES OF ROTATION, 90
SPECIAL EQUIPMENT FOR CONOSCOPIC LIGHT, 91

EXERCISE 4.6   Setting Up the Universal Stage, 91
EXERCISE 4.7   Universal Stage Conoscopic Procedure, 93

METHOD FOR UNIAXIAL MINERALS, 93

EXERCISE 4.8   Conoscopic Method for Biaxial Minerals, 98
EXERCISE 4.9   Plagioclase Determination on a Five-Axis Stage, 104

Orientation of the Indicatrix, 104
Measurement of X to ⊥ (010), 104
Measurement of 2V, 104
Measurement of X to ⊥ (010) and Y to ⊥ (010), 105
Determination of Composition and Structural State, 105
Resolving Ambiguities, 108

EXERCISE 4.10   Structural State of Alkali Feldspar, 110
EXERCISE 4.11   2V in Topaz, 111
EXERCISE 4.12   Applications to Petrofabrics, 111

Chapter 5   Mineral Separation   **113**

EXERCISE 5.1   Hand Extraction, 113
EXERCISE 5.2   Crushing the Whole Rock Specimen, 114
EXERCISE 5.3   Cleaning, 115
EXERCISE 5.4   Magnetic Isodynamic Separation, 116

DENSITY SEPARATION, 120

EXERCISE 5.5   Choice of Specific Gravity Liquid and Diluent, 120
EXERCISE 5.6   Recovery of Heavy Liquids, 121
EXERCISE 5.7   Density Separation of Two Minerals Coarser than 200
Mesh, 123
EXERCISE 5.8   Density Separation for Fine-Grained Rocks, 123
EXERCISE 5.9   Mineral Crushing for Chemical Analysis, 126
EXERCISE 5.10   Rock Crushing for Chemical Analysis, 126
EXERCISE 5.11   Sample Splitting, 128
EXERCISE 5.12   Mineral Separation by Flotation, 129

Chapter 6   Powder Methods of X-ray Diffraction   **132**

A.   Camera Techniques, 132

EXERCISE 6.1   Sample Preparation for the Powder Camera, 132
EXERCISE 6.2   Preliminary Alignment of the Camera, 134
EXERCISE 6.3   Loading the Camera, 137
EXERCISE 6.4   Choice of Collimators, 139
EXERCISE 6.5   Exposure Time, 139
EXERCISE 6.6   Choice of X-ray Tube and Its Operating Conditions, 139
EXERCISE 6.7   Development of the Exposed Film, 141
EXERCISE 6.8   Measurement of the Film, 141

B.   Diffractometer Techniques, 144

EXERCISE 6.9   Diffractometer Specimen Preparation, 144

Method A, 144
Method B, 145

EXERCISE 6.10   Specimen Preparation for Avoiding Preferred
Orientation, 145

Method A, 146
Method B, 146
Method C, 147

EXERCISE 6.11   Specimen Preparation for Clay Minerals, 147

Powder Press Technique, 147
Smear on Glass Slide, 147
Suction on Ceramic Tile, 147
Centrifugation onto Ceramic Tile, 147

EXERCISE 6.12   Preliminary Alignment of the X-ray Diffractometer, 147
EXERCISE 6.13   Angular Calibration of the Goniometer, 154
EXERCISE 6.14   Adjusting the 2 : 1 Goniometer Setting, 155
EXERCISE 6.15   Choice of Divergence, Receiving, and Antiscatter
Slits and Centering the Primary Beam, 156

Divergence Slit, 156
Receiving Slit, 156
Antiscatter Slit, 157
Centering the Primary Beam, 157

EXERCISE 6.16   General Procedure for Running an X-ray
Diffractogram, 158
EXERCISE 6.17   Determination of Ideal Operating Conditions for the
X-ray Detector, 161
EXERCISE 6.18   Determination of Pulse Height Discriminator
Settings, 163
EXERCISE 6.19   Use of a Standard for Obtaining High 2θ Accuracy, 166
EXERCISE 6.20   Identification of an Unknown by the J.C.P.D.S.
Powder Data File, 168
EXERCISE 6.21   Indexing the Powder Diffraction Lines, 173
EXERCISE 6.22   Deductions Regarding the Lattice Type, 178
EXERCISE 6.23   Special Powder Diffraction Techniques, 179

Chapter 7   Application of X-ray Powder Data to Specific Mineral Groups   180

A.   Feldspars, 180

EXERCISE 7.1   Determination of Intensity Ratio, 181
EXERCISE 7.2   Determination of Bulk Composition of Alkali
Feldspar, 182
EXERCISE 7.3   Determination of Composition of Sanidine from
Volcanic Rocks, 186
EXERCISE 7.4   Determination of Temperature-Structural State of
Alkali Feldspar, 186

EXERCISE 7.5   Determination of the Triclinicity of Alkali
Feldspar, 188
EXERCISE 7.6   Determination of the Structural State of Alkali
Feldspar, 189
EXERCISE 7.7   Determination of Plagioclase Structural State, 197
EXERCISE 7.8   Determination of Plagioclase Composition, 204

B.   Garnet Group, 207

DETERMINATION OF GARNET COMPOSITION, 207

EXERCISE 7.9   Measurement of the Unit Cell Edge *a* in Å
of a Garnet, 207
EXERCISE 7.10   Identification of the Garnet, 209

C.   Carbonates, 214

EXERCISE 7.11   Determination of Percentage of Dolomite in
a Carbonate Rock, 214
EXERCISE 7.12   Determination of the Amount of Ca Replaced
by Mg in Calcite, 216
EXERCISE 7.13   Determination of Carbonate Minerals and
Quartz in Rocks, 217

D. Olivine, 219

EXERCISE 7.14   Determination of Composition of Common Olivine, 219

E. Cordierite, 221

EXERCISE 7.15   Determination of the Structural State of Cordierite, 221

F. Pyroxene, 224

EXERCISE 7.16   Determination of the Pyroxene Composition
in the Range Enstatite to Hypersthene, 224

Method A, 224
Method B, 225

G.   Clay Minerals, 225

EXERCISE 7.17   X-ray Diffraction Preparation of Clays, 226

Oriented Mounts, 226
Unoriented Mounts, 226

EXERCISE 7.18   Routine X-ray Diffractometer Procedure, 227

H.   Summary of Applications to Other Minerals, 231

Aluminum silicates, 231
Amphiboles, 231
Analcite, 231
Apatite, 232
Aragonite, 232
Arsenopyrite, 232
Carbonates, 232

Chlorite, 232
Feldspars (Plagioclase), 232
Ilmenite, 232
Loellingite, 232
Micas, 232
Nepheline, 233
Pyroxene (Clinopyroxene), 233
Pyrite, 233
Pyrrhotite, 233
Scapolite, 233
Serpentine, 233
Sphalerite, 233
Spinel, 233
Topaz, 234
Tourmaline, 234
Vesuvianite (Idocrase), 234

Chapter 8   Specific Gravity Determination   **235**

EXERCISE 8.1   Density Determination by the Archimedes' Principle, 236

Method A, 236
Method B, 239
Method C, 241

EXERCISE 8.2   Density Determination with a Pycnometer, 241
EXERCISE 8.3   Density Determination with Heavy Liquids, 243
EXERCISE 8.4   Density Determination with a Pycnometer and
Heavy Liquid, 244

Chapter 9   Refractive Index Determination   **247**

EXERCISE 9.1   Routine Method for Refractive Index Determination, 247
EXERCISE 9.2   Special Method of Orientating Grains
for Refractive Index Measurements, 250

Method A, 250
Method B (*for uniaxial minerals only*), 250
Method C, 253

EXERCISE 9.3   Matching Refractive Index of Solid and
Immersion Oil, 253
Relief, 253

Method A   Becke Line, 254
Method B   Oblique Illumination, 256
Method C   Use of Phase Contrast Equipment, 256
Method D   Use of Wavelength Dispersion, 259

EXERCISE 9.4   To Check the Refractive Index of an Immersion
Liquid, 261

Method A, 261
Method B, 262

Method C, 263
Method D, 263

Chapter 10   X-ray Fluorescence Spectrometry   **264**

A.   Qualitative Analysis, 264

EXERCISE 10.1   To Make Pressed-Powder Self-Supporting Specimen
Discs of 30.4-mm Diameter, 265

Method A, 265
Method B, 268

EXERCISE 10.2   Alignment of the Spectrometer, 269
EXERCISE 10.3   Adjustment of the Crystal Holders, 271
EXERCISE 10.4   Selection of Optimum X-ray Tube and its
Operating Conditions, 274

For Light Element Detection, 274
For Heavy Element Detection, 274

Relationship Between Kilovolts and the Production of $K\alpha$ Radiation, 275

K Spectrum, 275
L Spectrum, 276
M Spectrum, 277
Fluorescent Yield, 280

EXERCISE 10.5   Angular 2θ Limits of the Spectrometer, 280

Minimum 2θ Angular Limit, 280
Maximum 2θ Angular Limit, 280

EXERCISE 10.6   Choice of Dispersing Crystal, 281

Basic Combination of Three Crystals, 281
Additional Crystals for Specific Elements, 282

EXERCISE 10.7   Routine Procedure for X-ray Spectroscopy
Scanning, 282

B.   Quantitative Major Element Analysis, 286

SPECIMEN PREPARATION, 286

EXERCISE 10.8   Specimen Preparation, 290

Method A (suitable for Mg to Fe analysis), 290
Method B (suitable for Mg to Fe analysis), 293
Method C (suitable for Na to Fe analysis), 294
Method D (suitable for Na to Fe analysis), 295

EXERCISE 10.9   Determination of Optimum Detector Voltage for All
Elements Excluding Sodium, 296
EXERCISE 10.10   Determination of Pulse Height Analyzer Settings
for All Elements Excluding Sodium, 299
EXERCISE 10.11   Setting of Counter HV and Pulse Height
Analyzer for Sodium, 302

EXERCISE 10.12   Routine Instrumental Settings for Major
Element Analysis, 303
EXERCISE 10.13   Background Determination, 308

Method A, 308
Method B, 308

EXERCISE 10.14   Avoidance of High Count Rates and Dead-Time
Determination, 310

DETERMINATION OF X-RAY DETECTOR DEAD TIME, 310

EXERCISE 10.15   Matrix Corrections Only for Specimens Made
by Method A of Exercise 10.8, 313
EXERCISE 10.16   Direct Measurement of Mass Absorption
Coefficient, 317

Specimen Preparation, 317
Measurement, 319
Calculation, 319

EXERCISE 10.17   Composition Determination in Mineral Isomorphous
Series by Mass Absorption Coefficient Determination, 320

Plagioclase, 321
Orthopyroxene, 321
Sphalerite, 321

C.   Quantitative Trace Element Analysis, 321

EXERCISE 10.18   Trace and Minor Heavy Element Analysis with
Mass Absorption Coefficient and Fluorescent Radiation Intensity, 321

Method, 321
Calculation, 322

EXERCISE 10.19   Determination of Rubidium and Strontium in a
Rock, 323
EXERCISE 10.20   Trace Element Analysis by Matrix Calibration
Against Al_2O_3, 324

Method, 324
Calculation, 326
Determination of Relative Absorption R, 326

EXERCISE 10.21   Trace Element Analysis Without Matrix
Correction, 329

D.   Some Special Techniques, 330

SPECIAL DILUTION METHODS, 330
ANALYSIS OF VANADIUM IN THE PRESENCE OF TITANIUM, 330
THE MACROPROBE ATTACHMENT FOR THE X-RAY SPECTROMETER, 331

EXERCISE 10.22   Determination of Chlorine in Silicate Rocks and
Minerals, 331

EXERCISE 10.23   Determination of Phosphorus in Silicate Rocks and Minerals, 332

Chapter 11   Atomic Absorption Spectrophotometry   **333**

SPECIMEN PREPARATION, 333

A.   Silicate Rocks and Minerals, 333

EXERCISE 11.1   The HF-H$_2$SO$_4$ Method, 333
EXERCISE 11.2   The Na$_2$CO$_3$ Fusion Method, 334
EXERCISE 11.3   Fast and Complete Decomposition for Rocks, Refractory Silicates, and Minerals, 334
EXERCISE 11.4   Decomposition with a Teflon-Lined Bomb, 335

Method A, 337
Method B, 337

EXERCISE 11.5   The NaOH Fusion Method for Silica and Alumina Determination, 338
EXERCISE 11.6   Dissolution of Limestone, 338
EXERCISE 11.7   The Basic Instrument and Adjustments, 339

Preliminary Adjustments, 340
Flame Types and Lighting Instructions, 342

Air-Acetylene Gas Mixture, 342
Nitrous Oxide-Acetylene Gas Mixture, 343
Nitrous Oxide-Propane Flame, 344
Air-Propane and Air-Coal Gas Flame, 344
Air-Hydrogen Flame, 344
Nitrogen/Entrained Air-Hydrogen Flame, 344
Argon/Entrained Air-Hydrogen Flame, 345

EXERCISE 11.8   Standard Operating Procedure, 345

Fitting Lamps, 346
Adjusting the Photomultiplier, 346
Preliminary Settings, 346
Atomic Absorption Mode, 346
Flame Emission Mode, 347

EXERCISE 11.9   Making Standard Solutions, 348

Major Elements, 348
Minor Elements, 348

EXERCISE 11.10   Analytical Conditions, 349
EXERCISE 11.11   Removing Layer Silicates from Quartz and Feldspar, 352

Chapter 12   Determination of Chemical Components Not Attainable by the Foregoing Methods   **354**

A.   FeO and Fe_2O_3 Determination, 354

   EXERCISE 12.1   Determination of FeO in a Rock or Mineral, 354

      Method A, 354
      Method B, 356
      Method C, 358

STANDARDIZATION OF 0.01 N FERROUS AMMONIUM SULFATE, 359
FeO DETERMINATION, 359

      Method D.   Ferrous: Ferric Ratios by an Electron Microprobe, 361

B.   Total Water, Carbon Dioxide, and H_2O- Determination, 361

LOSS ON IGNITION AT 1000°C, 361

   EXERCISE 12.2   Simultaneous Determination of Water and Carbon Dioxide, 361
   EXERCISE 12.3   Determination of Total Water, 365
   EXERCISE 12.4   Determination of H_2O-, 367
   EXERCISE 12.5   Water Determination by Infra-Red Spectroscopy, 368

C.   Fluorine, 369

   EXERCISE 12.6   Fluorine Determination in Rocks and Minerals, 369

      Method A, 369
      Method B, 372

   Total Chlorine in Silicates, 373

Chapter 13   Display of Data   **374**

A.   Graphical, 374

   EXERCISE 13.1   Triangular Variation Diagrams: General Description, 374
   EXERCISE 13.2   Graphical Representation of the Metamorphic Paragenesis of Carbonate Rocks, 376
   EXERCISE 13.3   The ACF Diagram for Representing Metamorphic Paragenesis, 378

      General Description, 379
      Method, 379
      Plotting of Minerals on the ACF Diagram, 382
      Use of the ACF Diagram, 382

   EXERCISE 13.4   The A'KF Diagram for Representing Metamorphic Paragenesis, 385

General Description, 385
Method, 385
Plotting Minerals on the A'KF Diagram, 388

EXERCISE 13.5   The Use of ACF and A'KF Diagrams, 390
EXERCISE 13.6   The AFM Diagram for Representing Metamorphic
Paragenesis, 391
EXERCISE 13.7   Classification of Basaltic Rocks, 393
EXERCISE 13.8   Alkali Versus SiO_2 Diagram for Volcanic Series, 394
EXERCISE 13.9   The Solidification Index for Volcanic Series, 395
EXERCISE 13.10   The Crystallization Index, 396
EXERCISE 13.11   The Differentiation Index, 398
EXERCISE 13.12   Other Types of Variation Diagram, 399
EXERCISE 13.13   Calculation of Pyroxene Analyses in Terms of
End Members, 401

Method A, 401
Method B, 405

EXERCISE 13.14   Relationship of Pyroxene Composition with
Metamorphic Grade, 406
EXERCISE 13.15   Classification and Display of Amphibole
Structural Formulas, 406
EXERCISE 13.16   Relation Between Garnet Chemistry and
Metamorphic Grade, 411
EXERCISE 13.17   Weathering Index of a Silicate Rock, 414
EXERCISE 13.18   The Alkalinity Ratio for Igneous Rocks, 414

B.   Norm Calculations, 414

EXERCISE 13.19   Rules for Calculating the C.I.P.W. Norm, 414
EXERCISE 13.20   Rules for Calculating the Niggli Molecular
Norm (Catanorm), 419
EXERCISE 13.21   The Mesonorm for Metamorphic and Granitic
Rocks, 423

C.   Statistical Analysis of Data, 425

EXERCISE 13.22   Frequency Distributions, 426

Histograms, 426
Populations and Samples, 427
Mean, Variance, Standard Deviation, and Coefficient of Variation, 428
Interval Estimation, 430

THE CONFIDENCE INTERVAL FOR THE DIFFERENCE BETWEEN TWO
POPULATION MEANS, 434

EXERCISE 13.23   Simple Linear Regression, 435

Chapter 14   Thermal Analysis Techniques   **438**

A.   Differential Thermal Analysis, 438

EXERCISE 14.1   Choice of Sample Holders and Thermocouples, 439
EXERCISE 14.2   Choice of Heating Furnace, 442
EXERCISE 14.3   Sample Pretreatment, 442
EXERCISE 14.4   Loading of Type SH-11 BR Sample Holder, 443
EXERCISE 14.5   Loading the SH-11 BP Sample Holder, 444
EXERCISE 14.6   Loading the Sample Holder SH-8 BE, 444
EXERCISE 14.7   Selection of Heating Rate, 444
EXERCISE 14.8   Temperature Calibration, 445
EXERCISE 14.9   Operating Procedure, 445
EXERCISE 14.10   Reporting the Conditions of an Experiment, 447
EXERCISE 14.11   Identification of the Mineralogy of a Specimen
by the DTA Thermogram, 448

DTA Thermograms of Mineral Groups   **452**

Silica Minerals, 452
Feldspars, 452
Feldspathoids, 453
Amphiboles, 453
Epidote Group, 453
Tourmaline, 453
Zircon, 453
Zeolites, 453
Carbonates, 453
Ore Minerals, 454
Oxides and Hydroxides, 455
Clay Minerals, Micas, and Chlorites, 456
Carbon and Graphite, 457

EXERCISE 14.12   Qualitative and Quantitative Estimation, 457

B.   Thermogravimetric Analysis, 458

EXERCISE 14.13   Routine Operation of the TGA Apparatus, 460
EXERCISE 14.14   Results and Interpretation of TG Curves, 464

C.   Thermoluminescence, 465

EXERCISE 14.15   Apparatus and Procedure for Glow Curve
Plotting, 466
EXERCISE 14.16   The Artificial Glow Curve, 470
EXERCISE 14.17   Thermoluminescence Calibration Curve, 471
EXERCISE 14.18   Other Applications of Thermoluminescence, 472

References   **473**

Appendix   Selected List of Suppliers of Equipment and Materials   **485**

Index   **491**

LABORATORY HANDBOOK OF
PETROGRAPHIC TECHNIQUES

THIN-SECTION PREPARATION 1

The standard methods of rock and mineral thin-section preparation are so widely used that they need no description here. Adequate accounts may be found in Johannsen (1918) p. 572 to 604, Holmes (1921), p. 231 to 249, Winchell (1937), p. 71 to 74, and Kerr (1959), p. 3 to 9, but the most thorough and up-to-date treatment of the subject appears in Hartshorne and Stuart, (1970), p. 219 to 251. Only a few new techniques which are considered to improve thin-section quality and especially to make the sections generally applicable for transmitted and reflected light study, involving final polishing rather than covering, are discussed in this chapter.

Porous and friable rocks present difficulties in the making of good-quality thin sections. Methods of successfully overcoming them are reported first. The old way of boiling in Canada balsam should, by now, be in general disuse.

EXERCISE 1.1

Cutting the Hand Specimen

The rock or mineral specimen is first sawed into a small wafer in the required orientation. A large variety of saws in a considerable range of quality and price is available on the market. The choice of saw naturally would have to be dictated by the money to be spent. Only a few of a large number of suppliers are given here:

Ward's Natural Science Establishment Inc., P.O. Box 1712, Rochester, New York, 14603.

Colorado Geological Industries Inc., 1244 East Colfax Avenue, Denver, Colorado, 80218.

Kyoto Scientific Specimens Co. Ltd., 378, Ichinofunairicho, Kawaramachi-Nijo Minami, Nakagyoku, Kyoto, Japan.

These three suppliers market a comprehensive list of geological materials. The following are sources of higher quality geological preparation equipment, including saws, and their products are highly recommended:

Buehler Ltd. and Adolph I. Buehler Inc., 2120 Greenwood Street, Evanston, Illinois, 60204.

Buehler-Met AG., Postfach 4000, Basel 23, Switzerland.

Cutrock Engineering Co. Ltd., 35 Ballards Lane, London N.3, England.

1. If the rock is firm and coherent, saw it into a wafer approximately 3 mm (1/8 in.) thick and trim it to a square about 24 × 24 mm; a larger specimen, however, may be left as a rectangle 24 × 34 mm. These measurements are given only as a guide but they represent useful maximum and optimum sizes. The width of 24 mm as an upper

1

limit is critical. Otherwise the rock slab will overlap the width of the glass slide on which it will be mounted and this is not permissible.

2. If the rock is friable, the wafer need not be sawed to a thinness of 3 mm at this stage and can be left considerably thicker, even up to 10 mm.

3. Although a large-diameter diamond blade may be used initially for sawing large handspecimens into smaller slabs, the final trimming is best done with a 127 × 0.38 mm (5 × 0.015 in.) diamond saw blade because the cut is considerably smoother than that of a 127 × 1.4 mm (5 × 0.055 in.) blade. The thinner blade also conserves material, hence is preferable if several thin sections are to be made from a large slab.

EXERCISE 1.2

To Impregnate a Porous or Friable Rock

Two methods are proposed. The first is suitable for all but the most friable rocks. The second should be reserved for weathered, decomposed, or very fragile rocks that otherwise would be impossible to section.

METHOD A   (*for most friable rocks*)

1. Dry the specimen for 30 minutes at 95°C.

2. In any suitable container, such as a small beaker or watch glass, mix equal portions of Araldite AY 105 epoxy and hardener 935F (supplied by Chemical and Engineering Co. Inc., 221 Brook Street, Media, Pennsylvania, 19063). One volume of freshly mixed epoxy and hardener is immediately dissolved in 3 volumes of toluene, stirred occasionally, and is then ready for use. The container may be lined with aluminum foil, which can be discarded later. Without the foil lining the epoxy, once cured, will be permanently stuck to the container.

3. Immerse the dry specimen in the diluted epoxy. Cover the epoxy container with a bell jar and evacuate with a vacuum pump until bubbling in the specimen ceases. Break and restore the vacuum several times. This helps to expel all trapped air from the specimen and to fill all the pore spaces with epoxy.

4. Curing time is about 1 hour at 100°C and 15 minutes at 150°C. Remove the specimen and let it cure in an oven or let it cool to room temperature and cure overnight. Once hard it is ready for further grinding and mounting and may be treated like any firm and coherent rock specimen.

(Method after Moreland, 1968.)

DISCUSSION

This epoxy medium gives excellent results on very friable materials such as carbonaceous chondrites and decomposed granite and has a refractive index of 1.55.

METHOD B   (*for extremely friable rocks*)

1. Dry the sample thoroughly.

2. A suitable amount of Epon 828 (Shell Chemical Co.) mixed with its catalysing agent, Shell Chemical curing agent "Z," 10 to 15 ml altogether, is heated to about 60°C in a temperature-controlled oven. The sample is heated separately to the same temperature, then immersed in the resin. The epoxy recommended here may be substituted equally well by that of method A.

3. A standard molding die with a close-fitting plunger, similar to that illustrated in Exercise 10.1 (Fig. 10.1), is also heated in the same oven. The specimen, of course, must fit inside the die. If necessary, it may be trimmed.

4. The resin container is then transferred to a bell jar and evacuated until all bubbling ceases in the sample. If the specimen is very porous, the container should be reheated and again evacuated until all bubbles disappear.

5. The specimen and enough resin to cover it completely are poured into a small rubber bag, such as the cut-off finger of a rubber glove or a small surgical finger cap, which is then tied securely with twine just below the epoxy level to exclude air. Excess twine and rubber are cut off and the bag is put into the press die into which a few millimeters of mineral oil have been poured. Enough oil is then poured in to cover the bag.

6. The plunger is fitted and pressed to 12,000 psi. The oil will squeeze out past the plunger until the rubber bag seals the cavity. At this point the pressure rises sharply and thereafter drops slowly. Pressure may have to be increased if the system leaks.

7. After 1 hour impregnation is complete for specimens of 1 or 2 cm^3.

8. The specimen is removed from the bag, and excess epoxy is wiped off. Curing is done overnight at about 60°C. The specimen is then ready for sawing and thin sectioning.

(Method after Sinkankas, 1968.)

DISCUSSION

This method is recommended for friable materials that are scarce and when the operator wishes to make a good thin section from what is perhaps the only small fragment available. If the specimen is friable but not scarce, method A is simpler to operate. The cured Epon epoxy has a refractive index of 1.591 (Hagni, 1966) and subsequent care must be taken to remember the high refractive index of the medium (c.f. Canada Balsam 1.54).

EXERCISE 1.3

Mounting the Specimen

Choice of Glass Slide. *Slide size.* The older 3×1 in. (25×76 mm) slide, although convenient for teaching purposes and offering a large area for labeling, is inconvenient because it cannot be mounted on the universal stage or the electron probe microanalyzer.

A better and more generally useful slide size is approximately 2×1 in. or 26×50 mm, with variation from country to country to 27×46 mm. It has room only for a very small label.

METHOD A

1. The specimen surface to be mounted is ground on an 850-rpm cast-iron lap, first in a water slurry of 100 carborundum; it is then washed, and reground with 600 alumina in a water slurry. The second grinding is repeated until the specimen surface is flat and all pits and imperfections are removed. Frequent washing of the specimen and lap to remove any grit is necessary to obtain a good finish.

2. The thoroughly washed specimen is placed, polished surface upward, on a 95°C hot plate with a clean glass slide, numbered side downward, alongside. The glass slide can be numbered by scratching the specimen number near the short edge with a diamond-tipped or hardened-steel-pointed marker.

3. After 30 minutes equal portions of Araldite Ay 105 and hardener 935F (obtained from Chemical and Engineering Co. Inc., 221 Brook Street, Media, Pennsylvania, 19063) are mixed in a small aluminum dish on the same hot plate and stirred until it clears (not more than 40 seconds). This plastic, so prepared, must be applied immediately because it thickens in 3 to 5 minutes.

4. A thin even coat of the plastic is spread on the polished side of the specimen with a wooden spatula. The glass slide is then slowly lowered onto the specimen from a 45° angle. Any trapped air bubbles are immediately removed by rubbing the back of the glass slide with a soft object such as a sharpened wooden dowel.

5. The slide is now turned over, the specimen quickly moved gently to its final central position on the slide, and left 30 minutes on the hot plate to cure completely.

6. After removing and cooling, traces of plastic sticking to the glass can be removed gently with a razor blade.

(Method after Moreland, 1968.)

DISCUSSION

Method A is highly recommended. Epoxy mounting is superior to any other mounting media. The Araldite AY 105 is to be preferred because its refractive index is 1.55, which is close to the conventional Lakeside 70C and Canada Balsam of 1.54. Epon 828 (Shell Chemical Co.) may be used equally well but it has the disadvantage of having a refractive index of a 1.59. Nevertheless it may be preferable for some petrographic work, but 1.55 is of greater general advantage for the distinction of common rock-forming minerals—quartz, alkali feldspar, and plagioclase.

METHOD B

Lakeside 70 cement (obtainable from Cutrock Engineering, Ward's, or Buehler) is used in this method.

1. Heat the specimen and the glass slide at 140°C on hot plate with a temperature gage. Accurate heat control is essential. Below 135°C the cement is viscous; above 150°C bubbling will occur.

2. Touch the end of a Lakeside 70 bar to the hot specimen and the glass slide. The melted cement will flow and make an even film coating. Invert the glass slide over the specimen slowly from a 45° angle. Any air bubbles trapped must be removed immediately by rubbing the glass slide with a pointed wooden dowel.

3. Invert the slide so that the rock slab rests on top and remove the assembly from the hot plate. Position the rock specimen gently in its final central place on the slide and allow it to cool at room temperature for several minutes.

4. Excess cement may be scraped off with a razor blade.

METHOD C

1. Dry the polished specimen and a glass slide on a hot plate and allow both to cool.

2. Cover a watch glass with a sheet of aluminum foil. Mix the following at room temperature on this foil: 10 parts Epon No. 828 (supplied by Shell Chemical Co.,

P.O. Box 2392, Church Street Station, New York, New York. 1000) with 1 part triethylene tetramine (supplied by Union Carbide Corporation, Chemicals Division, P.O. Box 6112, Cleveland, Ohio).

3. At room temperature apply this epoxy to the polished surface of the specimen. Lower the glass slide from a 45° angle onto the specimen. Remove any air bubbles by rubbing the glass with a pointed wooden dowel.

4. Turn the slide over and move the specimen gently to its final central position. Allow the slide to stand without movement for at least 2 hours, preferably overnight, when the epoxy will be completely cured. This epoxy has a refractive index of 1.59.

(Method after Hagni, 1966.)

EXERCISE 1.4

Sawing and Grinding

At this stage of the operation the procedure varies from laboratory to laboratory, depending on manpower and/or money available. Several automatic and semiautomatic machines will bring the slides down nearly to their final thickness. Nevertheless, a large number of operators, whether by choice or because of lack of funds, may wish to complete this stage of the procedure manually.

Once the specimen is mounted on the glass slide it is less easy to hold during subsequent sawing and grinding. Some form of holder will help to produce high-quality sections. The holder shown in Fig. 1.1 is useful for reducing the section to a thickness of about 60 μ. The holder is attached to a vacuum source, such as a small vacuum pump, by a piece of flexible tubing through a bottle liquid trap. Do not attach the holder directly without a bottle liquid trap; otherwise water may be sucked back into the vacuum pump. The vacuum holds the glass slide firmly onto the holder

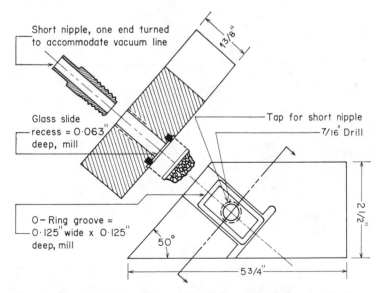

Fig. 1.1   Design of handholder for sawing and grinding glass-mounted slabs down to approximately 60 μ. (After Copeland, 1965.)

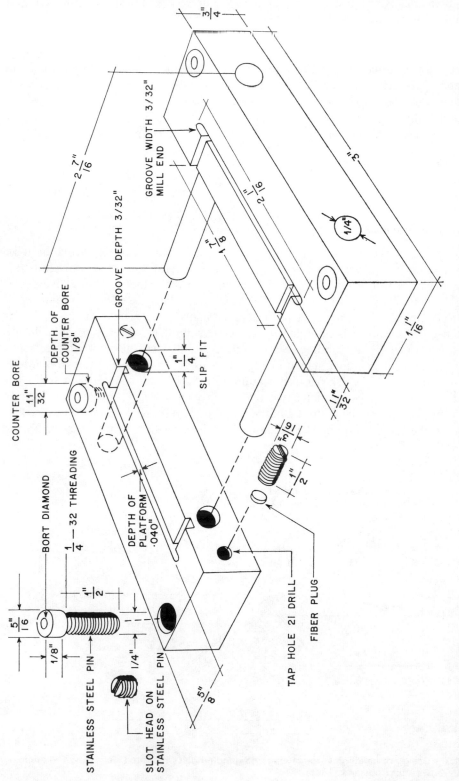

Fig. 1.2 Holder with diamond wear points for grinding thin sections. The rock slab cannot be ground beneath the diamond plane. (After Coch an and Jenson, 1965.)

COUNTER BORE

DEPTH OF COUNTER BORE 1/8"

$\frac{11"}{32}$

GROOVE DEPTH 3/32"

GROOVE WIDTH 3/32"
MILL END

$\frac{3"}{4}$

$2\frac{7}{16}"$

BORT DIAMOND

$\frac{1}{4}$ — 32 THREADING

DEPTH OF PLATFORM ·040"

SLIP FIT

$\frac{1"}{4}$

$3"$

$2\frac{9}{16}"$

$1\frac{7}{8}"$

$\frac{1}{4}"$

$\frac{11"}{32}$

$1\frac{1}{16}"$

$\frac{5"}{16}$

$\frac{1"}{2}$

1/8"

STAINLESS STEEL PIN

1/4"

SLOT HEAD ON STAINLESS STEEL PIN

$\frac{5"}{8}$

$\frac{3"}{16}$

$\frac{1"}{2}$

TAP HOLE 21 DRILL

FIBER PLUG

against the O-ring. The back of the glass slide may be smeared with a thin coating of vaseline to help form the vacuum which is applied during the complete operation. The holder and slide are fed by hand through the diamond saw. The size and shape of the holder, which may be aluminum or brass, promotes easy handling. After sawing the holder continues to be used for grinding the specimen down to 60 μ (any quartz in the rock will show first-order orange or pink). Final grinding should then be carried out with the slide removed from the holder and held by hand.

Since the end point in grinding is difficult to judge and the section must be frequently observed and occasionally removed for observation of interference colors under crossed polars, a holder that can prevent grinding beyond a certain selected thickness would be of great benefit. Such a holder made of two brass blocks joined by $\frac{1}{4}$ in. stainless steel pins has been designed by Cochran and Jensen (1965) and is illustrated in Fig. 1.2.

The petrographic glass slide with its mounted rock slab is accommodated in a rectangular depression. The small groove around the platform facilitates milling of the platform.

Four commercial bort diamonds of approximately $\frac{1}{4}$ carat are mounted, one in each end of screwlike stainless steel pins. The exposed diamond surface should be 2 to 3 mm in diameter. The diamonds are mounted with their flatter surfaces upward and with no sharp points protruding. If peen mounting cannot be done, the diamonds may be brazed in minimum heat into shallow drilled holes.

The pins are adjusted with a screwdriver until the diamonds protrude far enough to give the desired final thin-section thickness. Then the diamond-tipped screws are locked by set screws with a fiber plug to prevent damage to the screw thread.

The rock slab cannot be ground beneath the diamond plane. The sawing and grinding of the mounted slab is, of course, more conveniently done by some form of automatic or semiautomatic machine. Several models are available.

The MICRO-TRIM (manufactured by Microtec Development Laboratory, P.O. Box 1441, Grand Junction, Colorado 81510) is the most advanced automatic machine known to me (Fig. 1.3).

In this machine the thin sections are held by vacuum on a heavy metal chuck which takes up to seven 27 × 46 mm slides or may be changed for others that take up to five 1 × 3, five 2 × 2, or four 2 × 3 in. slides. The vacuum chuck slowly rotates automatically, the rate being controlled by a selector dial, against a fast-spinning diamond blade (for initial rough trimming of the thick sections), then against a diamond impregnated grinding wheel (for grinding down to 0.035 to 0.045 mm final thickness), and, if required, against a buffing plate (for automatic polishing of the thin sections). The blade, grinding wheel, and buffing plate are rapidly interchanged. The section thickness is set by a micron-dial of 6 μ increments. The machine is compact and takes up a bench space of only 22 × 18 in.

Ingram model 103 saw and Ingram model 303 grinder (both supplied by Ingram Laboratories, Griffin, Georgia, or by Wards Natural Science Establishment Inc., P.O. Box 1712, Rochester, New York 14603) are suitable. A wide variety of equipment is available from Cutrock Engineering Co. Ltd., 35 Ballards Lane, London N.3, England, from Buehler Ltd. and Adolph I. Buehler Inc., 2120 Greenwood Street, Evanston, Illinois, 60204, or from Buehler-Met AG, Postfach 4000, Basel 23, Switzerland. Buehler now has an automatic machine comparable to the MICRO-TRIM.

Fig. 1.3 The MICRO-TRIM automatic thin-sectioning machine. (Photograph by courtesy of Microtec Development Laboratory.)

The procedure, whether manual or by machine, is as follows:

1. Saw the mounted slab with a diamond saw 0.38 mm thick to a thickness of 0.5 mm if an automatic machine is used or to a final thickness of 1 to 2 mm if held by hand. The final thinness will depend on skill. The section should be fed slowly through the saw to prevent plucking of the specimen from the glass slide.

2. The section is further reduced either on an automatic machine or hand-held on a lap. A diamond impregnated wheel is recommended. If a machine is used, the micro-screw adjustment is positioned so that the specimen just barely touches the wheel. The specimen is slowly advanced against the wheel and from time to time removed and examined under crossed polars for thickness. Grinding is stopped at 40 μ (0.04 mm) at which point quartz shows first-order yellow-to-orange birefringence.

3. Further grinding is done by hand, first on a 850 rpm horizontal lap, with 600 alumina in a water slurry. The slide is held with one hand while the slurry is applied sparingly with the other. After a few seconds, when the slurry has become semidry, grinding is started with light pressure. After 15 to 20 seconds the slide is rinsed in clear water and examined under the microscope for thickness. The grinding proceeds until the desired thickness is obtained.

4. The final step is done on a glass plate with 1500 alumina in a water slurry. This hand grinding provides excellent control over the final thickness which should normally be 30 μ (0.03 mm) with standard birefringence colors. An advantage of hand grinding is that plucking is almost entirely eliminated.

Conventional Covering

If slides are to be covered in the conventional way, they can be employed in normal petrographic and universal stage work but will be of no use in the study of the opaque constituents under reflected light or in the electron microprobe. The steps in covering are the following:

1. The section is washed free of grinding powder and then laid section uppermost on a warm hot plate (at about 140°C) to dry. A cover glass of slightly larger size than the rock section is also placed on the hot place.

2. Fresh Canada Balsam is smeared over the surface of the section. As soon as it is cooked (begins to smoke) the cover glass is laid on top from a 45° angle and any air bubbles are carefully expelled. Trapped air bubbles are released by gently rubbing the cover glass with a pointed wooden dowel. The slide is immediately removed and allowed to cool.

3. Excess balsam around the edge of the cold cover glass is removed by trimming with a razor blade or dissolved with xylol. The whole section is then washed with kerosene. The thin section is ready for use and may now be labeled.

Polished Thin Sections

The polished thin section has the following advantages over the conventional covered section:

1. Petrographic study can be made of both transparent and opaque minerals.

2. Reflectivity and microhardness can be determined.

3. Areas of petrographic interest can be marked for subsequent microprobe study.

4. Microchemical tests can be performed.

5. Oil immersion lenses of high power can be used.

6. High resolution is obtained with oil immersion objectives; hence polished sections are superior for fine-grained rock study.

7. Better photomicrography is possible.

Because polished thin sections can be repolished if the surface is marred, they therefore compare more than favorably with covered thin sections in petrographic work.

Proper grinding before polishing is critical to the production of high-quality polished sections. Woodbury and Vogel (1970) recommend that resin-bonded diamond grinding disks are superior to wet, loose, abrasive techniques in terms of time, convenience, and superior quality of surface.

Polishing can be conveniently carried out, for example, on the Buehler "AB Whirlimet attachment" with three "AB petro thin slide holders" of appropriate size 27×46 mm or circular cross-section slabs of 1 or $1\frac{1}{4}$ in. diameter.

If the AB petro thin slide holder is not available, a plastic or perspex holder, illustrated in Fig. 1.4, can be made. It is recommended that a holder of some kind be

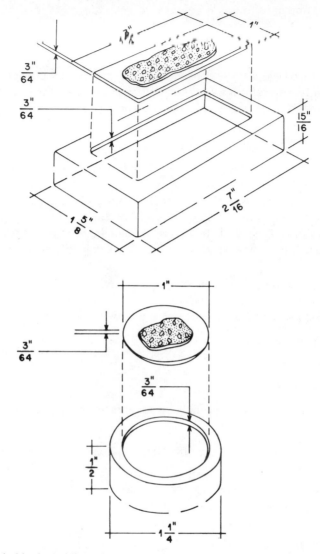

Fig. 1.4   Slide holder for holding thin sections during polishing on a lap. (After Moreland, 1968.)

used if good polishing is to be attained. Different laboratories use different polishing techniques. Four sequences are given here. Sections for polishing must NOT be mounted with Canada Balsam; either Lakeside 70 or epoxy is suitable.

METHOD A

If the rock section contains both opaque and nonopaque minerals, two steps are recommended. Step 1 can be eliminated if the rock is completely free of opaque minerals.

1 (*rough polish for opaque minerals*).   An AB Texmet lap cloth is needed (supplied by A. Buehler, Ltd., 2120 Greenwood Street, Evanston, Illinois). The cloth is impregnated with 3 μ diamond abrasive. The lap is lubricated with a 1 : 1 mixture of Buehler 40-8140 mineral spirits and 40-8142 polishing oil. Apply enough to give the cloth a slightly damp appearance. One application will last several days. The speed of the lap is 600 rpm. The specimen is moved in a circular direction opposite to the lap rotation. Use moderate pressure and polish for $1\frac{1}{2}$ minutes. Rotate the specimen through 180° and polish for another $1\frac{1}{2}$ minutes. This step removes the pits and scratches. After each step the specimen is carefully cleaned with water and liquid soap applied with a soft camel's hair brush. Alcohol should be used if the specimen is slightly water soluble.

2 (*for polishing the nonopaque minerals*).   Polytex Supreme lap cloth is recommended (supplied by Geoscience Instrument Corp., 110 Beckman Street, New York 10038). The polishing medium is Linde grade 0.3A alumina abrasive. A light slurry is prepared by sprinkling a small amount of the abrasive on the lap, adding water or alcohol if needed, and mixing with the fingers. While polishing it is essential to keep the lap wet. A dry lap causes scratches; 600 rpm is the speed required. Use light pressure and move the specimen in a circular motion, opposing the lap rotation. After 2 to 3 minutes rotate the specimen through 180° and continue polishing for 2 to 3 minutes more. If the specimen begins to grab, stop and add more water.

Periodic microscopic examination will indicate when a condition is reached in which most of the scratches have been eliminated. Then stop.

3 (*final polishing*).   Use AB microcloth (from A. Buehler) and Linde type 0.05B high-purity alumina abrasive. The lap speed is 163 rpm. Make the slurry as in step 2. Remove most of the slurry with a laboratory wiping cloth. The final polish should be done on an almost dry lap. The specimen holder becomes a necessity in this step. Move the specimen slowly (1 to $1\frac{1}{2}$ seconds per revolution) against the lap rotation direction. Do this 15 times, rotate the specimen through 180°, and repeat for 15 revolutions. This will be enough for most specimens. If some pits or scratches remain, the process may be repeated.

4. Wash carefully with soap and water (or alcohol) and clean in an ultrasonic cleaner, first with warm soapy water (or alcohol), then with clean water. Then dry carefully. Label the slide.

(Method after Moreland, 1968.)

DISCUSSION

Experience alone produces good-quality polished sections. Prolonged polishing is of no advantage and can cause a loss of the softer minerals. Therefore one should not strive for a better and better polish. It is better to tolerate some pits and scratches than to lose all the softer minerals. Some pits may be natural. Therefore do not attempt to eliminate all imperfections.

METHOD B (*for rocks that do not contain especially hard minerals and all components have approximately similar hardness*)

All at wheel speed 160 rpm.

All materials may be obtained from A. B. Buehler at any one of its addresses.

1. *Rough polishing*

Diamond paste, 6 μ
Lapping oil
Nylon cloth   for 10 minutes

2. *Fine polishing*

Alpha alumina, 0.3 μ
Distilled water
Nylon cloth   for 5 minutes

3. Gamma alumina, 0.05 μ
Distilled water
Microcloth   for 2 minutes

(Method after Woodbury and Vogel, 1970.)

METHOD C (*for rocks displaying great variation in hardness of the minerals*)

All at speed 160 rpm.

All materials are obtainable from Buehler.

1. *Rough polishing*

Diamond paste, 6 μ
Lapping oil
Texme cloth   for 10 minutes

2. *Fine polishing*

Alpha alumina, 0.3 μ
Distilled water
Texme cloth   for 5 minutes

3. Gamma alumina, 0.05 μ
Distilled water
Silk cloth   for 2 minutes

(Method after Woodbury and Vogel, 1970.)

DISCUSSION

The above times in methods B and C are for a new cloth and new charge of abrasive, full pressure, and a 160-rpm speed with three 3/4 × $1\frac{1}{4}$ in. samples. The times must be varied naturally with different materials, worn cloths, and so on.

Cleaning between each step is essential. The ideal should include light scrubbing with a soft brush and soap solution, followed by ultrasonic cleaning to prevent abrasive carryover from the preceeding step.

A technician with relatively little experience can quickly and inexpensively produce high-quality polished thin sections by this method.

METHOD D

Specimens are best prepared down to a final thickness of 60 μm before polishing begins. Such sections are easier to polish mechanically than those that have been ground down to 30 μm. The individual glass slides are mounted in circular Perspex (lucite) holders and inserted in a polishing machine. The polishing sequence is as follows

1. Hypress diamond compound on Hyprocell-paper lap, 6 μm.
2. Hypress diamond compound on Hyprocell-paper lap, 3 μm.

3. Hypress diamond compound on Hyprocell-paper lap, 1 μm.

4. Hypress diamond compound on Hyprocell-paper lap, $\frac{1}{4}$ μm.

(Method after Long, 1967.)

The diamond compounds may be obtained from Griffin and George, Ltd.,57 Uxbridge Road, London, W.5, England, and the paper laps from Engis Ltd., Park Road Trading Estate, Maidstone, Kent, England.

EXERCISE 1.7

Polished Thick Sections

Because attaining a good polish on a thin section is something of an art not given to all laboratory workers and because the normal thin-section thickness of 0.03 mm may be too thin for Vickers Hardness determination (see Chapter 2C), it may be preferable for most purposes to proceed as follows:

1. When the rock has been sawed into thin slabs (Exercise 1.1), make two 25 × 25 mm slabs about 5 or 6 mm thick, working so that one surface of each slab is separated from the other by the same saw cut. These two surfaces will be closely similar. Mark them for identification.

2. Using one of the slabs, grind the marked surface and mount it on a glass slide (Exercise 1.3). Complete the thin section and cover it in the conventional way (see Exercise 1.5).

3. Place a bakelite ring form 3.82 cm in diameter (obtained from A. B. Buehler) on a sheet of glass on which a thin film of vaseline has been smeared.

4. Lay the rock slab (marked side down) inside the ring to lie flat on the vaseline layer (it may be necessary to cut off the corners of the slab to gain entry into the ring area).

5. Mix EITHER by volume 10 parts Ciba Araldite MY 753 with 1 part Ciba Hardener Hy 951 in a watch glass (both obtainable from Ciba Ltd., Basle, Switzerland) OR by volume 5 parts Epoxide resin with 1 part Epoxide hardener. These two materials may also be obtained in sealed packs ready for mixing in the correct proportions, under the trade name AB Epo-Mix Epoxide (all obtainable from A. B. Buehler).

6. Pour the mixed epoxy into the bakelite ring form to cover the specimen and fill the ring just to the brim.

7. Allow to set for 2 to 3 hours in an oven at 60°C or overnight at 40°C.

8. Any epoxy protruding above the top of the ring form may be ground off level with the ring top with carborundum on a lap.

9. The rock slab surface and the epoxy surface which will have a vaseline smear on it are now ground and polished by the same recipe given in the various alternatives of Exercise 1.6.

DISCUSSION

The polished rock slab is permanently set in the bakelite ring and is thick enough for Vickers Hardness determination (Chapter 2). It can, however, be used only with incident light, hence is suitable only for the opaque mineral identification. The thin section made from the same rock surface must be used for the transparent mineral determination with transmitted light. Both thin section and polished section can be given the same number and stored together. Together they allow a complete study of all the minerals, both opaque and transparent. It may well be that time spent in completing the two preparations is less than time spent in making one good polished thin section.

EXERCISE 1.8

Grain Thin Sections

The foregoing methods were formulated for large and coherent rock and mineral sections. Frequently, however, a thin section of normal thickness of an assemblage of mineral grains or crystals in various orientations is required.

Use is made of two mounting media—an epoxy such as Hysol epoxi-patch (supplied by Ward's Natural Science Establishment) or Epon No. 828 or Araldite Ay and Lakeside 70.

1. Put a glass slide on a hot plate at approximately 120°C. Smear on Lakeside 70 cement and sprinkle the grains over it. Allow the slide to cool and wait for the cement to set.

2. Grind the Lakeside surface to a flat smooth surface until it transects the broadest central section of the majority of the grains.

3. This flat and semipolished surface is now mounted on a second petrographic slide which has been smeared with hot epoxy mixed with its hardening agent. Heating the epoxy to around 100°C makes the smear more liquid; hence the bubbles are more easily removed. Heating also encourages the epoxy to set in as short a time as 6 hours for Hysol Epoxi-patch and 30 minutes for Araldite Ay 105 (see Exercise 1.3).

4. The sample, which is now sandwiched between two slides, is allowed to set overnight.

5. On the following day it is placed on a hot plate and heated to 120°C. The Lakeside 70 will melt but the epoxy will remain set and the first glass slide can be removed.

6. The exposed side is then ground to the desired thickness and either covered in the conventional way or finished as a polished thin section.

(Method after Adams and Arvidson, 1969.)

EXERCISE 1.9

Removal of Thin Section

To separate a thin section from its glass slide after it has been ground to its final thickness it is preferable initially to have mounted the rock slab with Polyvinyl acetate (" Mowilith " obtainable from Farbwerke Hoechst, Frankfurt a.M., Germany). This material is more easily removed than Lakeside 70 or epoxy.

For mounting, the slide and slab are heated to 35°C on a hot plate. Polyvinyl acetate is smeared on the slide and the rock slab is attached to it. Hardening is speeded up by running cold water over the slide.

Once complete the slide is inverted on a wire gauze and immersed in acetone for a few hours. The gauze prevents the freed rock slice from curling.

(Method after Zeidler, 1968.)

AIDS IN THIN-SECTION STUDY 2

A. STAINING TECHNIQUES

As explained in the preface, this book does not include the standard optical methods of mineral identification and petrographic study. These methods are already abundantly available in a large number of excellent textbooks, a few of which are mentioned here: Johannsen (1918); Wahlstrom (1969); Kerr (1959); Heinrich (1965); Hartshorne et al. (1970); Winchell (1937); Winchell and Winchell (1951); Deer et al. (1962a, b, c; 1963a, b; 1966); Williams et al. (1954); Moorhouse (1959); Spry (1969), and Harker (1950).

The study of rocks and the identification of minerals both in hand specimen and thin section may be greatly facilitated by various staining techniques, which unfortunately usually find no place in standard determinative mineralogy and descriptive petrography books such as the majority listed above. To remedy this deficiency the techniques that are considered to be helpful in petrography are described in this chapter. The list is not exhaustive and includes only those that apply to rock-forming minerals. Staining tests have been devised for a large number of minerals of economic significance. They are excluded. The reader is referred to Reid (1969) for a comprehensive tabulation which incorporates the staining of opaque minerals.

The staining of hand specimens of rocks requires no special equipment. The hand specimen may be conveniently sawed to give a flat surface, which is then ground to a good finish in successive grades in the same way as a rock slab is prepared for mounting on a glass slide. This surface is ideal for staining.

Staining is employed occasionally to identify, hence separate, grains from a crushed rock in mineral separation, in which case the grains themselves are completely treated with the reagents.

Most frequently, however, staining is performed on uncovered thin sections, which are sections that have been ground to the final stage and are now ready for the application of a cover glass. After staining the cover glass may be attached to the surface in the usual manner or with a cold-setting adhesive. Alternatively, a finished polished thin section may be stained if difficulty has been encountered in distinguishing the mineral constituents.

Thin-section staining need have no special equipment other than photographic developing dishes or the small plastic containers that are obtainable in any supermarket for holding sandwiches. Nevertheless, it is always convenient to prepare in advance a special holding rack for thin section staining if this is to be a technique regularly used in your laboratory. (See Fig. 2.1.)

The rack is made of $\frac{1}{4}$-in. perspex sheet. It consists of a base plate with a window and two triangular end pieces joined above by a length of perspex rod $\frac{1}{2}$ in. in diameter. The window is recessed around its edge to the depth of the glass slides and holds four

15

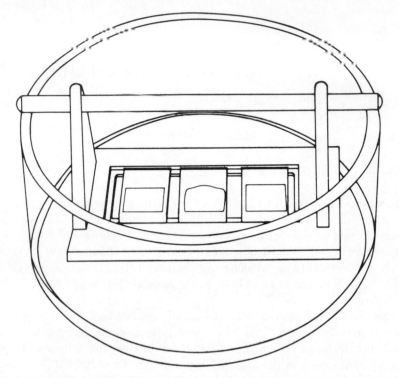

Fig. 2.1 Perspex rack for thin-section staining. Actual dimensions will depend on the slide sizes used. (After Dawson and Crawley, 1963.)

sections firmly in place. The thin sections are placed *face down* in the window and weighted down with a piece of sheet lead. The sections are left in this position on the rack during all steps of the staining procedure until they are removed from the wash water at the very end. The window can alternatively be cut for four small or three large thin sections. The triangular side pieces are cemented to the recessed base by carbon tetrachloride.

Feldspars

The following methods are designed to distinguish between alkali feldspar, plagioclase, quartz, cordierite, and feldspathoids.

We begin with the best known and most widely used method:

EXERCISE 2.1

Staining Rock Slabs and Thin Sections for K-Feldspar, Plagioclase and Quartz

1. Porosity causes difficulty in staining. It should be eliminated at once by soaking the rock specimens, on which a flat surface has been sawed and roughly polished, in molten paraffin for 15 minutes. Excess paraffin is wiped off. The flat surface is then polished with 400 abrasive and dried. Do not touch the polished surface; otherwise some paraffin may adhere to it and prevent etching. Sprayed-on plastic also eliminates porosity.

2. Prepare the following reagents:

Hydrofluoric acid, concentrated 52% HF. CAUTION. It causes painful burns. Five percent barium chloride solution in water. Saturated sodium cobaltinitrite solution in water.

Rhodizonate reagent: dissolve 0.05 g rhodizonic acid potassium salt in 20 ml distilled water. Make it fresh in a small dropping bottle because the solution is unstable.

3. Prepare the following apparatus:

Plastic vessels about 2-cm deep and in diameters slightly less than the size of the specimens to be etched. Plastic cover vessels large enough to cover the etching vessel and the specimens.

Plastic or paraffin etching vessel to fit thin sections.

Ribbed vessel to hold the sodium cobaltinitrite solution.

If the thin-section rack is used (Fig. 2.1), several larger dishes will be required for holding the various solutions.

METHOD FOR ROCK SLABS

1. In a well-ventilated fume cupboard put conc. HF into the etching vessel to about 6 mm from the top.

2. Place the rock specimen across the top of the etching vessel, polished surface down.

3. Cover the whole with a larger inverted plastic vessel to prevent drafts. Let it stand for 3 minutes. If the acid is less than 52% or weakened by age, experience may indicate a lengthening of this time to 5 minutes or thereabouts.

4. Remove the specimen, dip it in water and then twice quickly in and out of the barium chloride solution.

5. Rinse it briefly in water and immerse it face down for 1 minute in the sodium cobaltinitrite solution.

6. Rinse gently in tap water, tilting it back and forth, until the excess sodium cobaltinitrite solution is removed. The K-feldspar is now bright yellow. If it is not, remove the etch residue by rubbing the surface under water; then dry and etch it again for a longer period. Continue from step 4.

7. Rinse briefly with distilled water and with a dropping bottle cover the surface with rhodizonate reagent. Within a few seconds the plagioclase will become brick red. When the color is sufficiently intense, rinse the specimen in tap water to remove excess stain.

DISCUSSION

This procedure produces a brilliant yellow (K-feldspar) and red (plagioclase) which appear grainy under magnification. For fine features such as zoning less intense colors may be desired and the above method may be modified accordingly. Quartz remains unstained, but its surface may be somewhat dull due to frosting with the HF.

The etching time will vary somewhat with different rock types, but over-etching should be avoided because thick etch residues flake off the surface.

Staining of somewhat weathered granite slabs is not satisfactory by the standard method. The following modification is proposed by Lyons (1971):

1. The rock surface is immersed in 52% HF for 45 seconds (badly weathered granites require a longer bath; very fresh ones, a shorter bath.

2. Rinse and dry at 80°C in an oven until the surface is light gray and powdery.

3. Fume the surface over the HF bath for 3 minutes.

4. Immerse in saturated sodium cobaltinitrite solution for 1 minute.

   This procedure produces the following results: plagioclase is light gray, K-feldspar, bright yellow, and quartz, medium gray to glassy. The dark minerals look corroded and are easily identified.

METHOD FOR UNCOVERED THIN SECTIONS

1. Etch the rock surface by leaving it face down for 10 seconds over HF at room temperature. For weaker acid etching up to 30 seconds may be necessary or alternatively the acid may be warmed. Do not rinse. Rinsing causes an uneven stain.

2. Immerse the slide in the saturated sodium cobaltinitrite solution for 15 seconds. The K-feldspar stains light yellow.

3. Rinse briefly in tap water to remove excess reagent.

4. Dip the slide quickly in and out of the barium chloride solution.

5. Rinse briefly, first with tap water, then with distilled.

6. Cover the rock surface with the rhodizonate reagent from a dropping bottle. When the plagioclase becomes pink, rinse the slide in water.

7. Allow the slide to dry; then apply the cover slip in the conventional way.

DISCUSSION

Under the microscope the K-feldspar is evenly stained pale yellow, but occasionally the stain is distributed over the K-feldspar as small, evenly spaced, intense yellow spots. The plagioclase is pink. The pink color is usually not so even as the yellow and is often more intense along the cleavages and twin planes. Quartz remains unstained.

   Even though the staining may be uneven and pale, there will be no difficulty in identifying every plagioclase and K-feldspar grain. The micas also take up some of the stain and look darker. Ideally, thin-section staining should be pale. Darker stains make the differences more obvious but they also make the crystals almost opaque and this interferes with normal optical identification. Pale staining does not affect the optical properties of the minerals at all.

NOTE.   Pure albite does not stain red, but if the plagioclase contains at least 3% An it will take up the red stain.

(Method after Bailey and Stevens, 1960.)

EXERCISE 2.2

Staining Rock Slabs for K-Feldspar, Plagioclase, and Quartz

METHOD FOR ROCK SLABS

1. Etch rock slabs and hand specimens by submerging them completely in conc. HF for 15 to 20 seconds. Dip them in water to remove the acid and, while still wet, immerse them in the saturated sodium cobaltinitrite solution (prepared as in Exercise 2.1, step 2) for 1 to 2 minutes.

2. Rinse and allow to dry. The K-feldspar stains bright yellow, the plagioclase, chalky white, and the quartz, dull gray. This method can be used in the field with a small squeeze bottle of HF. Several drops are spread over the rock surface with the bottle tip. The sodium cobaltinitrite is then applied and after 1 minute, rinsed.

(Method after Nold and Erickson, 1967.)

EXERCISE 2.3

Staining Rock Slabs and Thin Sections for K-Feldspar, Plagioclase, and Cordierite

METHOD FOR ROCK SLABS

1. Prepare a smooth surface with 400 to 800 grit on a lap. If the rock is porous, fill the surface with molten Lakeside Cement or impregnate it with epoxy as in Exercise 1.2.

2. Etch the surface for 10 to 15 seconds in conc. 52% HF acid.

3. Dip once in water.

4. Immerse in saturated sodium cobaltinitrite solution for 1 minute.

5. Remove excess sodium cobaltinitrite with a gentle water rinse.

6. Dry under a heat lamp.

7. Immerse in W/V 5% barium chloride solution for 15 seconds.

8. Dip once quickly in water and dry gently with compressed air.

9. Immerse for 15 seconds in amaranth solution [1 oz (28.35 g) F, D, and C, Red No. 2, 92% pure coal-tar dye, in 2 liters of water].

10. Dip once quickly in water.

11. Direct a gentle stream of compressed air to sweep off the excess amaranth solution.

12. When milky white areas suggest albite (which does not stain), repolish the slab. Repeat steps 1 to 3, dip in calcium chloride solution, dry, and proceed as in steps 4 to 11.

METHOD FOR THIN SECTIONS

1. Etch uncovered thin sections for 15 seconds in HF vapor.

2. Immerse in cobaltinitrite solution for 15 seconds.

3. Rinse briefly under tap.

4. Immerse a few seconds in barium chloride solution.

5. Dip in distilled water.

6. Immerse 1 minute in amaranth solution.

7. Dip once in water.

8. Sweep away excess amaranth with compressed air.

9. Dry and cover.

(Method after Ruperto et al., 1964.)

DISCUSSION

Plagioclase stains red, K-feldspar, yellow, and quartz is unstained. Other minerals that stain are benitoite, celsian, cordierite, dolomite, hydrogarnet, pectolite, vesuvianite, witherite, and wollastonite.

The deep red of stained cordierite contrasts well with the lighter stain of plagioclase

Sanidine and anorthoclase do not respond so well to sodium cobaltinitrite as
K-feldspar of plutonic rocks.

The color of amaranth stain on sodic plagioclase can be intensified if the section
is first dipped in calcium chloride solution, but this is not effective on pure albite
(Ford and Boudette, 1968).

EXERCISE 2.4

Staining for Anorthoclase

Anorthoclase may be misidentified for plagioclase in the routine modal analysis on
stained thin sections.

1.   After staining with sodium cobaltinitrite, rinse, dry, and immerse in approx. 15%
solution of baruim chloride for 15 seconds.

2.   Rinse, dry, then immerse in concentrated amaranth solution for about 1 second.
The anorthoclase stains grayish red-purple. Anorthoclase phenocrysts stand out from
the yellow-stained K-feldspar matrix. Thus anorthoclase is indistinguishable from
sodic plagioclase by this method.

(Method after Ford and Boudette, 1968.)

DISCUSSION

It is recommended that detailed optical and X-ray studies be made to identify anorthoclase before modal analysis. The staining tests may assist in the identification, but
alone they are inconclusive.

Cordierite

Cordierite is often difficult to distinguish from plagioclase feldspar in thin section.
Hence a staining test which helps to identify them will be of great benefit to the
petrographer.

Whenever cordierite is suspected, staining is desirable because of the possibility
of mistaking cordierite for plagioclase in metamorphic rocks.

EXERCISE 2.5

For Distinguishing K-Feldspar, Plagioclase, and Cordierite

1.   Prepare the following reagents:

HF concentrated.
Sodium cobaltinitrite saturated water solution.
Barium chloride 5% by weight in distilled water.
Amaranth, Biological Stain C.1.16185, 14.2 g per liter of distilled water.

IMPORTANT NOTES. (a) The principal cause of failure is weak HF. Acid that has deteriorated to a point at which plagioclase fails to stain is still strong enough to produce a satisfactory K-feldspar stain. This causes the operator to believe that the acid is still good and to arrive at the conclusion that his technique is wrong. Even freshly purchased HF may have deteriorated. It is therefore of the utmost importance to have fresh strong HF.

(b) If 20 g sodium cobaltinitrite is added to 60 ml distilled water, there will be a residue of undissolved reagent. The solution remains effective as long as this residue is present, and more reagent may be added as it is used up by decomposition in the solution. It can be kept for further use in a tightly capped bottle.

(c) The amaranth solution will keep fresh for at least a year.

2. Prepare the following apparatus:

Plastic box for HF slightly larger than the thin sections, with a plastic cover for opening when etching is not in progress.

Plastic cover with platinum wire hooks for supporting the thin section over the HF or a perspex holder (see Fig. 2.1).

Hot plates with sensitive temperature control in low-temperature range.

Flexible tubing attached to a compressed air source.

100-ml beaker
250-ml beaker
Teflon-coated forceps
Three 30-ml beakers
Petri dish
Stop watch

3. Conventionally mounted thin sections (with Lakeside 70, Canada Balsam, or epoxy) may be used. They must be uncovered. Wash the surface with detergent and then with distilled water. A clean, freshly prepared thin section needs no washing.

4. Dry on a hot plate at approximately 56°C in a current of air from a fan or compressed air source. A sheet of paper toweling on the hot plate helps to prevent excessive heating of the thin section and absorbs moisture from the back.

5. Fume 20 seconds in proximity over HF (if the HF is weak, this time may have to be increased considerably, even up to 40 seconds, based on experience). Mild heating of the acid will counteract the effect of reduced HF content. Tentatively, the temperature of the acid should be increased by as many degrees above 20°C as the HF percentage is decreased below 50.

6. Immerse 15 seconds in a 30-ml beaker of sodium cobaltinitrite solution.

7. Wash briefly with tap water and then with distilled water from a wash bottle. The time is not to exceed 15 seconds. Water should be streamed over the thin section from the edge to avoid damaging etch films on mineral surfaces. Washing time here and subsequently should be kept to a minimum to avoid leaching the films.

8. Dry thoroughly on a hot plate as in step 4.

9. Fume 7 seconds over HF as in step 5 (again this time may have to be extended if the acid is not fresh)

10. Immerse with agitation in a 30-ml beaker of barium chloride solution.

11. Dip twice briefly in a 100-ml beaker of fresh distilled water.

12. Press section face down on folded Kimwipe (or any good quality, highly absorbing tissue), backed by a paper towel to absorb free moisture. Pools of moisture that remain on the surface can cause concentration of $BaCl_2$ where they dry off and result in excessively dense amaranth stain.

13. Dry on a hot plate as in step 4.

14. Immerse in a 30 ml beaker of amaranth solution for 15 sec.

15. Dip briefly four times in a 250-ml beaker of fresh distilled water to remove most of the amaranth red solution. Excessive washing will remove the water-soluble amaranth stain. Conversely, if the stain is too dense, it can be reduced by additional washing. High-density spots can be reduced by application of water from a wash bottle.

16. Blow off remaining liquid with compressed air, resting the edge of the thin section on a paper towel to absorb liquid driven to this edge by the air stream.

17. Slides may be covered at room temperature by Permount or similar mounting medium to secure the cover slip. Any cold setting epoxy may also be used.

(Method after Boone and Wheeler, 1968.)

DISCUSSION

K-feldspar will stain yellow. Plagioclase and cordierite stain orange to reddish. The stain on the cordierite is lighter than on the plagioclase and there is sufficient difference to allow a distinction between the two. If required, a final HF etching at the end after step 15 will bleach the plagioclase.

EXERCISE 2.6

For Distinguishing K-Feldspar, Plagioclase and Cordierite

NOTE.   Only EPOXY adhesive can be used to mount thin sections because NaOH dissolves Lakeside 70 and Canada Balsam cement.

1. Prepare the following reagents:

Concentrated HF.

Trypan blue basic solution.

Dissolve 0.2 g trypan blue (Diamine blue 3B) powder in 25 ml of methanol (Methyl alcohol). Mild heating and stirring are necessary to disperse the powder effectively because it tends to form small polygranular clumps. Add methanol if necessary to maintain the solution volume. Add 15 ml of 10% NaOH aqueous solution to the trypan blue-methanol mixture and heat to about 50°C for staining.

2. Use *only EPOXY*-mounted thin sections. Etch the surface of the thin sections as for cobaltinitrite staining (Exercise 2.5, Nos. 4, 5). Stain for K-feldspar or plagioclase first (Exercise 2.5, Nos. 6, 7, or 14, 15), if desired, then re-etch the surface for 7 seconds at room temperature over HF fumes. Do not rinse or disturb the re-etched surface.

3. Immerse in warm trypan blue basic solution. Immersion time varies from 30 seconds to 1 minute. Withdrawing and rinsing the section for inspection followed by re-immersion is *not* recommended.

4. Rinse, dry, and cover with cover slip at room temperature with Permount adhesive or any cold-setting epoxy.

(Method after Boone and Wheeler, 1968.)

DISCUSSION

Staining time cannot be indefinitely long because plagioclase begins to take on the blue stain in addition to the cordierite. Initial tests may have to be run to determine the best staining time at which cordierite is blue but plagioclase is still unstained. All Mg-minerals, for example, chlorite and hornblende, take up the blue stain, but this is not detrimental to optical studies. Fe-rich cordierite may not take up the blue stain, and cordierite strongly pinitized and of yellowish color in thin section will not accept it.

The foregoing method works best if combined only with sodium cobaltinitrite staining (omitting the amaranth plagioclase staining) to give a striking contrast between yellow K-feldspar and blue cordierite.

It can also be combined with the amaranth plagioclase stain (omitting the K-feldspar staining) to give reddish plagioclase with blue cordierite.

It does not work well if all three successive items are attempted because the sodium cobaltinitrite cannot withstand the two successive HF etches.

Micas

There is usually no difficulty in distinguishing the common micas in thin section. It is impossible, however, to distinguish muscovite from paragonite. Hence paragonite is frequently misidentified as muscovite. The following method allows a distinction to be made in thin section.

EXERCISE 2.7

Thin-Section Staining for Paragonite

1. The uncovered thin section is exposed to fresh hydrofluoric acid vapor for 3 minutes.

2. The thin section is placed on a plastic tray and allowed to dry for 2 or 3 hours in the draft of the fume cupboard.

3. Immerse the thin section for 4 minutes in a well-mixed saturated sodium cobalt-initrate solution (6g/10 ml water).

4. Rinse gently but thoroughly in tap water.

5. Dry and cover in the usual way.

(Method after Laduron, 1971.)

DISCUSSION

The muscovite is stained yellow, whereas paragonite remains unstained. The muscovite stain works better in sections cut ⊥ the mica foliation. Lamellae lying parallel to the thin section do not stain so well. The yellow color is best seen with a daylight filter.

Neither margarite nor pyrophyllite take the stain, but these minerals cannot be confused with paragonite.

The presence of a feldspar grain inhibits the staining of surrounded or included muscovite, which may be mistakenly identified as paragonite.

Sometimes a fringe of muscovite, surrounding feldspar, is unstained, but farther away the same crystal is stained.

This is not a serious drawback, however, for if a stained section shows that unstained micas occur only in contact with feldspar it can be assumed that paragonite is absent from the rock. As soon as unstained white micas occur away from feldspar crystals, paragonite is confirmed.

Feldspathoids

1. Prepare the following reagents:

Phosphoric acid (U.S. Pharmacopoeia 85%).
0.25% methylene blue in demineralized water.

2. Spread a thin film of the syrupy phosphoric acid over a thin section with a glass rod.

3. After 3 minutes dip gently in water to remove the acid.

4. Immerse in the methylene blue solution for 1 minute.

5. For preservation of the stains dip the section in water a few times to remove excess dye and immediately cover the section with a drop of thick glue solution (Durofix plus amyl acetate 1 : 3) with a cover slip on top. Any cold setting epoxy may also be used.

(Method after Shand, 1939.)

DISCUSSION

Nepheline, sodalite, and alacime are stained deep blue. Melilite stains paler blue, and leucite is not affected by the color.

Because zeolites are also stained by this method and any colloidal weathering products will take up the dye, it is necessary to use fresh rock and double check the identity of the stained minerals by optical methods.

Carbonates

Because of their chemical variability, much emphasis has been placed on the distinction of the various carbonates by staining techniques both on hand specimens and uncovered thin sections. Staining techniques for carbonates are not exclusively useful to the sedimentary petrologist, for many metamorphic, some igneous rocks, and many mineralized zones often contain significant amounts of carbonate minerals.

Thin-Section Carbonate Staining

Each step should be performed in the given order until a definite identification is made.

1. Etch the uncovered thin section with 1.5% HCl for 10 to 15 seconds.

Result. Both calcite and ferroan calcite show considerable etch. Dolomite and ferroan dolomite show negligible etch.

2. Stain with 0.2 g Alizarin red S dissolved in 100 cc of 1.5% HCl (hereafter called ARS) + 2.0 g potassium ferricyanide dissolved in 100 cc 1.5 HCl acid (hereafter called PF), mixed in the ratio ARS : PF = 3 : 2, for 30 to 45 seconds.

Result. Calcite is very pale pink to red. The calcite surface parallel to the *c*-axis is more deeply stained than a basal section. This feature assists petrofabric analysis. Basal sections are pale pink, prismatic sections, red. Ferroan calcite is very pale pink–red and pale blue–dark blue. The two superimposed produce mauve-purple-royal blue. Dolomite takes no stain. Ferroan dolomite becomes deep turquoise, depending on the ferrous iron content.

3. Stain with ARS solution (prepared as in step 2) for 10 to 15 seconds.

Result. Calcite and ferroan calcite become very pale pink–red. Dolomite and ferroan dolomite take on no color.

(Method after Dickson, 1965.)

DISCUSSION

Distilled water must be used in making up all solutions and for washing surplus stain from the thin sections.

Step 1 removes all grinding dust. The etched section is then immersed in an acidified mixture of the two stains (step 2). Each stain works independently and there is no mutual interference. If the partly stained section is immersed in an acidified solution of alizarin red S for a few seconds, the color differentiation of the carbonate will be increased, but it is important not to exceed 15 seconds of staining (step 3).

Then wash rapidly and carefully with distilled water, and dry quickly because the stain is water soluble. Do not touch the stained surface, for the stain is easily rubbed off.

To protect the surface a layer of Durofix is painted on with a camel's hair brush. The Durofix is diluted with amyl acetate 1 : 3, which evaporates rapidly, and the remaining dry surface has an RI = 1.54.

Staining Thin Sections for Dolomite and Brucite

The presence of dolomite and brucite in marble may be easily detected by this method.

1. Make up the following reagents:

1 : 20 HCl.

0.2% alizarin red S in 0.2% HCl (2 ml HCl per liter of water).

2. Etch the thin section for 3 to 5 seconds in 1 : 20 HCl.

3.   Wash with water.

4.   Apply the alizarin red S reagent drop by drop on the etched area. When a strong pink color has developed on the calcite, wash the sample gently with demineralized water and allow it to dry at room temperature.

(Method after Haines, 1968.)

DISCUSSION

This treatment stains brucite purple and calcite pink and leaves dolomite unstained.

Microscopic examination at × 10 with reflected light provides the best contrast. Because the calcite stain is on the surface layer only, use extreme care not to remove the stain in washing.

This test is best confined to marble in thin and polished sections. It is not good on hand specimens because the red liquid flows into cavities.

EXERCISE 2.10

Staining Thin Sections for Dolomite

1.   Make up a trypan blue solution by dissolving 0.2 g trypan blue in 25 ml methanol by heating. Replenish methanol lost by evaporation. Add 15 ml 30% caustic soda (30 g NaOH in 70 ml water).

2.   Immerse the thin section, which must have been made with an *Epoxy* cement, in the boiling reagent solution.

(Method after Friedman, 1959.)

DISCUSSION

Dolomite is stained blue and calcite remains colorless. The stain fades with time but can be restored by immersion in dilute sodium hydroxide.

Lakeside 70 and Canada Balsam are dissolved by NaOH; therefore only epoxy mounted slides can be treated in this way. Friedman (1959) also lists 23 other dyes that are suitable for staining dolomite.

EXERCISE 2.11

A Complete Staining Scheme for Carbonates

This scheme is illustrated, step by step, in Fig. 2.2 and a summary of the results is given. The following notes supply the details of each step and the results stage by stage.

1.   *Prestaining.*   Etch in dilute HCl (8 to 10 ml HCl in 100 ml distilled water). If the specimen effervesces briskly in the cold acid, allow it 3 minutes (aragonite, calcite, witherite). If the specimen does not react or reacts slowly, heat the acid until it reacts vigorously. Allow only 30 seconds to 1 minute of this reaction time (siderite, dolomite, magnesite, rhodochrosite, ankerite, cerussite, smithsonite, strontianite).

2.   *Staining.*   Hand specimens, grains, and uncovered thin sections are all treated similarly. Use only small quantities of reagents and discard after a single use. Do not reuse; otherwise inconsistent results will be obtained. Freshly etched rock specimens

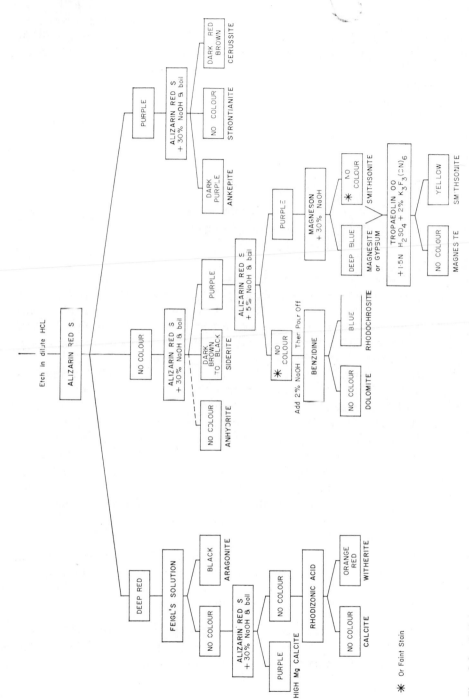

Fig. 2.2 Staining scheme for carbonates in hand specimen and thin section. (After Warne, 1962.)

27

are required for each staining stage. All stains except that produced by magneson, are stable. The faded magneson color may be regained by wetting the previously colored surface with dilute sodium hydroxide solution.

3. *Poststaining.* Any liquid once used should be thrown away. Wash the stained surface several times by adding distilled water gently. If the stain had been applied in an alkaline medium, the final washing is best done with a dilute sodium hydroxide solution. In this case the use of water would change the pH radically and perhaps alter the color.

4. *Details of Staining*

ALIZARIN RED S (SODIUM ALIZARIN SULFONATE)

a. *Solution.* Dissolve 0.1 g alizarin red S in 100 ml of 0.2% cold HCl acid (2 ml HCl diluted with 998 ml distilled water).

b. *Method and results*

   i. The etched rock surface is covered with the solution and held cold for 5 minutes (calcite will stain darkly in 2 to 3 minutes). Pour off the solution and wash carefully by decanting. Calcite, witherite, high Mg calcite, and aragonite stain deep red. Ankerite and Fe dolomite, strontianite, and cerussite stain purple; anhydrite, siderite, dolomite, rhodochrosite, magnesite, smithsonite, and gypsum do not stain.

   ii. If the staining is repeated with equal volumes of the solution and 30% sodium hydroxide solution and boiled for 5 minutes, each of the above unstained minerals except anhydrite, will become stained. The siderite stains dark brown to black; the others become purple.

   iii. Further differentiation is possible if the specimens are boiled for 5 minutes in a liquid of equal volume of the solution and 5% sodium hydroxide solution. Dolomite and rhodochrosite remain unstained, but magnesite, gypsum, and smithsonite stain purple. The stains do not affect organic matter in coal, hence are useful for carbonate determination in such rocks.

FEIGL'S SOLUTION

a. *Solution.* 1 g Ag_2SO_4 is placed in a solution of 11.8 g $MnSO_4 \cdot 7H_2O$ in 100 ml water and boiled. After cooling the suspension is filtered and one or two drops of dilute NaOH solution are added. The precipitate that forms is filtered off after one or two hours. The solution keeps if stored in a dark bottle.

b. *Method and results.* The etched rock is immersed in cold solution. Aragonite turns black. Calcite, witherite, and high Mg calcite are unaffected. The blackening is a slow process and usually takes 10 minutes.

RHODIZONIC ACID SOLUTION

a. *Solution.* Dissolve 2 g of disodium rhodizonate in 100 ml distilled water.

b. *Method and results.* The etched specimen, after washing with distilled water, is immersed in the solution for 5 minutes. Witherite stains orange-red. The color depends on the age of the solution. (Sodium potassium rhodizonate made in the same way produces the same staining.)

BENZIDINE SOLUTION

a. *Solution.*   Dissolve 2 g pure benzidine in 100 ml distilled water containing 1 ml 10 N HCl acid.

b. *Method and results.*   The etched and washed rock is immersed in a dilute solution (1 to 3%) sodium hydroxide for $1\frac{1}{2}$ minutes. The specimen is then exposed to the air for $1\frac{1}{2}$ minutes and covered by the cold benzidine solution. It produces an immediate blue stain on rhodochrosite but dolomite is completely unaffected.

MAGNESON (*P*-NITROBENZENE-AZORESORCINAL) SOLUTION

a. *Solution.*   Dissolve 0.5 g of the dye in 100 ml 0.25 N (1%) sodium hydroxide solution.

b. *Method and results.*   The etched and washed rock is covered with equal quantities of the magneson solution and 30% cold sodium hydroxide solution. Magnesite stains blue in 1 minute and deep royal blue in 3 minutes. Smithsonite is only very faintly stained even after 5 minutes. If the stained specimen is washed several times by decantation with distilled water and strongly heated, but not boiled, in 30% sodium hydroxide solution, the deep stain of rhodochrosite will remain but the faint smithsonite stain will disappear.

TROPAEOLIN 00 SOLUTION

a. *Solution.*   Dissolve 0.01 g tropaeolin 00 in 100 ml distilled water.

b. *Method and results.*   The etched and washed specimen is covered by a liquid composed of 1 part tropaeolin 00 solution, 1 part 1.5 N sulfuric acid, and 3 to 5 parts freshly prepared 2% solution of potassium ferricyanide (made by dissolving 2 g potassium ferricyanide in 100 ml distilled water). After 5 minutes the liquid is poured off and smithsonite will be stained yellow; magnesite is unaffected.

POTASSIUM FERRICYANIDE TEST

a. *Solution.*   Combine equal parts of 2% HCl acid and potassium ferricyanide (prepared as in the tropaeolin 00 test).

b. *Method and results.*   Immerse the sample in the solution. Ankerite and ferroandolomites stain dark blue in a cold solution. Dolomite, generally, and siderite, always, require heating before the color appears. The reaction is rapid but dolomite often needs 5 minutes.

POTASSIUM HYDROXIDE-HYDROGEN PEROXIDE METHOD

a. *Solution.*   Make a hot concentrated solution of caustic potash to which is added a little hydrogen peroxide from time to time during staining.

b. *Method and results.*   The etched rock is immersed for 5 to 10 minutes in the solution. The surface of the rock is then washed in distilled water and dried in air. Siderite is stained brown.

AMMONIUM SULFIDE AND COPPER SULFATE TEST

b. *Method and results.* A polished rock surface is placed in 10% HF acid for 2 minutes to etch the surface, washed thoroughly in running water, dried, and then immersed for 1 minute in ammonium sulfide solution. Remove, wash, dry, and place the specimen in a 10% copper sulfate solution. Ankerite turns gray-black. Sodium sulfide may be substituted for ammonium sulfide. Ferroandolomite and ankerite are stained almost black, but siderite is hardly stained.

(Method after Warne, 1962.)

Opaque Minerals

Because this book is primarily concerned with rock-forming minerals and because space is not available, the vast number of staining and microchemical techniques applicable to opaque minerals and other less common silicates must be omitted. The most complete compilation of staining methods has been made by Reid (1969). Another very complete list of tests available for opaque and ore minerals is given by Short (1940). Hosking (1958, 1964) has also compiled a variety of methods for identifying and distinguishing ore minerals, and Smith (1953) describes a wide range of microchemical tests applicable to mineralogy in general.

B. USE OF ULTRAVIOLET LIGHT

Fluorescence under ultraviolet radiation, and the afterglow (called phosphorescence), although on the whole capricious and unpredictable, when present can be used to provide information about crystal growth, zoning, and inclusions that may not otherwise be discernible. It is also a guide to identification.

A frequently used ultraviolet (UV) source is the mercury arc lamp in a quartz envelope which has appreciable emission down to a wavelength of 185 nanometers (nm). Ordinary glass in layers a few millimeters thick is transparent to wavelengths of 300 nm and above.

Shortwave UV light ranges from 185 to 300 nm and cannot penetrate a thin section cover glass. Longwave UV light ranges usually from 300 to 400 nm and can penetrate a cover glass. However, because it is the shortwave UV light that excites most fluorescence in minerals, it is recommended that observations for fluorescence and phosphorescence be conducted only on uncovered polished slabs or thin sections. Shortwave UV light alone is obtained by placing a special shortwave filter in front of the mercury arc lamp to absorb the longer wavelengths of both UV and visible radiation. It is important initially to ascertain that the bulb does actually produce shortwave UV radiation, for the majority of commercially available UV bulbs produce only longwave UV and some blue light. A shortwave UV filter in front of this kind of bulb would be useless.

By far the best known supplier of UV lights is Ultraviolet Products Inc., San Gabriel, California 91778, whose products are marketed under the trade name MINERALIGHT, and may be purchased from most geological supply houses; for example, from Colorado Geological Industries Inc., 1244 E Colfax Avenue, Denver, Colorado 80218. The most useful model for laboratory work is MINERALIGHT model MPR 2, which has two externally clamped tubular bulbs that can be adjusted in position above the polished surface on either side of the microscope objective. A shortwave UV filter can be fitted on one or both of the lamps and the fluorescent colors of the polished section studied through the microscope (petrological or binocular). A wide range of models is available for hand specimen and field use.

WARNING.   Ultraviolet radiation, especially of short wavelength, will cause temporary or permanent blindness either by looking directly into the lamp or at a reflection from a surface. Hence never observe UV fluorescence except through a glass (spectacles or the microscope system).

Because of the capricious nature of fluorescence, a complete list of mineral responses is not given here. The reader is referred to Gleason (1960) for a comprehensive list of mineral, rock, and gemstone UV reactions, and to Smith (1953) and Przibram and Caffyn (1956) for less complete details. The majority of the minerals that fluoresce do so because of a trace impurity (the activator or spike). The activator may commonly be manganese, uranium, or rare earths, but other elements are known to cause fluorescence, and the reason for it in some cases may not be chemical but physical. A few minerals always fluoresce, and for them the UV procedure is a definite step in their identification.

The following is a list of minerals that *ALWAYS FLUORESCE*, and UV irradiation should be a standard procedure if their presence is suspected. The common fluorescent colors are given.

CUPROSCHEELITE	Yellow with a faint tinge of green, brassy yellow; rarely blue.
FLUORITE	Commonly blue, white, cream, yellow, green. All colors are best seen under long wave UV.
HYDROZINCITE	Brilliant blue under short wave UV. Bluish-white and white under a very strong longwave lamp.
MALAYAITE	Yellow, very slightly greenish. This mineral is tin-sphene and always fluoresces. Sphene never does.
PETROLEUM	Most petroleum shows white to yellowish to blue fluorescence; the color varies with the gravity.
SCAPOLITE (WERNERITE)	Brilliant orange to yellow. Best under long wave UV. Sometimes phosphoresces for long duration.
SCHEELITE	Brilliant blue to bluish-white under shortwave. No fluorescence under longwave. The range of color is due to variable molybdenum content.

Mo-free	bright lustrous blue;
0.05% Mo	faint blue;
0.48% Mo	white fluorescence;
0.96 to 4.8% Mo	increasingly yellow.

Above 4.8% Mo no further change in color (Smith, 1953).

URANIUM MINERALS	Most uranium secondary minerals show a strong lemon-yellow to light-green fluorescence. Species included are autunite, gummite, meta-torbernite, schroekingerite, torbernite, uranophane, and several others.
WILLEMITE	Bright green; better with shortwave UV. Often phosphoresces, sometimes brilliantly and persistently.

The following minerals will fluoresce if they contain the correct activators. In their pure state they are nonfluorescent, but it may be said that they commonly do fluoresce: **Anglesite** (yellow-orange); **aragonite** (orange, yellow, green); **brucite** (blue, white, greenish-white); **calcite** (red to pink, orange, white, yellow, blue); **calomel** (red to pink, orange); **cerussite** (bright yellow); **chabazite** (deep green); **chlorapatite** (yellow); **colemanite** (bright white to cream, pale yellow); **diamond** (blue, green—about 15% of all diamonds fluoresce); **eucryptite** (begonia rose pink); **hackmanite** (sulfur-bearing sodalite: orange, reddish, pinkish—phosphoresces for a long time and thereafter daylight color becomes deep purple to raspberry, fading slowly to its original color); **idrialite** (pale green); **leadhillite** (dull or pale yellow, orange); **manganapatite** (yellow, orange, pink); **nepheline** (orange and red); **opal** (brilliant pale yellow and yellow-green, due to uranium trace content); **pectolite** (orange-yellow); **phosgenite** (yellow-orange); **selenite** (variety of gypsum: yellow); **serpentine** (yellow variety only: yellow to white); **sodalite** (orange to yellow); **stolzite** (bright greenish-white); **thernardite** (pale orange to yellow); **trona** (white); **witherite** (blue, white, cream); **wollastonite** (yellow to cream); and **zircon** (golden yellow).

There are, of course, several famous localities, such as Franklin, New Jersey, where almost every mineral fluoresces.

These lists are only for the most generally occurring fluorescence. The reader is referred to Gleason (1960) for a complete listing and for applications to specialized fields such as gem testing and chemical tests involving untraviolet reactions.

It is recommended that polished sections be observed routinely under ultraviolet light in the hope that some of the minerals in the study locality will fluoresce. If none does, then little time is wasted, but if the locality is one fortunately endowed with the necessary activators, UV observation will add a valuable petrographic tool to the standard methods. Care will be needed in making a preliminary identification of the fluorescing species, but once their identity is established by other mineralogical methods their UV colors and properties will become a rapid identification technique; for example, in the tin-tungsten mineralized region of Southeast Asia, extending from Yunnan through Thailand to Indonesia, the systematic use of UV light on polished rock surfaces is invaluable for rapid identification of scheelite and malayaite.

C. OPAQUE MINERAL IDENTIFICATION

The identification of minerals that are transparent in thin section is fully aided by a large number of good reference books. To mention only a few, Kerr (1959), Heinrich (1965), Winchell and Winchell (1951), and Deer et al. (1962, 1963, 1966), contain all the necessary data.

Books on optical mineralogy, however, usually fail rather badly, when it comes to those minerals that are opaque. As a result, many petrographers are content to record "opaque minerals" as the only description of opaque components or what is even worse to make a guess and list "magnetite" and "ilmenite" simply because they are the most common opaque minerals in igneous and metamorphic rocks. Petrographers in general should by now be aware that the incident light microscope is of great applicability to petrology and is not the sole prerogative of the ore mineralogist and applied geologist. Opaque minerals invariably are present, at least in accessory amounts, in most rocks, and to encourage their proper identification and study the methods are briefly given in this section.

When incident light is used, the conventional thin section with cover glass is of little use. The preparation necessary is that the rock or mineral surface should be flat and polished. An ideal preparation is a polished thin section, which is equally useful in a transmitted light and an incident light microscope. Polished sections can therefore be used equally for opaque as well as transparent mineral study.

Two primary parameters that characterize opaque minerals are reflectivity and hardness. Methods of measuring them are now described separately.

EXERCISE 2.12

Determination of Reflectivity

A suitable microscope for measuring reflectivity and for general transmission and incident light study is the Vickers model 74 (Vickers Instruments, Haxby Road, York, England). It must have an external lamp fitted with an iris diaphragm and should be of high intensity and stabilized. The ideal lamp for this purpose is a 100-W, 12-V quartz-halogen lamp power pack (Vickers Instruments).

A continuous spectrum interference filter, ranging from 400 to 750 nm and adjustable by sliding a thermally insulated holder calibrated directly in wavelengths on a vernier scale, is attached to the lamp projection housing of the microscope (Vickers Instruments). It can be slid out of the light path to allow white light into the microscope.

The vertical ocular of the microscope is replaced by a microphotometer. The most suitable model is a Vickers digital microphotometer, which constitutes a photomultiplier and amplifier and a control unit with a four-figure digital readout. The complete microscope assembly with digital photometer is illustrated in Fig. 2.3. Viewing is carried out through the inclined ocular.

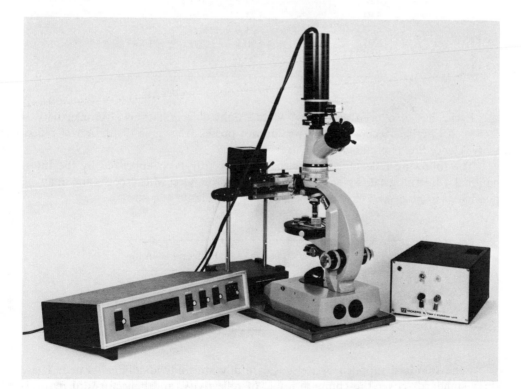

Fig. 2.3 Vickers reflecting microscope complete with microphotometer, high intensity lamp, and interference monochromater. (Courtesy of Vickers Instruments, Haxby Road, York YO37SD England.)

The vibration direction of the polarizer must be made to coincide with the symmetry plane of the illuminator. Otherwise there will be an appreciable loss of light.

The polished carborundum SiC standard used has a given reflectivity of 20.2% for "white" plane polarized light (2850°K).

The polarizer is put in the 0° position and the reflectivity values of the mineral to be studied are determined in comparison with the SiC at each of the two extinction positions 90° apart—its two principal directions—unless of course the mineral is cubic and any direction will suffice.

With SiC as standard, and with the microscope not fitted with an ocular iris diaphragm, the error due to glare in measuring a high reflectivity unknown is substantial. If, however, an ocular iris is fitted and made coincident with the field iris, the glare error will be greatly reduced. Errors may be reduced to less than 5%.

Carborundum is the most satisfactory standard. It grows in plates parallel to the basal plane and crystals are uniaxial.

Spectral reflectivities from an SiC basal plate determined by the National Physical Laboratory are the following:

nm	R%	nm	R%
Nanometers $(10^{-9}\ m)$			
400	21.8	546.1	20.4
420	21.5	589.3	20.2
440	21.2	620	20.0
460	21.0	660	19.9
480	20.8	700	19.8
500	20.6		

Pyrite has also been used for "white" light determinations. Its reflectivity is 54.5%. All pyrite specimens, however, do not possess the same dispersion of reflectivity. Hence SiC is preferable.

The following reflectance standards have recently been approved by the International Mineralogical Association Commission on Ore Microscopy and may be purchased from Zeiss, 7082 Oberkochen, West Germany.

	Reflectivity in air
1. Black glass, NG 1	4%
2. Black silicon carbide	20%
3. Tungsten carbide	47%

Each standard supplied by Zeiss is carefully and individually calibrated. These three standards cover the common range of reflectivity, and henceforward pyrite is discarded.

The reflectivity of a specimen depends on the quality of polish. Overpolishing produces a hardened layer with lower reflectivity. The polished surface must be flat, and the slide must also be flat on the microscope stage. The actual method is outlined step-by-step:

1. The standard is placed on the microscope stage and a flaw-free area is brought into focus with white or selected monochromatic light.

2. The field and ocular iris diaphragms are adjusted so that with no more than one-quarter of the field illuminated the digital readout from the photometer is approximately equal to the reflectivity of the standard. (The digital readout system may be replaced by a galvanometer system).

3. A small box painted matte black internally and with an aperture at the top is substituted for the standard and a second reading is taken. This reading equals that caused by primary glare. It is offset by adjusting the zero setting control until the galvanometer or digital readout reads zero.

4. The standard is then replaced, and the iris is adjusted slightly until the readout exactly equals that of the standard provided by the supplier. All subsequent galvonometer or digital readings obtained on selected areas of unknown minerals give reflectivities directly. No further adjustments to the equipment are needed.

5. The standard is replaced by a rock or mineral polished section and the microscope is carefully focused on a flaw-free area of the mineral under study.

6. The galvanometer or digital readout gives the reflectivity of the specimen, provided no settings of the diaphragms, etc., have been changed. If the mineral is isotropic, the reflectivity can be taken at any stage setting. If the mineral is anisotropic, two reflectivity readings are taken at each of its extinction positions.

(Method after Bowie, 1967.)

DISCUSSION

This method does not correct for secondary glare. Secondary glare can be minimized, however, by selecting a standard not too dissimilar from the unknown in reflectivity. Because secondary glare is proportional to the reflectivity of the specimen, this error is reduced to a minimum by using an ocular iris and making it coincident with the field iris. Alternatively, secondary glare can be eliminated by drawing up a calibration graph for any particular wavelength of light by using a series of five or more standards covering a wide range of reflectivity. If pyrite were used alone as the standard, the measured reflectivity of carborundum would be too low and that for silver too high because of the secondary glare.

For measurement of reflectivity in oil use oil immersion objectives. Use the recommended oil. One with $n = 1.515$ at 589-nm wavelength at 20°C is satisfactory.

EXERCISE 2.13

Choice of Microindentation Hardness Method

We have a choice of using the Vickers hardness or the Knoop hardness test.

Vickers hardness is determined by producing an indention with a pyramid-shaped diamond and averaging the lengths of both diagonals. The apical angle between opposed pyramid faces is 136° and the depth of penetration of the diamond is one-seventh of the length of the indentation diagonal.

The Vickers hardness number is found by the formula

$$HV \ (kp/mm^2) = \frac{1854.4P}{d^2},$$

where P is the load in grams and d is the mean diagonal length in microns (0.001 mm).

Knoop hardness uses a diamond whose base is the shape of a rhombus. The ratio of the two base diagonals of an indentation is 7 : 1. The depth of penetration is only one-thirtieth of the long diagonal. Knoop hardness is found by the formula

$$HK \ (kp/mm^2) = \frac{14230P}{l^2},$$

where P is the load in grams and l is the length in microns (0.001 mm) of the long diagonal of an indentation.

The minimum permissible thickness of a mineral layer for accurate hardness determination is 10 times the depth of penetration of the diamond (see Leitz instructions for the Miniload tester). For a polished thin section of normal 0.03-mm thickness the maximum permissible penetration of the diamond is therefore 0.003 mm (3 μ). This makes the maximum length of the indentation diagonal

for Vickers hardness	0.021 mm (21 μ),
for Knoop hardness	0.090 mm (90 μ).

From these values the following table can be used to indicate the MINIMUM hardness that can be determined accurately on a 0.03-mm thick polished thin section. In every case the maximum permissible indentation diagonal is 0.021 mm for Vickers and 0.090 mm for Knoop.

	Vickers	Knoop
Load in Grams	HV Minimum Hardness	HK Minimum Hardness
5	21	8.8
10	42	17.6
20	84.1	35.1
50	210	87.8
100	420	175.7
200	841	351.4

The softest mineral likely to be encountered in a rock is molybdenite of HV 20 and HK 12. Its hardness can be determined accurately on a polished thin section only if a 5-g load is used. On no account may there be a heavier load. However, the softest common accessory opaque mineral to be found in polished thin sections is magnetite of HV 560 and HK 514. Other minerals such as ilmenite and pyrite are harder. Therefore under normal conditions, provided nothing softer than magnetite is measured on a rock, a polished thin section may be used for accurate hardness determination with a 100-g load. Each grain being measured must have a width at least twice the length of the identation diagonal.

All grains being measured on polished thin sections must therefore exceed a diameter of 0.042 mm for Vickers determination and 0.180 mm for Knoop determination, and the indentation should be at least one-half the indentation-diagonal length from the edge of the grain.

If minerals softer than HV 420 are to be commonly measured, it may be preferable to choose a polished thick section so that a greater depth of penetration can be tolerated and a longer diagonal can be attained, but the grains then must have an increased surface diameter.

Many petrologists have the fear that polished thin sections are too thin for hardness determination. Our discussion shows that this fear is without basis, and if the above table is employed to make the proper selection of load the complete range of hardness down to HV 21 and HK 9 can be accurately specified.

A note on the choice of Vickers or Knoop hardness might be useful here. Theoretically, Vickers hardness is identical to Knoop hardness, but in practice divergences are found, the causes of which are not understood and depend on the measuring forces as well as on the minerals being measured. The smaller the measuring forces, the higher the divergences. Correlation tables accordingly do not exist between Vickers and Knoop hardness determinations.

The majority of workers have come to consider that the Vickers diamond gives greater precision; hence most opaque mineralogy books quote only Vickers hardness. The Knoop diamond is presently in disfavor. For the reader who is interested, however, Knoop values for 127 opaque minerals are listed and described by Robertson (1961).

EXERCISE 2.14

Determination of Indentation Hardness

The ideal instrument for hardness determination is the Leitz Miniload-Pol hardness tester (Ernst Leitz Gmbh, D6330 Wetzlar, Germany) (Fig. 2.4). This instrument has a rotating microscope-type stage and incorporates a polarized analyzer so that anisotropism and extinction positions can be measured on crystals to be tested. It should be noted that a Knoop diamond in this instrument is a standard fitting, but of course it may be readily interchanged for a Vickers diamond. The Leitz standard nonpolarizing Miniload hardness tester is also suitable, the only differences being the absence of an analyzer and a rotating stage.

A Vickers microindentation hardness tester may be attached directly to the Vickers M74 incident light microscope by making use of a special diamond indenter objective and a remote control pneumatic cylinder for releasing the diamond at the required force.

Figure 2.5 shows a typical microindentation made by the Vickers diamond as it would be seen through the microscope.

The length of the indentation diagonal is taken as the average of both diagonals in Vickers diamonds and the length of the long diagonal in Knoop diamonds. The procedure is simple:

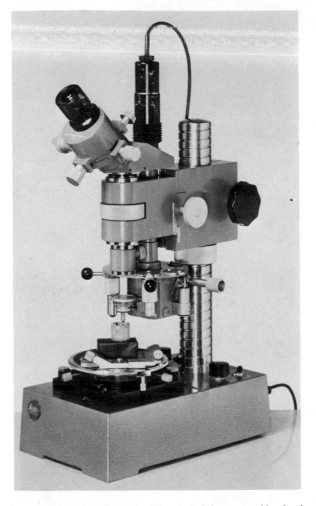

Fig. 2.4 Leitz Miniload-Pol hardness tester, with a rock slab mounted in plastic on the stage ready for testing. (Photograph by courtesy of Ernst Leitz, Wetzlar, Germany.)

1. Position the selected load P on the indenter. The load may be 200, 100, 50, or fewer grams. The choice will depend on the hardness of the mineral to be tested—200 g for a very hard mineral, 50 g for a very soft mineral; 100 g is the most commonly used load. The indenting body is put in its raised position. The choice of load will also be selected according to the table in exercise 2.13 if a polished thin section is being used.

2. The polished thin section, or thick section mounted in plastic, is laid on top of a layer of heat-softened cedukol (supplied by E. Merck Ag, 61 Darmstadt, Germany), or Tissuemat (supplied by Fisher Scientific Co., 633 Greenwich Street, New York, New York) which has been placed on the Leitz 3-point support (code 3539a); then pressed into a firm and horizontal position by means of the Leitz hand-press (code 799876). The soft embedding material is allowed to harden and the 3-point support is set on the hardness-tester stage.

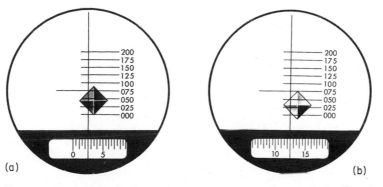

(a)   (b)

Fig. 2.5 Measuring a Vickers indentation. The micrometer eyepiece scale is aligned parallel to an identation diagonal by swiveling the eyepiece. One corner of the indentation is set on the 000 line (a) by means of the micrometer screw. The micrometer screw reading (a) is 2.5μ. Now turn the screw until the opposite corner lies on a scale division (b), the reading is 12.5μ. The length of the diagonal is therefore 75μ (a) $+12.5$ (b) -2.5 (a) $=85\mu$. Knoop indentations are measured in the same way, except only the long diagonal is measured. (Courtesy of Ernst Leitz, Wetzlar, Germany).

3. Using a low-power objective, center the crystal to be measured on the cross wire. Then center more exactly with the high-power objective and *focus carefully.* Sharp focus is most essential under high power to ensure the correct working distance.

4. Now rotate the indenter into position and release it onto the specimen. Indentation time is normally 15 seconds, but the exact time is not critical.

5. Raise the indenter and with the high-power objective and the calibrated eyepiece micrometer measure the two diagonals of the indentation and average (Fig. 2.5). Use this value (d) and the load (P) to compute the hardness.

$$\text{HV (Vickers hardness)} = \frac{1854.4P}{d^2},$$

$$\text{HK (Knoop hardness)} = \frac{14230P}{l^2}.$$

For Knoop only the length (l) of the long diagonal is measured.

DISCUSSION

Hardness should always be calibrated against stainless steel polished plates of known and measured hardness; for example, HV 300 and 800 (Ernst Leitz). The hardness tester must be set up on a solid concrete or stone bench to eliminate vibration effects.

The diagonals of the indentation are best measured with an airpath n.a. 0.85 objective with illumination of wavelength 500 nm.

The hard layer (10 to 20 μm) which is produced by some polishing methods may make the determined HV too high for light loads. If, however loads of 100 g or more are used, the hard surface layer will be unimportant. If a 100-g load is used, tests show that a similar hardness is obtained from the same mineral polished by a large variety of techniques.

The shape and quality of the indentation can be described as perfect (p), slightly fractured (sf) fractured (f), concave-sided (cc) or convex-sided (cv).

Roughly, hardness correlates with Mohs' scale:

Mohs' scale	HV	HK
1	40	50
2	50	72
3	110	119
4	190	215
5	550	616
6	750	707
7	1100	
8	1600	
9	2000+	

(Bowie, 1967; Robertson, 1961)

Reliable hardness determination can be performed only if the specimen is rigidly fixed to the instrument; for example, if the specimen is leveled on the instrument base by means of a volume of modeling clay, the nonrigidity of the clay will cause the determined HV numbers to be unreliably high. Polished sections mounted with epoxy give the best results. A very complete discussion of the microhardness of opaque minerals can be found in Young and Millman (1964).

Water absorption from room air onto the polished surface of a mineral can reduce the Vickers hardness by as much as 32% from the value obtained when the surface is dry (Westbrook and Jorgensen, 1968). Reproducibility of HV therefore requires that the polished sections be kept in a dessicator and the hardness tester in an air-conditioned or dehumidified room (if the climate is humid).

Systematic Scheme of Opaque Mineral Identification

If the mean reflectivity value of a mineral in white light has been determined according to Exercise 2.12 and the microhardness according to Exercise 2.13, these values can be used to identify the mineral in Figure 2.6. A similar chart with comprehensive tables has been produced by McLeod and Chamberlain (1968).

The metals and metalloids with metallic bonding are soft but have high reflectivity. At the other extreme the metallic oxides with ionic bonds are hard and have low reflectivity. Between them lie the sulfides and sulfo salts.

The following data will help in the identification of opaque minerals that occur most commonly in igneous and metamorphic rocks [data from Bowie (1967) and Bowie and Font-Altaba (1970)]. The white plane polarized determinations are in air relative to pyrite of 54.5% reflectivity. The VH numbers are with a 100 g load and n.a. 0.45 objective.

CHROMITE $FeCr_2O_4$. Cubic. Isotropic with red internal reflections. HV (100-g load) perfect indentation 1288–1561. Color dark gray. HK (200 g load) 1069–1152. Reflectivity in air, 12.1% for white light.

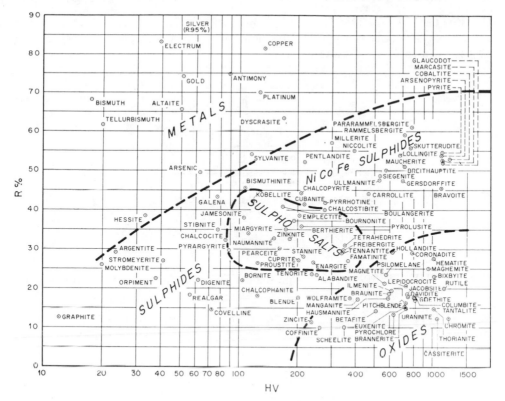

Fig. 2.6 Reflectivity (R) in white light and Vickers microhardness numbers (HV) for the common opaque minerals. (After Bowie, 1967.)

λ nm	Reflectivity % Air
470	11.9 to 13.8%
546	11.5 to 13.3%
589	11.5 to 13.3%
650	11.2 to 12.7%

Range due to composition: high%—Fe, Cr,
low%—Mg, Al.

Hutchison (1972) showed that for Alpine-type chromite (Al increasing reciprocally with Cr decrease) the reflectivity at 590 nm varies linearly with composition: Al cations per unit cell = $27.93 - 1.82 \times$ reflectivity % ranging from 10 to 14. This relationship, however, cannot be applied to stratiform chromite in which Fe increases as Cr content falls.

HEMATITE Fe_2O_3. Trigonal. Strongly anisotropic, with pale green polarization colors in the 45° position. Color in polished section is white. Reflectivity 27.5% (25.0 to 30.0). Hardness only for coarsely crystalline material; HV 1009 (920 to 1062). HK (300 g load) 533 to 794.

ILMENITE $FeTiO_3$. Trigonal. Light brown color in polished section. Moderately anisotropic. Pale greenish gray polarization colors in the 45° position. Reflectivity, 19.4% (17.8 to 21.1) in white light.

	Reflectivity %			
	In Air		In Oil	
λ nm	Re	Ro	Re	Ro
470	17.3	20.6	5.4	7.7
546	17.0	20.1	5.4	7.5
589	17.4	20.2	5.8	7.6
650	18.0	20.4	6.1	7.7

Hardness: HV 611 (519 to 703) with a slightly fractured indentation. HK (25 g load) 703 to 1291.

MAGNETITE Fe_3O_4. Cubic. Isotropic. Color, gray. Reflectivity, 21.1% in white light.

	Reflectivity %	
λ nm	In Air	In Oil
470	19.7	8.0
546	19.8	8.0
589	20.0	8.1
650	20.0	8.1

Hardness: HV 560 (493 to 607) with an indentation perfect to slightly fractured. HK 429 to 574.

PYRITE FeS_2. Cubic. Isotropic. Color. yellowish white. Reflectivity, 54.5% in white light.

	Reflectivity %	
λ nm	Air	Oil
470	46.5	32.8
546	54.0	40.2
589	55.4	41.6
650	55.9	42.4

Hardness: HV 1165 (1027 to 1414) with a fractured indentation. HK 719 to 834.

RUTILE TiO_2. Tetragonal. Color, gray. Strongly anisotropic, usually masked by internal reflections. Reflectivity, 20.2% in white light. Hardness, HV 1139 (1074 to 1210).

SPINEL $MgAl_2O_4$. Cubic. Isotropic.

λ nm	Reflectivity % Air
470	6.9
546	6.8
589	6.8
650	6.7

Hardness: HV 1505 to 1650; indentation perfect.

The foregoing list, of necessity, is admittedly brief, because of available space. For a complete listing and more comprehensive details of opaque mineral identification the reader is referred to the following publications:

Ramdohr (1969), Schouten (1962), Uytenbogaardt (1968), and Robertson (1961).

GRAIN SIZE, MODAL ANALYSIS, AND PHOTOMICROGRAPHY

3

This chapter describes certain ancillary macroscopic and microscopic techniques which are all too frequently dismissed in a few words in or completely omitted from standard optical mineralogy and petrography textbooks.

A. GRAIN-SIZE DETERMINATION

Although students of petrology make almost daily use of the polarizing microscope and may be completely conversant with all its accessory operations, surprisingly few have an accurate perception of the order of size of the crystals they are studying in thin section. This is because they have never calibrated their fields of view in terms of a millimeter scale for the microscope ocular in combination with the range of objectives. This calibration is an essential preliminary to regular use of the polarizing or any other microscope.

EXERCISE 3.1

To Calibrate the Eyepiece Micrometer

Because the modern trend is toward the general use of a binocular tube in the polarizing microscope, it is good practice to fit one eyepiece hole with a plain ocular and the other with an ocular incorporating a crosswire and eyepiece micrometer. The Zeiss Standard WL Research Microscope (Carl Zeiss, Oberkochen, West Germany) or the Leitz Ortholux-POL (Ernst Leitz, Wetzlar, Germany), among others, can be so fitted.

The procedure of calibration is as follows:

1. Place on the microscope stage a stage micrometer, which can be purchased from Leitz (usually 2 mm long with 200 divisions; each division = 0.01 mm) or from Zeiss (usually 5 whole mm with 1 mm subdivided into 100 intervals of 0.01 mm). Every department should possess one stage micrometer (one is enough because it is not used again after the microscope has been calibrated). Focus on the micrometer and place its engraved scale exactly along and parallel to the crosswire which includes the eyepiece scale (Fig. 3.1)

2. With both micrometers in focus, note the equivalence of one large eyepiece division in terms of millimeters on the stage micrometer. Note also the power of the objective. Thereafter, if the same eyepiece continues to be used in the microscope with this particular objective, the micrometer equivalence will be permanent.

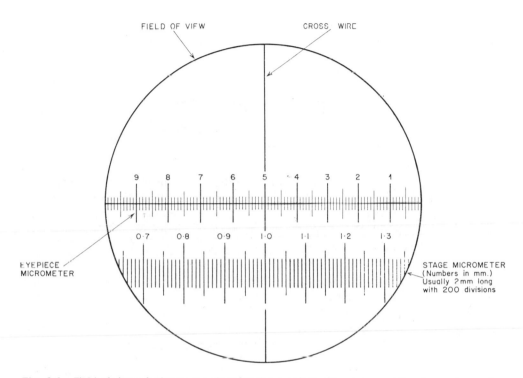

Fig. 3.1   Field of view of microscope showing the eyepiece micrometer and the stage micrometer both in focus. In practice the stage micrometer is manipulated to lie directly underneath the eyepiece scale, but it is shown here displaced to the south for clarity of illustration. In this arrangement one large eyepiece micrometer division measures 0.08 mm for this particular combination of objective and ocular.

3. Repeat the calibration for all the objectives of the microscope and keep the calibration on a small card near the microscope for future work.

4. The stage micrometer will never be needed again, provided the microscope in use does not require a change of lenses.

EXERCISE 3.2

To Calibrate the Field of View

For routine work the eyepiece micrometer may be completely unnecessary and a simple eyepiece incorporating a crosswire will be sufficient. It is essential, however, to know the diameter or radius width equivalence of the field of view as seen through this ocular combined with every objective of the microscope.

    The calibration is done as follows:

1.   Focus on the stage micrometer. Position the micrometer to lie exactly along the crosswire and note the length equivalence in millimeters from the center cross to the edge of the field of view.

2.   Repeat this operation for each objective belonging to the microscope; for example, the Zeiss Standard WL fitted with a ×12.5 ocular has the following calibration:

Using Objective	Radius of field (mm)
×2.5	3.5
×10	0.875
×25	0.35
×40	0.22

3.   Prepare a small card with this kind of calibration for your microscope and always keep it handy for the future estimation of grain size. The length of a grain can be compared with the field radius. Of course, if the field radius for one objective is known, that for another can be calculated; for example, given 0.875 for a ×10 objective, the field radius for a ×40 objective $= 0.875 \times \frac{10}{40} = 0.219$ mm.

4.   Should the ocular be subsequently changed, a new calibration must be done, for no two oculars have exactly the same width of field. Generally, however, although oculars give different magnifications and make objects look larger or smaller, the radius or diameter of the field of view may not be changed: for example the same Zeiss microscope fitted with a ×6.3 eyepiece gives an identical field calibration to that in (2) above, but the objects look about half their size to the eye.

EXERCISE 3.3

Grain Size of Rocks

Using the eyepiece calibrations of exercises 3.1 and 3.2, we can now focus on a rock thin section and estimate the particular and average grain sizes. The grain size of a rock is described qualitatively as follows:

	Average Grain Size
Coarse-grained rocks	> 5-mm diameter
Medium-grained rocks	1 to 5 mm
Fine-grained rocks	< 1 mm

We also use the prefix *micro* (e.g., microgranite or microdiorite). To deserve such a name the grain size should be between 0.5 and 0.05 mm, although in practice it is not generally so precisely restricted. The foregoing grain-size ranges follow the general practice of Moorehouse (1959) and Williams et al. (1954) and are preferred to older practices that set the medium/coarse limit at 10 mm (Johannsen, 1931).

B.   MODAL ANALYSIS

The petrographer frequently wishes to measure the relative amounts of the various mineral components of a rock. This is called modal analysis. To do so he has of necessity to examine a flat-sawed surface, a nearly flat outcrop surface, or a thin section. On these flat surfaces he will estimate or measure the relative area occupied by the individual mineral species and then relate it to the relative volume. But the point cannot be overstressed that the relative area occupied by any mineral species on a particular planar surface is not always equal to the modal (volume) percentage of that mineral in the rock mass.

This point can be illustrated by taking two extreme cases, shown schematically in Fig. 3.2

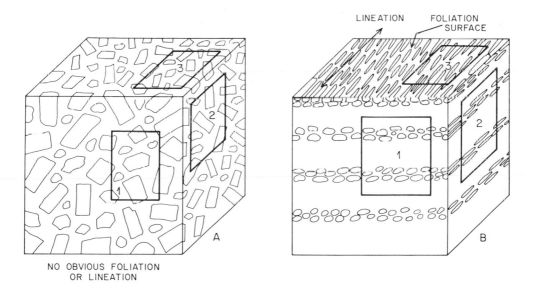

Fig. 3.2 Schematic block diagrams of A an unfoliated unlineated (isometric) rock and B a strongly foliated and lineated rock and three areas selected for modal analysis. See text for description.

Diagram A shows a cube sawed from a rock such as a porphyritic granite which has no apparent foliation or lineation. Because the crystal components are randomly oriented and evenly distributed, a large enough area selected for areal analysis on each of the three orthogonal faces (1, 2, 3) should give similar results, in each case rather close to the true mode of the rock. An average of the three would be the closest analytically possible value to the true mode.

Diagram B, on the other hand, shows a strongly foliated and lineated rock, such as a banded amphibolite gneiss, with layers alternatively rich in amphibole and plagioclase. A glance at this diagram shows immediately that an areal analysis on surface 3 will in no way be similar to that on surfaces 1 and 2, for 3 lies entirely on a mafic layer. Areal analyses on faces 1 and 2, cut perpendicular to the foliation, are the only ones suitable for modal analysis of the rock as a whole. The foliation surface, of nonuniqity, gives information of only one of the banded layers and cannot be related to the entire rock. Equally, we could imagine an area 4 parallel to 3, lying entirely on the felsic layer under the cube which would be devoid of mafic minerals.

EXERCISE 3.4

Selection of Area for Modal Analysis

1.   Carefully observe the hand specimen of the rock, noting in particular its average grain size and whether it is equigranular or porphyritic (porphyroblastic). Note also whether the rock has a homogeneous texture (Fig. 3.2A) or whether the minerals are segregated into bands (Fig. 3.2B). These observations will decide the minimum area that will have to be analyzed and the choice of plane on which the counting will be carried out.

2.   The problem of selection of a suitable area for counting is analogous to deciding on the unit cell of a crystalline lattice. The unit cell is the smallest possible volume representative of the whole crystal in terms of chemical components and geometrical distribution. In modal analysis we can select a cubic or orthorhombic cell for convenience so that all faces will be squares or rectangles.

3.   For a uniform unfoliated rock (Fig. 3.2A) there is no special direction to control our selection of unit cell edges or faces. Hence the specimen can be sawed randomly to give three mutually perpendicular planar surfaces. Rectangular unit cell areas can be inked on the surfaces (Fig. 3.2A), and each of these rectangles must be large enough to be representative of the whole planar face in terms of grain-size distribution and mineral content. It is impossible to give hard and fast rules for optimum unit cell size that would apply to all rocks because every rock specimen is, in a way, unique. The student must use his own discretion, but the following discussion should be a guide:

a.   If the rock is uniform in grain size, nonporphyritic, and of uniform texture, with a maximum grain size of 3 mm, a unit cell face need be no bigger than the size of a conventional thin section (24 × 24 mm). For such a rock the complete modal analysis may be done on thin sections under a microscope with a stage point counter.

b.   If the rock has uniform grain size but is extremely coarse-grained (pegmatite), clearly a thin section would be a useless unit cell face, for it would contain only part of a single crystal. If the grain size averages 2 cm, a unit cell face of at least 20 × 20 cm would have to be selected. This face would have to be analyzed by a point counter designed for large rock slabs or even carried out on a flat rock outcrop surface.

c.   If the rock is strongly porphyritic with phenocrysts up to 20 mm long set in a groundmass of grain size smaller than 3 mm, again a unit cell face cannot be a thin section, for one thin section may be entirely of groundmass or chosen to include too much phenocryst; it could not in any case be representative of the rock because it is too small. The proper choice of a unit cell face will depend on the distribution of the phenocrysts. It is suggested that the minimum length of the unit cell edge should be long enough to pass through at least two phenocrysts of the same mineral (Fig. 3.7). If there is any doubt, it is always best to err on the bigger side. A larger-than-necessary unit cell will give accurate results. A unit cell that is too small makes the exercise statistically meaningless, hence a waste of time. The modal analysis is performed on a large sawed slab.

4. If the rock has a uniform isometric fabric (Fig. 3.2A) in which unit cell faces 1, 2, or 3 are considered identical, modal analysis can be performed on one unit cell face alone. The areal analysis then becomes the modal analysis. If there is any doubt about the equivalence of the three mutually perpendicular faces, at least one specimen from a suite of similar rocks should be areally analyzed on all three faces. If they give similar results, the rest of the suite may be counted on a single unit cell face and the areal analysis becomes the mode.

5. In rocks that are foliated and banded (Fig. 3.2B) more care has to be taken in the selection of the unit cell areas and faces. The most useful faces are sawed perpendicular to the foliation. If the rock is also lineated, one of these faces should be parallel to the lineation, the other perpendicular to it (Fig. 3.2B). Much care has to be taken in outlining the unit cell face when the rocks have a segregation of the particular minerals.

6. A face lying on the foliation plane (Fig. 3.2B, No. 3) is generally of no use in modal analysis, for it is confined exclusively to a single segregated layer.

7. Faces selected at right angles to the foliation are suitable for determining bulk modal analysis of the whole rock, but care must be taken that the area selected for analysis does equal justice to the segregated bands. In Fig. 3.2B faces 1 and 2 are suitably chosen because each extends over two complete mafic and two complete felsic layers. Greater accuracy will be attained by extending the dimension of the unit cell face perpendicular to the layering to include a greater number of bands, but care must always be taken to ensure an equal weightage to the different types of layer.

8. Unless the banding is on a microscopic scale, the modal analysis of such rocks will of necessity have to be performed on hand-specimen slabs, for thin sections will normally be too small to include a suitably chosen unit cell face.

9. If the modal analysis is to be performed one layer at a time, only the mafic parts of areas 1 and 2 (Fig. 3.2B) are areally analyzed and averaged. The felsic layers are similarly analyzed. Individual layers may have a unit cell face small enough to permit their analysis to be done in thin section.

DISCUSSION

Because the majority of rocks tend to be (a) inequigranular to strongly porphyritic (porphyroblastic), (b) made up of at least four mineral components of inequal abundances, (c) often strongly foliated and/or lineated, and (d) lacking in any definite break in grain size between "megacrysts" and "groundmass," it can be generalized that modal analyses will usually have to be performed on hand specimens on which a flat surface has been sawed and polished. The identification of each grain may be simplified by using one of the staining techniques described in Chapter 2. Thin section modal analysis must be restricted to equigranular nonporphyritic (non-porphyroblastic) rocks that lack coarse layering or banding. This point is emphasized because most students and professional geologists simply cut off a portion of a rock, make a thin section from it, analyse it areally, and quote the result as the mode. This all-too-common procedure completely disregards the basic principles of sampling and the suitable choice of a representative modal unit cell. The numbers so produced will be without value. A modal analysis is meaningful if it is done only after careful

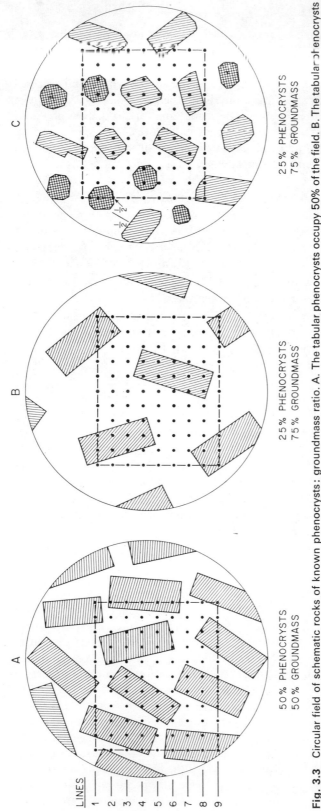

Fig. 3.3 Circular field of schematic rocks of known phenocrysts : groundmass ratio. A. The tabular phenocrysts occupy 50% of the field. B. The tabular phenocrysts occupy 25% of the field. C. The total phenocrysts, including prismatic and basal sections, occupy 25% of the field. The regular grids of 99 points are used to ascertain the mode. (See text, exercise 3.5, for discussion.)

50% PHENOCRYSTS
50% GROUNDMASS

25% PHENOCRYSTS
75% GROUNDMASS

25% PHENOCRYSTS
75% GROUNDMASS

observation of the specimen and the correct selection of an optimum modal unit cell, as outlined above. All that is needed is common sense.

It is hoped that the foregoing plea for rigor in modal analysis will not detract petrographers from performing only one analysis on a well-chosen thin section, which in their visual estimation appears to be representative of the rock, if time is not available for a complete analysis. It should not be forgotten that a single modal analysis is better than none at all.

Visual Estimation of Areal Mode

It is good petrographic practice to make a regular visual assessment of the relative proportions of mineral grains, or phenocrysts, that show on a flat outcrop surface, on a sawed hand specimen, or in a thin section as observed under low-power magnification. With practice and experience most petrologists become very good at estimating the mode at a glance. The beginner frequently overestimates the proportion of a colored mineral and underestimates the felsic minerals. No rules can be given for this exercise because the method per se is rather unscientific, analogous to asking a physicist to estimate the length of a metal rod without using a ruler. Some of us are good at estimating lengths, others rather poor; therefore estimation cannot be generally recommended as a proper scientific method. For interest Fig. 3.3 illustrates three circular areas with carefully measured ratios of simulated mineral phenocryst to undifferentiated groundmass, which might be seen under a low-power microscope. Figure 3.3A could be a thin section of a porphyritic rock containing phenocrysts of tabular feldspar. In this case the phenocryst : groundmass ratio is 50 : 50. B in Fig. 3.3 is a similar rock containing only 25% phenocrysts, and C, a rock composed of 25% phenocryst in different orientations, both basal and prismatic sections. Estimation is difficult because a few large and widely spread phenocrysts (B) often do not appear to have the same relative total area as a larger number of smaller phenocrysts (C).

EXERCISE 3.5

Principle of Modal Analysis by Point Counting

Whether the surface to be areally analyzed is a flat outcrop surface, a sawed hand-specimen slab, or a thin section, the method is universal—point counting; the principle is illustrated here with reference to Fig. 3.3.

1. Superimpose on the surface a transparent grid pattern. For an outcrop surface a fishing net gives an ideal grid. For a rock slab a suitable grid is a sheet of transparent graph paper. For a thin section a grid is obtained by moving the thin section in a controlled manner step by step along E-W lines, then jumping to another line in a N-S direction, thus allowing the crosswire of the microscope to trace out a grid. It is unnecessary, of course, that the grid steps be identical in both x and y directions, but x as well as y must be regular for statistical counting. Figure 3.3 has the same grid superimposed. Only the intersections of the lines, called points, have been shown.

2. Now identify the mineral under each point, tabulating the results as you proceed. In Fig. 3.3 the grid chosen has 99 points and in each illustration there are only two features to identify—phenocryst or groundmass. The results of this analysis are informative.

Consider Fig. 3.3A first. Along line 1 four points lie on phenocryst and seven on groundmass. Now count line 2, and so on to obtain the total:

	Points	Mode %	Value Used
Phenocryst	49	$\dfrac{4900}{99} = 49.5\%$	50%
Groundmass	50	$\dfrac{5000}{99} = 50.5\%$	50%

The result is remarkably close to the true value of 50 : 50 and we can assume that this was a well-chosen modal unit cell face. Consider the face carefully. It contains a total of about four phenocrysts and the edge of the modal unit includes at least two.

Now consider Fig. 3.3B. With the same modal unit the results are as followss

	Points	%
Phenocrysts	29	29%
Groundmass	70	71%

The true value is 25 : 75. Why the lack of accuracy in this case? Clearly our modal unit cell is too small, for it contains only about two phenocrysts. For this wide spacing of phenocrysts a modal cell of twice the edge length should have been selected.

Now consider C. We have the same size modal cell, but the grain size is smaller and the unit chosen looks like a suitable unit cell. The accuracy should be good. The actual results are the following:

	Points	%
Phenocrysts	$24\frac{1}{2}$	25%
Groundmass	$74\frac{1}{2}$	75%

NOTE.   If a grid is superimposed on a rock surface, it is inevitable that some points will fall exactly on the edge between two mineral species, in which case one-half point must be ascribed to each mineral.

DISCUSSION

A suitably chosen modal unit cell face should contain at least four or five, preferably more, of the largest phenocrysts of any one mineral. The length of a counting line should pass through at least two and preferably more crystals of any one mineral species. For coarse-grained rocks a wide spacing grid is suitable (Fig. 3.3A and B), but for fine-grained rocks the grid is chosen smaller because the modal unit cell is smaller. In Fig. 3.3C the grid spacing could be improved by having the points closer together.

EXERCISE 3.6

Modal Analysis of Rock Slabs

1.   The rock is sawed into a slab somewhat larger than the modal unit cell, as described in exercises 3.4 and 3.5. This slab may be about 10 × 8 cm or less in area and 1.5 to 2 cm thick, although thickness is of no importance. The face to be areally analyzed is ground and finished off with 600 alumina.

2.   The cleaned and dried surface is then stained for whatever minerals the operator wishes to accentuate. These will almost certainly include K-feldspar, plagioclase, and quartz but may involve the more appropriate stains described in Chapter 2.

3.   The modal area is outlined with a felt pen. From this point on, several techniques have been devised:

METHOD A   (*for porphyritic rocks*)

4.   Those phenocrysts more than 3 mm long (the choice of the lower size limit depends on the grain-size variation of the rock) are outlined in ink with a fine-pointed felt pen.

5.   Lay a sheet of tracing paper over the rock slab and, without moving the paper, trace the outline of every phenocryst, labeling each with its correct appropriate mineral identification (aided by the appropriate stain color).

6.   Remove the tracing and lay a sheet of transparent graph paper over it; then pin both to the table.

7.· Points of intersection of the lines on the graph paper can be counted visually for whichever feature they overlie—quartz, K-feldspar, plagioclase, phenocryst, or groundmass.

8.   If the modal area had been 10 × 8 cm and the graph paper used had divisions at every 5-mm interval, a total of 357 points would have been counted; therefore 1 point is equivalent to 0.28% (method after Nesbitt, 1964). Alternatively, instead of point counting the tracing the cumulative area of each phenocrystic mineral can be determined with a planimeter.

DISCUSSION

Larger areas or, alternatively, more than one slab may be counted, totaled, and converted to a percentage modal analysis. If x is the total number of counts for any one component and N is the total points counted, then x in terms of a modal % equals $100x/N$.

   By this method we are still left with a percentage labeled as groundmass, which is subsequently modally analyzed by using a thin-section analysis.

METHOD B

This alternative makes use of a macropoint counter designed by Smithson (1963) and illustrated in Fig. 3.4.

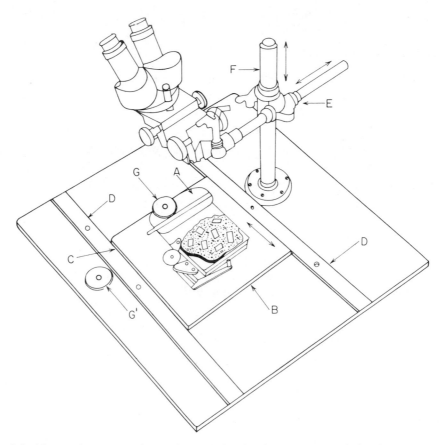

Fig. 3.4 Macropoint-counter for modal analysis of stained rock slabs (After Smithson, 1963.)

The components are fixed onto a plate of fiberboard. A normal binocular microscope is fixed for viewing and a mechanical stage is fixed to the base plate for translating the specimen.

The mechanical stage is simply a larger version of the common stage micrometer and in most workshops can be made of brass (see Fig. 3.4).

The drive for N-S and E-W movement of the stage is by a commerically available rack and pinion gear.

The mechanical stage (A) is mounted on a plate of clear plastic (B) that can be slid N-S. A flat spring (C) provides sufficient pressure to hold the plate fixed between its guide (D) during counting; E-W adjustment is allowed by loosening the clamp on the pipe (E) that holds the binocular microscope. Adjustments in these two directions allow the rock slab to be positioned under the microscope, which can be grossly focused by vertical adjustment on (F).

4. The rock slab is pressed into modeling clay on the mechanical stage and leveled.

5. The outlined modal area is positioned by movement of (**B**) and the microscope. If the area of the slab exceeds the translation of the mechanical stage, the slab can be repositioned as necessary for complete coverage.

6. The counting interval is varied by interchanging one notched wheel (**G**) for another (**G′**). The wheels are machined with different notch intervals around their peripheries and are quickly interchanged with a single locking screw.

DISCUSSION

When the macrostage described in this method is used, plagioclase staining is found to be unnecessary; K-feldspar stain, followed by a clear laquer spray, is enough to produce an obvious distinction between

    yellow K-feldspar,

    frosty white plagioclase,

    vitreous white quartz,

    yellowish-green biotite,

but by this method accessory minerals cannot be identified (Smithson, 1963).

METHOD C

This alternative makes use of a macropoint counting stage, illustrated in Fig. 3.5 and designed by Emerson (1958).

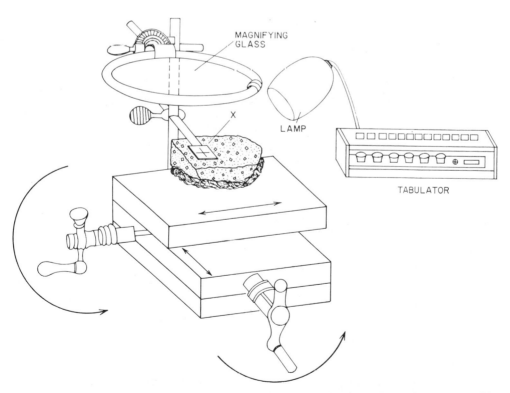

Fig. 3.5 Macropoint-counter for modal analysis of stained rock slabs (After Emerson, 1958.)

The stage is a standard dual cross-feed milling table with a maximum translation in each direction of 7 in. (18 cm).

The rock slab is leveled on the stage by pressing it into modeling clay attached to the stage.

As a reference point, a small glass plate with engraved cross hairs is mounted on a movable iron rod attached to a clamp stand so that it can be adjusted in height to ride over the surface of the sample. The cross hairs remain fixed as the rock sample translates under them. To aid the viewing a large low-power lens is placed above the cross wires, also attached to the clamp stand. Good lighting from a lamp is needed for rapid mineral identification.

4. The modal area of the stained rock slab is positioned with one of its corners under the engraved cross hair. The stage is translated a fixed amount each time and every mineral appearing under the cross hair is identified.

5. Counts can be recorded on a tabulator.

DISCUSSION

All minerals should be identifiable on the rock specimens.

The stage can handle a sample of 49 in.^{2} (7×7) or 324 cm^2, which is equivalent to 35 standard thin sections.

Rocks with grain sizes smaller than 3 mm are counted more conveniently on a microscope with standard equipment.

A porphyritic rock with fine-grained groundmass can be counted on this stage for phenocrysts (of various kinds) plus groundmass. The groundmass can then be determined separately on a thin section by the microscope method.

Phenocryst/groundmass ratio can also be determined by using a photograph of a flat outcrop surface. The actual photograph may be point-counted by this macro-stage (Emerson, 1958).

METHOD D

A macropoint counting stage, such as that described in method B, can be designed to be activated by the Swift point-counter controller (Schryver, 1968), which greatly facilitates translation of the specimen and tabulation. A binocular microscope is used with a zoom $7 \times$ to $50 \times$ lens to aid identification. This stage is now commerically available (James Swift and Sons).

DISCUSSION

Grains of diameter smaller than 0.1 mm cannot be identified. Initially, a comparison of the appearance of the rock slab under reflected light with some thin sections of the same rock under the polarizing microscope will be necessary to check the mineral identification. With experience, definite and accurate identification can be made with the binocular, provided, of course, that one is not expected to distinguish, for example, ortho- and clinopyroxene (Schryver, 1968).

EXERCISE 3.7

Modal Analysis of Thin Sections

1. The conventional thin section is stained for K-feldspar. Some workers prefer a plagioclase stain in addition. If other minerals such as cordierite are suspected, the staining technique should be modified (Chapter 2).

2. The thin section may now be covered.

3. Set up a polarizing microscope with a Swift automatic point-counter stage unit attached to its stage (Fig. 3.6). This unit, together with the electronic Control Box

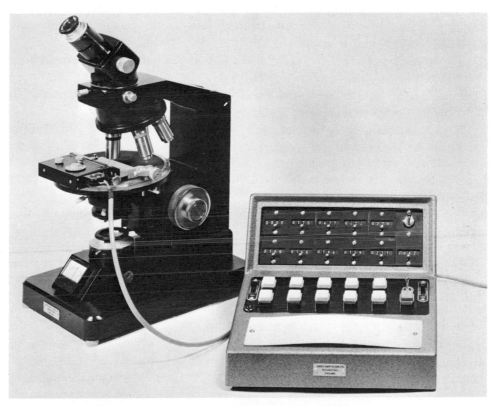

Fig. 3.6 The Swift automatic thin-section point counter showing the stage unit and the control box with counters plus total. (Courtesy James Swift and Sons, England.)

with 7 nonresetting counters (catalog 407), 14 nonresetting counters (414), or 11 resetting counters (411) may be purchased from James Swift and Sons Ltd., Joule Road, Houndmills Industrial Estate, Basingstoke, Hampshire, England; it can also be supplied with a wide range of screws to attach it properly to a wide range of well-known makes of polarizing microscopes. The stage unit has a maximum E-W traverse range of 27 mm, operated stepwise by pressing any one of the counter buttons on the control box. The 27-mm traverse limit is unfortunately rather small and requires that a thin section of normal length be counted in two square blocks by reversing the slide. It is to be hoped that future models of the stage will have a considerably longer traverse limit. The gears in the stage unit may be adjusted before beginning to allow a choice of E-W translation steps of $\frac{1}{3}$, $\frac{1}{6}$, $\frac{1}{10}$, or $\frac{1}{20}$ mm. Model A or model B, of improved design, permits the choice of horizontal traverse at $\frac{1}{2}$, $\frac{1}{5}$, $\frac{1}{10}$, and $\frac{1}{20}$ by simply turning a selector dial. The selection should depend on the grain size of the rock. A coarse-grained rock might be better suited to a $\frac{1}{3}$- or $\frac{1}{2}$-mm translation and a fine-grained one to a $\frac{1}{20}$-mm translation. The N-S translation range is 27 mm, effected manually by turning a knob, each click being $\frac{1}{2}$ mm. For fine-grained rocks, the Swift Eyescan attachment allows a crosslined graticule in the eyepiece to traverse at small intervals, depending on the primary magnification of the microscope optical system.

4. Firmly position the thin section in the stage unit. Connect the control box to the unit and note the reading in each of the counters. If the counters are not self-zeroing, the beginning numbers must be noted. The bottom right-hand counter gives the total. Each counter is labeled with the name of the mineral to be identified; for example, quartz, K-feldspar, or plagioclase.

5. The stage unit clutch is engaged and the unit pulled fully back to zero with the sliding knob.

6. Lock the microscope stage so that the E-W translation of the stage unit is exactly along the E-W cross wire.

7. Identify the grain under the cross wire. Experience proves that a × 10 objective is best for routine work. Press the appropriate button on the control box and the stage will automatically translate one step. Identify the grain under the microscope by pressing the appropriate button until one complete line of translation has been performed.

8. Now pull the zeroing knob on the stage unit back to zero and at the same time turn the N-S adjusting knob one notch to start a new translation line.

9. Once approximately 1000 to 2000 points have been counted (observe the totalizer) the operation is stopped. Note the final reading on each mineral counter, subtract from it the beginning reading, add up all the mineral totals, and convert the individual minerals count to a percentage.

DISCUSSION

To aid identification the staining will normally be enough of a guide, but plane-polarized and cross-polarized light may be chosen at will. Also, at any one point in a traverse the microscope stage may be unlocked and rotated to facilitate identification. Remember to return the stage immediately afterward to its locked zero position.

Because a large number of points is being counted (2000, approximately), one can afford to make a few mistakes in identification. A high speed can be achieved. Because this is a statistical analysis, there is no point in wasting time deciding if one difficult grain is, for example, K-feldspar or plagioclase. It is better to make a reasoned guess and pass on quickly and not sacrifice speed. It is preferable to analyse a large number of thin sections modally and rapidly with a few mistakes here and there than to analyze one very carefully with no mistakes.

As an example of an analysis on a nonresetting counter the record might look like the following; the resetting counter needs no beginning reading:

	Quartz	Plagio-clase	Ilmenite	Ortho-pyroxene	Clino pyroxene	Total
End	3612	2837	2761	2072	7855	9461
Beginning	3571	1752	2735	1654	7321	7354
Counts	41	1085	26	418	534	2107*

Total counts 41 + 1085 + 26 + 418 + 534 = 2104.

Although the areal analysis of this thin section may be precisely quoted, there is spurious accuracy here, for if we were to take another thin section of the same rock it would be impossible to obtain 51.6% for the plagioclase. Because rocks are inherently variable mineralogically and texturally, we should not claim superior homogeneity by quoting a precision that is too high. Hence we might be tempted to quote the above mode as

quartz, 2%;
plagioclase, 52%;
orthopyroxene, 20%;
clinopyroxene, 25%;
ilmenite, 1%.

Chayes (1953), however has argued convincingly in favor of retaining and quoting the first decimal in modal analysis even when as few as 1500 to 2000 points have been counted. The tenths place should not be rounded out, particularly when (as is usually the case) many modal analyses of individual specimens will be pooled in statistical analysis.

EXERCISE 3.8

Modal Method for Porphyritic Rocks

It is common for a rock to consist of larger crystals (phenocrysts or porphyroblasts) which can be identified in the hand specimen and of a finer grained groundmass which cannot (Fig. 3.7).

1. Prepare a flat stained slab of the rock and outline on it an area large enough to be a modal unit cell (Fig. 3.7). By using one of the hand-specimen analytical methods (exercise 3.6) measure quantitively the area covered by each component.

PHENOCRYSTS   quartz a% of the total area,
                  feldspar b%,
                  biotite c%,
GROUNDMASS D%
                  $D = 100 - (a + b + c)$.

---

* Frequently the final totalizer reveals a small error, usually because one of the buttons has been badly depressed and more than one count has registered, but the two totals should check approximately as a similar amount.

Mode: quartz $\% = \dfrac{41 \times 100}{2104} = 1.9\%$;

       plagioclase $\% = \dfrac{1085 \times 100}{2104} = 51.6\%$.

       orthopyroxene $\dfrac{418 \times 100}{2104} = 19.9\%$;

       clinopyroxene $\dfrac{534 \times 100}{2104} = 25.4\%$;

       Ilmenite $\dfrac{26 \times 100}{2104} = 1.2\%$.

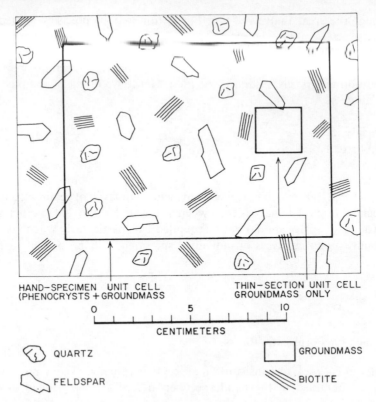

HAND–SPECIMEN   UNIT CELL
(PHENOCRYSTS + GROUNDMASS

THIN–SECTION UNIT CELL
GROUNDMASS ONLY

0   5   10

CENTIMETERS

QUARTZ

FELDSPAR

GROUNDMASS

BIOTITE

Fig. 3.7 Diagrammatic sketch of a porphyritic rock. The hand-specimen large unit cell area is first modally analyzed for all phenocrysts plus groundmass by exercise 3.6. A smaller unit cell is then selected to represent the groundmass and modally analyzed in a thin section by method 3.7.

2. Prepare a thin section from an area free of phenocrysts. Now analyze it modally by the thin-section method (exercise 3.7). Suppose it is made entirely of the same minerals as the phenocrysts, with the following percentages:

GROUNDMASS   quartz $x\%$
                  feldspar $y\%$
                  biotite $z\%$
                              ————
                              100

Then the overall modal composition of the rock, including phenocrysts and ground-mass, is

$$\text{quartz} = a + \frac{xD}{100}\ \%,$$

$$\text{feldspar} = b + \frac{yD}{100}\ \%,$$

$$\text{biotite} = c + \frac{zD}{100}\ \%.$$

DISCUSSION

Alternatively, the phenocrysts/groundmass ratio could be determined by counting modally, using exercise 3.6, a photograph of a large flat outcrop surface of the rock.

EXERCISE 3.9

Modal Analysis by X-ray Diffraction

This method will be more easily understood after reading Chapters 6 and 10.

If a rock is ground to a fine-grain size so that the particles do not exceed 5μ and a small representative amount of the powder is made into an X-ray powder diffraction preparation, each mineral component of the rock will give its own characteristic X-ray powder pattern. If certain diffraction lines can without doubt be ascribed to any one of the mineral components, and to that component alone, the intensity of the identified lines can be related to the concentration of that mineral in the rock.

Theory. $I_x = KV/\mu$, where I_x = the measured intensity of a diffraction line of any crystalline component (x) in the rock sample, V = the volume fraction of that component, μ = the linear absorption coefficient of the specimen, and K = a constant that depends on the incident X-ray intensity and goniometer geometry. The specimen must be "thick" enough to attenuate the incident beam completely.

$$I_x = \frac{KX}{\rho_x A_x},$$

where ρ is the apparent density of the specimen and ρ_x = the true density of the component being estimated.

$$X = \frac{\rho x}{\rho} = \text{the weight fraction of the component being estimated,}$$

$$A_x = \frac{\mu}{\rho} = \text{the mass absorption coefficient of the specimen.}$$

If the same diffraction line is measured on a standard sample for which X is known and has the value S, then

$$I_S = \frac{KS}{\rho_S A_S}.$$

Because $\rho_S = \rho_x$, we can combine the two equations above to obtain

$$X = \frac{A_x I_x S}{A_s I_s}.$$

Generally a pure mineral can be selected as the standard, in which case $S = 1$.

METHOD

The method works only for fine-grained preparations in which the particle size is $<5\mu$. A natural clay may be used directly without any grinding. Any fine-grained rock, such as basalt or hornfels, somewhat ground up, would also be suitable. Such fine-grained rocks are difficult to analyse modally by conventional point-counting techniques, and this X-ray method may be the general answer to their analysis.

If the constituent particles exceed 5μ, microabsorption effects cause the above equations to be invalid.

Any diffraction peak may be used. Therefore in a rock composed of a mixture of minerals lines that do not interfere and can with certainty be assigned to one mineral species are used. A suitable line for each mineral present can be selected after a diffractometer trace has been obtained. Cu or Co $K\alpha$ radiation is suitable.

1. Samples must be ground and homogenized in a Wig-L-Bug or similar mixer. The grain size must be less than 5μ. One sample is made of the rock to be measured, and one sample of each of the pure mineral components of which the rock is known to be made is also prepared.

2. *Specimens for diffractometry* are prepared by lightly pressing the powder into 2×1 cm rectangular metal holders. If certain fibrous or platy minerals become preferentially oriented in preparation, the method is not useful, for peak intensity is not then linearly related to mineral concentration. Orientation effects may be reduced by pressing the front surface of the specimen against ground glass or coarse paper instead of against a polished surface.

3. Run a complete diffractogram of the rock powder by using the selected radiation and identify the various lines as belonging to the various mineral components. The identification is aided by running the prepared specimens of the individual minerals and comparing the 2θ positions of the lines.

4. Select one peak for each mineral constituent, both in the rock specimen (X) and in the individual pure mineral preparations (S) which can with certainty be ascribed wholly to that particular mineral, and set the goniometer exactly on the precise 2θ setting for one peak of each mineral constituent in turn.

5. Take a large number of counts (fixed time, 10 seconds or more) at this 2θ setting for the rock preparation and for the appropriate pure mineral standard. Then move the goniometer to either side of the peak to measure the background.

6. Subtract the background value from the peak value for each mineral component in the rock and for each individual pure mineral standard. All counting times should be identical for all peaks and background measurements. The values so obtained are I_x and I_S, one I_x and one I_S being determined for each of the minerals in the rock. If no other lines lie close to the one being measured and the background is flat, there is no problem in obtaining an accurate peak height of the line above background. For complex mixtures of minerals, as in rocks or clays, the background is often complex. It will have to be judged on a diffractogram traced several degrees on each side of the peak. In such a case integrated peak intensity is obtained as a better value than peak height. Two methods are possible:

a.   Measure the area under the peak above the background line;
b.   Calculate the product of the maximum peak height and the peak width at half maximum height (all above background). Whatever method is used, it must be used throughout.

Alternative (b) is perhaps the best measure of line intensity. A diffractometer with a proportional counter may be used, but a scintillation counter is preferable. A pulse height analyzer is also a good addition for measuring I_x and I_S for the mineral components.

7. *Specimens for mass absorption measurements* are prepared by pressing the rock powder or the individual pure mineral component powders into a hole 1.27-cm in diameter in a $\frac{1}{8}$-in. Perspex holder. Pressures varying between 200 and 2000 kg/cm^2 are necessary to ensure a self-supporting sample. The sample must be of uniform thickness. Each sample should weigh 0.07 to 0.10 g generally. Many rock or mineral powders will not press into a self-supporting disc. They should be mixed with a known amount of filter-paper pulp and pressed.

8. A monochromatic beam of high intensity is obtained by putting a single crystal slab, for example, quartz, lithium fluoride, or oriented powder specimens such as graphite, into the diffractometer sample holder and setting the detector at the correct 2θ position corresponding to the reflecting plane. Instead of the diffractometer, the spectrometer may be used by placing a Cu metal disc in the sample chamber (if Cu radiation was used in step 3) or Co if appropriate. The goniometer is then set at the appropriate 2θ angle for the dispersing crystal to receive the first order Cu$K\alpha$ or Co$K\alpha$ radiation.

9. The intensity of the beam is measured first by counting for 10 seconds. Then the rock or mineral disc is placed between the counter and its collimator and the intensity again counted for the same time. This is done for the rock and for each of the mineral standards. When a sample disc has been diluted by adding filter paper pulp, a correction has to be made for the dilution. This is done by running a specimen with the same amount of filter paper but without any powder added. The reduction in X-ray intensity because of that thickness of filter paper alone has to be added to the value obtained for the disc prepared by mixing paper pulp and rock powder. Mass absorption coefficient can also be determined with a spectrometer by using the attenuation of the Mo$K\alpha$ Compton-scattered primary radiation, after the manner of Reynolds (1963; 1967) (see exercise 10.18).

10. *Calculation of Mass Absorption Coefficient.* $I = I_0 \exp(-\mu l)$, where $I_0 =$ monochromatic primary beam intensity, $I =$ attenuated beam intensity, and thickness $= l$. Therefore measurement of I and I_0 determine μl. The weight of the sample (the holder is weighed before and after filling) and its area 1.27 cm^2 gives its mass per square centimeter, ρl, and the mass absorption coefficient $A = \mu/\rho = \mu l/\rho l$. Ratios of I/I_0 as low as $3 \times 10^{-5}(\mu l = 10)$ can be used to give A accurately and can be measured with a reproducibility of about 1%. The main error is deviation in thickness l.

11. *Final Calculation.* The equation $X = A_x I_x S/A_s I_s$ can be applied to artificial mixtures and natural mixtures (rocks). The quantities A_s, I_s, A_x, and I_x have been determined experimentally; hence X, the weight fraction of the component, is obtained. In the described procedure all standards were of pure minerals; hence $S = 1$. Our equation therefore is $X = A_x I_x/A_s I_s$.

(Method after Norrish and Taylor, 1962.)

DISCUSSION

Analyses made by using different lines of a diffraction pattern give good agreement even when they have large angular separation.

The accuracy of the method will be limited by the nature of the mineral in the rock powder, its orientation, and microabsorption variability. Even with the limitations the method gives results as good as with other modal methods and avoids the necessity of calibration curves or the addition of an internal standard.

The method is particularly valuable for modal analyses of fine-grained rocks, provided the presence of any mineral can be definitely proved by the presence of its

X-ray diffraction pattern and any one X-ray peak assigned without doubt to a single mineral component.

A method of rock modal analysis which uses the electron microprobe has been described by Keil (1965).

EXERCISE 3.10

Modal Analysis of Granite by X-ray Diffraction

Peak height determinations on diffractograms can be as accurate as stained thin-section point counting for modal analysis of granitic rocks. An advantage in using the powder method for modal analysis is that K-feldspar triclinicity can be determined at the same time.

1.   Crush the granite to $<5\mu$ grain size and homogenize the powder.

2.   Make a diffractogram of the rock powder.

3.   It is convenient to measure the K-feldspar-plagioclase-quartz ratio first. The heights of the plagioclase $\bar{2}01$ peak, quartz 100, orthoclase $20\bar{2}$, plus microcline 111 give the plagioclase : quartz : K-feldspar ratio. The heights of the orthoclase 111 and microcline 111 give the orthoclase-microcline ratio. Figure 3.8 displays the ratios determined from artifical mixes.

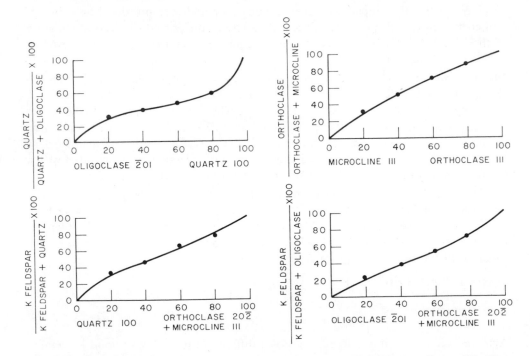

Fig. 3.8   Relative intensity of quartz, orthoclase, microcline, and plagioclase X-ray diffraction peaks. (After Bradshaw, 1963.)

4.   From Fig. 3.8 the K-feldspar : oligoclase, oligoclase : quartz, and quartz : K-feldspar ratios give three equations in three unknowns from which the percentages of K-feldspar, oligoclase, and quartz can be determined. From the other graph the microcline : orthoclase ratio can also be calculcated.

DISCUSSION

The results fall within $\pm 6\%$ of the absolute value. They are less accurate than point counting but well within acceptable limits.

EXERCISE 3.11

Modal Analysis by Mineral Separation

For rocks of simple mineralogy it is often possible to make a clean separation of the minerals after crushing the rock. The weight of the separated minerals may be converted to a volume to obtain the mode. An amphibolite composed of only three minerals—plagioclase, hornblende, and ilmenite—would be ideal for this method.

1. Note the average grain size of the rock.

2. Crush, first in a jaw crusher, then in a percussion mortar, a volume of the rock large enough to be representative of the whole rock mass. The final grain size after crushing should be slightly smaller than the rock grain size observed in step 1.

3. Wash the crushed rock to remove the fine dust and dry.

4. Separate the rock powder into its various components by using an electromagnetic separator, heavy liquids, or both.

5. After separation clean and dry the separates.

6. Weigh each mineral separate and convert its weight to a volume by means of its known or measured specific gravity.

7. Convert the individual mineral separate volumes to a percentage mode.

DISCUSSION

Although in theory this method is ideal, in practice there are weaknesses. During crushing of the rock a particular mineral, for example, feldspar, may crush more readily into a fine dust than the other components and much of the dust may be lost. It is not by any means always easy to effect a perfect separation, and grains of one mineral may contain inclusions of another even after crushing and separation. The method is also time consuming. Nevertheless it may be usefully employed in the modal analysis of rocks of simple two- or three-component mixtures which can readily be separated by their magnetic properties or by density.

C. PHOTOMICROGRAPHY

Taking a good photograph through a microscope is in essence simple, provided a few basic rules are observed. There must be a careful choice of the correct combination of objective and ocular. The combination at present in use in your microscope may appear perfect for visual observation. However the camera, unlike the eye, does not accommodate and requires a perfectly flat field. The lighting system must be well aligned and the optical system, clean. A stage micrometer should be photographed along with a batch of photographs to provide the enlargement scale. Careful focusing and correct exposure (based on a previous calibration) ensure a good negative.

Choice of Objective and Eyepiece Combination

1.   Decide on the object field of view required in the photomicrograph. For coarse-grained and porphyritic rocks perhaps the largest possible object field will be required. On the other hand, one might wish to photograph a small feature in detail and will choose a small object field.

$$\text{Object field} = \frac{\text{eyepiece field-of-view index}}{\text{objective magnification}}.$$

The field size may therefore be increased by (a) choosing an eyepiece of larger field-of-view number (not its magnification), (b) by choosing an objective of lower magnification, or (c) both; for example a Zeiss combination of a $10 \times$ objective with an $8 \times$ Kpl (compensating flat field) eyepiece of field number 18 will give an object field of $18/10 = 1.8$-mm diameter.

2.   Choose the correct objective. For good photomicrography it is essential to use a flat-field objective. Others should not be considered because the whole field will not be simultaneously focused. The best on the market are of German manufacture, and among them Leitz (Ernst Leitz, Wetzlar, West Germany) and Zeiss (Carl Zeiss, Oberkochen/Wuertt, West Germany) are outstanding in quality. FLAT-FIELD ACHROMATS are objectives suitable for black-and-white photomicrography. FLAT-FIELD APOCHROMATS are more expensive and are fully color-corrected. They are essential for color photomicrography and are the best overall for black and white. Because polarized light is being used, all objectives must be strain-free; they are designated P or POL, meaning suitable for use with polarized light.

ZEISS Recommended Objectives. Use PLANACHROMATS Pol for black-and-white photomicrography. Suitable magnifications are $\times 2.5$, $\times 10$, $\times 40$, and $\times 100$ (oil immersion). The 40 and 100 are correctly designed for a 0.17-mm cover glass. The lower power objectives can be used with or without a cover glass. Use PLANAPO-CHROMATS Pol for color and black-and-white photomicrography. Suitable magnifications are $\times 4$, $\times 10$, $\times 40$, and $\times 100$ (oil immersion). Only the $\times 4$ may be used without a cover glass. All the necessary information is engraved on the objective metal barrel:

Pol

Plan 40/0.65

160/0.17

meaning an objective suitable for polarized light, planachromatic, $\times 40$ magnification, of numerical aperture 0.65, to be used with a microscope tube of 160-mm length and a cover glass 0.17 mm thick. The Planapochromatic objectives are similarly labeled with Planapo instead of Plan. The cover glass thickness quoted is that recommended to give optimum results. When none is quoted, the objective may be used on covered or uncovered thin sections. For reflected light photomicrography all objectives give a flat field and are for use without a cover glass. They are designated EPIPLAN POL. Compared with dry objectives, immersion objectives result in an enhanced contrast because much reflection is eliminated.

LEITZ Recommended Objectives. The objectives must be designated Pl (PLANA-CHROMAT) and P (strain-free for polarized light) or (P) strain-free within limits. NPl objectives are also recommended. They are for a reduced field and are less expensive. A suitable range of objectives is Pl(P) 1, Pl P 2.5, NPl P 6.3, NPl P 16, NPl P 25, NPl P 40, and NPl P Oel 100 (oil immersion). All are achromats and all must be combined with a Periplan eyepiece. The ×1, ×2.5, and ×6.3 may be used without a cover glass. The others must have a 0.17 ± 0.05 mm cover glass. The information is engraved on the objective barrel:

$$170/0.17$$
$$NPl\ P\ 40/0.65$$

meaning for use with a 170-mm tube and a 0.17-mm cover glass, and new Plano polarizing (strain-free) ×40, numerical aperture 0.65. It is understood that Leitz Plano lenses are achromats. No apochromats are available. The Leitz PLANO 1/0.04 objective is extremely useful for petrographic photomicrography. If it is properly combined with a Periplan GW 6.3× widefield eyepiece, it covers a field of view of 28-mm diameter and produces a plane image over the entire field. A special condenser must be used in conjunction with this objective to illuminate the larger field; it has a dovetail attachment suitable for the Leitz Ortholux microscope. One feature needs a special note: it incorporates an iris diaphragm. My experience shows that this diaphragm should always be fully open; otherwise the photographs produced will be unsharp. For incident light the recommended flat-field objectives are Pl R 2×, Pl 3.2×, Pl 8×, Pl Oel 8×, Pl 16×, Pl Oel 16×, Pl 32×, Pl 80×, and Pl 160×. All are achromats. There is one Apochromat Pl APO Oel 160×. All must be used with a Periplan eyepiece and give a fully flattened field even with an eyepiece up to field-of-view index 28.

3. The correct eyepiece should be combined with a properly chosen objective. COMPENSATING FLAT-FIELD eyepieces must always be used in conjunction with FLAT-FIELD objectives.

ZEISS Eyepieces. Only the KPl eyepieces (compensating flat-field) should be used. They reproduce a perfectly flat aerial image and lateral chromatic aberration is compensated. W engraved on the eyepiece means wide-angled. The choice is (field-of-view number in parenthesis) KPl 8× (18), KPl 10× (16), KPl 16× (10), KPl 20× (8), and KPl 25× (6.3). Hence for the largest possible object field with Zeiss equipment a combination of KPl 8× and a Planachromat ×2.5 gives an object field of 18/2.5 = 7.2 mm.

LEITZ Eyepieces. Periplan, GF Periplan, or GW Periplan eyepieces are suitable. The selection will depend on the tube diameter. All are clearly labeled Periplan. The Periplan series is for tube diameter 23.2 mm and the choices are 6.3× (18), 8× (16), 10× (18), and 25× (8) (field-of-view index in parenthesis). GF Periplan are for tubes of 23.2 mm and come in the following magnifications: GF 10× (18), GF 12.5× (18), GF 16× (15), and GF 25× (10). The GW Periplan series is for tubes of 30 mm and the following are available: GW 6.3× (28) and GW 10× (24), both of which are suitable for photomicrography because of their large fields. Hence the largest possible object field with Leitz equipment is obtained by using the Pl 1 objective in combination with the Periplan GW 6.3×, with object field of 28/1 = 28 mm. Clearly, then the Leitz equipment is ideal for photomicrography of coarse-grained and porphyritic rock textures. Much contradictory advice is to be found in photomicrography textbooks; for example, Schenk and Kistler (1962) state that plano objectives must always be used with normal compensating eyepieces for photomicrography. This is untrue, for both Leitz and Zeiss recommend the combination of Plano objectives with plano eyepieces.

EXERCISE 3.13

Defining the Scale of a Photomicrograph

It is bad practice to present a photomicrograph print with the caption stating that the print is ×35 or a similar magnification. Such a statement is ambiguous, for one is uncertain if it means ×35 linear or areal, and there is always a question in the reader's mind whether ×35 was the correct magnification of the negative or the original print. The blockmaker may have reduced the print size for publication; hence ×35 is no longer the correct magnification, though it still appears in the caption.

The best scale for a photomicrograph is an inked line on the original print whose length represents a measured length of 1/10 or 1 mm on the original rock. Such a line scale immediately indicates the grain size of the photomicrograph and is not subject to mistake during final reduction in block making for publication.

Since many factors contribute to the final scale of the original print, such as magnification of the objective, eyepiece, camera adaptor, camera, and enlarger, it is best not to compute the enlargement. The following method is recommended as a standard procedure:

1.   Select a series of thin sections that will be photographed under identical magnification. Do not mix magnifications.

2.   Take an exposure of each thin section in turn, each under identical magnification settings. Then focus on a stage micrometer and photograph it in an identical manner under identical magnification.

3.   This length of developed negative then includes several photomicrographs and the correct millimeter scale for the set. If other photomicrographs under greater or lesser magnification are required, repeat the procedure and finish by photographing the stage micrometer under magnification identical to the thin sections.

4.   When exposing the photographic paper under the enlarger during printing, once the correct enlarger setting has been established, use border masking as is usual practice. Now slide the strip of negative through the enlarger until the negative of the stage micrometer is centered. By masking the area of the print just exposed expose only one margin of the photographic paper to the image of the stage micrometer. This will be facilitated if the sheet of paper is somewhat larger than necessary, with a wider than normal margin on one end.

5.   Develop, wash, and fix the print. It will consist of the photomicrograph with the correct millimeter scale on one of its margins

6.   Thus any photomicrograph will always have its scale in millimeters. For publication or mounting for display, however, the white margin may be trimmed off and the millimeter scale inked onto the photomicrograph in one corner. The caption should state what the scale divisions mean (mm or 1/10 mm).

The Simplest Photomicrography System

1. Mount a camera with its lens set at infinity as close as possible above the microscope eyepiece and open the camera stop as far as you can. Many camera manufacturers supply an attachment for this purpose. Preferably a vertical tube should be attached to the microscope for holding the camera.

2. If the microscope has been carefully focused with relaxed eye accommodation and the camera set at infinity, the photomicrograph will be sharply focused. This system may cause slight vignetting of the negative because the camera lens will usually have a wider angle than the eyepiece. Best results are therefore obtained with wide-angle eyepieces. Any doubts regarding critical focus can be eliminated by using a single-lens reflex camera; the focus can be seen on the ground glass of the camera. Use the fine focus adjustment of the microscope for final focusing.

3. A photoelectric exposure meter is convenient for determining the correct exposure.

If working conditions are always identical (objective, eyepiece, position of condenser iris, lamp, and film speed), exposure measurements are not required. A few test exposures may be made at a variety of speeds; for example, $\frac{1}{50}$, $\frac{1}{25}$, $\frac{1}{10}$, $\frac{1}{5}$, $\frac{1}{2}$, 1, 2, 5, 10 seconds. The correct exposure is selected as that giving the best negative. After this test the optimum time determined can always be used.

Another simple and accurate method is direct measurement of the finder image brightness of the reflex camera, preferably by means of a photoelectric exposure meter which has a small acceptance angle and measuring surface (e.g., the Zeiss IKON "IKOPHOT CD"). No calibration photographs are required if the following procedure is used:

a. Point the camera with lens set at f 2.8 at a bright wall. Set the meter to the correct film speed and take a meter reading of the wall with f 2.8-1/50 second settings on the meter.

b. Then take a reading with the meter pointing at the ground glass of the camera finder image and read the f stop corresponding to 1/50 second; it may be f. 1.4.

c. When taking the photomicrographs with the camera fixed on the microscope, all readings should be done with the meter pointing at the camera finder image. The correct exposure time is the value corresponding to f 1.4.

(Method after Mollring, 1968.)

Photomicrography with a Photomicrographic Camera

The microscope in front of the camera stops down the lens to a large extent so that the focusing elements of the camera (ground-glass screen or image splitter) are not ideally suited and critical focusing may not be possible.

Much better results are obtained by removing the camera lens and attaching the camera body to the microscope with a photomicrographic camera adaptor which is available with or without a photoelectric exposure meter. Both Zeiss and Leitz supply these attachements. If this instrument is used, it is best to have one that incorporates the photocell. The procedure is as follows. these instructions are for the Zeiss photomicrographic camera but are similar to those for Leitz and other equipment. The Leitz equivalent is the microattachment for the Leica camera complete with Microsix-L exposure meter. It has a built-in vibration absorber which ensures perfect sharpness of the photomicrographs.

1.   Fit the microscope with the correct eyepiece/objective combination and make sure that the correct condenser is used and that the lighting is centered to ensure even illumination of the whole field.

2.   Select the correct setting on the adaptor focusing barrel; for example, on the Zeiss adaptor CL is the setting for the CONTAX camera.

3.   Remove any unnecessary substage lenses. The fewer the number of lens components the light has to pass through the better. Those that are used must be clean.

4.   Check the plunger positions. Make a note of the position for viewing, measuring, and taking the exposure. The Zeiss attachment has

fully out	for measuring
middle (click stop)	for viewing and focusing
fully in	for taking

5.   Push the plunger to the viewing position.

6.   Carefully focus the cross hairs of the photomicrography attachment; this is often engraved as a double line on the ground-glass screen. If it is, the double line must be sharp and resolved. Once the ground-glass screen is focused the barrel focusing device MUST NOT be further moved.

7.   Focus on the thin section with the microscope focusing device. NEVER use the photomicrography attachment for focusing the thin section.

8.   Push the plunger fully in (taking position).

9.   Switch on the photocell measuring meter. Allow approximately 10 minutes for it to equilibrate.

10.   Put its scale to 0–0.

11.   Adjust the meter reading exactly to ZERO by using the Nullpunkt knob.

12.   Now put the scale to 0–100 μA.

13.   Adjust the meter reading to exactly ZERO by using the Kompensation knob.

14.   Now check that the scale reading is ZERO on all sensitivity scales (0–0, 0–1, 0–10, 0–100, and 0–1000). If not, a slight adjustment of Nullpunkt when on scale 0–0 and a slight adjustment of Kompensation when on scale 0–1000 will ensure the proper settings. DO NOT play violently with the knobs; otherwise it will be difficult to obtain their correct relative settings. The knobs should be adjusted only when the instrument is properly warmed up, as indicated by the stability of the meter needle.

15.   Put to scale 0–1.

16.   Pull plunger out (measuring position).

17.   Note the deflection on the scale. If it is less than 10, put on scale 0–10. If the reading is still less than 10, put on scale 0–100. In this way you will obtain a suitable scale and a reading of more than 10.

18. The correct exposure for the most widely used black-and-white film PAN-ATOMIC-X daylight film ASA 32 Din 16 is the following (all times in seconds):

Scale Used	Scale Deflection			
	Between 5 and 10	Between 10 and 25	Between 25 and 50	Between 50 and 100
0–1	$\frac{1}{10}$	$\frac{1}{25}$	$\frac{1}{50}$	$\frac{1}{100}$
0–100	1	$\frac{1}{2}$	$\frac{1}{5}$	$\frac{1}{10}$
0–100	10	5	2	1
0–1000	100	50	25	10

19. Having selected the correct exposure time, put the plunger to the middle (viewing) position and check the focusing and photo composition.

20. Put plunger IN (taking). Set the correct camera speed and, using a cable or time delay, expose. For exposures greater than 1 second use the B position on the camera.

21. The above table is suitable for all films of ASA 32 Din 16 speed ratings. Adjustments should be made for other films; for example, if the film speed is ASA 50 Din 19, all the times given in the table should be halved. If the film is ASA 15 Din 13, all given times should be doubled.

DISCUSSION

Panatomic-X is one of the best black-and-white films, and Kodachrome is a favorite for color photomicrography. Agfa CT 18 is popular also. Kodachrome A film is balanced for photoflood lighting (3400 K); therefore a photoflood bulb may make an ideal light source for the microscope when this film is used. Care must be taken to check the film's color balance; that is, whether it is balanced for daylight or tungsten light. All steps should be taken to see that the lighting is appropriate for the film. Other combinations may be used with light-balancing filters; for example the Kodak Wratten filters. Color-compensating filters may also be needed to adjust slight color imperfections in subsequent films.

More elaborate and expensive instruments such as the Zeiss Photomicroscope or the Leitz Orthomat fully automatic camera are available for photomicroscopy. These instruments should enable the operator to obtain high-quality photomicrographs, but with normal cameras the simpler methods described in exercise 3.15 should with care produce results of professional quality.

EXERCISE 3.16

Low-Power Photomicrography

1. Insert a rock thin section into the film carrier of a photographic enlarger and project its image onto white cardboard.

2. An open 4 × 5 in. sheet film holder, in which a white card has been inserted, is placed flat on the cardboard and moved about until the desired feature is projected onto the card.

3. The thin section is carefully focused on the card in the holder. The position of the holder is marked.

4. All lights are turned off. A loaded film holder is substituted and the exposure made by turning the enlarger light on for the correct exposure.

5. For a photograph under crossed nicols two sheets of polaroid at right angles to one another are placed one on top of and the other beneath the thin section before it is placed in the enlarger.

DISCUSSION

Because of the many variables, the operator will have to experiment with the exposure time. A trial negative can be exposed by releasing a cover on top of the negative an inch at a time every three seconds after the lighting of the enlarger. This will produce a film with strip exposures of 3, 6, 9, 12, and 15 seconds from which the ideal exposure time may be determined.

Scale is best determined by replacing the slide with a microscope stage micrometer on which a millimeter scale is ruled and photographing it on the next film exposure without changing the height of the enlarger. The thin-section film should always be kept with the film of the stage micrometer as representing the same enlargement.

(Method after Atchley, 1958.)

EXERCISE 3.17

Photography of Irradiated Specimens

1. A polished slab of rock or polished thin section is irradiated for 2 to 5 minutes by 50 kV X-rays at 2 mA (about 30,000 R/min) in the specimen chamber of an X-ray spectrometer.

2. The irradiated surface is quickly placed in contact with a sheet of film in a light-tight container or darkroom; 4×5 in. packets of Polaroid ASA 3000 black and white or color ASA 75 used in a Polaroid 4×5 in. film holder are suitable. The film is kept in contact with the rock surface for several seconds to several minutes, based on experience, to give the correct exposure.

3. The film is then developed, fixed, and washed.

DISCUSSION

This method can register low-intensity thermoluminescence of the minerals, and the different intensities of thermoluminescence can be correlated with different minerals. It is useful for displaying zoning.

(Method after Colucci, 1970.)

ROTATION METHODS FOR THE 4
POLARIZING MICROSCOPE

Convention: In this chapter the principal optical parameters are designated as follows: Biaxial: The vibration directions are X (fastest), Y (optic normal), Z (slowest) and the corresponding refractive indices are $N\alpha$ (smallest), N_β (intermediate), $N\gamma$ (largest). $2V$ is the optic axial angle, BXa the acute and BXo the obtuse bisectrix. Uniaxial: The vibration directions are O (ordinary) and E (extraordinary) and their corresponding refractive indices are $N\omega$ and $N\varepsilon$.

Many students of mineralogy and petrology have an aversion for the universal stage. This is because many instructors are purists and insist that the student learn rotation methods entirely by the orthoscopic method. I can only repeat the sentiments of Tuttle and Bowen (1958):

> The conoscopic method permits more accurate results and is far less tedious than the extinction methods. It is surprising that this method is not more widely used in conjunction with the universal stage.

In normal thin-section work we are constantly employing conoscopic light for interference figure and sign determination. Now that modern universal rotation methods need no longer rely entirely on extinction methods and the orthoscopic system, I strongly recommend that wide use be made of conoscopic light and interference figures in rotation methods. In this way the student will not continue to be discouraged from taking fuller advantage of them.

Orthoscopic methods of course are excellent, though tedious and time consuming, and most workers would agree that they are more precise. Nevertheless, conoscopic methods are more direct, more easily understood, and may attain a precision as high as the orthoscopic methods if care is taken.

The purpose in this chapter is to promote the faster and less dreary conoscopic methods. Orthoscopic methods are described in Wahlstrom (1969), Hartshorne and Stuart (1970), Emmons (1959), Muir (1967); their application to plagioclase determination is covered in Slemmons (1962).

The following exercises, then, are confined to conoscopic determinations, which deserve wider use in view of their greater simplicity.

Some Preliminary Exercises

To enable us to use the spindle stage it may be necessary to calibrate the conoscopic field of view of the microscope. This should be done as a standard procedure on any polarizing microscope. Another useful exercise is to calculate or obtain by a graphical method the value of $2V$ if $N\alpha$, N_β, and $N\gamma$ are known.

Calibration of the Micrometer Eyepiece for the Measurement of BXa Interference Figures

METHOD A

This method is based on the equation

$$\frac{d}{N_\beta} = \frac{\sin V}{A},$$

where $d = 2D/2R$,

  $2D$ = distance between the melatopes (emergence points of the optic axes) in micrometer scale divisions,

  $2R$ = diameter of the complete interference figure, measured in the same way,

  N_β = refractive index of the mineral for the Y vibration direction,

  V = half the true optic axial angle,

  A = numerical aperture of the objective of the microscope.

NOTE.   A standard eyepiece should be used throughout and it should not be changed.

1. Obtain an acute bisectrix interference figure from a crystal in a thin section. The optic plane must be vertical. This is confirmed by the fact that in the 90° position the isogyre cross lies exactly underneath the cross wires.

2. Rotate 45° exactly from the extinction position. Measure in ocular micrometer scale divisions the complete diameter ($2R$) of the figure.

3. Measure in ocular micrometer scale divisions the distance between the two melatopes (optic axes) ($2D$).

4. Calculate $d = 2D/2R$.

5. If the mineral is known, then look up its N_β in a mineralogy textbook; otherwise it will have to be measured by an immersion method.

6. If the objective in use has a numerical aperture of 0.85, then, using Fig. 4.1, obtain $2V$ directly from the graph.

(Method after Tobi, 1956.)

DISCUSSION

The chart is for NA 0.85 objectives only. When another objective of NA $= A'$, is used the chart can still be consulted after multiplication of the computed d value by $A'/0.85$ (easily done with a slide rule); $2E$, the optic axial angle measured in air, can also be obtained if required.

  The chart may be applied to oblique interference figures in which BXa is not vertical, but the optic plane must be vertical, in which case the distance between the melatopes should be measured in two parts, that is, from the center of the interference figure to each of the melatopes. From these two distances, $2D'$ and $2D''$, the values $2V'$ and $2V''$ should then be read separately from Fig. 4.1 and afterward $2V$ can be computed: $2V = 2V' + 2V''$.

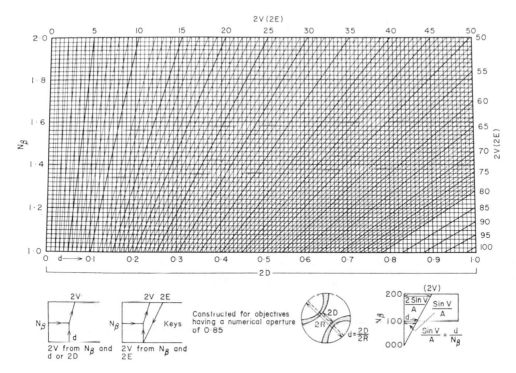

Fig. 4.1 Chart for the measurement of the optic axial angle in BXa figures when the optic plane is vertical. (After Tobi, 1956.)

METHOD B

1. Buy an aragonite (001) and barite (100) oriented thin section from Wards, Krantz, or any other supply company. These sections give centered *BXa* figures. Aragonite has $N_\beta = 1.680$, $2V = 18°$; barite has $N_\beta = 1.638$, $2V = 37°$.

2. Using the eyepiece containing the micrometer and the objective that will constantly be needed on this microscope for arriving at interference figures (usually $\times 40$ NA 0.85), obtain under conoscopic light and crossed polars an interference figure from aragonite.

3. In the 45° position measure accurately on the ocular micrometer the separation of the melatopes (optic axes), which is $2D$ in eyepiece micrometer divisions.

4. From the formula $2D/2 = N_\beta K \sin 2V/2$ (Winchell, 1946) we can obtain K, which is Mallard's constant,

$$K = \frac{2D/2}{N_\beta \sin 2V/2},$$

since we are using aragonite for which $2D$ has just been measured; N_β and $2V$ are known and do not vary significantly. In the above formula $2D/2$ and $2V/2$ are half the separation of the melatopes and half the optic axial angle, respectively.

5. The same procedure can be repeated by using the *BXa* figure of barite and determining K. The value of K should be identical. If it is not exactly, the average K is used as Mallard's constant for this particular combination of eyepiece and objective.

6. For any interference figure of unknown $2V$ simply measure $2D$ and N_β and obtain $2V$ from this relationship

$$\sin \frac{2V}{2} = \frac{2D/2}{N_\beta K}.$$

If N_β cannot be measured, an average value can be found in any mineralogy textbook if the mineral is known.

DISCUSSION

For convenience a chart is available (Winchell, 1946) to convert $2D$ to $2V$ if K and N_β are known. Please note, however, that K is *only* for a specific combination of objective and eyepiece and must be redetermined if any change in the optical system is made.

EXERCISE 4.2

To Derive $2V$ from $N\alpha$, N_β, and $N\gamma$

When any three of the variables V, $N\alpha$, N_β, and $N\gamma$ are known, the following equations may be readily solved with sin tables and a desk calculator to determine the unknown:

$$\sin^2 V\gamma = \frac{1/N_\alpha{}^2 - 1/N_\beta{}^2}{1/N_\alpha{}^2 - 1/N_\gamma{}^2}.$$

and

$$\sin^2 V\alpha = \frac{1/N_\beta{}^2 - 1/N_\gamma{}^2}{1/N_\alpha{}^2 - 1/N_\gamma{}^2},$$

where $V\gamma$ = half the true optic angle of a positive biaxial mineral,
$\qquad V\alpha$ = half the true optic angle of a negative biaxial mineral,
$\qquad N\alpha$ = the lowest index of refraction,
$\qquad N_\beta$ = the intermediate index of refraction,
$\qquad N\gamma$ = the greatest index of refraction.

(Mertie, 1942).

In the absence of a calculator the nomogram (Fig. 4.2) may be consulted. A straight-edge, preferably a transparent one, is placed on the value for $N\alpha$ along the top and on the value for $N\gamma$ on the bottom. The value for N_β is projected on a vertical line to the intersection with the straight-edge, and the value for $2V$ is indicated by a horizontal line from this intersection. The optic sign is also indicated.

DISCUSSION

The nomogram is based on the $\sin^2 V\gamma$ function and, in common with most types of nomogram, it does not give accurate estimations of V for crystals with small birefringences or with small optic angles.

Parker (1956) describes a graphical method of high accuracy for deriving $2V$ from the refractive indices using the stereographic projection. However, the calculations that use the two equations given above require less time than drawing a stereogram.

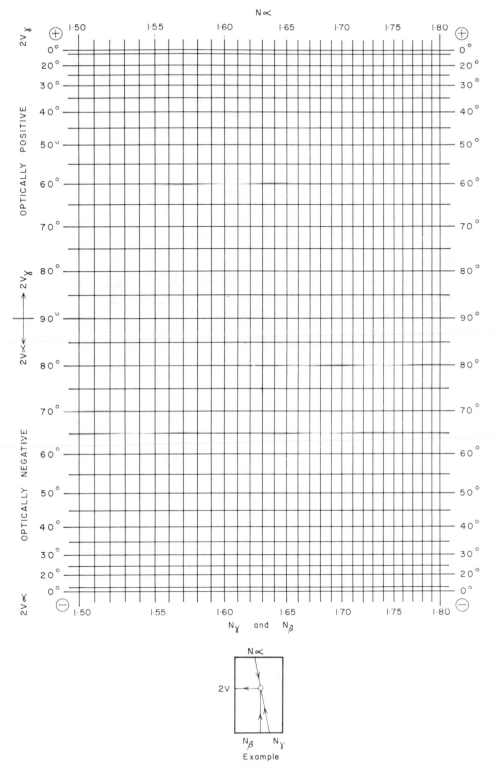

Fig. 4.2 Nomogram for determining V, the half optic angle, and the optic sign from measured values of $N\alpha$, N_β, and $N\gamma$. (After Mertie, 1942.) See the small diagram for an example of the use of the nomogram.

THE SPINDLE STAGE

The spindle stage has certain advantages over the universal stage. One is its simplicity of contsruction and manipulation: there is only one axis, which is horizontal, and this is combined with the vertical microscope stage axis. With these two movements, any desired line in the crystal may be rotated into the plane of the microscope stage and the plane of the polarisers, either N-S or E-W. Hence $N\alpha$, N_β and $N\gamma$ may be determined. $2V$ and sign may also be readily determined. Being of simple construction, the immersion oils are changed very readily. One most important additional advantage over the universal stage is that no correction to angular rotations is necessary.

EXERCISE 4.3

Constructing and Setting up a Spindle Stage

Design by Wilcox (1959). Figure 4.3 illustrates the stage. The base plate is stainless steel 0.04 in. thick and has a circular glass window. A cover glass will sit on supporting blocks above the crystal, with a drop of immersion oil underneath to fill the space.

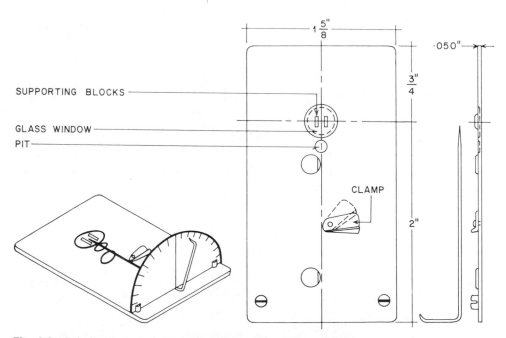

SUPPORTING BLOCKS

GLASS WINDOW

PIT

CLAMP

Fig. 4.3 Spindle stage as designed by Wilcox. (After Wilcox, 1959.)

The supports for the cover glass are 0.2 in. thick. The spindle is $\frac{1}{32}$-in.-diameter piano wire $2\frac{1}{2}$ in. long.

Design by Jones (1968). Figure 4.4 illustrates the stage. It is a plate with two holes drilled to take two screws (A) exactly located to screw into the microscope stage. A metal protractor is fixed at right angles to the base plate and a rotatable knob is

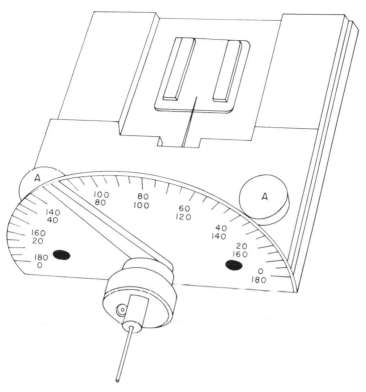

Fig. 4.4 Spindle stage designed by Jones (1968).

inserted exactly at the protractor center. A hole of sufficient diameter to take the removable crystal holder must have been drilled through the knob axis. The knob requires a locking screw for the spindle.

The immersion cell is made of a standard glass slide $26 \times 26 \times 1$ mm thick. Two thin strips of similar glass slide are mounted onto this glass plate with epoxy cement so that a 22-mm square cover glass can rest on them.

The center of the axis for the spindle must coincide with the center of the space between the slide and the cover glass; that is, it must be $1\frac{1}{2}$ mm above the base plate on which the 1-mm glass slide rests.

Design by Roy (1965). Roy (1965) has described a spindle stage that is suitable for conoscopic observations in which the direct determination of the optic constants and $2V$ are possible. This stage is similar to the two illustrated in Figs. 4.3 and 4.4.

The Hartshorne Spindle Stage. This stage is the only commercially available one known to me. It is fully described by Hartshorne and Stuart (1970, pp. 367 to 369) and may be obtained by ordering "Hartshorne's Rotation Apparatus" from McCrone Research Associates, 493 East 31st Street, Chicago, Illinois 60616, and 2 McCrone Mews, Belsize Lane, London, N.W.3, England. This apparatus has a fine adjustment for both longitudinal and lateral positions of the needle which greatly facilitates centering the crystal in the microscope field of view. An elegant and inexpensive spindle stage is coming into production at the Virginia Polytechnic Institute, and Dr F. D. Bloss is completing a brief text describing its straightforward use.

Procedure for Attaching the Crystal

1. The fragment, which may be a small crystal removed from a hand specimen with a chisel or spatula blade or a small cleavage fragment of any mineral (e.g., feldspar), should be 0.03 to 0.3 mm in diameter. Too small a fragment will be difficult to handle; 0.2 mm would be a good average size.

2. *Choice of Adhesive.* Wilcox (1959) recommends a mixture of four parts of common water-soluble carpenters' glue (e.g., Lepage's liquid glue) and one part of crude (blackstrap) molasses. Jones (1968) recommends Weldwood furniture cement. Araldite AY 103 thoroughly mixed with hardener HY 951, in proportion about 10 to 1, is also a good adhesive. Durofix (obtainable from Rawlplug Co. Ltd., Cromwell Road, London, S.W.7, England) diluted with an approximately equal volume of amyl acetate is also a satisfactory cement.

3. A drop of adhesive is placed on a flat surface and the spindle end is touched to it to bring a minimal quantity onto the spindle. The spindle is touched onto the grain to attach it as nearly parallel to the spindle as possible. While the cement is still soft the crystal can be nudged into an ideal orientation and allowed to set.

4. After the glue has hardened the spindle is clamped into the protractor scale.

5. The coverglass is placed on the supports and a drop or two of the chosen refractive index oil is introduced under the cover glass.

6. An alternative method (Jones, 1968) places a drop of the immersion liquid $(RI \simeq N_\beta$ of the mineral) near the center of one edge of the cover glass. Quickly invert the cover glass and place it on the two glass spacers so that the hanging drop is directly over another drop of identical oil which has been placed on the slide (before the spindle is introduced). These drops are promptly caused to coalesce with a needle touched to them in the space between the slide and the cover glass. The spindle is inserted through the hole in the base plate so that the far end (away from the crystal) protrudes through the knob. The glass cell is laid in place and the spindle pushed slowly back toward the cell until the crystal penetrates and comes to rest within the immersion oil bubble. The spindle is then locked in the knob. The size of the immersion oil drop must be arranged to suit the size of the crystal.

7. The spindle stage is attached to the microscope stage. It can be screwed on or simply held in place by the stage clips of the microscope.

8. The grain must be centered. This is done best under a high-power objective with the Bertrand lens in and the tube raised.

9. At any stage the immersion oil may be changed. The cell is easily cleaned by wiping with tissue paper and washing with a solvent or detergent. The crystal attached to the spindle can be cleaned by immersion in iso-octane.

10. A device called the "Hartshorne mounting apparatus," available from McCrone Associates, greatly facilitates the precise mounting of small grains on the spindle.

EXERCISE 4.4

Measurement of $N\omega$ and $N\varepsilon$ or $N\alpha$, N_β, and $N\gamma$ on a Spindle Stage

The following method assumes that the lower polar is polarized N–S.

A symmetry axis of the optical indicatrix (X, Y, or Z for biaxial and O or E for uniaxial) lies horizontal N–S when an isogyre lies E–W across the center of the conoscopic field. There are three such settings of the spindle axis and the microscope stage, corresponding to X, Y, or Z, for biaxial minerals, (Fig. 4.5). The detailed procedure is as follows:

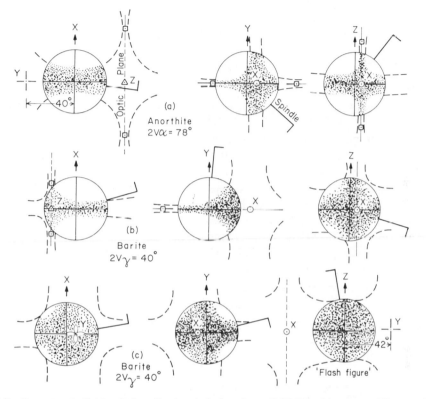

Fig. 4.5 Conoscopic fields of view (in the circles), using a 0.65 NA objective and Bertrand lens, with crystals in position for measuring their principal indices of refraction. In each case the isogyre is set symmetrically E–W across the field of view, thus making a principal axis N–S and horizontal. (After Wilcox, 1959.)

1. With a high-power objective (say NA 0.65 or 0.85), crossed polars, a Bertrand lens, and conoscopic light, rotate the crystal about the spindle and microscope stages until an isogyre is maneuvered into a position symmetrically along the E–W cross wire, that is, so that the cross wire divides the isogyre into two mirror-image halves. Examples with a 0.65 NA objective and Bertrand lens are shown in Fig. 4.5. A correct selection of objective may give narrower and sharper isogyres that are more readily and accurately positioned E–W; for example, a 0.85 NA objective will give narrower isogyres than a 0.65 objective and for certain figures a 1.25 NA oil immersion objective may have advantages. Once the isogyre is placed precisely E–W note the readings on the spindle and microscope stages.

2. Remove the analyzer, conoscopic converger, and Bertrand lens. Using the Becke line method, compare the refractive index of the crystal with that of the immersion oil.

3. Restore crossed polars and conoscopic light. Repeat the same procedure for the second and third positions at which an isogyre is set symmetrically E–W, taking care not to duplicate a position on the microscope stage at 180° from one already found. Figure 4.5 illustrates the three settings of spindle and microscope axes for anorthite and more settings for barite.

4. Refractive index liquids may be changed by drawing off the former liquid with a piece of blotting paper and replacing it with a new liquid until matches can be made with α, β, and γ.

(Method after Wilcox, 1959.)

DISCUSSION

Accuracy of orientation is obtained by the following procedure:

The substage assembly must be carefully centered, the Bertrand lens centered on the cross wire, and the polars accurately crossed and parallel to the cross wires.

The grain must be accurately centered on the cross wire. Becke line determinations are best made with a ×10 or ×20 objective rather than with the higher power used for the interference figures.

Occasionally, on aligning the isogyre with the E–W cross wire, one may find that the image obtained appears to be a centered figure of special orientation, which must be confirmed by unlocking the microscope stage and giving it a 360° spin to see the figure in both the 90 and 45° positions. This figure can be treated with normal thin-section methods. The optic sign may be determined. After checking the figure return the microscope stage to its original position.

Truly centered interference figures are special cases:

    a. A centered optic axis figure gives β in all positions of the microscope stage.

    b. A truly centered *BXa* figure gives β and α for $a + ve$ mineral at the extinction positions.

    c. A truly centered *BXo* figure gives β and γ for $a + ve$ mineral at the extinction positions.

    d. A truly centered optic normal figure gives both α and γ at the extinction positions.

Uniaxial crystals need be oriented only for a flash figure when $N\omega$ and $N\varepsilon$ are determined at the respective extinction positions. Of course, a check must be made to determine whether the crystal is indeed uniaxial.

The orientation is achieved by Lommel's rule which states that the optic axis lies in the quadrant from which the brushes of the flash figure leave the field on slight rotation of the microscope stage. The sign is determined by comparing $N\omega$ and $N\varepsilon$.

WARNING. For a biaxial crystal of small to moderate 2V, when the acute bisectrix is horizontal regardless of the position of the optic plane otherwise, the interference figure behaves much like the flash figure of an uniaxial mineral.

The conoscopic method is faster than the orthoscopic, but less accurate. However, it gives results sufficiently accurate for all but the most exacting work.

EXERCISE 4.5

2V Determination on the Spindle Stage

METHOD A

When the Two Optic Axes Are Accessible.

    1. Make certain that the microscope lens system, including the Bertrand lens, is centered.

2. Find the reference azimuth, that is, the microscope stage reading when the spindle axis is exactly N–S, with the spindle tip pointing S. Note the reading.

3. Set the microscope stage exactly 180° from this reference.

4. Immerse the crystal fragment in an oil of similar refractive index (± 0.02), use conoscopic illumination, and rotate about the spindle axis until the narrowest and densest part of the isogyre (the optic axis or melatope position) lies approximately on the N–S cross wire.

5. Rotate *both* polars of the microscope the same amount to swing the portion of the isogyre near the melatope parallel with the N–S cross wire. Now refine the setting of the spindle arm (rotation of microscope stage) so that this portion of the isogyre lies exactly on the N–S cross wire (B in Figure 4.6). Record the spindle arm

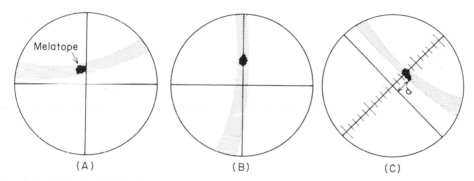

(A)   (B)   (C)

Fig. 4.6 Method of adjusting the spindle stage by using conoscopic light (A) initial position; (B) setting of the malatope on the N–S cross wire; (C) measurement in the 45° position. See exercise 4.5 (method A). (After Wilcox and Izett, 1968.)

reading for this optic axis OA_1 and note whether the melatope is north or south of the cross wire.

6. Rotate microscope stage 45° and measure d in ocular micrometer scale divisions from cross to melatope. It is good to rotate the stage 180° and repeat the d measurement. If the spindle stage axis can be rotated 180°, repeat the d determinations. In this way we can get four values of d which should be averaged. Convert the measured d to ψ min by the equation

$$\sin \psi \min = \frac{d}{N_\beta K},$$

where K (Mallard's constant) is known for this combination of objective and eyepiece with micrometer (exercise 4.1), and N_β of the mineral has already been measured (exercise 4.4) or is known approximately. If K has not already been determined, use method B, which follows.

7. Plot OA_1 on a Wulffnet. Count ψ min north or south from the center as noted in step 5. (This direction is used literally because Bertrand lens inversion has been canceled by rotation of the microscope stage 180° from the reference azimuth, as in step 3.) Thence count east along the small circle an angle equal to the spindle arm reading of step 5. This is the optic axis position OA_1. If the spindle axis reading exceeded 90°, then at 90° the primitive circle will be reached. In this case continue counting beyond 90° in from the primitive circle at a point *diametrically* opposite along an identical small circle. As an example, the following data are given:

Ferroaugite			Reference azimuth $= 0°$ $\beta = 1.71$ Mallard's constant $K = 18.0$	
	Microscope Stage Reading	Spindle Arm Reading	d Ocular Micrometer Divisions	ψ (min)
OA_1	180°	25°	$7\frac{1}{2}$ N	14°N
OA_2	180°	76°	4 S	$7\frac{1}{2}$°S

$$2V_z = 55° \pm 1°$$

and the plotting on the stereonet is given in Fig. 4.7.

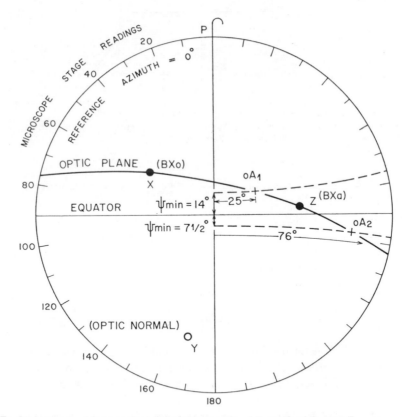

Fig. 4.7 Stereogram of ferroaugite; data from exercise 4.5, method A, step 7.

8. Return the microscope stage to 180° from the reference azimuth and repeat steps 3 through 7 for the second optic axis OA_2. Finally rotate both polars to their cardinal positions and with the appropriate accessory plate determine the optic sign.

9. A great circle drawn through OA_1 and OA_2 defines the optic plane. The optic sign will determine BXa and BXo, and $2V$ can be measured along this great circle (Fig. 4.7).

When Only One Optic Axis Is Accessible

Steps 1 through 7 are performed as before; that is, go as far as plotting OA_1.

8a. With the polars in their cardinal positions rotate the spindle arm and microscope stage to place an isogyre symmetrically along the cross hair parallel to the analyzer direction. In this position the optic symmetry axis is horizontal and in the plane of the polarizer. (Record readings of microscope stage and spindle arm). Now align this axis with the accessory slot of the microscope by a 45° rotation of the microscope stage and determine whether it is X, Y, or Z with an accessory plate. Orient and identify the other two symmetry axes similarly.

9a. Plot the symmetry axes on the stereogram: subtract the reference azimuth from the microscope stage reading: starting at the north point, count off the remainder counterclockwise around the primitive circle and count inward along the small circle an angle reading of the spindle arm. Plot the symmetry axis at this point.

10a. Draw in the great circle through OA, X, and Z and read off V.

	Microscope Stage				
Cummingtonite			Reference azimuth = 5° β = 1.648		
			Mallard's constant $K = 29.0$		
	Reading	Corrected for Reference Azimuth	Spindle Arm Reading	d Ocular Micrometer Divisions	ψ (min)
OA_1	185	180	139	$12\frac{1}{2}$N	15 N
X	143	138	91		
Y	59	54	128		
Z	76	71	23		

See Fig. 4.8 for plotting.

(Method after Wilcox and Izett, 1968.)

DISCUSSION

$2V$ precision is thought to be $\pm 1°$ when two optic axes are accessible and $\pm 3°$ when only one optic axis is accessible, provided care is taken to align the microscope. Abnormal broadening or distortion of the isogyre in the vicinity of the melatope indicates that the crystal is not optically homogeneous, in which case the margin for error is greater.

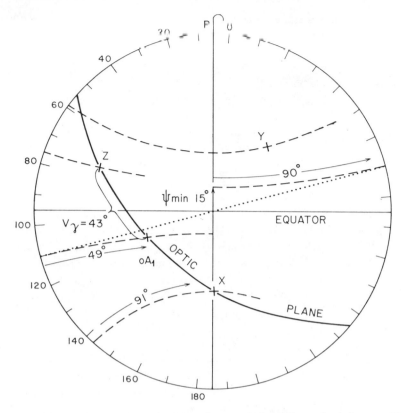

Fig. 4.8 Stereogram of cummingtonite; data from exercise 4.5, method A, step 10a. (After Wilcox and Izett, 1968.)

METHOD B

This method is an alternative to method A and follows directly from it. Figure 4.9 shows a stereographic plot in which the optic axis is oriented in a vertical plane lying 45° from the spindle axis by rotation about this axis; R is the amount of rotation about the spindle axis required to move the optic axis from the plane defined by the microscope and spindle axes into another plane through the microscope axis lying at a known angle λ to the spindle axis, (Fig. 4.9).

$$\tan \psi \min = \sin R \cot \lambda,$$

$$\tan \psi = \frac{\tan R}{\sin \lambda}.$$

Steps 1 through 5 of method A (exercise 4.5) are performed. Then complete the following steps:

6′a. Rotate a selected angle λ (20, 25, 35, 45, or 60°) exactly, either clockwise or counterclockwise, on the microscope stage and clamp. (The amount of rotation preferably should be as large as possible but still allow the subsequent placement of the melatope on the cross wire).

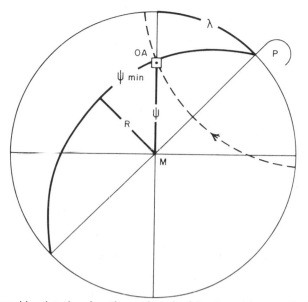

Fig. 4.9 Stereographic plot showing the optic axis *OA* oriented in a vertical plane lying 45° from the spindle axis by rotation about the spindle axis. Solid lines are great circles, dashed lines are small circles, *P* is the spindle axis, and *M* is the microscope axis. Ocular cross wires are N–S and E–W. See text, exercise 4.5, method B. (After Noble, 1968.)

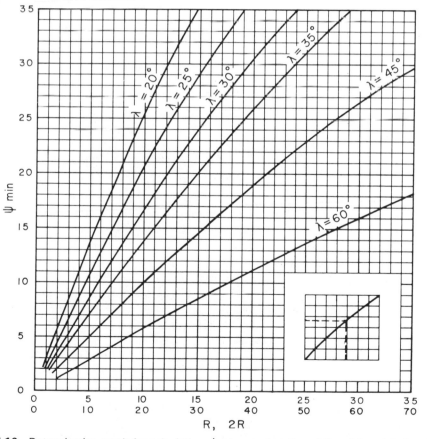

Fig. 4.10 Determinative graph for calculating ψ min as a function of *R* or *2R* and λ. A method of using the graph is illustrated in the small inset. (After Noble, 1968.)

6′b.   Return the melatope precisely to the N–S cross wire by rotation about the spindle axis in conjunction with simultaneous rotations of the nicols. Record the spindle-axis rotation

6′c.   If λ is small or if the optic axial plane lies at a small angle to the spindle axis, rotate the same number of degrees on the microscope axis in the opposite direction from the zero point and again place the melatope on the N–S cross wire. Record the reading on the spindle-axis scale (when the optic axial plane makes a small angle with the spindle axis, the accuracy of the $2V$ determination depends mainly on the accuracy of the ψ-min determination. In such cases double settings are an advantage even if angles of λ greater than $30°$ are used. If $\lambda = 45°$ can be used, a second optic axis can be set with little or no additional rotation of the nicols, provided the ocular cross wire can be rotated precisely $45°$ in the microscope tube).

6′d.   Determine R, the absolute difference between the spindle scale reading obtained in 6′b and that noted in 5, or determine $2R$ from the readings recorded in 6′b and 6′c.

6′e.   Determine ψ min from R or $2R$ and λ, using Fig. 4.10, or from the equation $\tan \psi \min = \sin R \cot \lambda$. Record ψ min and whether the melatope is north or south of the E–W cross wire.

(Method after Noble, 1968.)

DISCUSSION

This method has the advantage over method A of determining ψ min without an ocular micrometer calibration.

METHOD C

If an objective of numerical aperture 0.85 is used, at least one optic axis can be seen in the field of view.

Occasionally two optic axes can be seen in the field of view; $2V$ can be calculated from the separation of the optic axes by one of the methods based on Mallard's formula (exercise 4.1). The distance between the point of emergence of each optic axis and the center of the field of view are measured with a micrometer ocular and the measurements converted to $2V$ by calculation or by graphs provided by Winchell (1946) or Tobi (1956) (Fig, 4.1).

When only one optic axis is visible, the angle V that includes the microscope axis (M) (Fig. 4.11) is divided into two parts: Ψ is the angle between the optic and microscope axes and Φ is the angle between the microscope axis and the bisectrix. The angles are named $\Psi\alpha$, $\Phi\alpha$, and $\Psi\gamma$, $\Phi\gamma$ if $V\alpha$ or $V\gamma$ respectively, includes the microscope axis. Figure 4.11 illustrates the case for $V\gamma$.

The angle Ψ is determined by a method based on Mallard's formula (exercise 4.1, method A): Φ is calculated from θ and either ξ or R, θ is the angle between the spindle axis and Y, whereas ξ is the angle between X or Z and the spindle axis. Both are measured directly on the microscope stage when the principal axis being considered is normal to the microscope axis: R is defined as the rotation about the spindle axis that places X or Z normal to the microscope axis, measured from the position of the spindle at which the optic plane is vertical; ξ and R are designated $\xi\alpha$ and $R\alpha$ or $\xi\gamma$

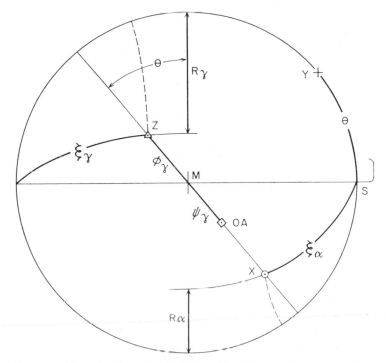

Fig. 4.11 Stereographic projection of a biaxial crystal, with its optic plane parallel to the microscope axis *M*. The angles are referred to in the text (method C, exercise 4.5). Spindle axis *S* is E–W. (After Noble, 1965a.)

and $R\gamma$, depending on whether X or Z has been made normal to the microscope axis; $\phi\alpha$ and $\phi\gamma$ are related to $\xi\alpha$, and $\xi\gamma$, and θ by the equations

$$\sin \phi\alpha = \frac{\cos \xi\alpha}{\sin \theta}, \tag{1}$$

$$\sin \phi\gamma = \frac{\cos \xi\gamma}{\sin \theta} \tag{2}$$

and $R\alpha$, $R\gamma$ and θ by equations

$$\cot \phi\alpha = \tan R\alpha \cos \theta, \tag{3}$$

$$\cot \phi\gamma = \tan R\gamma \cos \theta. \tag{4}$$

Because

$$\tan \phi\alpha = \cot \phi\gamma, \tag{5}$$

$\phi\alpha$ can be calculated from $\xi\gamma$ or $R\gamma$ or vice versa.

The parameter ξ becomes insensitive as θ approaches 0°, whereas R becomes insensitive as θ approaches 90°. Thus ξ should be measured when the spindle axis lies at a low angle and R used when the spindle axis makes a high angle to the optic plane.

METHOD D

1. Under conoscopic illumination make the crystal optic plane vertical and E–W by rotation about the spindle and microscopic axes. Measure θ on the microscope stage scale. Record the spindle-stage reading.

2. Place the optic plane exactly 45° from the polarizer vibration directions by rotation on the microscope stage.

3. Measure Ψ by a method based on Mallard's formula (exercise 4.1, method A). Using a compensator, determine whether $\Psi\alpha$ or $\Psi\gamma$ has been used.

4. Place either X or Z normal to the microscope axis by rotating about the spindle axis. Determine ξ from the reading on the microscope scale or R from the spindle axis scale. Identify the principal axis with a compensator.

5. Compute $\phi\alpha$ or $\phi\gamma$ from equations 1 to 5.

6. Compute V by adding Ψ and ϕ, (If, for example, $\Psi\alpha$ has been measured and $\phi\gamma$ has been calculated, $V\alpha = \Psi\alpha + 90 - \phi\gamma$.)

(Method after Noble, 1965a.)

DISCUSSION

In some cases, because of unfavorable orientation of the crystal or high dispersion, it is difficult to place X or Z accurately normal to the microscope axis.

The equation $\cot R\alpha \cot R\gamma = \cos^2 \theta$ allows calculation of the amount of rotation about the spindle axis necessary to orient the crystal such that X and Z may be measured. The calculation is based on the rotation required to orient Y and Z or Y and X, respectively.

The accuracy of this method with care is ± 1 to $2°$.

If the method is to be used frequently, then, rather than perform the computations, charts in Noble (1965a) may be referred to for obtaining ϕ.

UNIVERSAL STAGE

Universal stages with four or five axes are available from both Zeiss and Leitz and from several other manufacturers. The major difference between Leitz and Zeiss stages is that in the Leitz a circular glass plate supports the slide; one glass hemisphere is placed on top of the slide and the other under the glass plate. In Zeiss stages the central glass plate has been eliminated and the slide is placed directly on the lower hemisphere.

Nomenclature of Axes of Rotation

The best system of nomenclature is the following:

Inner vertical axis	A_1.
Inner horizontal east-west axis	A_0 (or A auxiliary)

(this axis is present only in five-axes stages and not in four-axes stages).

Middle horizontal north-south axis	A_2.
Outer vertical axis	A_3.
Outer horizontal east-west axis	A_4.
Microscope vertical axis	A_5.

The two types of stage, as seen from above, are shown diagrammatically in Fig. 4.12. With this system all axes with odd numbers are vertical and the numbers progress outward; A_1, A_3, A_5. The even numbers are horizontal axes; A_0 (if present) is E–W, A_2 is N–S, and A_4 is E–W. They also progress outward as the number increases.

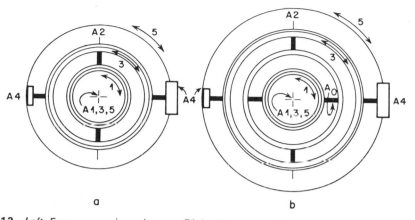

Fig. 4.12 *Left*. Four axes universal stage. *Right*. Five axes universal stage, showing the relative position of the axes. (After Hartshorne and Stuart, 1970.)

Special Equipment for Conoscopic Light

1. Special condensers must be fitted to the microscope. These are the Zeiss UD 0.6 condenser, Leitz conoscopic light condenser, or Leitz UT4 and UT5 condenser attachments which are screwed above the conventional lower condensing system.

2. Special small upper glass hemisphere segments. These are Zeiss small segments of RI 1.555 and 1.649 and Leitz small upper segments of RI 1.516, 1.557 and 1.649.

3. Special high-power objectives, numerical apertures in parenthesis: Zeiss UD 20(0.57)C UD 40(0.65)C and Leitz UMK 32(0.60) and UMK 50(0.60).

EXERCISE 4.6

Setting Up the Universal Stage

1. Align and center all the optical parts of the microscope: each objective that will be used with the universal stage, the condensing system, the Bertrand lens, and the light. A low-voltage high-intensity lamp is ideal for universal stage work.

2. Make sure that the polarizer and analyzer are at 90° and that their vibration planes are parallel to the cross wires.

3. With the microscope stage horizontal, remove the central ring from the microscope stage. Raise the objective (or lower the stage, depending on microscope design).

4. Place the universal stage on the microscope stage with the drum (axis for rotating A_4) on the right. Hold it loosely in place by means of the two set-screws, which must not be tightened until final centering is complete.

5. Place the central glass (Leitz) or the lower hemisphere (Zeiss) in position and set all the scale readings for A_1, A_0, A_2, and A_4 at zero; A_0 and A_2 are adjusted by means of metal arcs, which can be raised, and lowered after the set-screws have been tightened.

6. Turn A_3 until 270° coincides with the zero on the vernier scale, thus making A_2 normal to A_4. All the scales are now properly set.

7. To center the stage place a thin section on the universal stage and by rotating the outer stage A_3 find its center of rotation—that is, the point in the field of view about which all the grains on the slide rotate. Now center the universal stage with small lateral movements so that this center of rotation coincides with the cross of the eyepiece. Tighten the set screws. Some newer stages have two centering screws that move the stage along the 45° diagonals. These facilitate centering. In this case the set screws may be tightened at stage 4, but at stage 7 the centering screws alone are needed.

8. It is necessary to see that the horizontal axes are parallel to the cross wires. Fix an upper glass segment on the stage and focus on the top of the hemisphere. This usually means focusing on a particle of dust lying on the segment surface. It is necessary to fit a Leitz auxiliary (559001) objective to the microscope for this operation because there is insufficient distance for normal objectives to focus. After loosening its set-screw rotate on A_4 and notice whether the particle of dust moves parallel to the N–S cross wire. If it does not, the microscope stage should be turned a little until rotation on A_4 causes all dust particles to move exactly parallel to the N–S cross wire. The microscope stage is now clamped and its vernier reading taken to give the true zero for A_5.

9. Place on the glass plate (Leitz) or the glass top of the lower hemisphere (Zeiss) of the universal stage a drop of liquid of about the same refractive index as the glass plate or microscope slide. Nujol (RI = 1.48) is commonly used. Place the slide on it so that its length is normal to the line joining the set-screws of the upper hemisphere. If the slide contains a thin section complete with cover glass, proceed to step 10. If an immersion slide is to be prepared, crush the crystal grains, put one or more on the glass slide, center the crystals, place the cover glass on top, and allow the selected immersion oil to fill the space between the glass slide and the cover glass.

10. If a permanent thin section was used, place on top of its cover glass a drop of the same immersion oil used under the slide. If, however, it was a crushed immersion preparation, put a drop of the *same* immersion oil that was used to immerse the grains on top of the cover glass. This is to ensure that the immersion oil around the grains will not be contaminated by a different oil.

11. Carefully screw the upper hemisphere (small size) in place, watching the mount to see that the centered crystal is not broken by too much pressure or displaced from the center. Tighten the screws only gently. Do not exert excessive pressure.

12. The lower hemisphere of the universal stage using a central glass plate (Leitz) must now be fixed. Place a drop of the same liquid used in step 9 on top of the glass plate, on the upper flat surface of the lower hemisphere. Carefully push the hemisphere into the well beneath the stage, fixing it with a slight turn to engage it in the clips, or by giving it firm pressure if it is held by a friction ring.

13. Choice of hemispheres. The refractive index of the hemisphere glass should be as close as possible to N_β of the mineral. The top hemisphere must have the same RI as the lower hemisphere; for example, if plagioclase is being studied, use a hemisphere of RI 1.555 (Zeiss) or 1.557 (Leitz).

14. The height of the specimen must now be adjusted. The grain centered on the stage must lie on the plane of the horizontal axes. If the grain is above or below the axes, it will not remain centered in the field of view on rotation on the horizontal axes A_2 or A_4 but will plunge away from the center. An adjustment for height is possible by means of a threaded ring situated around the lower hemisphere underneath the stage. By rotating this ring clockwise and anticlockwise the height of the thin section can be adjusted on the stage until rotation on A_4 allows the centered grain to remain centered and not plunge away. (Viewed from underneath the stage, a clockwise rotation of the threaded ring raises the thin section). The stage is now set up and ready for use.

Universal Stage Conoscopic Procedure

It is imperative that the optical system be in perfect alignment and that the universal stage and specimen be exactly centered so that with all axes at zero rotation on A_1 will not disturb the position of the crystal. The crystal must also be carefully adjusted for height, coinciding exactly with the A_4 axis, so that it will not appear to plunge north or south out of the central position on rotation on A_4.

The choice of crystal for study in thin section is important. If we are using a normal thickness thin section of 0.03 mm, the diameter of a suitable crystal should be at least three or four times its thickness; that is, 0.1 mm diameter would be the minimum size of a suitable crystal. Thin sections that are a little thicker than normal give better results by conoscopic light because the isogyres are more clearly defined.

Preliminary adjustments are better done without a Bertrand lens by removing the ocular, but for final precise adjustments the ocular should be replaced and the interference figure viewed through a Bertrand lens.

METHOD FOR UNIAXIAL MINERALS

1. With all axes in their zero positions (A_3 at 270°), turn the crystal on A_1 to extinction; an isogyre will be seen (under the Bertrand lens) parallel to a crosswire through the center of the field. The optic axis (isogyre cross and center of any isochromatic rings) may or may not be visible. The image will be similar to those on Fig. 4.13.

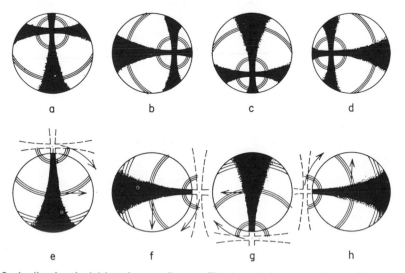

a   b   c   d

e   f   g   h

Fig. 4.13 Inclined uniaxial interference figures. The isogyre is shaded black. The isochromatic rings are shown as thin circular lines. Their presence depends upon the birefringence and thickness of the mineral fragment. Arrows show direction of movement of optic axis and isogyres upon stage rotation. (After Hartshorne and Stuart, 1970.)

2. If the axis lies outside the field, rotation on A_5 (microscope stage) will reveal the direction in which it lies as shown in e, f, g, and h of Fig. 4.13. Return A_5 to its correct zero position.

3. Rotation on either A_2 or A_4, whichever is appropriate, will bring the optic axis into a central position if it does not make an angle too high with the normal to the thin section.

4. In petrofabric work the angle made by the optic axis with the plane of the section needs to be measured accurately; therefore it is proper to bring the appropriate isogyre into a N–S position so that A_4 may be used to measure this angle. The angle of rotation of A_4 has to be corrected thus:

$$\frac{N\omega \text{ (ordinary RI of mineral)}}{\text{RI of glass hemispheres}} = \frac{\text{sine of observed angle on } A_4}{\text{sine of true angle in the mineral}}.$$

If a large number of readings are to be taken on the same mineral from many crystals in a rock thin section the above angle conversion is simplified because in the equation

$$\text{sine of true angle} = \text{sine of observed angle on } A_4 \times \left(\frac{\text{RI of glass}}{N\omega \text{ of mineral}}\right),$$

the factor in parentheses is constant for any one mineral. It can be computed and used to multiply the sine of the observed rotation on A_4 to give the sine of the true angle between the normal to the thin section and the optic axis of the mineral. Alternatively diagrams in Tröger et al. (1971, p. 152) or in the pocket on the back of Emmons (1959) may be used for a graphical conversion.

5. The angles recorded on A_1 and A_4 (corrected) may now be used to plot the direction of the optic axis on a Wulff net or Schmidt equal-area stereographic net. Figure 4.14 contains a stereographic plotting of two examples to show the method that employs the upper hemisphere of the stereogram. Use the following conventions for recording angles: rotations on A_1 are clockwise from the vernier zero; hence a counterclockwise rotation of 60° is actually recorded as $360 - 60° = 300°$. Rotations on A_4 are clockwise looking horizontally west onto the A_4 rotating knob. Hence, if the stage is tilted away from the operator 20° toward the north, this is recorded as 20°. If the stage is tilted 20° forward toward the south, this is recorded as $360° - 20° = 340°$.

6. The plotting of optic axis 1 (Fig. 4.14) is as follows:

a. Rotate the stereographic paper overlay the same amount as the A_1 rotation in the same direction. For $A_1 = 40°$ the overlay is rotated clockwise 40°.

b. Trace in the N–S diameter of the stereographic net.

c. For an A_4 rotation of 30° the optic axis is plotted 30° from the center toward the front of the stereographic net.

The plotting of optic axis 2 (Fig. 4.14) is by rotating the overlay 300° clockwise (60° counterclockwise), tracing in the N–S diameter from the net, then plotting the axis at 40° away from the center toward the back. These directions are for plotting the optic axes on the upper hemisphere of the stereogram. Some operators prefer to use the lower hemisphere, in which case the positions of the optic axes will be diametrically opposite those shown in Fig. 4.14. Whether you choose upper or lower hemisphere plotting, it is of the utmost importance to be consistent. The distances plotted outward from the stereogram center are not the actual measured A_4 reading but the true angle as computed in step 4 above.

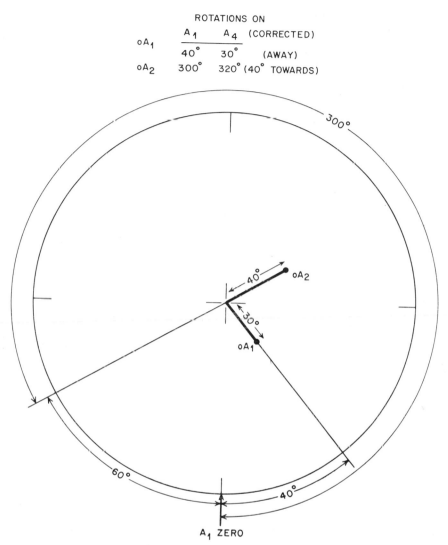

Fig. 4.14 Stereographic projection of the plane of a rock thin section showing the plotting of two uniaxial optic axes OA_1 and OA_2 in which the optic axes have been set vertically. See text of exercise 4.7 for description.

7. With mounts of separate crystal fragments in an immersion oil (as distinct from a thin section) a comparison of $N\omega$ of the mineral may be made with the RI of the immersion oil when the optic axis is centered.

8. When the optic axis figure of any mineral is centered, the sign may be determined with an appropriate accessory, comparing the E radial vibration (parallel to the microscope accessory slot) with the O tangential vibration (normal to the microscope slot).

9. It may not always be possible to set the optic axis vertically in which case it must be set horizontally with the equatorial plane of the indicatrix vertical. This will occur when the optic axis of a crystal forms a small angle with the plane of the thin section. Proceed as in step 10 on.

10. Rotate on A_1 until an isogyre is centralized on the N–S cross.

11. Rotate on A_4 in the direction in which the isogyre becomes broader. Continue rotation on A_4 until a diffuse black cross is centered on the cross wire. If this does not happen by rotating fully on A_4 in both directions, A_4 should be setback to zero and the section rotated through 90° on A_1 to set another isogyre exactly centered along the N–S cross wire. Then tilt on A_4 in whichever direction the isogyre broadens. In this case a position will be found in which a broad cross is centered on the cross wire.

12. Note the readings on A_1 and on A_4 when the cross is centered. In this position we know for certain that the optic axis is horizontal N–S. Step 13 explains how this may be confirmed.

13. The direction of the optic axis is revealed by the direction in which the diffuse cross breaks up; that is, the direction in which the isogyre brushes recede from the field of view as the microscope axis A_5 is slightly rotated one way and another from its zero setting. Return A_5 to its zero setting after this test. A further confirmation of the N–S horizontal setting of the optic axis is confirmed by tilting the optic axis up or down out of its horizontal setting by rotation on A_4. This is done by turning 45° clockwise on A_5 and tilting up or down on A_4; as the optic axis tilts up or down from its horizontal setting, the interference colors will go down the scale in the Michael-Levy chart. Maximum interference color is obtained when the optic axis is horizontal in this 45° position. Return A_5 to its zero setting and A_4 to its setting when the isogyre cross is centered.

14. For a microscope whose polarizer is N–S the optic axis now lies in the plane of the polarizer. The reading of A_4 (obtained in step 12) must now be corrected. The correction is by use of the extraordinary refractive index:

$$\text{sine true angle} = \text{sine observed } A_4 \text{ angle} \times \frac{\text{RI of glass}}{N\varepsilon \text{ of mineral}}.$$

15. A special case may be encountered when the optic axis actually lies in the plane of the thin section; that is, the section includes a prismatic section of the mineral. In such a case the interference figures will vary somewhat from the above steps. They are described as follows:

16. With all axes at their zero positions rotate on A_1. It will be impossible to obtain the effect of step 10. Instead, a diffuse cross will immediately be centered on the cross-wire at a certain position of A_1. In this case the optic axis is horizontal, either E–W or N–S. We shall have to confirm the direction.

17. If the optic axis lies E–W, tilting on A_4 in either direction will cause no change—the cross will persist. Therefore A_4 is parallel to the optic axis.

18. If the optic axis lies N–S, tilting on A_4 will cause the cross to break up into a single isogyre which remains centralized N–S, becoming narrower as the tilt on A_4 is increased. The cross appears only when A_4 is set at zero. This determines that the optic axis is N–S and horizontal.

19. There is no correction for such an axis lying on the thin section plane because the A_4 reading is zero.

20. The position of the optic axes obtained in steps 9 through 19 may be plotted by referring to examples given in Fig. 4.15.

21. The optic axis OA_1 (Fig. 4.15) is plotted as follows:

a. Rotate the stereographic paper overlay clockwise a distance 50° (= A_1 rotation).

b. Trace a N–S diameter from the stereographic net.

c. A_4 is 30°; the optic axis is plotted 30° from the extremity of the primitive circle in along this N–S diameter.

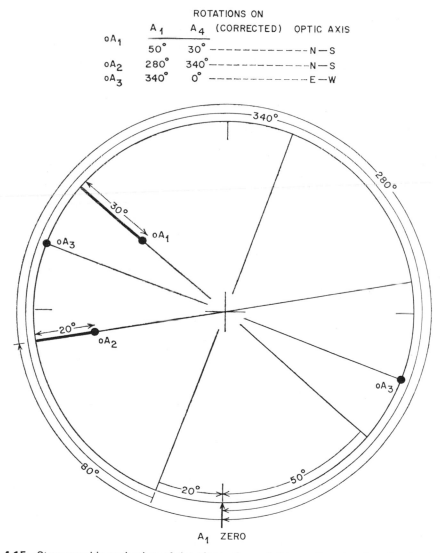

ROTATIONS ON

	A_1	A_4 (CORRECTED)	OPTIC AXIS
oA_1	50°	30°	N–S
oA_2	280°	340°	N–S
oA_3	340°	0°	E–W

Fig. 4.15 Stereographic projection of the plane of a rock thin section showing the plotting of three optic axes which have been set horizontally. The upper hemisphere is used. See exercise 4.7 for description.

OA_2 is plotted by clockwise rotation of the overlay 280° (counterclockwise 360° − 280° = 80°); A_4 is 340° (= 360° − 20). Plot OA_2 20° in from the front intersection of the N–S diameter with the primitive circle. OA_3 is plotted by clockwise rotation of 340° (counterclockwise 20°). Sketch in the N–S diameter. Since the optic axis is horizontal E–W, sketch in the E–W diameter. Plot $O.A_3$ where this diameter cuts the primitive circle (two points because it is horizontal).

22. When the optic axis is set horizontal (steps 12, 17, and 18), the refractive indices $N\omega$ and $N\varepsilon$ may be compared with that of the immersion oil. If the optic axis is horizontal N–S and the microscope polarizer is N–S, uncross the nicols and remove the Bertrand lens. The Becke line gives a determination of $N\varepsilon$. Rotate 90° on A_5 and the Becke line gives a determination of $N\omega$.

23. Optic sign may be determined by rotating the optic axis parallel to the accessory slot of the microscope by a 45° rotation on A_5. The accessory can determine (under crossed polars) whether the direction parallel to the slot (extraordinary) is faster or slower than the ordinary vibration which now is perpendicular to the slot.

EXERCISE 4.8

Conoscopic Method for Biaxial Minerals

Biaxial minerals are easily recognized because generally the isogyres are curved and sweep across the field of view in directions oblique to the cross wires. The chances are small that symmetrical sections of the indicatrix (normal to BXa, BXo or to the optic axial plane) will be presented in random orientations in a rock thin section or a crushed grain preparation. The following procedure is for randomly oriented crystals. The introductory remarks in Exercise 4.7 should be heeded before beginning.

1. With all axes at their zero positions (A_3 at 270°) rotate on A_1 to study the movement of the isogyres. Select one and let it intersect the center of the field and lie as nearly as possible to the N–S cross wire (Fig. 4.16a).

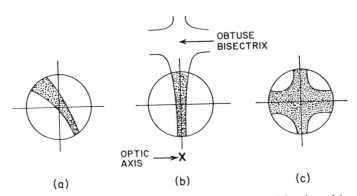

Fig. 4.16 Conoscopic adjustment of a biaxial symmetry plane: (a) selected isogyre made to pass through the field center by rotation on A_1; (b) rotation on A_2 and A_1 brings the symmetry plane vertical, rotation on A_4 retains the isogyre N–S; (c) exploration of symmetry plane by rotation on A_4 centralizes the bisectrix (in this case BXa). (After Hartshorne and Stuart, 1970.)

2. By small alternate movements of A_2 (tilting the isogyre to the center) and on A_1 (making the isogyre straighter) it is possible to align the isogyre exactly on the N–S crosswire (Fig. 4.16b). In this position we have a vertical N–S indicatrix symmetry plane. Determine by rotation on A_4 that the isogyre will remain exactly aligned N–S.

If it departs slightly from the ideal N–S position, small, fine adjustments on A_2 and A_1 must be made.

3.   From the readings of A_1 and A_2 the great circle representing this first symmetry plane may be plotted on a stereogram. Its pole will be either X, Y, or Z. The tilt on A_2 has to be corrected by the formula

$$\frac{\beta(\text{or mean index of the mineral})}{\text{RI of glass hemisphere}} = \frac{\text{sine of observed angle}}{\text{sine of true angle}}$$

or by the graph in the back pocket of Emmons (1959). The value of β taken here will be an average value that can be found in any mineralogy textbook. The plot is shown in Fig. 4.17A.

4.   In the example given the indicatrix plane of symmetry is vertical and N–S; its pole is horizontal and E–W when A_1 is at 40° and A_2 30° up to the right, down to the left. The tilt of A_2 was on a metal arc raised on the right. Hence it is best that we record the tilt as A_2 30° (right-up). Rotate the zero arrow on the stereographic net overlay $A_1 = 40°$ clockwise. Sketch in lightly the N–S and E–W diameters from the Wulff net; N–S represents the axis A_2. Using a dark line, draw in the great circle tilted at 30° to the right of the N–S diameter. This is the plane of symmetry. Its pole is plotted inward 30° from the west point of the primitive circle. All these plotting directions are for the upper hemisphere of the stereographic projection. If the lower hemisphere were used, the great circle would lie 30° on the west side of the center and the pole 30° in from the east point of the primitive circle. Use either upper hemisphere throughout or lower hemisphere throughout. Do not mix hemispheres. I alway use the upper hemisphere and the illustrations are so constructed.

5.   With A_1 and A_2 in the settings found in step 2 rotate on A_4 both ways from its zero setting. Observe what happens to the straight N–S isogyre. It may broaden on rotation of A_4, in which case continue rotation on A_4 until a crossed isogyre is exactly centered on the cross wire (see the example in Fig. 4.16c). Note the exact setting of A_4 when the isogyre cross is centered. This is the setting when BXa, BXo, or the optic normal Y is centered and vertical. Use the equation in step 3 to correct the rotation on A_4.

6.   With the corrected value of A_4 plot the position (BXa, BXo, or Y) on the same stereogram (Fig. 4.17B).

7.   We now have to determine whether the symmetry plane, which is set vertical, is the optic axial plane. This is done by rotating the microscope stage exactly 45° clockwise so that the plane is still vertical but now N.E.–S.W. In this setting rotate fully on A_4 from its two extremes. If the vertical plane is indeed the optic axial plane, then at one specific position of A_4 a centered optic axis figure will be obtained (Fig. 4.18).

8.   The value of A_4 to center the optic axis is corrected by the equation in step 3 and used to plot the optic axis on the stereogram as in the example in Fig. 4.17B. For the great majority of cases one would expect to be able to center only one of the optic axes; the second optic axis will generally lie too close to the plane of the thin section to be set vertically. When the optic axis is centered as in Fig. 4.18, it can be used for sign determination of the mineral. An accessory plate is put in the microscope slot and the birefringence color is observed in area A (Fig. 4.18). If the microscope slot is parallel to the A_4 axis, then in area A we are comparing the length of the accessory plate with Y and the width of the accessory plate with BXa. In this way we

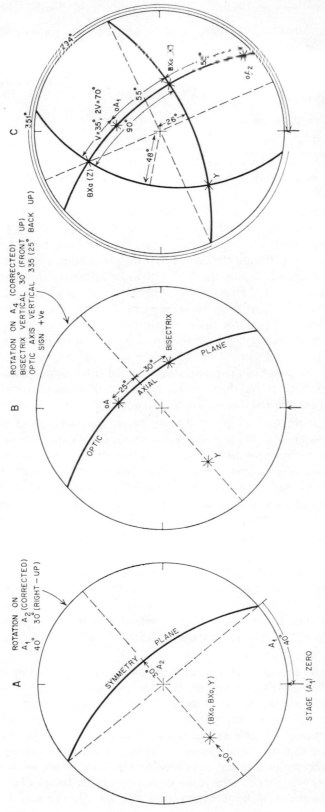

Figure 4.17 Stereographic plot of the optical properties of a biaxial mineral by the conoscopic method on a universal stage. The upper hemisphere is used.

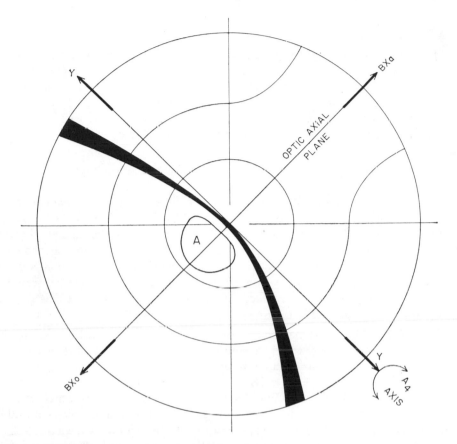

Fig. 4.18 Centered optic axis figure. The reading on A_4 allows plotting of the optic axis. The curvature of the isogyre indicates the direction in which BXa lies. Isochromatic rings may not be present if the mineral has low birefringence.

can tell whether BXa is faster or slower than Y. If BXa is faster than Y, the mineral is $-ve$; if slower, it is $+ve$. If $+ve$, the BXa is Z, if $-ve$, BXa is X.

9. It cannot be predicted for certain what the first symmetry plane will include; perhaps only a bisectrix or optic normal (BXa BXo or Y) or one or more optic axes as well. Whatever it does contain, the values of A_4 must be corrected and recorded and if possible the optic sign determined.

10. Suppose that the first plane did contain a bisectrix and a single optic axis (Fig. 4.17A,B) and that from the optic axis figure the sign was determined to be $+ve$. Now proceed as follows (see Fig. 4.17C). Since the optic axis is $30 + 25°$ on one side of the bisectrix the second optic axis may be plotted an equal distance of $55°$ on the other. This would make the bisectrix BXo. Plot the BXa bisectrix a distance of $90°$ along the optic axial plane great circle from BXo. Now V may be measured as $90 - 55° = 35°$ and $2V = 70$. The sign was determined as $+ve$; hence BXa is Z and BXo is X. Figure 4.17C shows the complete plot, including the other optical symmetry planes through $BXo(x)$ and Y and through $BXa(z)$ and Y. Therefore, if the first plane chosen is the optic axial plane, we could obtain from it the sign and the $2V$. Notice in Fig. 4.17C that rotation on A_4 could not have made BXa vertical, for a tilt of 60° is not possible on the universal stage.

11.   If we were lucky enough to have chosen the optic axial plane as our first symmetry plane, there will be no need to proceed, for as shown in Fig. 4.17 the complete optical orientation of the mineral has been obtained. It is unlikely, however, that one will always be able to select the optic axial plane first go. Therefore, having plotted the first plane of symmetry, in this case not the optic axial plane, we now wish to plot and measure the second plane, then the third plane, and one of them will be the optic axial plane and will allow $2V$ to be measured.

12.   In the five-axis universal stage the vertical setting of a second optic plane is done by means of the A_0 axis. From step 3 onward tilt on A_0 until the isogyre cross is exactly centered on the cross wire. Now we have the mineral oriented so that two optical symmetry planes are vertical, one N–S, the other E–W, and the third is horizontal. The bisectrices BXa, BXo, and Y lie N–S, E–W, and vertical. In the example shown in Fig. 4.17 the rotation on A_0 to center the cross is 30° (front-up), and as shown on Fig. 4.17C it would be the BXo bisectrix, which is vertical. BXa is horizontal N–S, and Y is horizontal E–W. Axis A_4 can now be used to measure the angle between BXo and an optic axis directly; hence $2V$ can be determined. With the five-axis stage no stereographic plotting is really necessary.

13.   With the four-axis stage, assuming that the first plane (steps 1 to 3) was not the optic axial plane, that plane would have given us readings for A_1 and A_2 when it was set vertically and would also certainly have given a reading for A_4 when an isogyre cross was set vertically. In Fig. 4.17B imagine that the plane is not the optic axial plane but that our 30° tilt on A_4 did allow us to plot the bisectrix (which could be BXa, BXo, or Y). Now rotate the stereographic net overlay until this bisectrix and the normal to the symmetry plane (marked as Y in Fig. 4.17B, but now known not to be Y, for the plane is not the optic axial plane) both lie on a great circle. Draw in this great circle as shown on Fig. 4.17C. Note where it cuts the primitive circle and read off the rotation on A_1 necessary to put this great circle (symmetry plane) N–S (in our example 4.17C it is $A_1 = 294°$). Now note the tilt on this great circle from the vertical (measured from the center of the stereogram along the diameter which is perpendicular to the diameter joining the intersections of the great circle with the primitive circle). In our case (Fig. 4.17C) it is 26° (right-up).

14.   Now we can proceed to place the second plane in a vertical and N–S position, equipped with the values obtained from the stereogram; in Fig. 4.17 $A_1 = 294°$ and $A_2 = 26°$ (right-up). Return all stage settings to their original zero settings (A_3 at 270°), set A_1 at the value found from the stereogram (294°), and tilt A_2 to the value found (26° right-up). This should give us an isogyre exactly centered along the N–S cross wire as in Fig. 4.16b. Check that the isogyre is exactly symmetrically set on the cross wire. Final adjustments to A_1 and A_2 as in step 2 may be necessary. Repeat steps 3 on for this plane of symmetry. If the first plane was not the optic axial plane, this one may prove to be, in which case $2V$ may be determined.

15.   Two planes of symmetry have been plotted on the stereogram. Plot the normal to this second symmetry plane. Now we have all three bisectrices plotted on the stereogram. It may be that one of the two symmetry planes studied may allow a centering of two bisectrices when rotated on A_4, in which case the two bisectrices are plotted on the stereogram. As shown in Fig. 4.17C, the complete optical indicatrix is plotted.

16. It may be found that neither of the two planes investigated reveals optic axes and that the diffuse central cross is that presented by the flash figure (optic normal), the optic axial plane being horizontal. The stereogram should be studied to determine whether the third symmetry plane can be accessible on the universal stage. In Fig. 4.17C the example chosen gives the following settings for A_1 and A_2 to set each of the three planes vertical.

	Rotation on	
	A_1	A_2 (corrected)
Plane 1	40°	30° (right-up)
Plane 2	294°	26° (right-up)
Plane 3	351°	48° (left-up)

The third plane may be accessible on the universal stage, though a tilt of 48° is high. If it is accessible, it is examined by setting A_1 and A_2 to their appropriate values and measuring that plane like the others. It may also be that the diffuse centered figure on planes 1 and 2 may have been a BXo figure, the optic axes with a very small $2V$ lying outside the angular tilt of the stage.

17. In any event, if we have not succeeded in centering an optic axis, there are two possible lines of action:

a. Abandon this crystal and repeat the whole operation on another differently oriented crystal. The chances are that it would be more favorably oriented.
b. Use the Berck (1923) and Dodge (1934) method to determine $2V$ and sign on the present crystal.

Alternative (a) is preferable wherever possible because the steps in the Berek and Dodge procedure are somewhat complicated. The Berek-Dodge method is given in exercise 4.9, step 4, with a determinative graph in Fig. 4.19. It is also fully described in Emmons (1959, p. 29–38). I recommend that a suitable crystal be found on the thin section such that one of the accessible symmetry planes will permit an optic axis to be centered. The results obtained will be more accurate.

18. For immersed grains the principal refractive indices α, β, γ can be compared with the immersion oil when the bisectrices are centered vertically; for example in Fig. 4.17C, when the optic plane is set vertically N–S and A_4 is set 30° up to the front, X is vertical and Z lies horizontally N–S. Uncross the polars. For a microscope whose polarizer is N–S the Becke line allows a comparison of γ with the immersion oil. Rotate the microscope stage A_5 through 90° and the Becke line allows β to be measured. Again, any one crystal may not allow the horizontal setting of all three principal directions X, Y, and Z. If two can be measured, as well as $2V$, the third can be calculated with the equation

$$\tan^2 V = \frac{N^2\gamma(N_\beta{}^2 - N^2\alpha)}{N^2\alpha(N^2\gamma - N_\beta{}^2)}$$

or by using the graphical method in exercise 4.2.

Plagioclase Determination on a Five-Axis Stage

The anorthite content and structural state of plagioclase may be determined rapidly on a five-axis univeral stage. This method is approximately twice as fast as the well-known Slemmons (1962) method.

Orientation of the Indicatrix

1. Select a grain with (010) (albite twin lamellae) nearly vertical; in any case (010) must make an angle of more than 60° with the slide. Sharp clear albite twin boundaries indicate that (010) is nearly vertical. The grain should have (001) (the basal cleavage), preferably at an angle of more than 50° to the plane of the thin section. A sharp thin 001 cleavage plane would indicate that (001) makes a high angle to the slide.

2. Lock A_2 exactly parallel to A_4.

3. Place the principal axis (X or Y) lying nearest to (010) in one twin individual exactly parallel to A_2 by rotation about A_1 and A_0. For this step any subindividual of a polysynthetic twin having (010) as a composition plane may be employed. It does not have to be the very albite-twinned subindividual used to orient $\perp$(010) in later steps.

4. Lock A_1 and A_0, having put Y or X parallel to A_2. Unlock A_3 and rotate 90° about A_3 to the zero position. Lock A_3.

5. Place Z parallel to A_4 by rotating about A_2. Lock A_2. Record the reading on the A_2 scale.

Measurement of X to $\perp$(010). The angle X to $\perp$(010) may be measured if Y and Z have been made parallel to A_2 and A_4, respectively.

6. Place a (001) cleavage plane parallel to A_4 and A_5 by rotation about A_3 and A_4. Record the reading on the A_4 scale, which is the complement of X to $\perp$(001); that is, X to $\perp$(001) = 90° − rotation on A_4.

Measurement of $2V$. If Y has been made parallel to A_2 in the orientation procedure, $2V$ is measured directly as follows:

7. Unlock A_3 and place A_2 parallel to A_4 by rotation about A_3. Lock A_3.

8. Put A_5 in the 45° position and make one or, if possible, both optic axes vertical in turn by rotation about A_4. Since X is vertical when A_4 is at zero, $V\alpha$ equals the amount of rotation about A_4 to reach the optic axis. Repeat for the second optic axis if it can be obtained. Repeat the readings in the other 45° position for A_5 and average the $V\alpha$ values obtained.

If Y has been made parallel to A_3, $2V$ is determined indirectly by the Berek-Dodge method as follows:

9. Unlock A_3 and place X and Z 45° from A_4 by rotation about A_3. Lock A_3.

10. Place Y 50, 60, 70, or 75° from A_5 by rotation in either sense about A_4. Lock A_4.

11.   Measure the extinction angle from X' (fast vibration) to A_4 and determine $2V$ and sign from Fig. 4.19.

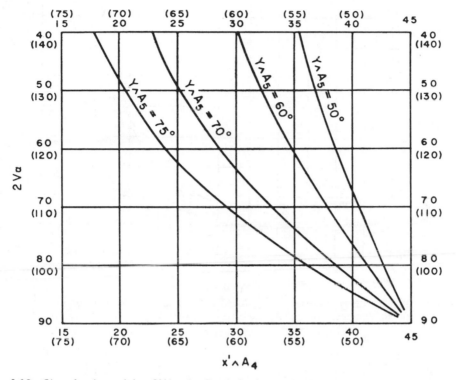

Fig. 4.19   Chart for determining $2V$ by the Berek-Dodge method when X and Z are both normal to A_3 and 45° from A_4. (After Noble, 1965b.)

Measurement of X to $\perp$ (010) and Y to $\perp$ (010)

12.   Unlock A_3. Place $\perp$(010) parallel to A_4 by rotation about A_2 and A_3. For this the composition plane of an albite twin is set vertical and N–S. This is accurately done by adjusting the crystal until interference colors in adjacent albite-twinned individuals match exactly across the twin plane throughout a wide rotation about A_4. Comparison of colors should be made in both the zero and 45° positions; the insertion of a first-order red plate sometimes increases the sensitivity in the 45° position. This method of orientation is very precise.

13.   Read the angle separating A_2 and A_4, which is either Y to $\perp$(010) or X to $\perp$(010), directly on the scale of the A_3 axis. Record the reading on the scale of the A_2 axis. The parameter $|\Delta A_2|$ is the absolute value of the difference between this reading and the reading recorded in step 5. In step 12 rotation about A_2 is necessary to place $\perp$(010) parallel to A_4. If the angle Y to $\perp$(010) is measured directly on the A_3 scale, the amount of rotation is termed $|\Delta A_2|x$. If X to $\perp$(010) is measured directly, the amount of rotation is termed $|\Delta A_2|y$.

Determination of Composition and Structural State.   Composition is determined by plotting $|\Delta A_2|x$ versus Y to $\perp$(010) in Fig. 4.20 or X to $\perp$(010) versus $|\Delta A_2|y$ in Fig. 4.21.

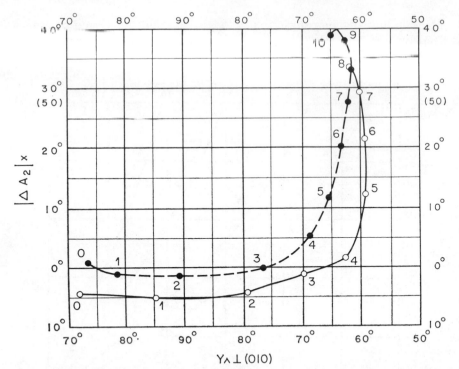

Fig. 4.20 Plot of $|\Delta A_2|x$ versus Y to $\perp$ (010) for plutonic (dashed line) and volcanic (solid line) plagioclases. Composition, from An_0 to An_{100}, is given by the numbers 0 to 10. (after Noble 1965b.)

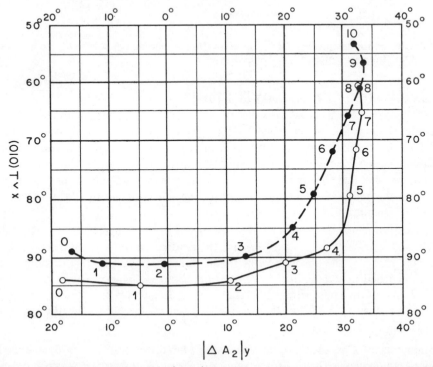

Fig. 4.21 Plot of X to $\perp$ (010) versus $|\Delta A_2|y$ for plutonic (dashed line) and volcanic (solid line) plagioclases. Composition, from An_0 to An_{100}, is given by the numbers 0 to 10. (after Noble, 1965b.)

106

In addition, for sodic plagioclase X to $\perp(001)$ is plotted versus Y to $\perp(010)$ (Fig. 4.22). This plot may be also used to check the composition of more calcic plagioclases.

If the data fall on either the plutonic or the volcanic curve, interpolation is made between the points that represent 10% increments of An content. If the point falls between or slightly outside the curves, lay a ruler along similar An% of the volcanic and plutonic curves and interpolate.

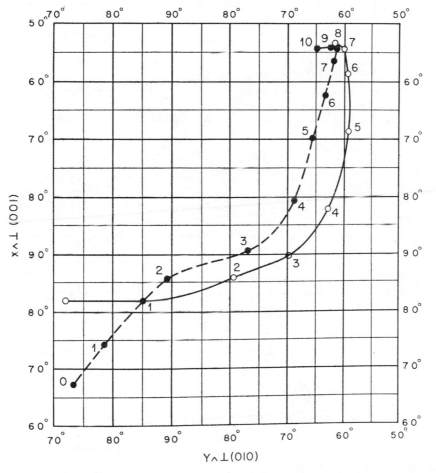

Fig. 4.22 Plot of X to $\perp$ (001) versus Y to $\perp$ (010) for plutonic (dashed line) and volcanic (solid line) plagioclases. Composition from An_0 to An_{100} is given by the numbers 0 to 10. (after Noble, 1965b.)

Structural state is indicated by the closeness to the plutonic or volcanic curves. The position of $\perp(010)$ with respect to the indicatrix is the most sensitive optical indicator of structural state of plagioclase more calcic than about An_{35}. For more sodic plagioclase (albite + oligoclase) $2V$, shown in Fig. 4.23, is the most sensitive.

The plot of X to $\perp(001)$ versus Y to $\perp(010)$, as shown in Fig. 4.22, is a sensitive indicator of the structural state of albite.

The plot of X to $\perp(010)$ versus Y to $\perp(010)$, as shown in Fig. 4.24, may also be useful for structural state and composition determination of most plagioclases.

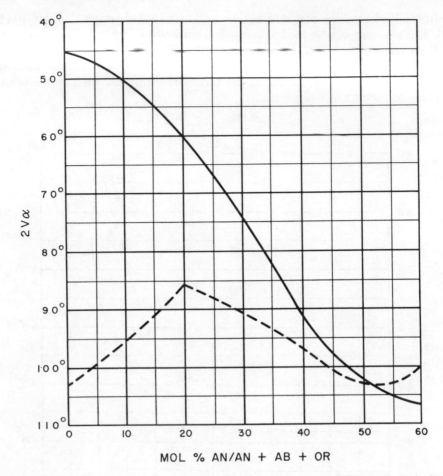

Fig. 4.23 Plot of $2V\alpha$ for plutonic (dashed line) and heated plutonic (solid line) plagioclases of sodic and intermediate composition. (after Noble, 1965b.)

Because of possible scatter, several grains from the same rock should be determined.

For sodic plagioclase the composition indicated by a given value of Y to $\perp(010)$ is greatly affected by small errors in X to $\perp(010)$. If the structural state indicated by the indicatrix orientation disagrees with the more reliable $2V$, the $2V$ is more likely to give the correct determination.

Resolving Ambiguities

14. Make (001) vertical by rotation about A_4 and note qualitatively the position of Y with respect to (010) and (001).

15. For sodic plagioclases note qualitatively the position of X with respect to (010) and (001) from the sense of rotation about A_2 and A_4 required to make (010) and (001) vertical.

16. If necessary, determine qualitatively the position of X with respect to [100] and $\perp[001]/(010)$.

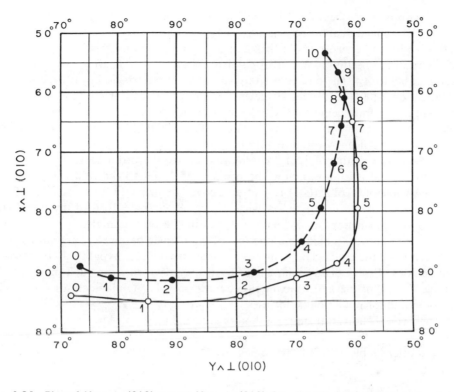

Fig. 4.24 Plot of X to $\perp$ (010) versus Y to $\perp$ (010) for plutonic (dashed line) and volcanic (solid line) plagioclases. Composition, from An_0 to An_{100}, is given by the numbers 0 to 10 (after Noble, 1965b.)

DISCUSSION

Values of Y to $\perp$(010) from 70 to 90° may represent either of two compositions. The correct one may be determined by noting whether Y lies within the acute or the obtuse angle between (010) and (001). For all plagioclases more sodic than An_{20} and An_{13} Y will lie within the obtuse angle between (010) and (001), whereas for more calcic compositions Y will lie within the acute angle.

The relative relief of the grain against the immersion oil may also indicate the correct composition.

In most cases the position of Y with respect to (010) and (001) shows immediately on which side of the ordinate X to $\perp$(001) = 90° on the graph (Fig. 4.22) the measured value of X to $\top$(001) is to be plotted.

In sodic plagioclase X is very close to (010), equivalent to the ordinate $|\Delta A_2|x = 0$ in Fig. 4.20. The side of this line on which $|\Delta A_2|x$ is to be plotted may be determined by noting qualitatively the relationship between X and (010) and (001). The position of X relative to (010) and (001) is obtained from the sense of rotation about A_2 and A_4 made to place (010) and (001) vertical. For albite, sodic oligoclase, calcic oligoclase, and sodic andesine with Y to $\perp$(010) greater than 82° $|\Delta A_2|x$ plots above the line $|\Delta A_2|x = 0$ if X lies within the acute angle formed by (010) and (001) and below the line if X lies within the obtuse angle.

For compositions near An_{30} the position of X relative to [100] and $\perp[001]/(010)$ must be determined qualitatively before $|\Delta A_2|y$ is plotted versus Y to $\perp(010)$ or X to $\perp(001)$ is plotted versus Y to $\perp(010)$. For such compositions X is nearly parallel to (010). Thus the projection of X on (010) is virtually parallel to A_3 after $\perp(010)$ has been made parallel to A_4. The direction [100] is the line of intersection of (010) and (001). The direction $\perp[001]/(010)$ may be determined if a Carlsbad or albite-Carlsbad twin is present in the grain being studied. The interference colors in the Carlsbad or albite/Carlsbad twinned subindividuals will match when $\perp[001]/(010)$, which lies 26° from [100], is made vertical by rotation about A_4.

The measured value of $|\Delta A_2|x$ is plotted above the ordinate $|\Delta A_2|x = 0$ if X lies to the left of (010) and below the ordinate if X lies to the right of (010). The measured value of X to $\perp(001)$ is plotted above the ordinate X to $\perp(001) = 90°$ if the projection of X on (010) lies within the acute angle formed by [100] and $\perp[001]/(010)$ and below the ordinate if X lies in the obtuse angle.

Extinction angle techniques alone, based on a single parameter, are *not* accurate for determining the plagioclase composition from a volcanic, hypabyssal, or epizonal plutonic rock. Errors of as much as 13% An may develop from such methods, for example, by use of the $\perp(010)$ extinction curves on volcanic plagioclases which have been prepared for plutonic plagioclases.

The five-axis universal stage method is accurate to $\pm 3\%$ An.

(Method after Noble, 1965b.)

EXERCISE 4.10

Structural State of Alkali Feldspar

The optic axial angle of alkali feldspars distinguishes high-temperature modifications from low-temperature forms if the chemical composition is known. The chemical composition may be determined by the method described in Chapter 7A; $2V$ should be determined by a universal stage method in thin section with conoscopic light, in grains mounted on a spindle stage, or in an immersion mount on a universal stage. Tuttle and Bowen (1958) claim a $\pm 1°$ accuracy for the conoscopic method on K-feldspars.

Figure 4.25 shows the relation between $2V$, composition, and structural state. On the basis of $2V$ and the position of the optic plane, four series can be recognized. Gradations exist between these series. The high sanidine-high albite and low sanidine-high albite series are high temperature modifications and occur in volcanic and high-level epizonal and hypabyssal rocks. The orthoclase and microcline series occur in plutonic rocks.

A useful way of studying a suite of rocks is to plot a histogram that shows the number of crystals analyzed in which $2V$ was determined to have discrete values ranging from 40° to 90°. A volcanic rock suite may be expected to show a mode on the histogram at 45°, a high-level epizonal granite at 55 to 60° $2V$, and a mesozonal or catazonal plutonic granite a mode at 80° $2V$; $2V$ measurements on K-feldspars are therefore of extreme importance in distinguishing granite plutons that have cooled rapidly from those that have cooled slowly. Volcanic rocks can be readily distinguished on the basis of their K-feldspar $2V$.

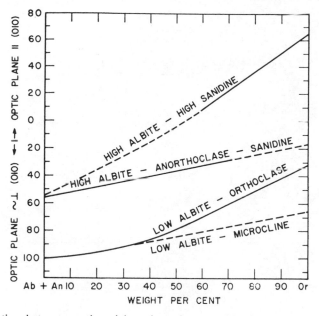

Fig. 4.25 Relation between optic axial angle and composition of the alkali feldspars. (After Tuttle and Bowen, 1958.)

2V in Topaz

Ribbe and Rosenberg (1971) have shown that the 2V of topaz is a sensitive indicator of its fluorine content:

$$\text{weight } \% \text{ fluorine} = 3.91 + 0.24 \times 2V_\gamma.$$

The 2V can be determined by the conoscopic method on a universal stage.

Applications to Petrofabrics

An oriented thin section of a foliated rock may be studied on a universal stage by using conoscopic light to enable the collection of the orientation of a particular mineral parameter on a large number of crystals; for example a large number of quartz crystals in a thin section may have their optic axes set vertically or horizontally (exercise 4.7) and each optic axis may be plotted on a stereographic net. In marble the calcite optic axes may be plotted likewise.

It is common in petrofabric work to plot the normals to all mica cleavages on a stereographic net. For this the universal stage is used without crossed nicols. Each mica crystal in turn is rotated on A_1 until its cleavage is exactly parallel to the E–W cross wire. The crystal is then tilted on A_4 until the image of the cleavage plane becomes as narrow as possible. This makes the cleavage plane exactly vertical and E–W. The pole to the cleavage plane is then plotted. The example shown in Fig. 4.26 illustrates the method. It may happen that several mica crystals may be so oriented in the thin

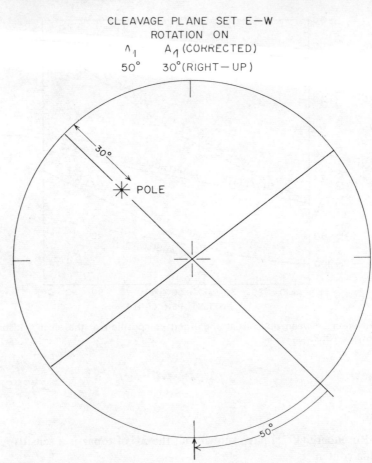

Fig. 4.26 Stereographic plotting of the pole to a mica cleavage plane which has been set E–W and vertical. The tilt on A_4 has to be corrected before plotting. The upper hemisphere is used.

section that their cleavage cannot be set vertically, in which case the polars have to be crossed, the universal stage adjusted so that a *BXa* figure is centered (exercise 4.8), and the position of *BXa*, which is, in fact, the normal to the cleavage plane, plotted on the stereogram.

The net result of all these methods is a stereogram that contains a large number of points.

MINERAL SEPARATION 5

It is frequently necessary to separate individual mineral grains from a whole rock for independent study. A separated mineral is commonly required for optical study in immersion mounts, and for X-ray diffraction and chemical analysis.

The method of separation selected must depend on the amount of separated mineral required and on the grain size of the mineral in the rock. The following is an outline of some of the possible procedures, but they may be modified in the light of individual requirements and the operator's experience.

EXERCISE 5.1

Hand Extraction

1. First make a careful study of the rock both in hand specimen and thin section with a view of selecting the best procedure for separating the selected mineral from the rock matrix. Features such as ease with which the mineral may be identified from its matrix and distinguishing physical properties should be noted.

2. If the mineral to be separated forms large discrete crystals—phenocrysts or porphyroblasts, for example, of K-feldspar or garnet—the simplest procedure may be to detach from the hand specimen with a chisel and hammer or dental pick a few parts of a conveniently placed larger crystal. The crystal fragments so detached can be further purified under a hand lens or binocular microscope and impurities chipped away with a small chisel on a steel anvil. If there is any doubt about the identification of the minerals, the grains or the whole rock surface may be stained by one of the methods in Chapter 2. Staining, for example, may help to distinguish K-feldspar from plagioclase. Once the separation is complete the stain should be removed from the grains by immersing in 1 N hydrochloric acid for 10 minutes at room temperature, then washed and dried.

3. If the separation is for optical immersion or X-ray diffraction studies, a high degree of purity of the separation will not normally be required, and this kind of rough hand extraction may be all that is necessary.

4. At any stage during hand extraction the hand specimen may need to be broken into smaller chips. This is conveniently done with a geological or sledge hammer on a specimen placed in a depression on a heavy metal plate or rock monolith resting on the workshop floor. Alternatively the hand specimen can be compressed between the hardened steel jaws of a manually operated hydraulic rock-splitting machine (supplied by Cutrock Engineering Co. Ltd., 35 Ballards Lane, London, N.3, England, or Wards Natural Science Establishment Inc., P.O. Box 1712, Rochester, New York 14603). When applying hydraulic pressure to the specimen, the metal guard should

113

be placed around the specimen holder and the operator should wear a plastic face visor as a protection against flying splinters.

5. Further percussion in a steel mortar (of 15 cm external diameter) with a steel pestle (supplied by Griffin and George Ltd., Ealing Road, Alperton, Wembley, Middlesex, England) may also help in the extraction of discrete grains. A percussion action is to be preferred, for it tends to break the specimen along grain boundaries; a grinding action in the mortar produces too much fine powder which is of no use to the operator. Before making use of the grains their purity should be checked under a binocular microscope.

EXERCISE 5.2

Crushing the Whole Rock Specimen

If individual grains cannot readily be hand extracted, the alternative may be to crush the whole rock specimen and then effect a separation of one mineral by means of a distinctive density or magnetic property.

1. Carefully note the average grain size of the mineral to be separated. This should be done preferably on a thin section. The common grain diameters in igneous and metamorphic rocks may range from less than 1 to 6 mm.

2. The complete rock should now be crushed to a final particle size between one-quarter and one-tenth the average grain size of the rock. This will ensure that the individual particles have a good chance of being monomineral. Crushing to a particle size that is too fine is not recommended; the aim should be only as far as necessary for monomineralic grains to be released.

3. Break up the hand specimen into smaller chips by pounding the specimen with a hammer against a heavy metal plate or rock monolith or by using a hydraulic rock splitter. The chips as a rule should be less than 6 × 6 cm.

4. Adjust the jaw spacing on a jaw crusher (supplied by Glen Creston, The Red House, 37 The Broadway, Stanmore, Middlesex HA7 4DL, England, or Fritsch, D 6580 Idar-Oberstein, West Germany) so that it is as small as the final particles required. Laboratory jaw crushers can accept rock fragments up to 60 × 60 mm and can crush them down to a final particle size of about 1 mm diameter. The setting of 1 mm will normally be ideal. Replaceable jaws can be obtained in manganese steel or stainless Cr-Ni steel. Before crushing it is essential to clean the jaw crusher thoroughly. To do this one jaw is removed to allow both jaws to be scrubbed clean with a wire brush. A vacuum cleaner set to a blowing action is useful for blowing all dust from the crusher. Please note, however, that rock dust constitutes a severe health hazard. The operator must always wear a respirator or at least the kind of mouth and nose gauze protector worn by dentists and surgeons.

5. The rock chips are fed into the jaw crusher and the crushed material is collected in a clean receiving metal container. The jaw crusher is thoroughly cleaned after each use. The place of a jaw crusher may be taken by a laboratory roller crusher (supplied by Chas. W. Cook and Sons Ltd., 97 Walsall Road, Perry Barr, Birmingham 22B, England).

6. Sieve the crushed rock for 10 minutes through a bank of U. S. Standard Sieves

(A.S.T.M. E 11 series) (supplied by Endecotts Test Sieves Ltd., Lombard Road, London S.W. 19, England). A suitable bank would consist of the following:

Mesh No.	Aperture (mm)
LID	
18	1.00
40	0.420
100	0.149
170	0.088
230	0.063
270	0.053
Bottom receiver	

As standard practice collect the grains that pass through the 100 mesh sieve and settle on the 170 mesh. These grains therefore will fall in the range of 0.15 to 0.09 mm diameter and for the vast majority of rocks should be monomineralic. The -100 to $+170$ mesh range is ideal for many purposes; for example, grains of this size are perfect for immersion mounts with a $\times 10$ microscope objective for optical study.

7.   Rock particles that collect on sieves coarser than the 100 mesh can be put in a steel mortar and pounded with a steel pestle until on resieving they pass through the 100 mesh. Occasionally a steel piston-type percussion mortar with a close-fitting steel cylindrical pestle of 28 mm diameter (supplied by Griffin and George or Gallenkamp, P.O. Box 290, Technico House, Christopher Street, London E.C. 2, England) may be more effective for crushing than the bellshaped mortar and pestle. Particles that pass the 170 mesh may be discarded as too fine.

8.   Although -100 to $+170$ mesh is the most frequently chosen range for mineral separates, it is essential to check it as a wise choice. A binocular observation of the grains collected on the 170 mesh should confirm that the grains are monomineralic and not composite. If the rock was fine grained, it may be necessary to collect the grains that pass the 170 mesh and collect on either the 230 or even the 270 mesh sieve. After each sieving the bank of sieves should be cleaned thoroughly with a soft camel's-hair brush.

EXERCISE 5.3

Cleaning

The crushed and sieved grains must now be thoroughly washed free of the fine dust produced during grinding. This is an important step that must not be neglected, for dust-coated grains are difficult to separate by subsequent processes.

1.   Place the grains (sieved to -100 to $+170$ or finer, if selected) in a beaker.

2.   Run tap water forcibly into the beaker through a fine jet to stir up the contents. When the beaker is nearly full, decant the water, together with slime, into the sink. Repeat this washing operation as often as necessary until the supernatant water added remains clear and all traces of slime have been removed. Pour off the clear water.

3.   Transfer the wet grains to a watch glass and place overnight or for several hours in a drying oven at about 80°C.

4. If the desliming of the grains is required in a hurry, acetone may be used in place of water, forced into the beaker through the fine jet of a wash bottle. After the grains are clean the acetone evaporates quickly.

5. The clean grains are now ready for separation by a suitable selection of methods, usually by magnetic or density separation, separately or consecutively, or even by hand picking under a binocular microscope.

EXERCISE 5.4

Magnetic Isodynamic Separation

A large variety of minerals may be separated from one another on the basis of their magnetic susceptibility. The ferromagnesian components can usually be quite perfectly separated from the salic minerals. Feldspars, however, cannot be separated from quartz. It is essential to know exactly which minerals the crushed sample contains and to ascertain from Fig. 5.3 if any of these mineral components has a distinctive magnetic susceptibility that will enable them to be separated from the others.

1. Initially the dry grains must have the magnetic components removed before being fed into the isodynamic separator. These strongly magnetic grains will be magnetite belonging to the rock and particles of metallic iron that have been introduced by the crushing operations.

2. A Frantz LD-3 dry ferromagnetic separator is set at a slope of about 35° (Fig. 5.1). The powder is streamed slowly into the ferrofilter funnel, as the funnel is slowly

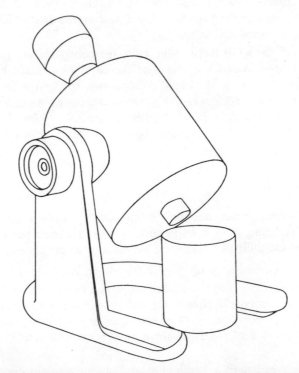

Fig. 5.1 Frantz model LD-3 dry ferromagnetic separator for removing the highly magnetic grains (magnetite and iron metal fragments) before the sample can be fed to the isodynamic separator.

rotated by hand. The highly magnetic particles attach themselves to the magnetic central part of the funnel, and the grains which pass into the metal container are then free of magnetite and iron metal contaminants.

3. After all the powder has been fed through, the funnel is lifted upward about 1 cm to a click stop, then fully rotated, pushed fully down again, rotated, lifted, and rotated again. This frees any trapped nonmagnetic grains.

4. The collector is removed. A cork is pushed into the lower open end of the funnel and the complete funnel assembly is pulled up out of the ferrofilter body. On its removal the magnetic grains fall onto the cork from which they can be released onto a white card and observed under a binocular microscope for identification. If this highly magnetic fraction contains both magnetite grains from the rock and iron fragments from the jaw crusher, the iron fragments may be manually separated, leaving the magnetite from the rock as a separate for mineral study if required. The ferrofilter funnel should be cleaned with a test-tube brush after every use.

5. If a Frantz LD-3 dry ferrofilter is not available, the operation may be done with a weakly magnetized hand magnet to pull the strongly magnetic grains away from the powder that has been spread on a card.

6. The powder is now ready for separation in the Frantz Isodynamic Magnetic Separator, model L1 (supplied by S. G. Frantz Co. Inc., 339 East Darrah Lane,

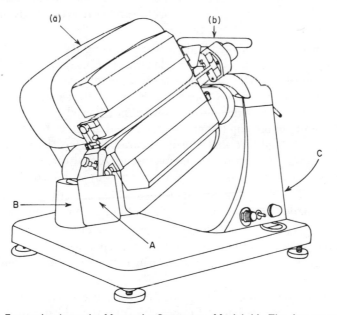

Fig. 5.2 The Frantz Isodynamic Magnetic Separator, Model LI. The important controls are (a) forward slope, set on a scale at the back; (b) side tilt, set by means of the wheel on the right; and (c) the electric current in amperes, set by an adjustable potentiometer and read on an ammeter. This controls the strength of the magnetic field. Final separation is into two cups: A more magnetic and B less magnetic.

P.O. Box 1138, Trenton, New Jersey 08606) (Fig. 5.2). The isodynamic separator consists of an inclined vibrating plane, which is set up according to definite settings of forward slope and side tilt. The forward slope is adjusted on a goniometer scale set at the back of the instrument (Fig. 5.2) and the side tilt is selected by rotating the black metal wheel on the right. Once slope and tilt have been selected the settings are locked. A small aluminum filter funnel is screwed fully clockwise into the top end of the inclined plane and filled up with the dry mineral powder. The principle of operation is that the magnetic field is directed up-slope and pulls the more magnetic particles up the inclined vibrating plane toward the front .Toward the lower end of the inclined plane a longitudinal divider separates the more magnetic grains, which collect in the front cup, from the less magnetic grains, which collect in the cup lower down slope. The amperage on the electromagnet is set at a value that would make a clear distinction between one mineral in the powder and the others. Amperage is selected first at low values and minerals are progressively separated from the mixture in the order of highly magnetic first and then onto the most weakly magnetic.

7.   The Frantz isodynamic separator can separate dry materials as fine as 300 mesh (0.053 mm diameter). Finer particles give trouble and should be avoided (Muller, 1967).

8.   Figure 5.3 lists the magnetic susceptibilities of common minerals. All names on the left of the table are for a side tilt of 15° and a forward slope of 25° (data after Rosenblum, 1958). The best results are for grains sieved −100 to +170 U.S. mesh, washed, deslimed, and dried. The table is also applicable for grains up to 65 mesh and down to 300 mesh. Figure 5.3 gives some data for a side tilt of 25° and a forward slope of 15° for mineral names tabulated at the right of the table. (Data after Flinter, 1959.) I doubt Flinter's identification of staurolite and have accordingly omitted it from the compilation.

9.   Although magnetite has already been removed from the powder, it is customary to run the sample through the isodynamic separator at 0.1 amp with a forward slope of 25° and side tilt of 15°. The vibrator is set to a middle value on its operating scale and the filter funnel rotated slowly counterclockwise to effect a steady stream of grains down the inclined plane. After all the powder has flowed through, magnetite will have collected in the upper cup (front container) and all other minerals in the lower. This magnetite can be added to that already separated in steps 1 to 4. This initial run at 0.1 amp is necessary to remove all the magnetite. A first run at a higher amperage would otherwise cause clogging of the magnet poles by attracted grains.

10.   The inclined plane is removed and cleaned with a camel's-hair brush. The space between the magnetic bars of the separator should also be brushed clean.

11.   The nonmagnetic powder collected in the lower cup (back container) is now fed through the filter funnel into the isodynamic separator set at a suitably higher amperage of, say, 0.3 amp. This operation will remove all the ilmenite to the front cup.

12.   The nonmagnetic powder may now be run through at 0.6 or 0.7 amp to separate, for example, the hornblende into the upper cup from the feldspar in the lower cup.

DISCUSSION

The foregoing procedure is standard. First the most magnetic grains are removed at a low amperage and then minerals of decreasing magnetic susceptibilities are progressively separated by increasing the amperage. Between each run the inclined plane should be detached and cleaned, together with the space between the poles of the magnet, with a soft-haired brush. A slow feed is preferred. If some separated samples

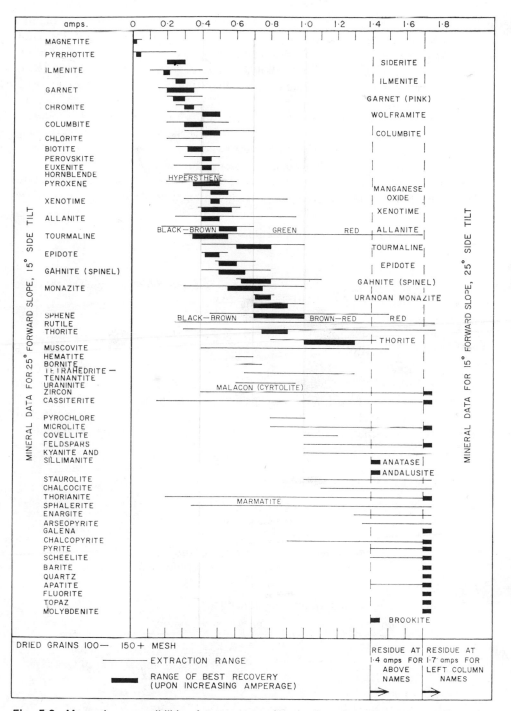

Fig. 5.3 Magnetic susceptibilities for common minerals. Data from Rosenblum (1958) and Flinter(1959) for the Frantz Isodynamic Magnetic Separator. Thick lines indicate the most frequent range; thin lines indicate the fuller range that may occasionally occur.

are found to be mixtures of more than one mineral, they may be rerun at a slightly different amperage, provided their magnetic susceptibilities shown in Fig. 5.3 are not identical; for example, it should be possible to separate chromite from ilmenite at an amperage of 0.25, but the separation will not be perfect because of overlapping fields. On the other hand, there would be no hope of separating biotite from pyroxene on an isodynamic separator.

The operator should select the amperage settings from Figure 5.3 but should be prepared to experiment by increasing or reducing the values slightly and rerunning the same powders until the best possible separation is effected. The separations should be checked frequently under a binocular microscope for purity of separation.

As an example of a rock that would give a good separation of all mineral components, an amphibolite composed only of ilmenite, hornblende, and plagioclase would require the following runs:

a.   At 0.1 A to separate any magnetite.
b.   At 0.3 A to separate ilmenite.
c.   At 0.7 A to separate hornblende.

The nonmagnetic residue would consist of feldspar plus any quartz that the rock may contain.

Density Separation

Choice of Specific Gravity Liquid and Diluent

A series of liquids can be prepared with specific gravities equal to the middle points of the specific gravity of each of the successive pairs of minerals to be separated. The average specific gravities of minerals are readily available in most mineralogy textbooks and need not be given here.

$$V_l = V_h \frac{G_h - G_x}{G_x - G_l},$$

where G_x = the specific gravity of the mixture desired,
  G_h = the specific gravity of the heavy end member liquid,
  G_l = the specific gravity of the light end member liquid (diluent),
  V_h = volume of the heavy member of the mixture,
  V_l = volume of the light member of the mixture (diluent).

(After Woo, 1964.)

The following heavy liquids are recommended:

		Density at 20°C	Solubility g/100 ml Water at 20°C
1.	Bromoform	2.8899	0.319
2.	Tetrabromoethane (acetylene tetrabromide)	2.9672	0.065
3.	Methylene iodide (Di-iodo methane)	3.325	1.42
4.	Clerici's solution	4.28 (can be 4.85 at 80°C)	1400 to 2000

The following diluents are recommended:

	Density at 20°C	Solubility g/100 ml Water at 20°C
1. Acetone	0.792	infinite
2. *NN*-Dimethyl formamide	0.93445	infinite

(After Muller, 1967; Woo, 1964; Benjamin, 1971.)

NN-Dimethyl formamide is the most satisfactory diluent because it has a low vapor pressure, approximately equal to that of the heavy liquids. Its use produces a stable diluted solution and low viscosity at low cost; it has the advantage of being infinitely soluble in water at room temperature, thereby facilitating final recovery of the heavy liquid.

Tetrabromethane is the most frequently used heavy liquid for minerals of specific gravity less than 2.95. Thallous formate can be used for minerals heavier than 2.95 and lighter than 4.95 but it is poisonous and expensive (Woo, 1964).

Dimethyl sulfoxide $(CH_3)_2SO$ is recommended as a diluent for bromoform instead of acetone. It is colorless, odorless, and less flammable than acetone and has a lower vapor pressure. Technical grade is similar in price to N.F. grade acetone (Meyrowitz et al., 1959).

Other diluents in use are benzene, carbon tetrachloride, and alcohol.

All heavy liquids should be kept in dark-colored bottles because they discolor and break down in light.

Clerici's solution may be made from a saturated solution of equal-weight thallous formate and thallous malonate in distilled water; 300 g of each salt in 40 g of water produce approximately 135 ml of solution of density 4.28 g/ml at 20°C and at 90°C up to 5.0 g/ml. This solution may be diluted with distilled water and is not light-sensitive like the others.

All heavy liquids, especially Clerici's solution which is extremely poisonous, must be used in a fume cupboard. Clerici's solution is a heavy-metal poison (thallium) that is cumulative in the human system and can be absorbed through the skin. Rigorous care in handling therefore is essential. This solution may be recovered after dilution by slowly heating in a sand bath. If overheated, however, it will turn dark colored.

The actual specific gravity of a diluted heavy liquid may be conveniently checked with a set of 21 glasses whose specific gravities range from 2.28 to 4.08 (obtainable from Gemmological Instruments Ltd., Saint Dunstan's House, Carey Lane, London E.C.2 England). Each glass has its specific gravity clearly marked, and the value marked on the particular glass which neither floats nor sinks in the liquid gives the specific gravity of the liquid.

EXERCISE 5.6

Recovery of Heavy Liquids

Heavy liquids are expensive; diluents are cheap. Hence it becomes necessary to recover the heavy liquid from a mixture after a mineral separation has been completed.

The method is based on the relatively insoluble nature of the heavy liquid in water. The most effective procedure (after Benjamin, 1971) is as follows:

Figure 5.4 illustrates the recovery apparatus. It is designed to give minimum wash-water volume consistent with the removal of the bulk of the diluents and to minimize losses.

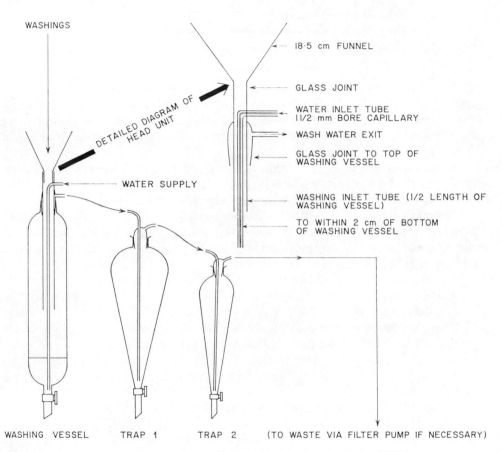

WASHINGS

DETAILED DIAGRAM OF HEAD UNIT

WATER SUPPLY

18·5 cm FUNNEL

GLASS JOINT

WATER INLET TUBE
11/2 mm BORE CAPILLARY

WASH WATER EXIT

GLASS JOINT TO TOP OF
WASHING VESSEL

WASHING INLET TUBE (1/2 LENGTH OF
WASHING VESSEL)

TO WITHIN 2 cm OF BOTTOM
OF WASHING VESSEL

WASHING VESSEL   TRAP 1   TRAP 2   (TO WASTE VIA FILTER PUMP IF NECESSARY)

Fig. 5.4 Apparatus for recovery of heavy liquids. (Design after Benjamin, 1971.)

The apparatus consists of a *washing vessel* which is a 2 liter cylindrical separating funnel, two trap vessels, which are 1 liter conical separating funnels, and trap heads, which are gas-washing jar heads.

The only nonstandard item is the washing-vessel head. Its construction is shown in Fig. 5.4. A cork can be bored to accomodate all the necessary tubing if glass-blowing facilities are not available. A water-flow meter can be inserted into the wash-water inlet line. Flow rates of 300 to 400 ml/min are ideal. The vessels are arranged in cascade fashion so that water will siphon through the system. To start the siphon the funnel inlet has to be closed with a cork. The waste tube may be fitted to a filter pump working at low flow rates to ensure steady operation.

Once the stabilized water flow has been established diluted heavy liquid washings are added via the funnel. If the flow rate is 350 ml/min, a washing time of 10 minutes is usually enough to remove the diluent completely from the washing. At the end of this time the recovered heavy liquid is run off from the bottom of the washing vessel. If any has carried over to traps 1 and 2 it can also be drained off. After filtering and dewatering it is ready for further use.

EXERCISE 5.7

Density Separation of Two Minerals Coarser than 200 Mesh

The basis of this method is that mineral grains of specific gravity lower than the liquid will float, whereas those of higher specific gravity will sink. The method may be illustrated by taking a common example.

A frequent problem is the separation of feldspars from quartz, which cannot be done by magnetic methods. If, for example, a granite had been crushed and separated into magnetic fractions according to exercise 5.4, the nonmagnetic residue left at the end would be expected to contain quartz of specific gravity 2.65, microcline of specific gravity 2.56 to 2.59, and a plagioclase in the oligoclase range of approximate specific gravity 2.65. Clearly, the quartz and oligoclase cannot be separated on a density basis, but the K-feldspar can be floated off by using a liquid of 2.60. This will be a useful separation because K-feldspar is one of the most informative minerals and such a separation can be used for X-ray diffraction work to indicate its composition and temperature state.

Since density separation is of frequent use in a mineralogical laboratory, it is advisable to set up some kind of separator (see Fig. 5.5).

A cheap electrically driven centrifugal pump continuously refills the vessel with the heavy liquid of selected specific gravity. If K-feldspar, for example, is to be floated from quartz and plagioclase, the liquid is made of S.G. 2.60 according to the recipe in exercise 5.5.

The funnel has a filter paper for catching the light minerals, and the heavy fraction has to be drained off at intervals. The pump need have a capacity of only 1.5 l/min. Grains of −100 to +200 mesh size are satisfactory in such a separation.

(Design after Østergaard, 1968.)

EXERCISE 5.8

Density Separation for Fine-Grained Rocks

Centrifuging to 1500 gravities will speed up the sink-float separation of grains in heavy liquids.

1. Crush sample to −200 mesh and sieve. Let it settle in a water-filled vessel 10 cm high for 2 minutes. Decant the suspension. Let the suspension settle for another 10 minutes; decant again. This procedure gives three solid fractions of approximate size limits.

$30 - 75 \; \mu$ after 2 min,

$10 - 30 \; \mu$ after 10 min,

$< 10 \; \mu$ in the remaining suspension.

Dry the two coarse fractions under a heat lamp. A final washing with acetone gives a powdery sample free of lumps.

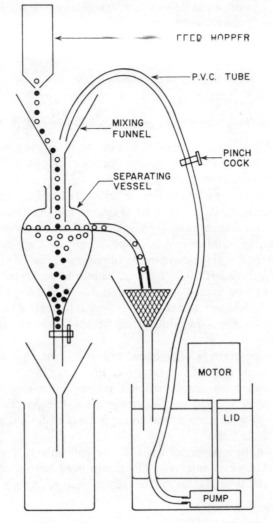

Fig. 5.5 A continuous density separator for mineral separation. (After Østergaard, 1968.)

2. Pour the size fraction whose largest grains are monomineralic onto a creased glazed weighing paper. Crush any lumps to form a free-flowing powder.

3. Pour the sample into a special centrifuge tube (see Fig. 5.6), which may be made by any glass blower. The 40 ml tube will hold 2 g or less and the 10 ml tube holds $\frac{1}{2}$ g or less.

4. Fill each tube to the same level with a heavy liquid mixture of the proper specific gravity. If a complete separation of a rock is attempted, it is best to start with the heaviest liquid and work to successively lighter liquids. In this way the bulk of the sample is always floating at the top of the tube where it can be stirred to free more heavy minerals.

5. Close the mouth of each tube, preferably with a plastic stopper, to prevent the liquid from evaporating and causing a consequent change in specific gravity.

6. Gently invert each tube several times to disperse the sample throughout the

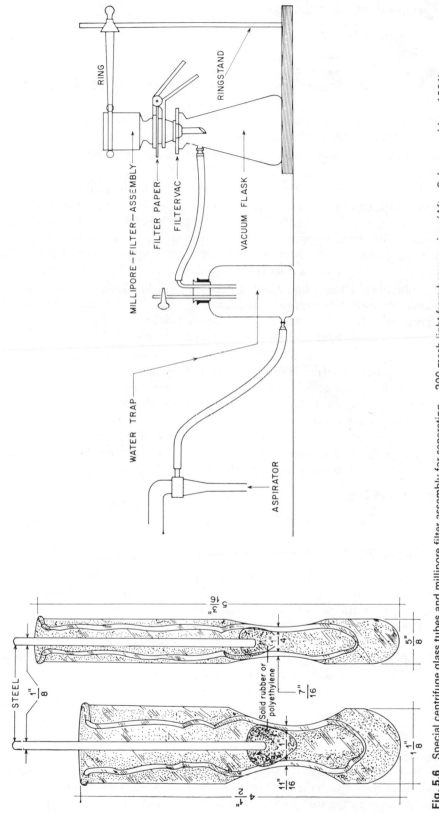

Fig. 5.6 Special centrifuge glass tubes and millipore filter assembly for separating −200 mesh light from heavy grains. (After Schoen and Lee, 1964).

125

liquid. Quickly place the tubes into a centrifuge and rapidly accelerate it to 1500 to 2000 gravities.

7.   Stop the centrifuge slowly after about 1 minute. Prolonged centrifugation is not good, for it hardens the floating cake of mineral grains and makes further separation difficult.

8.   Stir the floating cake to disperse it completely but avoid mixing it with the heavy sunk fraction. Quickly put on the centrifuge and rapidly accelerate to high speed.

9.   Three centrifugations are enough but more may be done until no further sinking is evident.

10.   Gently twist the specially shaped stopper through the floating cake. This loosens the cake so that it can be poured off, yet it causes no mixing of the light solids and heavy fraction. Seal off the lower bulb of the tube and pour the light fraction into a millipore filter. With the stopper held tightly against the constriction of the tube, rinse the upper portion with a jet of heavy liquid from a squeeze bottle until all the light fraction is on the millipore filter (Fig. 5.6). Repeat step 10 for each tube.

11.   After filtering the heavy liquid into a vacuum flask break the vacuum, remove the vacuum flask containing the heavy liquid, and insert a clean vacuum flask beneath the millipore filter. Wash the light fraction and the sides of the funnel with a jet of acetone from a squeeze bottle. Filter off the acetone washings.

12.   Carefully open the millipore filter holder and remove the filter paper. Scrape the grains sticking to the funnel onto the filter paper and dry under a heat lamp. Rinse the funnel with acetone and dry. Install a new filter paper and reassemble the filter. Pour off the heavy fraction from each tube and rinse it with heavy liquid. Change the vacuum flask and wash it with acetone as before.

13.   The heavy liquid filtrate can be reused after its specific gravity is checked and, if necessary, adjusted. The acetone washing should be saved in a large brown bottle for eventual recovery of pure heavy liquid by the method in exercise 5.6.

(After Schoen and Lee, 1964.)

EXERCISE 5.9

Mineral Crushing for Chemical Analysis

The mineral separate extracted from the rock by one or more of the foregoing methods may now be needed for chemical analysis by a classical wet method or by one of the rapid instrumental methods of Chapters 10 and 11.

The purity of the separated mineral grains should first be checked under a binocular microscope and any composite grains discarded. A selection of pure grains is placed in an agate mortar and ground with an agate pestle to a fine powder, which may then be kept in a plastic vial for subsequent analysis.

EXERCISE 5.10

Rock Crushing for Chemical Analysis

For preparation of a whole rock specimen for analysis the problem of sampling is critical. The final small sample chosen must be as nearly as possible representative of the rock mass from which the hand specimen was collected. The steps for obtaining such a small specimen for analysis are outlined as follows:

1. Carefully note the nature of the rock as it outcrops in the field. Particular attention must be given to the homogeneity of the rock type over the whole outcrop and to the grain size.

2. The hand specimen collected should be large enough to represent the rock unit or any variation of it. The size of the hand specimen collected should be controlled by the discussion in exercise 3.4. If the rock is coarsely porphyritic on outcrop, the smallest hand specimen collected should contain at least four or five or preferably more phenocrysts of any single mineral, which may require a hand specimen as large as 15 × 15 × 10 cm. On the other hand, if the rock is fine-grained and homogeneous, a hand specimen of only 5 × 5 × 6 cm may be adequately representative. On the other extreme, a coarse-grained pegmatite whose individual crystals may be up to 10 cm long would need an extremely large hand specimen as a representative sample.

3. The collected hand specimen is comminuted in the laboratory by using the methods in exercise 5.2. No part of the crushed material is discarded. The coarser chips are rerun through a jaw crusher until a final crush containing grains of maximum size 1 cm is obtained.

4. The complete crush is now pulverised. One of the best mills for this is the Tema laboratory disc mill (supplied by N. V. Tema, Riouwstraat 200, The Hague, Holland). Model TS250 is the most automatic and can be fitted with grinding barrels of 250, 100, or 10 cc net capacity. The barrels should be only half filled for efficient grinding. Grinding time is usually only 10 minutes for each load.

5. The complete rock crush is run through the Tema mill in as many charges as are required for the whole rock sample to be pulverized. The total rock powder is accumulated in a stainless steel container or beaker.

6. All laboratory mills cause some form of metal contamination of the rock. This is inevitable because of the lining material of the grinding elements. Prolonged grinding increases the contamination. Because of the short grinding time (usually 10 minutes), the Tema mill is one of those preferred for rock preparation. The grinding elements of the mill lined with either of the following materials may be obtained:

WIDIA	98.0 % W_2C
	2.0 % Co
COLMONOY	70.0 % Ni
	16.0 % Cr
	4.0 % B
	10.0 % max Fe, Si, and C
CHROME STEEL	1.7 % C
	0.3 % Si
	0.35 % Mn
	12.0 % Cr
	0.12 % V
	85.53 % Fe

Widia lining is to be preferred, but in view of its high cost it is seldom used in the larger vessels. Colmonoy lining is therefore the most common type. It should be noted that it will introduce nickel and chromium contamination. However, colmonoy lining is resistant to abrasion and tests on grinding coke have shown only 0.02 % contamination from the lining. After considerable use the lining of the grinding barrels may be replaced or renewed by the manufacturer.

7. Between each specimen it is essential to clean the elements of the grinding barrel thoroughly. Each element (inside of the barrel, grinding stone, and grinding ring) should be brushed clean and wiped with an acetone-soaked cloth. It is also good practice to keep a store of pure quartz sand, which can be ground in the mill barrel between each rock specimen. If ore specimens in particular have been ground, the intervening use of quartz sand is found to be especially satisfactory in removing all traces of the previous ore from the grinding elements.

8. The complete rock powder is now too bulky for our purpose and will have to be split to a smaller representative sample.

DISCUSSION

For ultrafine grinding and fast homogeneous mixing of smaller samples up to 100 ml volume the Mixer Mill M.280 (supplied by Glen Creston, The Red House, 37 The Broadway, Stanmore, Middlesex, HA7 4DL, England) is ideal. It can be supplied with a tungsten carbide grinding cylinder.

EXERCISE 5.11

Sample Splitting

Generally at this stage it will be found that the pulverized rock sample is too bulky for our particular purpose. It may have to be reduced to a smaller representative volume which is statistically equivalent to the whole sample.

The splitting of the larger sample should be done without bias.

The most convenient type of splitter is a small bench type gravity-operated two-way riffle pattern sample divider (supplied by Soiltest Inc., 2205 Lee Street, Evanston, Illinois, or by Gallenkamp, P.O. Box 290, Technico House, Christopher Street, London, E.C.2, England).

Two stainless-steel receiving containers are placed one on either side of the sample splitter and the powder is gently and evenly poured into the splitter from above. This effects an unbiased halving of the sample volume. If half the sample is still too much, it may again be split in half, and so on till a small representative sample of useful size is obtained.

The riffles should be cleaned with a suitable brush before each operation.

Larger bulk samples may be more rapidly reduced to manageable sizes with a six-way unbiased split on a motor-driven rotary sample divider (supplied by Pascall Engineering Co. Ltd., Gatwick Road, Crawley, Sussex, England). An accurate eight-way sample divider is available from Glen Creston (see above).

DISCUSSION

Sample division is an important laboratory procedure. It should always be kept in mind that the small powder sample selected for final analysis is considered to be representative of a larger hand specimen, which in turn is considered to be representative of a larger rock mass.

The final powder produced from a large hand specimen will always be too large in volume for our particular analytical need. Hence it must be accurately split in an unbiased fashion. Scouping up a few grains on the blade of a spatula is not be to recommended. This is bad sampling procedure. The whole sample should be progressively reduced to a suitable volume by using a riffle or rotary sample divider until the final sample split is of the required volume.

EXERCISE 5.12

Mineral Separation by Flotation

Occasionally two minerals that cannot be successfully separated by magnetic or specific gravity methods may be separated by flotation. A useful flotation cell is illustrated in Fig. 5.7. (After Partridge and Smith, 1971.)

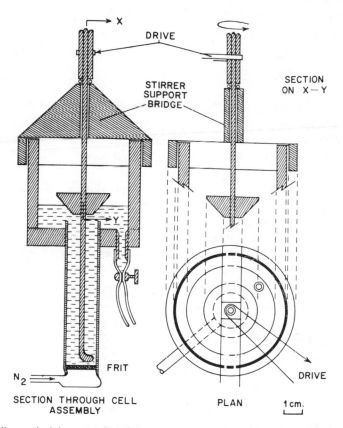

Fig. 5.7 Small sample-laboratory flotation cell. (After Partridge and Smith, 1971.)

The cell is made of glass to facilitate cleaning. Mineral particles are contained initially in a column of the solution closed at the bottom by a glass frit of fine ($<5\ \mu m$) pore size. The bed is maintained in a gently moving suspension by the variable-speed mechanical stirrer and a controlled flow of nitrogen from a bottle is passed through the frit. Particles exhibiting hydrophobic surface properties become attached to the bubbles that rise and are deflected outward by the conical perspex baffle before reaching the liquid surface. These particles can be collected at the end of the test by washing them out through the side tube.

Optimum conditions are obtained by trial and error. Gentle agitation and a 30 second float with a gas supply of 12 ml/min at 500 mm Hg gage was used for 2 g samples of −325 to +400 mesh hematite and quartz mixtures. Coarser magnetite (−65 to +100 mesh) required more vigorous agitation, a higher gas flow, and longer flotation time (Partridge and Smith, 1971). A variable speed motor is needed to drive the stirrer by means of a rubber band.

The following procedure is recommended by Herber (1969) for separating quartz from feldspar:

1. Prepare the following chemicals:
    a. Concentrated HF acid as a conditioning agent.
    b. One percent solution of Armac 12D (dodecylamine acetate) as collector agent* (obtainable from K and K Laboratories of California Inc., 6922

    Hollywood Blvd., Hollywood, California 90028).

2. Crush the granite (or other rock), sieve −120 to +230 U.S. standard and deslime with water until all dust is washed off. The magnetic fraction can be removed first by isodynamic magnetic separation.

3. Place about 30 to 40 g maximum of dry sample in a 500 ml beaker which contains 450 ml distilled water.

4. Add HF to the pulp until a pH of 2.5 is reached and allow the mixture to stand undisturbed for 5 minutes. Wash the acidified pulp into the flotation cell with HF diluted with distilled water to pH 2.5 and stir for a few seconds.

5. With the automatic stirrer at slow speed supply nitrogen through a tube at the bottom of the flotation cell and simultaneously add the Armac 12D collector agent drop by drop until a thick froth is obtained. As flotation occurs the feldspar-laden froth flows over the lip of the cell and is collected in a dish or beaker. Continue adding Armac 12D and HF diluted to pH 2.5 with distilled water to maintain the froth overflow until it no longer feels gritty.

6. Wash the flotation cell free of the tailings (material that did not float) and the tailings will be rich in quartz and relatively free of feldspar.

7. Decant and discard the scum from the wet feldspar concentrate which has been collected as overflow.

8. The procedure is repeated several times (up to five). For repeat runs steps 2 and 3 are deleted and instead of step 3,450 ml of HF already diluted to pH 2.5 with distilled water is added to the feldspar concentrate.

9. The feldspar concentrate can now be washed with distilled water and dried.

10. The K-feldspar can be separated from plagioclase by floating in tetrabromo-ethane of specific gravity 2.590 ± 0.003 g/ml.

DISCUSSION

With this method most samples should give a separation of oligoclase, K-feldspar, and quartz in which the amount of quartz remaining unseparated from the oligoclase is of the order of 1 or 2 %. Occasionally, however, for unknown reasons the method does not work so efficiently.

---

* In Herber's paper (1969) the collector agent is wrongly quoted as Alamine. He actually used Armac 12D [dodecylamine (distilled) acetate].

The flotation of feldspar from quartz can be performed with HF as a regulator, Aeromine as a promoter, and Aerofroth as a frother. For details see *The Mining Chemicals Handbook*: mineral dressing notes No. 26 (obtainable from American Cyanamid Co., Cyanamid International, Wayne, New Jersey 07470). This booklet contains details of many agents that are suitable for the flotation of a large variety of metallic ores and silicate minerals.

ICI (ICI House, 1 Nicholson Street, Melbourne, Victoria 3001, Australia) recommends that feldspar be floated from quartz by using HF as conditioner (kept below pH 3) and laurylamine acetate as the floating agent. The HF regulates pH, acts as a silica depressant, and activates the feldspar. A frother such as their Teric 402 must be added.

POWDER METHODS OF 6
X-RAY DIFFRACTION

The principles of X-ray powder diffraction are so well known that they need nothing but an introductory outline in this chapter. Readers unfamiliar with the techniques should refer to Azároff and Buerger (1958) or to Zussman (1967b) for a more complete discussion of the underlying theory.

A. CAMERA TECHNIQUES

Sample Preparation for the Powder Camera

1. Put a few crystals or crystal fragments of the mineral to be analyzed in a clean agate mortar and grind with an agate pestle until final traces of grittiness disappear. Undergrinding of the specimen produces spottiness of the X-ray film and overgrinding damages the crystal structure. A final grain size of around 300 mesh U.S. standard should be the outcome. In actual practice sieving to 300 mesh is not easy because the powder clings to the seive material. I recommend that the mineral for analysis be initially ground and sieved, the -100 to $+170$ mesh fraction alone being further used. An amount of this powder is placed in the agate mortar and ground by hand with the pestle for only a few minutes until all gritty feeling between mortar and pestle disappears. At this stage the powder should be the ideal grain size. Continued grinding to produce grains of 1000 Å or less will cause considerable broadening of the diffraction peaks because of lattice distortion and is to be avoided. There are other causes of peak broadening that are beyond our control, such as crystal zoning in an isomorphous series and natural crystal strain, without adding too fine grinding to make our record even less perfect.

2. The most convenient specimen holders are plastic specimen mounting capillaries (type PW 1951 supplied by N. V. Philips' Gloeilampenfabrieken, Eindhoven, The Netherlands). The capillaries of 0.3 mm internal diameter are preferable for more accurate work but are less easy to load than those of 0.5 mm internal diameter, which therefore may be preferable for routine identification work. Each capillary is about 8 cm long and has a wider neck 1 cm long to facilitate loading (Fig. 6.1). The secret of loading a capillary is to introduce only a very small amount of powder at a time. Tapping the open end with the spatula blade or rubbing it gently with a nail file helps the grains to move slowly and evenly down to the sealed end. Do not introduce more powder until the first grains are completely at the bottom of the capillary. Expel any

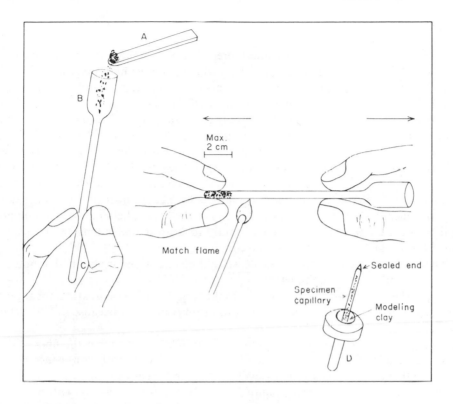

Fig. 6.1 *Left.* Loading of the capillaries. A = spatula blade, B = capillary neck, C = sealed capillary base. *Right*. Sealing the capillary after a length of $1\frac{1}{2}$ to 2 cm has been packed with powder. *Lower right:* The sealed specimen capillary attached to the camera plinth D by means of a spot of modeling clay.

trapped air bubbles from the bottom end by tapping its side. Now introduce more powder and repeat until a solid filling of powder about 1 to 2 cm long occupies the sealed end.

3. Hold the two ends of the capillary between the fingers and exert a gentle extension. At the same time touch the capillary to the small flame of a match or candle just 1 or 2 mm above the top of the powder level. The capillary will seal and the extension motion will draw off the top part from the sealed specimen.

4. The top part, if clean, may be reused.

5. The sealed specimen is now mounted into the central axial hole of the camera mounting plinth (Fig. 6.1D) and firmly stuck in place by a spot of softened modeling clay. The length of the capillary should be adjusted to coincide as exactly as possible with the axis of the plinth.

6. The metal plinth is pressed into the central axial holder of the Debye-Scherrer powder camera. The most commonly used camera has an internal diameter of 114.6 mm (PW 1024, Philips) which gives 1 mm on the film equivalent to $1° 2\theta$. For many purposes, however, the 57.3 mm camera is preferable. Exposure times are shorter and Straumanis (1959) has shown that with proper care the smaller diameter camera can give better results than the large cameras.

DISCUSSION

The following alternative powder camera preparation technique is recommended by Owen (1971). It produces good quality fibers which can give high precision.

A few milligrams of powdered sample (less than 200 mesh) are placed on a glass slide. A drop or two of collodion (supplied by BDH Chemicals Ltd., Poole, Dorset BH12, 4NN, England, or any chemical supply company) are added to the powder and the mixture is stirred rapidly with a sharp needle. After 20 to 30 seconds, as the collodion begins to set, the needle is lifted from the mixture. As it is slowly withdrawn a thin fiber of collodion-bound sample will form. If the fiber fails to form, the viscosity of the mixture is too low. The mixture should be continuously stirred until a fiber 0.05 to 0.2 mm in diameter and 2 to 5 cm long is formed. A fiber of such dimensions can be produced in less than 1 minute. The end of the fiber is held between thumb and forefinger of the free hand and pulled slightly to keep it under tension. Tension is applied for about 1 minute to allow the collodion to set.

The fiber, still attached to the needle, should be put aside for about 10 minutes to allow the collodion to set completely.

The earlier the needle is withdrawn from the sample collodion mixture, the thinner the resulting fiber. However, a fiber that is too thin is no good because it will curl on setting.

Two or three samples 10 to 15 mm long can then be cut from the fiber with a sharp razor blade. The thinness of the fiber gives improved resolution, hence higher accuracy. Of course the fiber thinness will necessitate longer exposure time. The fiber is then set axially on the camera plinth (Fig. 6.1D) and fixed firmly with modeling clay.

EXERCISE 6.2

Preliminary Alignment of the Camera

1. Fix the camera mounting bracket (PW 1012, Philips) to the X-ray tube shield (PW 1316, Philips) (Fig. 6.2).

2. Put the bracket adjusting device (PW 1020 Philips) onto the base plate of the camera mounting bracket and slide it toward the tube shield until the X-ray beam port fits into the opened (rotated) window shield of the tube shield. At this position the horizontal protruding bar at the base of the bracket adjusting device should press against and activate the microswitch at the tube shield's base (Fig. 6.2). If it does not, the protruding bar can be lengthened or shortened to make a convenient contact with the switch.

3. The X-ray generator is switched on, and with an intensity of about 30 kV, 10 mA, the image of the X-ray beam can be viewed directly by eye on the leaded-glass fluorescent screen of the adjusting device. Please note at this stage that the beam in view should be a horizontal slit. If the beam is circular, the camera mounting bracket will have been attached to the wrong outlet port and should be reassembled at one of the adjoining ports on the tube shield. Philips X-ray diffraction tubes give two horizontal-slit beams (ideal for powder diffraction) and two narrow pencil cones (ideal for single crystal work) in adjacent directions.

4. Adjust the screw A (Fig. 6.2) so that the camera mounting bracket allows the adjusting device to be the correct height for its nosepiece to engage comfortably in the opened (rotated) circular window of the tube shield.

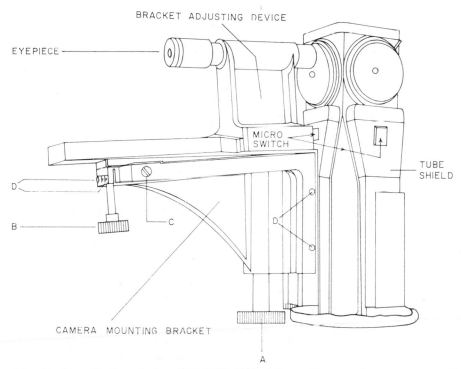

EYEPIECE

BRACKET ADJUSTING DEVICE

MICRO SWITCH

TUBE SHIELD

D

B

C

D

CAMERA MOUNTING BRACKET

A

Fig. 6.2 Bracket adjusting device (PW 1020 Philips) in position on the camera mounting bracket (PW 1012, Philips). The X-ray beam is now centered on the fluorescent screen by means of three adjustments: A for overall height, B for slope, C for sideways direction. All screws, D, are fully locked once the beam is centered on the cross wire.

5. Adjust screw **B** (Fig. 6.2) so that the X-ray beam lies horizontally along the center of the horizontal cross wire on the fluorescent screen.

6. Adjust screw **C** (Fig. 6.2) so that the X-ray beam is fully centered laterally in the field of view. Now lock screws **D** (Fig. 6.2).

7. Close the X-ray supply to this particular window. Remove the bracket adjusting device.

8. Place an open Debye-Scherrer powder camera (PW 1024, Philips), complete with its fine collimator of 0.5-mm diameter and appropriate outlet beam collector tube with leaded glass fluorescent screen, on the camera mounting bracket and insert the protruding collimator into the open window on the tube shield. Adjust the length of the bottom protruding bar of the camera so that it activates the microswitch on the tube shield. The camera is left open and no specimen or film has been mounted in it.

9. The camera body is attached to its metal base plate by clamping pressure on a single allen screw (Fig. 6.3A). This screw is loosened to allow the cylindrical camera body to be moved laterally in and out of its metal stand and rotated so that the line of the collimator axis can be tilted up and down from its horizontal position in relation to the base stand (Fig. 6.3).

10. Open the X-ray supply to the camera window. Take every precaution because the camera is not closed. Using a laboratory tongs, hold the green fluorescent screen (PW 1061, Philips) at the center of the camera body where the specimen would normally be mounted. Observe if any X-rays strike the screen from the collimator. A darkened room will facilitate this operation.

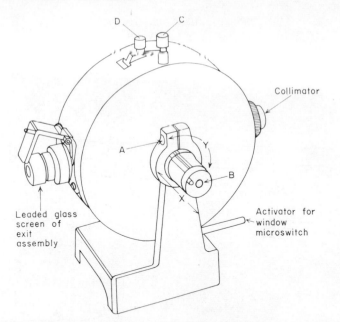

Fig. 6.3 Back view of the Debye-Scherrer camera (PW 1024, Philips), showing the allen screw (A) which tightens the camera body firmly onto the camera stand. This screw is loosened for preliminary adjustment of the camera; B is the rotation axial drive of the camera, to which the camera motor is attached during an actual exposure; C is the knob for centering the axis of rotation of the specimen capillary; and D is the adjustment for tightening the film close against the internal camera body.

11.   By gently pulling and pushing the cylindrical camera body in and out of the bracket mount (movement X, Fig. 6.3) and tilting the camera up and down on its axis (movement Y) the line of the collimator is gradually adjusted until the X-ray beam impinging on the screen is at its brightest.

12.   The allen screw A (Fig. 6.3) is gently tightened to fix the camera body more rigidly to its stand.

13.   Remove the fluorescent screen and put on the camera cover. A bright spot should appear on the leaded glass take-up shield of the camera. Increase the X-ray intensity to about kV 40, mA 20.

14.   A final sensitive adjustment of the position of the camera in its stand can now be made so that the fluorescent spot on the leaded glass screen will become its brightest and sharpest. At this stage the camera body is fixed rigidly to its stand by locking the allen screw.

DISCUSSION

Once this adjustment has been made the camera geometry should continue to be identical to that of the bracket adjusting device. There will be no further need for adjustment unless the camera is knocked or dropped and the firm attachment of the camera cylinder to its stand is altered in some way.

   To prevent this from happening the camera should always be held by its base, not by its cylinder, when being carried. Continual lifting of the camera by the cylinder may cause a rotation in relation to the stand and necessitate readjustment according to steps 1 to 14.

Loading the Camera

1. To maintain geometry identical to the bracket adjusting device make sure that the camera has been aligned on its base (exercise 6.2).

2. Place the bracket adjusting device on the base plate of the camera mounting bracket and align the X-ray beam horizontally in the center of the fluorescent screen (steps 1 to 6, exercise 6.2). This adjustment is necessary whenever the X-ray tube is replaced or whenever there is suspicion that the camera mounting bracket is out of alignment.

3. The specimen mounting plinth, to which the specimen has been attached (Fig. 6.1D) is pushed firmly into the central axial holder of the camera cylinder (exercise 6.1, step 6).

4. Screw the standard 0.5 mm collimator (Fig. 6.4A) into the camera body. Screw in the appropriate exit port assembly (Fig. 6.4B). Screw off the fluorescent screen (C) and leaded-glass terminal attachment from the exit port, and in its place push on the magnifying glass supplied with the camera.

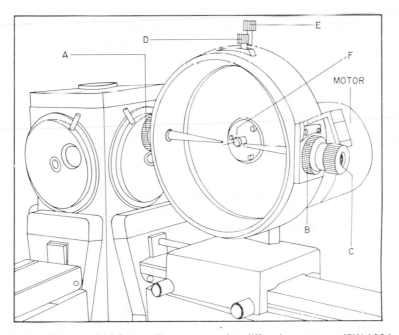

Fig. 6.4 Debye-Scherrer 114.6-mm diameter powder diffraction camera (PW 1024, Philips) A = collimator, B = exit port assembly, C = fluorescent screen and leaded glass, D = adjusting knob for tightening film against the camera wall, E = screw for centering the axis of rotation of the specimen, F = specimen mounting plinth pushed into the axial holder of the camera.

5. Place the camera on the specimen centering device bracket (PW 1021, Philips) so that the light from the device passes through the collimator and can be seen on the magnifying glass placed at the end of the exit port.

6. Be certain that the screw (Fig. 6.4E) is rotated fully up (counterclockwise) and that its lower end is not interfering with the rotation of the specimen.

7.   Turn the axis of the camera fully and regularly by hand, using the axial knob at the back of the camera (Fig. 6.3B), all the time observing the specimen capillary. If the capillary is perfectly horizontal, all is well. If it tilts out of the horizontal position, it should be adjusted in its modeling-clay mount on the plinth with a knife blade or pointed object until any rotation causes it to remain horizontal in the field of view, though not necessarily centered.

8.   If, as is normally the case, the specimen is not coaxial with the camera axis and moves up and down in the field of view on rotation, it must now be centered. Turn the camera axis (Fig. 6.3B) until the specimen moves up to its highest point in the field of view. At this stage hold the camera rotation knob firmly and slowly screw knob E (Fig. 6.4) clockwise until the specimen capillary is pushed down exactly to the center of the collimator field of view. Immediately turn screw E (Fig. 6.4) back up again.

9.   Continue rotating the camera axis. If the capillary is still not perfectly centered, repeat operation 8 as often as necessary to effect a perfect centering of the capillary.

NOTE.   After every operation of screw E (Fig. 6.4) the screw must be rotated counter-clockwise upward and not left in its down position; otherwise it will subsequently knock the capillary axis out of alignment. In this way the capillary will remain perfectly horizontal and centered in the field of view on a complete rotation of the camera rotation axis.

10.   Remove the magnifying glass and screw on the leaded-glass screen (C, Fig. 6.4).

11.   Carry the camera by its base to the darkroom. From now on care must be taken not to place the hands near the specimen capillary because no further chance will be given for centering. Extinguish the white light and use a suitable darkroom light (Kodak 6BR light brown filter for indirect lighting or Kodak 6B dark brown filter for direct lighting).

12.   Place the camera on the bench. Remove its collimator and outlet port assembly. Loosen screw D (Fig. 6.4) and push it fully to the left (toward the collimator side).

13.   Open a box of film. Ilford Industrial G X-ray film which comes in boxes of 75 flat films, 35×355 mm, is appropriate; each film is folder-wrapped in black paper (supplied by Ilford Limited, Ilford, Essex, England). Experience has shown that it is excellent for X-ray diffraction. However, it has a limited shelf-life of approximately 8 months. Old stock should not be maintained because the deteriorated film causes an uneven density and spotty appearance.

14.   Remove the paper folder from one of the films. Do not foget to close the box of unused films. Push the film, holding it always by its edges, fully into the slot of the film punch and cutter (PW 1022, Philips). Trim off the excess length of film with the cutter blade and press in the two punch knobs to make the holes for the collimator and exit port assembly.

15.   The film has emulsion on both sides so that either may be used. Lift the film by its edges and drape it around the inside circumference of the camera cylinder; its two free ends must come together at the top and be held apart by the two protruding metal spikes that descend from the top of the camera. Push the film firmly into the body of the cylinder.

16.   Push knob D (Fig. 6.4) fully to the right. This will tighten the film against the inner wall of the camera. Fully tighten screw D. Now screw in the collimator and the exit port assembly and push the camera lid tightly onto the camera body. The camera is now taken to the X-ray generator and is ready for exposure. The camera motor is fixed and locked onto the axial drive (B, Fig. 6.3) of the camera.

Choice of Collimators

1. The standard collimator has a diameter of 0.5 mm and should be used with its accompanying exit port assembly.

2. For rapid, less precise exposures a collimator of 1 mm diameter with appropriate exit port assembly may be obtained. Both collimators give a circular cross-section beam.

3. When capillaries have been prepared of clays and other minerals which have a large d spacing (up to 40 Å for CuK_α), a special collimator (PW 1024, Philips) should be used. It produces a narrow slit beam, and a small needle protrusion on the collimator fits into an appropriate slot in the camera body to ensure that the slit beam is horizontal and parallel to the capillary during use. This collimator is used in conjunction with a special short and narrow exit port assembly. Therefore it may be desirable to cut and punch the film with a special film punch (PW 1023 Philips) which makes smaller holes suitable for the smaller collimator and exit port assembly.

Exposure Time

Exposure time depends on the X-ray intensity, the film, the specimen preparation, the collimator, the crystallinity of the specimen, the choice of X-ray wavelength, and the alignment of the camera.

As a general guide, exposure time for a specimen mounted in a 0.3-mm plastic capillary should be of the order of 8 hours if the X-ray intensity is set at 40 kV, 20 mA, as standard procedure and a 114.6 mm diameter camera fitted with the normal 0.5 mm collimator and Ilford Industrial G film are used.

The 1-mm diameter collimator should cut the exposure time to approximately 3 to 4 hours.

The fine collimator for recording wider d spacings may require 8 to 12 hours exposure at 40 kV, 20 mA.

If, after development of the film, the lines are too dark, exposure time should be reduced. If the lines are too faint, exposure time should be extended. During exposure the camera motor should run continuously.

Choice of X-ray Tube and Its Operating Conditions

The two most commonly used X-ray tubes are the following:

Target Element		Wavelength Å			β Filter	(thickness mm)
	$K\alpha$	$K\alpha_1$	$K\alpha_2$	$K\beta$		
Cu	1.5418	1.5405	1.5443	1.3922	Ni	0.015
Co	1.7902	1.7889	1.7928	1.6207	Fe	0.012

The β filter gives $K\beta_1/K\alpha_1 = 1/100$ intensity.

The $K\alpha$ value $[=\frac{1}{3}(K\alpha_2 + 2K\alpha_1)]$ is used only when the $K\alpha_1$ and $K\alpha_2$ separate wavelengths are not resolved, at low angles of diffraction.

The X-radiation is always filtered by the appropriate choice of filter in the tube shield. For Cu radiation Ni is always used, and for Co Fe is always used. This filter greatly reduces the intensity of the $K\beta$ wavelength but does not eliminate it. It must therefore be remembered that a strong X-ray reflection may also register a weak β peak in addition to the α pattern. Even the "white" background radiation may register a broad diffraction peak because it has a maximum at about $\lambda = 0.5$ Å; for example, a strong reflection at $d = 3.5$ Å will occur at $25.45°$ 2θ with Cu$K\alpha$ radiation, but the β line may give a small peak at $22.94°$ and the "white" radiation, a broad peak at about $8°$ 2θ (Zussman, 1967b).

The $K\beta$ line occurs at a lower angle than $K\alpha$, and the $K\alpha_1$ at a slightly lower angle than the $K\alpha_2$.

Copper is the most useful radiation for all mineralogical work. However, when a mineral specimen contains appreciable iron, the background may be too high because of the iron fluorescing under the Cu$K\alpha$ radiation, in which case it is advisable not to use Cu. For minerals containing more than 10% iron the best X-ray tube is one with a cobalt anode. Co also gives better resolution because it separates the interesting peaks somewhat more than Cu, but the disadvantage is that the power rating of a Co X-ray tube is less good than that of a Cu tube.

The intensity of X-ray peaks from iron-bearing minerals is not linearly related to the amount of the mineral present in the specimen if Cu radiation is used because of the absorption of the Cu wavelength by the iron of the mineral (Norrish and Taylor 1962). It is therefore recommended that all X-ray diffraction work on iron-bearing minerals be performed using Co radiation. The advantages are that the background is less (because of the absence of iron fluorescence) and the calibration curve of the amount of iron-bearing mineral, for example, goethite, against peak intensity is linear.

A standard copper anode X-ray tube (PW 2103, Philips) has a rating of 2000 W, whereas the maximum rating of the Co tube (PW 2106, Philips) is only 1000 W. Both tubes should be used in tubeshield PW 1316 (Philips) and on any generator with a rating up to 2000 W.

The settings of kV and mA on the generator should never exceed the capacity of the generator and the capacity of the X-ray tube; for example, if a 1600 W generator is used (PW 1011, Philips), the product of kV × mA settings should not exceed about 90% of the generator capacity. In this case 50 kV and 28 mA (1400 W) might be considered an upper operating limit for a 1600 W generator. If a 2000-W Cu anode tube is used, an upper limit of, say, 60 kV and 30 mA (1800 W) may be used, provided, of course, that the generator rating is 2000 W. On the 1600 W generator the 2000-W tube would have to be run at a maximum of 50 kV and 28 mA (1400 W). The Co tube must never be run in excess of 90% of its 1000 W rating, whatever the generator.

Since both Cu and Co tubes listed above are rated at least 1000 W, it is safe practice to run exposures regularly at 40 kV and 20 mA. This is adequate for all powder camera work. The higher intensity beam that may be obtained from a 2000-W Cu tube is really required only when a diffractometer is used.

Should the product of kV × mA exceed the rating of the tube, tungsten from the filament will evaporate and contaminate the anode, thereby adding W wavelengths to the monochromatic spectrum of the tube. Eventually the filament will be burned out and the tube rendered useless. The life of a tube will be extended by running it regularly at less than 90% of its rating.

EXERCISE 6.7

Development of the Exposed Film

1. After the exposure is completed the camera is removed to the darkroom, which is illuminated by a Kodak 6B brown safelight.

2. Remove the camera lid. Loosen film-tightening screw D (Fig. 6.4). Remove the collimator and exit port assembly. Remove the film, holding it by its edges.

3. Develop the film in a dish, tank, or 35 mm daylight tank. Since it is of standard 35 mm width, it may be threaded into a Paterson or Kodak standard 35 mm developing tank. Once loaded in the tank, development and subsequent processing may be carried out in normal lighting.

4. Development time. For Ilford Industrial G X-ray film use Ilford Phen-X full-strength developer for 5 to 8 minutes at 20°C with 5-second agitation every minute. The developer may be replenished by adding Ilford Phen-X Replenisher according to instructions supplied in the packet. One clear sign that the developer should be discarded is slight opalescence shown by the film after fixation.

5. After development wash the film for about a minute or two in running water.

6. The film is now put in double-strength fixer for 5 minutes (Ilford Hypam rapid fixer or Kodak Rapid Fixer with hardener), then washed in running water for at least a half hour.

7. A final dipping in a weak solution of wetting agent (Kodak Photo-flo) before hanging the film to dry promotes even drying.

8. The dry film should have the specimen number and wavelength used (e.g., $CuK\alpha$) marked in its margin with a felt pen or with drawing ink.

EXERCISE 6.8

Measurement of the Film

The camera most commonly used has the Straumanis film arrangement (Fig 6.5)

1. Once the film is removed from the camera and developed and fixed, it can easily be reoriented. The arcs centering around the exit port of the camera are single low-angle reflections. Those centering around the inlet port are doublets, or high-angle, because at high angles the arcs are differentiated into $K\alpha_1$, $K\alpha_2$ doublets. Also, the intensity usually falls off from low angle toward high angle and some minerals show only faint high-angle arcs.

2. The following is a useful way of tabulating the results (Fig. 6.5). A millimeter scale is laid along the central axis of the film and without subsequent movement all readings are made. It does not matter where the zero of the rule lies. Lines are numbered progressively outward from the exit port.

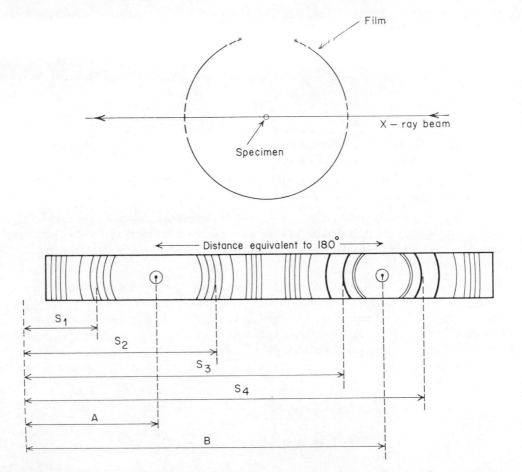

Fig. 6.5 Appearance of an X-ray powder diffraction film based on the Straumanis arrangement (taken on a PW 1024, Philips 114.6-mm diameter). A millimeter scale is laid along the film axis and absolute readings taken are S_1 and S_2 in millimeters. The estimated relative intensities of the lines are based on 100 for the most intense line.

	Line No.	Intensity		All in millimeters				
		I	S_1	S_2	$(S_2 - S_1)$	$(S_2 + S_1)$	θ	d
	1	.	.	.	.	.	.	.
	2	.	.	.	.	.	.	.
2θ	3	.	.	.	.	.	.	.
$<90°$	4	.	.	.	.	.	.	.
	.	.	.	.	.	.	.	.
	.	.	.	.	.	.	.	.
	.	.	.	.	.	.	.	.
	.	I	S_3	S_4	$(S_4 - S_3)$	$(S_4 + S_3)$	$(90 - \theta)$	θ d
2θ	.		.	.	.	.	.	.
$>90°$	.		.	.	.	.	.	.
	.		.	.	.	.	.	.
	$S_2 + S_1 = 2A$			$S_4 + S_3 = 2B$				

3. The intensity I is recorded thus: the most intense line is given a value 100 and all other lines are estimated visually for density in relation to this most intense line.

4. Starting from the left, all S_1's are recorded, followed by S_2's, pairing off with S_1's, then S_3 and S_4 in a similar manner. All sums $S_1 + S_2$ should be equal; similarly, $S_3 + S_4$ should be equal. Any large deviation from a constant will allow the operator to detect an error in reading the rule. For reflections $2\theta < 90°$

$$\frac{S_2 - S_1}{2B - 2A} = \frac{4\theta}{360} \quad \text{and} \quad \theta = (S_2 - S_1) \times \frac{90}{2B - 2A}.$$

For reflections $2\theta > 90°$

$$90 - \theta = (S_4 - S_3) \times \frac{90}{2B - 2A}.$$

The general relationship $\theta° = (S_2 - S_1) \times 180/\pi 4R$ relates θ to the radius R of the camera in millimeters.

5. The most convenient and most commonly used powder camera has a radius of 57.3 mm (diameter 114.6 mm), for in that camera 180 mm on the film is equivalent to a 180° angle subtended at the specimen. Hence direct measurement with a millimeter rule of the distance between a pair of arcs gives 4θ about the exit port or $360 - 4\theta$ about the inlet port. Corrections may be easily made; for example, if $2B - 2A$ were not exactly equal to 360 mm, the amount by which it differs $360 \pm x$ is the error. This error is used to correct all determined θ readings by adding or subtracting to θ, $\pm x/360 \times \theta$. In most cases this correction will be so insignificant that it will be omitted. The small diameter camera (57.3 mm) can give precision as high as the large camera (Straumanis, 1959); 90 mm on the film is equivalent to an 180° angle subtended at the specimen. It is cheaper, uses less film, and requires shorter exposures than the larger camera and will provide the cheapest and simplest high-precision results accessible to the average laboratory.

6. More accurate measurement and tabulation of the S_1, S_2, S_3, and S_4 values may be obtained by using a negatoscope (film viewer) or a Guinier viewer (supplied by Enraf-Nonius, Rontgenweg 1, P.O. Box 483, Delft, Holland).

7. *Error in method.* Differentiation of the Bragg equation $n\lambda = 2d \sin \theta$, given λ constant, gives $\delta d/\delta \theta = -d \cot \theta$; thus $\delta d \to 0$ as $\theta \to 90°$. If θ is near $90°$, the error in measuring it will become negligible. Higher accuracy is obtained by making use of high-order reflections. An additional help toward accuracy is that in this back reflection area all lines occur as $K\alpha_1 - K\alpha_2$ doublets in the region where θ approaches $90°$ (2θ approaches $180°$). Hence we use only one line of the doublet and the appropriate wavelength for α_1 or α_2 to convert θ to a d value. If the doublet is resolved, d values and all parameters can be calculated with a precision of about 0.01% or better. If the doublet is not resolved at high angles, the precision is more likely to be 0.1% or worse.

8. The values of θ or 2θ measured on the film are converted by the Bragg equation

$$d_{hkl} = \frac{\lambda}{2 \sin \theta_{hkl}}.$$

where d_{hkl} is the interplanar spacing between hkl planes, λ is the wavelength of X-rays giving rise to the reflection, and θ is the angle measured from the film. For a 114.6 mm diameter camera the distance $S_2 - S_1$ in millimeters equals 4θ in degrees. To avoid calculations that are tedious and may lead to mistakes it is better to use published charts and tables relating θ or 2θ to d for $K\alpha$, $K\alpha_1$, and $K\alpha_2$ wavelength; $K\alpha$ is used only if the $K\alpha_1$, $K\alpha_2$ doublet is not resolved. The following conversion charts and tables are available:

For copper radiation   Parrish and Mack (1963a),
                        Fang and Bloss (1966).
For cobalt radiation   Parrish and Mack (1963b).

B.   DIFFRACTOMETER TECHNIQUES

Because of the geometry of a flat specimen diffractometer, extremely good resolution can be obtained compared with the normal camera methods. A $CuK\alpha_1-\alpha_2$ doublet can be resolved at an angle as low as 40° 2θ, whereas in the powder camera this is not possible at such a low angle.

The powder diffractometer is therefore preferable to camera methods for high-precision work. The camera, however, may be a better choice for the identification of minerals, for it gives a complete record more readily and in a more compact form. The two diffractometers in most common use are

Philips vertical goniometer (PW 1050);
Philips horizontal goniometer (PW 1380).

Both can be fitted with a proportional detector (PW 1965, Philips) or a scintillation detector (PW 1964, Philips) and wired up to an electronic counting and measuring panel (PW 1350, Philips) which consists of a strip chart recorder, a high-tension supply for the detector, a ratemeter, a timer, a goniometer power supply, and a pulse-height analyzer (optional). For high-precision recording of the 2θ position of any reflection a step-scanning control (PW 1357, Phillips) combined with a digital printer (PW 4202, Philips) can be added. A suppressor (PW 4197, Philips) may be added to suppress electrical interference from the environment.

EXERCISE 6.9

Diffractometer Specimen Preparation

METHOD A

The simplest and most convenient method is to smear the fine mineral powder on a normal petrographic slide and slurry it with acetone.

1.   Prepare the mineral specimen as in exercise 6.1, step 1.

2.   Scatter a thin layer of the powder on one-half of one surface of a petrographic glass slide of approximate dimensions $25 \times 35 \times 1$ mm.

3.   From a wash bottle add enough acetone to the powder to produce a thin slurry. The powder will smear evenly and spread. Spreading may be assisted by a spatula blade. The amount of powder and acetone added should produce a nearly mono-granular layer of even thickness.

4.   Once the acetone has evaporated the powder layer will cohere. An area approximately 20 mm wide on the slide surface should be wiped free of powder. The final appearance of the preparation is shown in Fig. 6.6A.

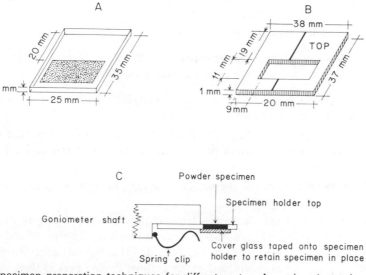

Fig. 6.6 Specimen preparation techniques for diffractometry: A = mineral powder slurry on a petrographic slide; B = aluminum specimen holder (PW 1062, Philips); C = specimen mounted flat against the goniometer shaft of the diffractometer in correct position (side view).

5. Once used, the glass slide may be washed free of powder and reused. If a permanent mount is required, a drop of Durofix, Peligom, Duco, or similar glue in the acetone will give a hard permanent final specimen.

6. The glass slide can be labeled on the back surface with the specimen name and number.

METHOD B

1. Invert an aluminum specimen holder (Fig. 6.6B) and place it in contact with a glass plate so that its well is sealed by the glass.

2. Fill the hole with the powdered specimen by pressing it in gently with a flat spatula blade. Smooth off any excess powder.

3. A thin glass plate, such as a small cut cover glass, is taped on top of the powder to cover the specimen well completely.

4. The specimen holder, together with the glass plate, is inverted right-side-up so that the free surface of the specimen powder is exposed on top after removal of the glass plate from the specimen holder. If the specimen surface has retained its flat smooth surface parallel to the aluminum surface, it is ready for use. The specimen powder surface may be smoothed flat if necessary by pressing the glass plate onto it before final removal.

5. The specimen holder is pushed into the goniometer shaft and held in place by the spring clip (Fig. 6.6C).

EXERCISE 6.10

Specimen Preparation for Avoiding Preferred Orientation

The methods in exercise 6.9 lead to preferred orientation of the mineral grains. The following methods have been designed to obviate this orientation and may be more useful when minerals of platy habit, for example, clays, are being prepared.

METHOD A

1. Obtain the following equipment:

1 can pressurized aerosol plastic spray (e.g., Blair spray-clear or Krylon);
1 roll of plastic food wrap (e.g., Saran wrap, or Handiwrap).

The mineral is first powdered ($<44\ \mu$) and sieved onto the bottom of a glass beaker, leaving no portion of the glass bottom exposed.

2. The powder is sprayed lightly with plastic from the aerosol, with the nozzle held about 20 cm away. The individual droplets of plastic clump the mineral grains into aggregates. Allow them to dry for a few minutes.

3. Brush the aggregates out of the beaker and separate the >115 mesh grain aggregates.

4. A small piece of plastic wrap is stretched taut over the opening in a standard aluminum sample holder (Fig. 6.6B) and taped in place. The plastic film is coated with a thin veneer of cement (Duco, Durofix, or other brand) onto which the grain aggregates are immediately sprinkled. After a short drying period the sample holder is tilted to allow excess aggregates to fall off.

5. The sample holder can now be inserted into the diffraction goniometer. To compensate for the thickness of the layer on top of the sample holder cover glass spacers can be inserted at the contact between the holder and the half-cylinder of metal (Fig. 6.6C) (Philips type) against which this holder is snugly held by a spring clip. This will ensure that the sample is correctly centered on the rotational axis of the goniometer.

6. The plastic wrap can be stored after removal from the sample holder as a future reference. The glass beaker cleans easily with acetone.

DISCUSSION

The plastic wrap itself produces distinct peaks in the 0 to 25° (2θ) region for Cu$K\alpha$. These peaks, however, are usually suppressed by absorption from the attached specimen layer. Background beyond 25° is extremely low.

Alternatively, the aggregates may be sprinkled directly onto a Duco or Durofix coated petrographic slide or packed very loosely into the standard specimen holder if the region 0 to 25° 2θ is to be studied in detail.

(Method after Bloss et al., 1967.)

METHOD B

1. The finely ground mineral powder is mixed with a hard setting plastic, for example, Lakeside 80, dissolved in dioxan (BDH chemicals Ltd. or other laboratory chemical supplier).

2. The solid mass, after setting, is then pulverized to yield roughly equidimensional particles containing flakes or fibers of the mineral specimen occurring in all orientations.

3. The pulverized particles are made into a regular specimen according to one of the methods in exercise 6.9.

(Method after Zussman, 1967b.)

METHOD C

The powdered specimen may be sprinkled onto a tacky adhesive surface like that of a glass slide coated with Durofix glue. This preparation provides less preferred orientation than a powder slurry with acetone (Zussman, 1967b).

EXERCISE 6.11

Specimen Preparation for Clay Minerals

Only four of the commonly used preparation techniques for clay mineral mounts fulfil the requirement of precision and accuracy; namely, powder press, smear, suction, and centrifugation. These four techniques therefore should be preferred.

Powder Press Technique. Approximately 150 mg of thoroughly mixed sample is loaded into an aluminum holder (Fig. 6.6B) from the back, with the surface to be exposed to the X-rays face down. For randomly oriented specimens the surface to be exposed is placed on filter paper and the sample tightly packed with low (10 psi) pressure. For partly oriented specimens the surface to be exposed is placed on a polished metal surface and the sample packed with high pressure (180 psi) in one high-pressure piston stroke, The pressure is released immediately.

Smear on Glass Slide. A spatula is made of a 12×30 mm rectangle of clear plastic 0.2 to 0.3 mm thick, mounted in a plastic holder so that the amount extending can be changed, depending on the consistency of the clay and flexibility of the plastic. The thick clay paste is placed along the edge of a 25-mm wide glass slide and spread across the slide in a thin even layer with a single stroke of the spatula.

Suction on Ceramic Tile. A clay suspension 2 to 3 cm deep is placed on an unglazed ceramic tile. The liquid portion is drawn through the tile by a vacuum from below (in less than 5 minutes), leaving the clay on the tile (available from Lordon Ceramics, 326 South Road, Croydon Park, South Australia 5008).

(Methods after Gibbs, 1965.)

Centrifugation onto Ceramic Tile. Kinter and Diamond (1956) have recommended the use of a porous ceramic bisque plate (e.g., an unglazed wall tile) on which the clay is deposited by centrifugation. The specimen is easily washed by passing distilled water through it with a suction pump and can be converted to other cationic forms by passing solutions of appropriate electrolytes followed by enough washings to remove excess salts. Although with glass slides 15 mg of clay per square inch is found to be sufficient, with porous plates 30 to 100 mg/in.^{2} is used. X-ray diffractions from the underlying ceramic material are largely cut out by the thicker layer of clay. Diffractograms are first done on the natural samples. Glycolation can then be done, if needed, by the vapor method of Brunton (1955) for at least $1\frac{1}{2}$ hours immediately before diffraction analysis.

EXERCISE 6.12

Preliminary Alignment of the X-ray Diffractometer

The diffraction goniometer has the ability to perform highly precise and reproducible measurements, but only if it is properly aligned and calibrated angularly. The method described in this section is for the Philips vertical goniometer (PW 1050), the one in most common worldwide use. The method may be modified for other makes according to the manufacturer's instructions. The goniometer is much more complicated than the camera, so that alignment is more involved, but optimum performance is possible only after alignment. Improper alignment results in loss of peak intensity, resolution, distortion of line profiles, increase of background, and incorrect angle measurements. The following account is largely after Parrish and Lowitzsch (1965) and the manuals of N.V. Philips' Gloeilampenfabrieken, Scientific and Analytical Department, Eindhoven, Holland.

The goniometer alignment should be done with a complete set of alignment tools (PW 1077, Philips), in addition to the single and double knife edges supplied with the X-ray diffractometry set (PW 1049, Philips).

1. Attach the support plate A (Fig. 6.7) to the face of the tube shield (PW 1316,

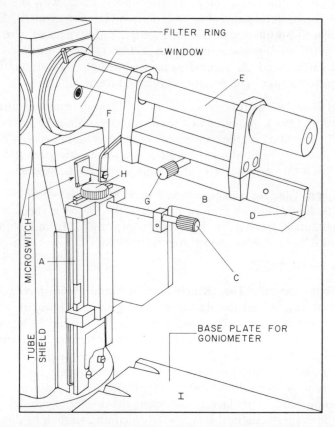

Fig. 6.7 Philips alignment tools (PW 1077) attached to tube shield (PW 1316). The various parts are referred to in the text of exercise 6.12. Only bar A is permanently fixed to the tube shield with its length precisely perpendicular to the top of the shield.

Philips) that gives a horizontal line focus from the X-ray tube (face 1 or 3). Plate A is permanently screwed onto the tube shield. Its length must be perpendicular to the top of the shield. This is done by removing the tube and using a machinist's set square on the shield's surface to align the length of plate A exactly perpendicular with the top of the shield before tightening the screws firmly on the plate. Replace the tube.

2.   Make sure that the hard fiber tip of screw C does not protrude from bracket **B** (Fig. 6.7).

3.   Fit bracket **B** onto the support plate A (Fig. 6.7). The bracket is delivered with a 6° take-off angle strip D attached. If the goniometer is to be aligned for a 3° take-off angle, strip D is replaced by the 3° strip which is also supplied with set PW 1077. A 6° take-off angle is to be preferred (Parrish, 1965) for when the line focus is used from the X-ray tube an increase in the take-off angle from 3 to 6° produces a 25% greater X-ray intensity with no loss of resolution.

4.   Position the monocular viewer E (Fig. 6.7) on the bracket and slide it toward the tube shield as far as possible. The front side of the viewer E should be in the hole of the mechanical shutter (rotated counterclockwise to the open position) and the activator arm F (Fig. 6.7) should close the microswitch on the tube shield. The switch is closed if a soft click is heard when it is pushed. If it is not closed, the pin on the activator arm F should be readjusted.

5.   Be sure that the plane at the bottom of the monocular viewer E coincides with the top plane of the bracket D, Tighten screw G (Fig. 6.7).

6.   Rotate the filter ring (Fig. 6.7) so that no filter lies in front of the window (no filter abbreviation is written at the top of the ring).

7.   Switch on the generator to approximately 40 kV, 24 mA for a normal Cu or Co tube (of 1000 W or more rating). Open the window so that X-rays will proceed from this particular window of the tube shield and look through the lens of viewer E.

8.   The image of the beam must be a horizontal slit. If it is a pencil beam, the wrong face of the tube shield is in use. Only faces 1 and 3 of tube shield PW 1316 which give horizontal slit beams may be used. Faces 2 and 4 give pencil beams that are unsuitable for a goniometer because they have poor resolution and lower intensities (Parrish, 1965).

9.   A bright line, the image of the X-ray focal plane, will be visible on the fluorescent screen of the viewer E, which may be focused by turning the viewer barrel.

10.   With screw H (Fig. 6.7), move the bracket and the viewer in a vertical direction until the center of the bright line coincides with the intersection of the cross wires of the viewer. If desired, the image of the cross wires can be readjusted when the front piece of the viewer is unscrewed one-quarter turn. The cylindrical center piece with the cross wire image can then be turned so that the cross wires are either horizontal or diagonal.

11.   Cut off the X-ray supply to this window.

12.   Loosen screw G and remove the viewer E from the bracket.

13.   Be sure that the counter arm of the goniometer (A of Fig. 6.8) has been properly and firmly fixed to the goniometer wheel.

14.   Make sure that the specimen holder shaft A (Fig. 6.9), with the spacer ring B and clamp C, has been fitted into the hollow goniometer shaft. Screws D and E should not be tightened. To prevent the shaft from dropping out screw in locking knob F completely; then unscrew it again about $\frac{1}{2}$ turn. (Fig 6.9.)

15.   Loosen screws G and make sure that clutch lever H (Fig. 6.9) is in the right-hand position.

16.   Turn the goniometer arm to its approximate horizontal position with the manual operating handle I (Fig. 6.9).

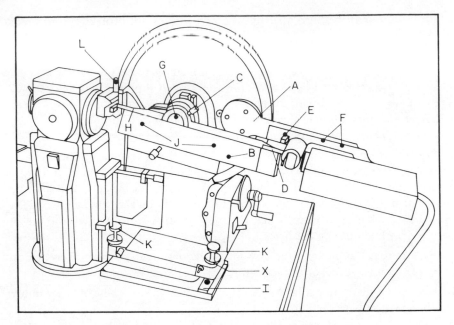

Fig. 6.8 View of the aligning devices (PW 1077) as attached to the Philips vertical goniometer PW 1050. See text of exercise 6.12 for description.

17.   The receiving parallel slit assembly (J, Fig. 6.9) should be positioned in the counter arm in such a way that channel K for the receiving slit is visible at one side of the assembly and channel L for the scatter slit is visible at the far side (Fig. 6.9).

18.   The lever M must be positioned between spring N and adjusting screw O. Turn screw O half way in. Be sure that the assembly is firmly seated in the counter arm.

19.   The divergence slit assembly P (Fig. 6.9) should be positioned on the cast-iron collar Q of the goniometer frame. If present, the pin in this parallel slit assembly should fit into the slot of the collar Q. Screw R (Fig. 6.9) should be tightened only to the extent that the divergence slit assembly is just movable in its collar.

20.   Position the aligning device B (Fig. 6.8) in the specimen holder. The aligning device must be straight against the specimen holder's back surface. Tighten screw C (Fig. 6.8).

21.   Make sure that the marking lines on the aligning device and the specimen holder are exactly in line. If not, readjust the adjusting screw on the aligning device (Fig. 6.8).

22.   Insert the aligning slit D (Fig. 6.8) for the receiving slit assembly into channel K (Fig. 6.9). Figure 6.8 shows it in place. The notched edge of the slit should be face down. The slit is in its correct position if it clicks and holds at its central notch.

23.   Loosen screw E (Fig. 6.8).

24.   Turn the specimen holder with the aligning device and the receiving slit assembly with its aligning slit until the top side of the aligning device coincides with the entire bottom surface of the aligning slit. Make sure that the receiving slit assembly is firmly seated in the counter arm and tighten screw E (Fig. 6.8).

25.   Be sure that the end of the adjusting slit reaches the marking line on the adjusting device. If not, loosen screws F (Fig. 6.8) and adjust.

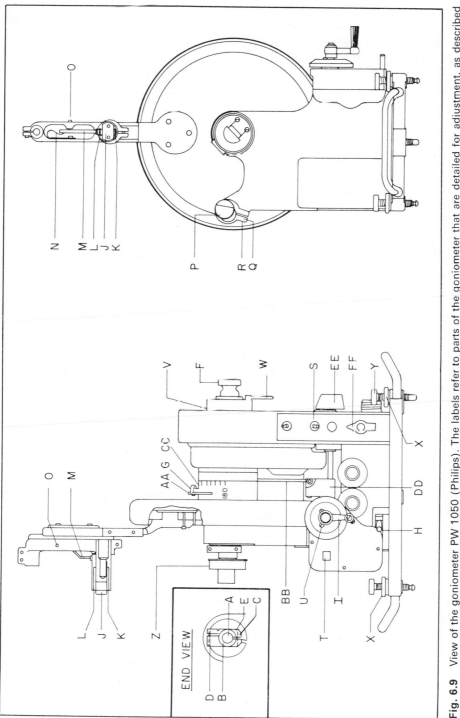

Fig. 6.9 View of the goniometer PW 1050 (Philips). The labels refer to parts of the goniometer that are detailed for adjustment, as described in exercise 6.12.

END VIEW

26. If the aligning device and the specimen holder assembly are firmly pushed into the hollow goniometer shaft, a plane over the vertical side of the device should coincide with the appropriate plane over the aligning slit. Adjustment is made with screw G (Fig. 6.8) until the planes coincide. This can be checked with a straight-edged bar touching the two vertical planes.

27. Insert the aligning slit H (Fig. 6.8) in the divergence slit assembly. If the divergence slit assembly shown in Fig. 6.10 is used, loosen screw A (Fig. 6.10).

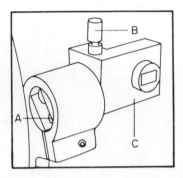

Fig. 6.10   Divergence slit assembly of the Philips vertical goniometer. See exercise 6.12 for details.

28. Turn the specimen holder with the aligning device and the divergence slit assembly with its aligning slit until the top side of the aligning device coincides with the whole surface of the aligning slit. If the divergence slit assembly shown in Fig. 6.10 is used, tighten screw A (Fig. 6.10).

29. Move the divergence slit assembly in its cast-iron collar Q (Fig. 6.9) until a plane over the vertical side of the aligning device coincides with the appropriate plane of the aligning slit.

30. Tighten screw R (Fig. 6.9) in the cast-iron collar. Check that the position is correct with a straight bar touching the two vertical planes.

31. Carefully turn the goniometer arm down to a position in which the top side of the aligning device just coincides with the bottom surface of the aligning slits in the divergence and receiving slit holders. *DO NOT TURN THE GONIOMETER ARM TOO FAR; OTHERWISE THE ALIGNING SLITS MAY BE BENT.* In this position be sure that the vertical planes over the aligning device and the slits still coincide.

32. Make sure that the goniometer switch S (Fig. 6.9) is in the off position.

33. Lock the goniometer in this position by pushing the clutch lever H (Fig. 6.9) to the left-hand position, thus locking the worm drive and the nonrunning motor. Adjust the angular 2θ reading T (Fig. 6.9) to 000° by loosening the three screws U (Fig. 6.9) on the front of the calibrated dial and turning the dial until 000° is indicated. In this position of the dial 999 and 000 alternatively must be partly visible on slight rotation in either direction. Set exactly on 000 (T, Fig. 6.9). Pull the dial forward on its shaft so that independent movement of the dial with respect to the degree indicator is obtained and turn the dial to 0. Slide the dial back and tighten screws U (Fig. 6.9) evenly and equally in turn. Make sure that the space between the dial and goniometer housing is 0.5 to 1 mm. This is conveniently done by inserting a card of appropriate thickness behind the dial while screws U are being tightened. Once tight, the card is pulled out. This ensures free subsequent movement of the dial.

34. Loosen screw V (Fig. 6.9) and make sure that lever W is in its midposition, protruding radially outward.

35. Push the specimen holder assembly into the hollow shaft and secure clamp C by first tightening screw E, then D. Tighten F (Fig. 6.9).

36. Check that the base plate I (Fig. 6.7) is loosely fixed on the table top of the X-ray generator.

37. Remove the locking strip I (Fig. 6.8) from the base plate.

38. Pull the base plate I (Fig. 6.7) away from the tube shield as far as possible.

39. Make sure that pins J (Fig. 6.8) in the aligning device are in the drawn-up position and locked in the horizontal slots.

40. Back off the knurled nuts X (Figs. 6.8, 6.9) and turn the leveling screws so far in that there will be some distance between the long bottom side of the aligning device and the strip on the bracket (between D of Fig. 6.7 and B of Fig. 6.8) when the goniometer is positioned on the base plate.

41. Position the goniometer on the base plate. Put right-hand leveling screw Y (Fig. 6.9) in the hard fiber cup supplied with the goniometer. Position the two left-hand leveling screws K (Fig. 6.8) on the left-hand grooves of the base plate I (Fig. 6.8).

42. Loosen the knurled screw B of the scatter case C (Fig. 6.10).

43. Turn the mechanical shutter disc of the tube shield to the left to open the window.

44. Move the goniometer toward the tube shield until the scatter case just stops short of touching the filter disc inside the shutter window of the tube shield. One of the V grooves of the base plate is provided with a pin. When the goniometer is adjusted for a beam take-off angle of 6°, make sure that the leveling screw touches this pin. If for any reason the goniometer should be removed later on, this pin can be used as a reference for relocation. Fix the locking strip I (Fig. 6.8) to the base plate so that it clasps the leveling screw.

45. To obtain a nearly correct position for the right-hand side of the goniometer adjust the perpendicular distance from the center of the locking knob F (Fig. 6.9) to the table top exactly 264 mm for a take-off angle of 6° with the PW 1316 tube shield. (For a take-off angle of 3° the distance should be 273 mm.)

46. Position the goniometer so that the aligning device B (Fig. 6.8) is above the take-off angle strip of the bracket D (Fig. 6.7), as shown in Fig. 6.8.

47. At this moment the following adjustments should be made:

    a.   The long bottom side of the aligning device should be at the correct distance from and in line with the strip on the bracket. This distance can be corrected by adjusting the two leveling screws K (Fig. 6.8) so that the pins J (Fig. 6.8), when released from their horizontal slots, are just below the surface of the aligning device. The best way to check correct alignment is to slide a straight object, for example, a specimen holder or glass slide, over the top surface of the aligning device. This object should just fail to touch the pins. If it touches a pin, the leveling screws have to be further adjusted until the object clears the tops of the pins when it is slid over the top surface of the leveling device. Both pins should just fail to touch the flat object.

    b.   The plane over the vertical side of the aligning device should coincide with the appropriate plane over the strip of the bracket. Alignment can be made by sliding the goniometer over the table top and by slight adjustment of the leveling screw Y (Fig. 6.9). The alignment can be checked with a straight bar touching both vertical planes.

The alignments (a) and (b) should be repeated a few times, for one step will alter the settings of the other.

NOTE.   If it is impossible to make B (Fig. 6.8) align with the strip on the bracket D (Fig. 6.7), the eight screws in the base of the tube shield (PW 1316) should be loosened and the tube shield rotated around its vertical axis slightly before all eight screws are firmly retightened.

48.   When the goniometer is positioned, tighten all screws on the base plate, including the screw on I (Fig. 6.8).

49.   Recheck points (a) and (b) of step 47.

50.   Tighten the nuts X (Fig. 6.8, 6.9) and recheck points (a) and (b) of step 47.

51.   Remove the aligning device B, the aligning slits D and H (Fig. 6.8), and the bracket B (Fig. 6.7).

DISCUSSION

Exercise 6.12, steps 1 through 51, should be only infrequently necessary for major periodic alignment if it is felt that adjustment is required. On the other hand, exercise 6.13 should be performed more often because it contains refinements of the alignment that need more frequent attention.

EXERCISE 6.13

Angular Calibration of the Goniometer

1.   Although Fig. 6.9 shows the goniometer arm without the detector attached, it should have been fixed to the arm as shown in Fig. 6.8.

2.   The hollow cast-iron cover with its metal cap is now slid into the receiving slit assembly J (Fig. 6.9) and loosely locked in its approximately correct position symmetrically about the groove K for the receiving slit and L for the antiscatter slit (Fig. 6.9).

3.   Push the radiation protection shielding (PW 1081, Philips), which is supplied with the goniometer, onto the central drum Z (Fig. 6.9) of the specimen shaft. A slit on top of the shield should engage in a pin on top of Z (Fig. 6.9). The shield should fit smoothly onto the top and around the divergence slit assembly C (Fig. 6.10). The rotatable shaft of the shield should be clipped onto the cast-iron sleeve which has been fitted around the receiving slit assembly in step 2. The cast-iron sleeve is now tightened by its locking nut.

4.   Remove the antiscatter slit from groove L (Fig. 6.9) if one has been fitted.

5.   Fit a single knife-edge into the goniometer specimen holder so that it coincides exactly with the specimen rotation axis. The knife edge is included with diffraction set PW 1049.

6.   With an X-ray intensity $CuK\alpha$ (nickel filtered) of 15 kV, 10 mA, a $1°$ divergence slit, and a 0.05 mm receiving slit allow the goniometer to scan at its lowest speed of $1/8°$ 2θ/min from, say, $\frac{1}{2}°$ below 000° to $\frac{1}{2}°$ above 000°. Allow the X-ray counting rate to be recorded on the strip chart recorder, running at its maximum speed of 1200 mm/hr. Figure 6.11 shows the experimental setup and the kind of record that should be obtained. Intensity rises regularly from a low I_0 to a high I_1, flattens out, and

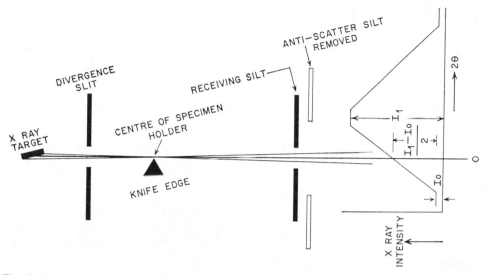

Fig. 6.11 Experimental set up for ascertaining the true 2θ zero setting of the goniometer, using the single knife-edge in the specimen position. The X-ray intensity output record is shown on the right. The midpoint on the curve indicates the exact 2θ zero setting. (After Parrish and Lowitzsch, 1965).

eventually falls off again. The true angular reading when the absolute intensity is $I_0 + (I_1 - I_0)/2$, that is, the exact midpoint of the intensity curve, gives the true zero setting of 2θ (Parrish and Lowitzsch, 1965).

7. The goniometer should be turned manually to the setting at which the X-ray intensity is exactly equal to this half value, and at this position the setting of T and I (Fig. 6.9) should be adjusted by unlocking screws U and resetting so that the 2θ reading is exactly 000°.

DISCUSSION

Any adjustment to the zero setting of 2θ necessary as a result of this exercise should be very slight if 2θ has been accurately set (see step 33 of exercise 6.12).

EXERCISE 6.14

Adjusting the 2 : 1 Goniometer Setting

1. Center the double knife-edge (supplied with diffraction set PW 1049) in the specimen holder.

2. With 2θ set on the zero found in exercise 6.13 and an X-ray intensity of similar settings, loosen allen screw V (Fig. 6.9). This is the larger screw on the collar closest to the goniometer body.

3. The specimen holder is now gently rotated by lever W (Fig. 6.9) until a maximum count rate is shown by the recorder pen deflection. This is a very sensitive adjustment. A slight movement of lever W from the correct 2 : 1 position causes the count rate to fall dramatically. When the count rate is at maximum, lock screw V (Fig. 6.9), but while tightening it make sure that the count rate remains at its highest. Check also that knurled nut F (Fig. 6.9) is fully tightened and that any tightening of it does not reduce the X-ray intensity.

DISCUSSION

The adjustments of exercise 6.13 and 6.14 are easy to make and should be done periodically to ensure correct 2θ and that the specimen is in the correct reflecting position.

EXERCISE 6.15

Choice of Divergence, Receiving, and Antiscatter Slits and Centering the Primary Beam

Divergence Slit

1. Ascertain the lowest 2θ angle at which the goniometer will be required to operate. This will depend on the specimen under measurement and the wavelength used. For specimens with a large d spacing operation will be required at a small 2θ angle and vice versa.

2. The following table lists the proper choice of divergence slit aperture for various 2θ ranges on the goniometer.

Divergence Slit Aperture	Recommended Goniometer 2θ Range	d-Spacing Range, Å for Cu$K\alpha$	d-Spacing Range, Å for Co$K\alpha$
$5'\,(\frac{1}{12}°)$	1.45° to 8.50°	61 to 10.4	75 to 12.0
$30'\,(\frac{1}{2}°)$	8.50° to 17.0°	10.4 to 5.2	12.0 to 6.05
$1°$	17.0° to 34.5°	5.2 to 2.6	6.05 to 3.02
$2°$	34.5° to 72.8°	2.6 to 1.3	3.02 to 1.51
$4°$	>72.8°	<1.3	<1.51

(Data from Parrish, 1965.)

3. From this table it can be seen, for example, that if a diffractogram is to be run only over the range Cu$K\alpha$ 17.0 to 34.5 2θ the ideal divergence slit to be put in slot H (Fig. 6.8) should be a 1° slit. For clay mineral diffractograms in the range 2θ 1.45 to 8.50 a smaller aperture slit of 5 minutes (1/12°) must be fitted. As 2θ goes to higher values on the diffractometer trace, one may persist with this narrow slit, but preferably the diffractogram can be momentarily stopped and the divergence slit replaced with a corresponding bigger aperture. On no account can a wider slit be used at lower 2θ angles than listed above.

Receiving Slit

The width of the receiving slit fitted has a major role in determining the resolution, intensity, and peak-to-background ratio of a diffractogram. The narrower the slit the greater the resolution, but of course the lower the intensity. Therefore the choice of receiving slit must be made by balancing the effect of peak intensity required against resolution. The most commonly used slit is 0.1 mm, and when used in conjunction with a scanning speed of 1° per minute and a time constant of 2 seconds gives rise to a good diffractogram. If a slower scanning speed of, say, 1/8° per minute is employed to gain maximum resolution of an individual peak, it may be desirable to combine this slow speed with a very narrow receiving slit of perhaps only 0.05 mm, provided of course that it does not reduce the peak intensity too much. If it does, then one may have no choice than to use the 0.1 mm slit. When a peak has very small intensity an even wider receiving slit of 0.20 mm may be necessary. For routine high scanning speeds of 1 or 2° per minute one cannot afford to fit a very narrow receiving slit; otherwise the peak intensities may be too drastically reduced.

Antiscatter Slit. A low, nearly uniform background can be obtained by the proper use of the antiscatter slit (the slit nearest the detector window). The receiver slit and the antiscatter slit together limit the radiation that enters the counter tube.

1. Set the 2θ exactly to get maximum count rate from a strong reflection; for example, the Silicon 111 peak at Cu$K\alpha$ 28.467 2θ.

2. Slowly rotate the screw O (Fig. 6.9) both in and out until the peak intensity reaches a maximum on the recorder. The peak intensity should now remain at nearly the same value when the antiscatter slit is removed. If the peak intensity rises (say, by more than 2%) when the slit is removed, the antiscatter slit is too narrow. It should be replaced by another slit of wider angular aperture. Antiscatter slits come in various widths of 4, 2, and 1° and smaller down to 1/30°. As normal practice it is not wise to have an antiscatter slit that is too narrow and a 2° slit should generally be fitted unless it is found that fitting a narrower slit does not reduce a peak by more than 2% of its total height.

Centering the Primary Beam

1. The choice of divergence slit has been determined on the lowest 2θ at which the diffractogram will have to begin. It is now desirable to center the X-ray beam with 2θ set at the lowest scanning angle to be used, particularly when working at small 2θ angles of less than 10°, because at such small angles the length of the specimen illuminated increases rapidly with a small decrease of θ.

2. Set 2θ to the smallest angle required on the diffraction scan. Center the fluorescent screen disc (PW 1061, Philips) exactly on the specimen holder so that its central radial line is parallel to the goniometer axis. Loosen screw D of the clamp C (Fig. 6.9). Loosen clamp F (Fig. 6.9). Pull out the specimen holder shaft approximately 5 mm, turn it until the fluorescent disc is perpendicular to the expected X-ray beam, and push the specimen holder assembly again fully into the holder shaft. This rotation will have to be done with the radiation shield removed. It can thereafter be refitted.

3. Switch on X-rays to this window—it will be necessary to hold the microswitch in the activated position, for with the shield door open the door lever will not activate the switch—set to kV 40, mA 20, and observe the X-ray beam as it strikes the screen.

4. The brighter strip of illumination should appear centered on the screen. If not, the image of the beam is centered by loosening knob B (Fig. 6.10) on the divergence beam assembly and tilting the assembly until the beam is centered. Tighten knob B (Fig. 6.10) and knob L (Fig. 6.8) after the beam is centered.

5. The X-ray supply is cut off, the fluorescent screen removed, and the specimen holding shaft returned to its normal horizontal position with screws D and F locked (Fig. 6.9).

6. Each time the specimen shaft has been removed and the primary beam centered according to steps 1 to 5 it will be necessary to follow immediately with an angular calibration of the goniometer (exercise 6.13) and an adjustment of the 2 : 1 goniometer setting (exercise 6.14).

DISCUSSION

Recentering the primary beam is required only when a change is made in the lower operating 2θ range for diffractograms and is especially important when operation will commonly be used at low 2θ values for clay minerals. Of course, after finishing with clay minerals it may be desirable to recheck the centering of the beam for normal mineral diffractograms beginning at a much higher 2θ. In every case any adjustment to the primary beam direction must be followed by exercises 6.13 and 6.14.

EXERCISE 6.16

General Procedure for Running an X-ray Diffractogram

1. Mount the specimen on the specimen shaft of the goniometer, firmly held by the spring clip. Close the shield door. Make sure that the space between the spring clip and specimen shaft is clean and free of specimen powder.

2. Select the most appropriate scanning rate by changing the gears on the goniometer. If a rapid record of low angular resolution simply for identification purposes is required, a fast scanning rate of 1 or even $2°$ per minute is selected. If, on the other hand, high 2θ precision is required over a smaller angular range, a slow scanning rate of $\frac{1}{2}$, $\frac{1}{4}$, or even $\frac{1}{8}°$ per minute is chosen.

3. Select the appropriate time constant (on ratemeter PW 1362) according to the scanning rate selected; for example, a scanning speed of $1°$ per minute should normally be combined with a 2 second time constant. For slower rate of $\frac{1}{2}$ or $\frac{1}{4}°$ per minute a longer time constant of 4 seconds may be preferable. For a $2°$ per minute rate the time constant would have to be 1 second. A longer time constant gives a smoother, more easily read diffractogram. Too short a time constant of less than 1 second will show too many random changes in the detector response; hence the diffractogram will not be smooth.

4. Choose the most appropriate divergence slit, receiving slit, and antiscatter slits according to exercise 6.15.

5. Choose an appropriate strip chart recorder speed to suit the scanning speed of the goniometer. A slow goniometer scanning speed should always be combined with a slow recorder speed and a fast goniometer speed with a fast recorder speed. A common goal is to achieve 2 cm per degree 2θ (or 1 in. per degree) for moderately slow scans of $\frac{1}{2}$ or $\frac{1}{4}°$ 2θ per minute, so that peak centers can be accurately located with the eye. There is no virtue in having a long chart distance per degree; indeed 1 in. per degree (or 2 cm per degree) is normal practice at the U.S. Geological Survey and the Carnegie Institute in Washington for high-precision lattice parameter refinement. The following combinations give ideal diffractogram scales of 2 cm/1°2θ:

Goniometer speed, degrees 2θ/min	$\frac{1}{8}$	$\frac{1}{4}$	$\frac{1}{2}$	1
Recorder speed, mm/hour	150	300	600	1200

The slowest scanning rates of $\frac{1}{8}°$ per minute should be reserved for highest 2θ precision determinations, whereas the fast rate of $1°$ per minute would be used for normal identification purposes requiring less precision, An even faster scan of $2°$ per minute is possible, and this would have to be combined with 1200 mm/hr giving $1°2\theta/cm$ on the diffractogram, which is perfectly adequate for rapid identification purposes.

6. With the detector voltage set at its appropriate operating high tension and the pulse height analyser (PW 4280) set with attenuator $Z = 2$, selector knob on threshold, lower level on zero, start the scaler (PW 4231) ensuring that preset counts and time (on PW 4261) are both set at INFINITY.

7. Now rotate the 2θ setting of the goniometer by arm I (Fig. 6.9), ensuring first that the clutch H is in its released right-hand position, over the 2θ scanning range. While this is being done note the deflection of the pen on the recorder and adjust 2θ manually exactly on what is discovered to be the strongest peak of the specimen. At this stage adjust the ratemeter (PW 1362) settings so that the pen of the recorder is deflected as far to the right as possible without going off scale. If necessary, this may be combined with a slight adjustment of milliamperes on the generator so that the deflection from this, the strongest peak, will fill 80 to 90% of the recorder scale. Turn the 2θ scale to about $1°$ below the peak value, engage the clutch, switch on the goniometer motor, and allow the goniometer to scan over this peak. See that the peak stays on scale. If not, make slight adjustments to the ratemeter and milliampere settings.

8. Return 2θ to the minimum value required on the diffractogram. This should be a whole number of degrees, for example, 27.00. Switch off the goniometer switch S (Fig. 6.9) and lock the clutch H (Fig. 6.9) in its left-hand position.

9. Turn the strip chart recorder manually so that the pen rests exactly on one of the chart lines. Check that the switch on the ratemeter (PW 1362) is in the ON position for the recorder motor. Check that the switch on the goniometer supply (PW 1361) is OFF. Now turn the goniometer switch S (Fig. 6.9) to ON. Put the goniometer selector switch on HIGH.

10. By turning on the switch on the goniometer supply (PW 1361) the goniometer scan and the strip chart recorder will begin simultaneously. Figure 6.12 shows a diffractogram obtained from SILICON. The lowest 2θ value selected was $27°$. The stepped line along the foot of the diffractogram indicates the one-half degree positions of the goniometer. These will coincide with the ruled lines of the chart paper only if both goniometer and chart were started simultaneously from a whole degree 2θ, as explained in Steps 5, 8, and 9.

11. At any stage during a diffraction run the ratemeter value may be changed to make full use of the chart paper width. The peak heights are measured in millimeters above the extrapolated baseline, as shown in Fig. 6.12. Provided a constant ratemeter setting is used throughout, the intensities are simply prorated arithmetically to peak height. If, however, a change in ratemeter setting is made between the most intense peak and another, the different value will have to be taken into account; for example, a peak recorded at a ratemeter setting of 1×10^3 will be four times more intense than if it were recorded at 4×10^3. Likewise a peak recorded at 4×10^2 will be 2.5 times more intense than if it were recorded at 1×10^3 counts per second full chart range.

12. A diffraction run can be made without any automatic stops on the goniometer by leaving screw G (Fig. 6.9) loose during the run, but this is a dangerous procedure,

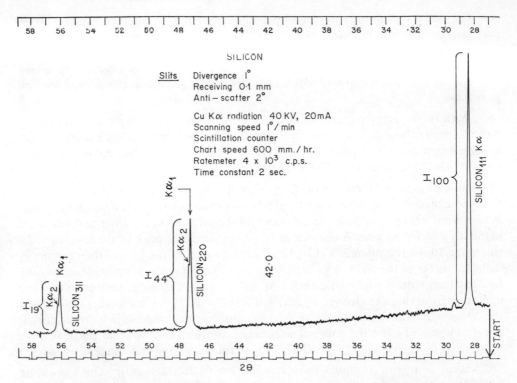

Fig. 6.12 Partial diffractogram of silicon powder. The most intense peak 111 is given an intensity of 100. The intensity of the subsequent peaks is given a realtive value prorated to their height above the base line.

for if the operator does not remain in the room the diffractometer will keep on increasing in angular position until the counter arm eventually strikes the tube shield. The result will be damage to the goniometer and its alignment. To prevent this kind of accident the automatic stops should be activated before a diffraction trace is initiated.

13. At step 8, when the 2θ is at its minimum position, say $27°$, decide over what angular range the diffractogram is to be run, say, only as far as 2θ $57°$, that is, over a range of $30°$. There is another locking nut similar to G (Fig. 6.9) around the back of the goniometer. Loosen this nut and rotate the zero line on the collar to which it is attached to coincide with the angular range selected, in this case $30°$, as shown on the collar CC (Fig. 6.9). Lock the nut to hold the collar zero against $30°$. This firmly locks lever **BB** (Fig. 6.9) onto this collar. With nut G (Fig. 6.9) loose, turn the collar so that the automatic stop lever AA (Fig. 6.9) just touches the front of the stop switch DD (Fig. 6.9). Tighten nut G (Fig. 6.9). It will now be found that when the diffractogram trace has run from its selected beginning of 27 for a selected range of 30 to $57°$ the lever **BB** will activate the back of the switch DD (Fig. 6.9), causing both the goniometer and the recorder to stop. This precaution ensures that no accident will occur. Once the goniometer has been automatically stopped by this device, it is restarted by releasing the nut G and moving switch EE (Fig. 6.9) in the way that is found necessary to light up the recorder once again. This automatic stop also stops the count on the scaler so that the start button needs to be reactivated to resume counting. If switch FF (Fig. 6.9) had been put to oscillate instead of high, then, instead of stopping when $57°$ is reached, the goniometer will continue to oscillate between 27 and 57 until manually stopped.

14. The 2θ peak position is best determined at the peak apex. With a good sharp peak visually bisect its extremity with the measuring hairline to obtain its 2θ position. The old practice of bisecting the $\frac{2}{3}$ peak height positions is arbitrary and cumbersome and should be abandoned. Use of the peak apex leads to higher precision. The recorded 2θ readings, as read from the diffractogram (Fig. 6.12), are noted and converted to dÅ values by using appropriate conversion charts or by the equation

$$d_{hkl} = \frac{\lambda}{2 \sin \theta_{hkl}},$$

where θ is half the angle recorded and λ is the wavelength used. It will be seen in Fig. 6.12 that the first peak is not resolved into a doublet and that λ is taken as $K\alpha$, whereas at higher 2θ values each peak is a $K\alpha_1$ and $K\alpha_2$ doublet. The 2θ position of $K\alpha_1$ should be converted to a d by using the wavelength $K\alpha_1$.

EXERCISE 6.17

Determination of Ideal Operating Conditions for the X-ray Detector

The diffractometer may be equally well operated with a proportional detector (PW 1965, Philips) or a scintillation counter (PW 1964, Philips). The following procedure should be carried out to ensure that correct high tension (on PW 4025, Philips) has been selected in conjunction with the appropriate attenuation on the amplifier (PW 4280, Philips).

The following description applies to the scintillation counter but is identical for the proportional counter except that the values will differ.

1. Put the silicon standard (PW 1062, Philips) or any specimen that is known to give an intense diffraction peak into the diffractometer. With the strip chart recorder running and counter high tension set to approximately 160 on the potentiometer (PW 4025, Philips) = 730 V for the scintillation counter (460 setting = about 1640 V for the proportional counter), attenuation switch at $Z = 1$ (attenuation is at steps of 2^Z; setting $Z = 1$ attenuation is 2), lower level setting on the pulse height analyser set at zero, and operating knob set to threshold (THR) (PW 4280, Philips), press the start button on the scaler (time at infinity). Note that the pen of the recorder deflects to the right. Adjust the ratemeter to make full use of the chart and adjust 2θ manually on the goniometer, whose motor is switched off, to get maximum deflection on the chart. If silicon$_{111}$ is used, then 2θ should be at 28.44 for Cu$K\alpha$ radiation. Lock the goniometer clutch when a maximum counting rate is obtained and the pen deflection is at about 60% of the scale. Adjustment of ratemeter and milliamperes will obtain a 60% deflection of the pen.

2. Now slowly drop down the high tension potentiometer setting until the pen deflection drops to nearly zero. Set the fixed time to 10 seconds. Perform a series of 10-second counts for various HV potentiometer settings by increasing the reading 10 divisions at a time. Plot the count rate against the HV setting as shown in Fig. 6.13. Do not proceed to a voltage that is too high beyond the plateau because it will spoil the counter. For the scintillation counter (PW 1964) the relation between the HV potentiometer reading on the power supply (PW 4025) and the actual voltage is found to be $263 + 3 \times$ the HV reading. Of course, this relation varies from laboratory to

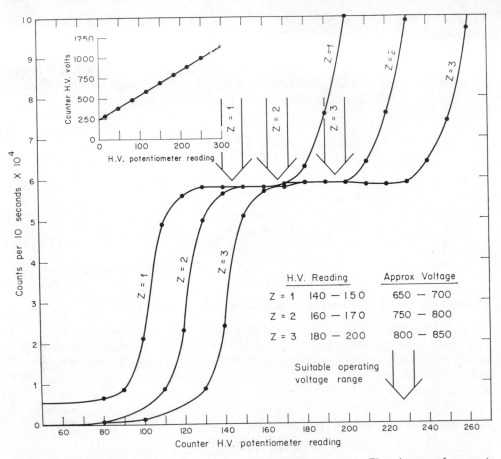

Fig. 6.13 Operating conditions for the Philips scintillation counter. The plots are for counts of 10 seconds for various counter HV settings on the silicon$_{111}$ peak, using CuKα radiation at 2θ 28.44°, 50 kV, and 26 mA. A ratemeter setting of 1×10^4 was used (slits are 1° divergence, 0.1 mm, and 2° antiscatter) to allow strip-chart deflections to stay on scale. The top left-hand graph relates HV potentiometer setting to true counter voltage. The selected operating voltage must be on a flat response curve plateau, preferably on the low-voltage side of its midpoint. Ideal HV settings are shown for various attenuations (2^z) of the amplifier. These settings should not be blindly followed but determined for any particular laboratory by each operator.

laboratory and can be checked by plotting the HV potentiometer reading against the voltage read from the scale. A similar plot is given at the top left corner of Fig. 6.13.

3. Switch the attenuation to $Z = 2$ and repeat the procedure to produce a plot of X-ray intensity against the HV potentiometer reading. Likewise produce a plot for $Z = 3$ (actual attenuation $= 2^3 = 8$). For copper radiation there is no point in going to higher attenuation values. Figure 6.13 shows the response curves for the three attenuation settings for CuKα radiation.

4. Select an appropriate operating voltage from the curves. The voltage should be such that the X-ray intensity is on a flat part of the curve plateau, preferably on the lower voltage side of the middle of the plateau. Note that the ideal voltage is suitable only for a chosen attenuation; for example an HV setting of 150 with $Z = 1$ would be no use if $Z = 3$ were used, for the X-ray intensity would not be on a stable plateau region, hence would be dependent on slight voltage fluctuations.

DISCUSSION

In some diffractometers, for example, the Philips with electronic panel PW 1352, the graph of counts per 10-second intervals against counter HV (Fig. 6.13) may be conveniently drawn directly by the strip chart recorder by allowing it to run continuously at its fastest speed of 1200 mm/hr; the starter button is pressed immediately after each 10-second count is completed and the HV potentiometer is quickly adjusted to its next highest setting.

Appropriate settings for the proportional counter for CuKα radiation are $Z = 3$ and HV = 460.

Whenever another wavelength is used, for example CoKα, the appropriate settings of Z and HV should be determined in the same way.

EXERCISE 6.18

Determination of Pulse Height Discriminator Settings

Although not essential for diffraction, the incorporation of a pulse height analyzer (PW 4280, Philips) in the electronic panel helps to give nearly perfect monochromatic $K\alpha_1 + K\alpha_2$ radiation in which $K\beta$ and white radiation are almost entirely eliminated. Its use therefore produces sharper diffraction peaks with better resolution. Proper use of the pulse height analyzer makes for good diffractograms of iron-rich minerals even when a Cu tube is used. There is no need therefore to use Co or Fe tubes if the pulse height analyzer is employed intelligently to discriminate against the unwanted fluorescent radiations from the specimen. It must be warned, however, that the pulse height analyzer must be properly set or the resulting resolution may be worse than if it were not used at all. Careless settings must be avoided. The procedure for selecting the settings is outlined below and should be carefully followed, step by step, until the operator becomes familiar with the method.

1. With the goniometer set on silicon$_{111}$, as in step 1 of exercise 6.17, and the pulse height analyzer set on THR with LOWER LEVEL on zero, allow the recorder to run continuously and adjust 2θ, the ratemeter, and milliamperes, if necessary, to get the pen response from the peak to fill most of the chart. Lock 2θ exactly on the value that is centered exactly on the CuKα intense reflection.

2. Select in turn the appropriate settings of Z and HV as determined in exercise 6.17 (see Fig. 6.13). Now, if the scintillation counter is used, set $Z = 1$ and HV = 140 (one of the appropriate combinations discovered in exercise 6.17).

3. Note the exact pen reading on the strip chart recorder. In our case assume that it is 60%. Now slowly turn the lower level potentiometer of the pulse height analyzer (knob on THR) until the recorder reading is reduced to exactly half its value. In our case LOWER LEVEL has to be turned to a value of 180 (1000 divisions = 5 V), at which time the recorder shows exactly 30%.

4. Repeat with the other possible combinations of Z and HV. By using $Z = 2$, HV 160 the chart deflection is reduced from 60 to 30% when the lower level potentiometer is turned to a value of 170. By using $Z = 3$ and HV = 190 the lower level adjustment necessary is 230 to reduce the intensity to half.

5. The *correct selection* should be made when the *value of the lower level* lies *between 200 and 500 divisions* to reduce the *recorder deflection* to its *half-value*. In our case the appropriate choice is obviously $Z = 3$, HV = 190. Unless this determination is made initially, the amplitude of the incoming X-ray pulses will not be fully accommodated within the capacity of the amplifier. Some modification is possible at any setting of Z, for a slight increase of HV at any Z setting will cause an increase in the value of LOWER LEVEL when the intensity is brought to its half-value. Also, if HV is allowed to remain constant, an increase in the Z value will cause LOWER LEVEL to go to a higher value. In any case whatever combination of Z and HV is chosen, HV must have been determined to give a suitable position on a plateau, as in Fig. 6.13.

6.   With LOWER LEVEL still at its value to give half-peak intensity we have now found the top of the pulse distribution curve. Now adjust the WINDOW potentiometer to a small reading of 10 divisions (500 divisions = $2\frac{1}{2}$ V). Turn the knob from THR to 1 to activate the window. This will immediately cause the pen deflection to reduce dramatically. Increase the deflection on the recorder by adjusting the ratemeter and the milliampere setting of the generator so that the pen deflection will fill at least 70% of the chart. It is not essential that WINDOW be exactly 10 divisions. It can be 9 or 8 but should not be wider than 10. An appropriate choice of WINDOW and ratemeter can be selected to give an approximately 70% chart deflection. It may be necessary to have the ratemeter running as low as 4×10^2.

7.   Turn the lower level potentiometer to lower values until the pen deflection goes down nearly to its zero value or as low as it can go to become stable.

8.   Select a preset 10-second time and adjust the recorder to its fastest speed of 1200 mm/hr. Allow it to run continuously.

9.   Start counting for 10 seconds. Mark the deflection of the pen by the value of the lower level potentiometer. As soon as the count stops, turn LOWER LEVEL to a value 10 divisions higher and repeat the 10-second count. Mark it and continue. Make as little delay as possible between each subsequent count. Between each count increase LOWER LEVEL by 10 divisions. This procedure is continued until a curve, illustrated in Fig. 6.14, is drawn out. Counting and plotting should continue till the pen deflection drops again to near zero and is stable. The plotted curve is the energy distribution curve of the Cu$K\alpha$ radiation being diffracted. It will be seen to be approximately Gaussian. A solid penned line should be drawn over the end of recorder pen deflections to outline the curve (Fig. 6.14).

10.   Draw a line to represent the baseline (background value). Draw a parallel line to indicate the maximum peak height. Bisect the distance between these two lines and draw in the line representing the half-maximum peak intensity. Note the two points at which the half-maximum line intersects the curve. Note the extrapolated value of LOWER LEVEL at each of these points. In the example given in Fig. 6.14 the values are 165 and 265 divisions, respectively. The difference between these two values is called the half-width value, in Fig. 6.14 it is $265 - 165 = 100$ divisions ($=0.5$ V).

11.   Adjust the helical WINDOW potentiometer to a value $2\frac{1}{2}$ times this half-width value. In Fig. 6.14 WINDOW is then set at 250 ($=1.25$ V). The choice of $2\frac{1}{2}$ times the half-value width is based on the assumption that the energy distribution curve is Gaussian, or normal. From statistical tables we can determine that the area of a normal curve between the two ordinates at half-peak height is 0.7580 of the total area, between the two ordinates at twice half-peak height 0.9809 of the total area, and between $2\frac{1}{2}$ the half-peak height 0.9964 of the total area. Thus by setting the WINDOW at $2\frac{1}{2}$ times the half-peak height width we arrange the window to accept 99.64% of the total energy from the monochromatic $K\alpha$ radiation. Further increase of the window will not take

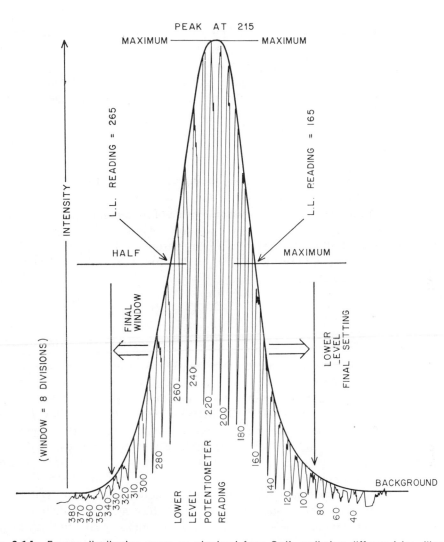

Fig. 6.14 Energy distribution curve as obtained from CuKα radiation diffracted by silicon$_{111}$ at 2θ 28.44°, kV 50, mA 26. Chart speed 1200 mm/hr. Ratemeter 4×10^2. WINDOW used for scan equals eight divisions. Counting time for each LOWER LEVEL setting 10 seconds. Final settings selected WINDOW $= 2\frac{1}{2} \times (265 - 165) = 250$ (1.25 V), LOWER LEVEL $= (265 + 165)/2 - 250/2 = 90$ divisions (0.45 V). Some recorders cannot run for short times of 10 seconds and the graph cannot be directly drawn out on the recorder. However, the graph can be redrawn manually by plotting counts per each 10 seconds versus the LOWER LEVEL value during each count.

in much more of this energy. Of course the purpose of the window is to discriminate against other energies such as $K\beta$ and white radiation. Therefore the window should not be made wider; otherwise its purpose will be lost.

12.   The maximum peak height value should be determined as the midpoint between the two half-peak height values. In our example (Fig. 6.14) this peak value will be

$$\frac{165 + 265}{2} = 215,$$

which is a more satisfactory value then just looking at the value of LOWER LEVEL at the top of the curve because the peak may not be absolutely symmetrical.

13.   Position the lower level potentiometer to a value of maximum peak value, as in step 12, minus half the value obtained for the window, as in step 11. The pulse height discriminator is now ideally set to receive the monochromatic Kα radiation from the diffractometer. The values of LOWER LEVEL, WINDOW, HV, and Z so determined are to be recorded and used always for CuKα diffraction. No single one of these values can be changed without affecting the other three.

DISCUSSION

Three of these determined values can be permanently maintained for all subsequent settings; that is, Z, LOWER LEVEL, and WINDOW. It is then recommended that for subsequent diffractograms the goniometer is set on an intense peak with these four settings and the amplifier knob is set at 1 to activate the window. To allow for any slight drift of voltage on the detector it is now recommended that the HV potentiometer be slightly adjusted up and down from the previously accepted value until the pen deflection is a maximum. This is a very sensitive adjustment and compensates for any electronic drift in the equipment. It is recommended always that the electronic panel of the diffractometer be switched on at least one-half hour before use to stabilize it and that periodically, say at least once an hour, a fine adjustment be made to the HV potentiometer to get maximum deflection from any one intense peak. In this way we ensure that the Kα energy spectrum will remain perfectly centered within the window.

Settings of Z, HV, LOWER LEVEL, and WINDOW need to be determined separately for different wavelengths, for example, CoKα and CuKα, and for the scintillation and proportional counters.

EXERCISE 6.19

Use of a Standard for Obtaining High 2θ Accuracy

For accurate calibration of a powder camera and especially a diffractometer a film or diffractogram should be taken of a pure standard substance. This must be a carefully selected pure chemical of known and highly refined structure determination lacking in isomorphic substitution and lattice variation.

Pure silicon powder ($a = 5.43050$ Å) is the most commonly used standard, but pure $KBrO_3$ (see exercise 7.2) is also frequently used. Silicon details are given in the J.C.P.D.S. powder data file, card 5-0565; the first six diffraction lines are tabulated here and their 2θ values are given for both Cu and Co radiation.

Line No.	Relative Intensity	Silicon		2θ	
		hkl	λ	CuK	CoK
1	100	111	α	28.467	33.177
			α_1	28.443	33.152
			α_2	28.515	33.226
2	60	220	α	47.345	55.577
			α_1	47.304	55.533
			α_2	47.428	55.663
3	35	311	α	56.174	66.279
			α_1	56.124	66.225
			α_2	56.275	66.386
4	8	400	α	69.197	82.496
			α_1	69.132	82.423
			α_2	69.328	82.640
5	13	331	α	76.453	91.857
			α_1	76.378	91.772
			α_2	76.602	92.027
6	17	422	α	88.124	107.704
			α_1	88.033	107.591
			α_2	88.308	107.929

There are numerous other weak lines on the silicon pattern; $\alpha[=\frac{1}{3}(\alpha_2 + 2\alpha_1)]$ is used only when a diffraction peak is not resolved into its $\alpha_1\alpha_2$ doublet.

1. Prepare a powder specimen of the pure silicon powder in a standard way (exercise 6.1 for the camera, exercise 6.9 for the diffractometer).

2. Take, develop, and measure an X-ray powder film, using the standard 0.5-mm collimator. Measure the film and tabulate the 2θ values (exercise 6.8).

3. Make a diffractogram of the silicon specimen from 2θ CuKα 28 to, say, 70° by using a 1° divergence slit (this may be replaced with a 2° slit above 2θ 35°), a receiving slit of 0.1 mm, and an antiscatter slit of 2°.

4. After the standard silicon scan has been completed the recorded 2θ positions of the peaks (Fig. 6.12) are compared with the theoretical positions given above. Any systematic error, either $+ve$ or $-ve$ is noted and used to correct the 2θ readings on any subsequent experiment, provided that the alignment of the goniometer has not been changed and the same specimen preparation technique is used for both standard and unknown.

5. If the difference between the recorded 2θ of the silicon peak differs too much from the ideal value (e.g., by as much as 0.10 or 0.20° 2θ), the 2θ drum of the goniometer I (Fig. 6.9) may be adjusted by loosening screws U (Fig. 6.9) and resetting the drum. The screws are then retightened.

6. Both standard specimen and subsequent unknown should be run with 2θ progressing in the same direction, both from low to high or both from high to low. The goniometer does not reproduce exactly the same angles for both directions of scan because of backlash. Hence scans proceeding from low to high should be compared *only* with scans proceeding from low to high. Similarly, scans from high to low should be compared with and calibrated against scans from high to low. Preferably also the unknown and the standard should be scanned at the same speed as well as in the same direction.

7.   As noted in Exercise 6.8 (step 7), the angular error in 2θ becomes less as 2θ approaches $180°$; therefore any error detected in the 2θ position of the silicon 111 peak at $CuK\alpha$ $28.467°$ 2θ should be used subsequently to correct only the lines in this 2θ neighborhood. Likewise, silicon peak 422 should be used for correction in the $88.124°$ 2θ neighborhood. It may well be found that the errors in these two neighborhoods are not identical.

8.   The best practice is to mix an unknown specimen with an approximately equal amount of the standard as an internal 2θ standard during the specimen preparation. The diffractogram or powder film will then contain the lines of both the standard and the unknown. Each standard line should be clearly labeled. This labeling may be facilitated by indexing the mixture by comparing it with a diffractogram of the pure standard, previously taken and kept on the laboratory wall for quick reference. Any systematic error in the position of a standard line may then be used to correct the 2θ values of neighboring lines of the unknown. The standard so mixed should have been chosen to prevent its peaks from overlapping with those of the unknown. Some of the peaks may overlap, but it is important that the peaks of particular interest to us be quite distinctly resolved from the standard peaks.

Swanson et al. (1966) recommend the following substances as alternatives to silicon as internal standards: high purity tungsten powder, silver powder, and cadmium oxide powder. Their 2θ angles for $CuK\alpha_1$ ($\lambda = 1.5405$ Å) are listed:

Line	W $a = 3.16504$Å	Ag $a = 4.08625$Å	CdO $a = 4.69558$Å
110	40.262	...	...
111	...	38.112	33.012
200	58.252	44.295	38.304
211	73.187	...	...
220	87.000	64.440	55.286
310	100.630	...	...
311	...	77.390	65.920
222	114.922	81.530	69.256

EXERCISE 6.20

Identification of an Unknown by the J.C.P.D.S. Powder Data File

Every mineral and crystalline substance, whether organic or inorganic, has a unique powder diffraction pattern. Isostructural compounds, however, may have identical diffraction patterns except for intensities; the CaF_2 pattern, for example, is nearly identical to that obtained from HoF_2. Nevertheless this should not be taken to mean that any single mineral can be identified solely on the basis of its powder pattern.

Minerals and other crystalline substances of fixed chemistry which are not character-ized by chemical substitution and isomorphism, for example, quartz and fluorite, can be readily identified by their X-ray powder patterns because of the rather constant nature of their crystal structure.

On the other hand, many minerals have no fixed chemistry; for example, olivine, which may range from forsterite to fayalite. Each member of the olivine series there-fore has its own particular diffraction pattern. This makes it somewhat more difficult to identify an olivine because it is unlikely that the olivine we have measured will have identical chemistry to the one published in the J.C.P.D.S. powder data file. Nevertheless there are certain similarities between all olivine diffractograms that should allow us to identify the unknown as an olivine. Feldspars also present difficulties because of their complex structural variation. Accordingly, for many minerals it will not always be possible to obtain a perfect match between the X-ray pattern of the unknown and one of the powder data file cards. Intelligent guesses will be necessary and the data file should not be used blindly. Other additional data must be sought. The unknown, for example, should always be crushed and identified as closely as possible in a microscopic immersion preparation. The X-ray diffractogram is then compared with the J.C.P.D.S. file to confirm the microscopic identification.

The following Joint Committee for Powder Diffraction Standards (J.C.P.D.S.) publications should be in every X-ray laboratory. They are usually revised and published annually and may be obtained from Powder Diffraction Standards, 1601 Park Lane, Swarthmore, Pennsylvania 19081.

1. Powder data file *data cards*, sets 1 through 21 (sets 1 to 5 and 5 to 10 may be purchased bound; the latter are supplied as separate 3 × 5 cards which must be kept in numerical order in a bank of card files).
2. Inorganic index for the powder diffraction file.
3. Organic index for the powder diffraction file.
4. Fink index for the powder diffraction file.
5. KWIC guide to the powder diffraction file.

The procedure is best outlined in an actual example.

1. An X-ray diffractogram of an unknown mineral has been obtained with $CuK\alpha$ radiation on a well-aligned diffractometer. It was measured and all 2θ values converted to d values by using $K\alpha$ for the first peak, which was not resolved, and $K\alpha_1$ for sub-sequent peaks, which were resolved into doublets. The results are tabulated:

Line	Intensity	2θ	d
1	94	28.30	3.153
2	100	47.01	1.931
3	35	55.76	1.647
4	12	68.64	1.366
5	10	75.86	1.253
6	16	87.38	1.115

These lines are followed by six more weak ones that were not measured.

2. The three most intense lines are tabulated from the diffractogram, the most intense first:

d values			Intensity		
1.931	3.153	1.647	100	94	35

In such a case as this, however, the difference between the first two lines is so slight that the possibility is that we may have listed them in their wrong relative order or that a subsequent X-ray preparation may even reverse their intensity orders. It must always be kept in mind that line intensities are not reproducible because of preferred orientation. Hence we should also give the tabulation

d values			Intensity		
3.153	1.931	1.647	94	100	35

3. Now look up the most recent Inorganic Index to the powder diffraction file (e.g., the 1972 edition). It will be found that the pages are headed in groups, beginning

$$999.99{-}10.0$$
$$9.99{-}\ 8.0$$
$$7.99{-}\ 7.0$$

and so on down to the smaller values. These headings are the range for the MOST INTENSE line of the diffractogram. According to our first tabulation in step 2, our most intense line is *d* 1.931. It should be found on pages headed 1.94–1.90. The tabulations on these pages list eight columns in decreasing order of intensity. The first is not numerically ordered but includes all diffractograms whose first most intense line is within the *d* range 1.94–1.90 (the page heading); this obviously includes our first value of 1.931. When the actual value is near the lower or upper limits of any page heading, it is wise to consult the adjacent page groups as well. It is the second column, which lists the second most intense peak, that is numerically arranged. Scan this column until a match is obtained with our second most intense *d* of 3.153. This will bring us to the entry

$$1.93 \quad 3.15 \quad 1.65 \quad 1.12 \quad 1.37 \quad 1.25 \quad 1.05 \quad 0.86 \quad CaF_2 \quad 4{-}0864.$$

Normally it will be found that there are several entries with an identical second-column value, but at this stage all others are eliminated by glancing left to column 1 and finding only the line that has the correct *d* spacing—in this case, 1.93. Then look along columns 2, 3, and 4, etc., to check that each entry matches our diffractogram. Even now we may be left with a possible choice. Therefore at this stage it is important not to use the index blindly. The preliminary identification under the microscope should enable us to make the correct choice between several alternatives. Usually most of them will be clearly wrong and, for example, may not be minerals at all. If we had tabulated the relative intensities as in the second alternative of step 2, the index entry should be searched under the page headings 3.19–3.15; a search of these pages down column 2 would bring us to the second most intense line of 1.931 and so to card 4–0864 in the same way. The various indices have been prepared with multiple entries to allow for the eventuality that the lines are not always consistently tabulated in the exact same order.

4. Of course, if the optical identification suggests the mineral identification, it may be confirmed by going directly to the Mineral Index in the Inorganic Index, in which mineral names are tabulated alphabetically; fluorite is listed thus:

   * FLUORITE Syn CaF_2 1.93   3.15   1.65   100   94   35   4-0864

Other indices that may be used are the Fink Index which works similarly to the Inorganic Index and the KWIC Guide which tabulates crystalline substances according to their chemistry and also includes an alphabetical mineral listing with the three most intense lines. It should be noted that because they are computer printed some of the older indices reveal an occasional serious omission which affects their usefulness; for example, the 1967 indices failed to carry some important fluorite entries. Information regarding omissions from the indices is usually available by applying to the publishers.

5. From the various index entries it is clear that card 4.0864 should be consulted. It is found in set 4 and because the cards are filed numerically it is easily located. Figure 6.15 shows this card as a sample illustration. The important information is as follows: All cards have identical formats; $A = d$ spacing of the most intense diffraction line, indicated by 100 below it; $B = d$ spacing of the second most intense line which has a relative intensity of 94, given below it; $C = d$ spacing of the third most intense line which also carries its intensity below it. $D = d$ spacing of the very first line on the powder diffraction pattern. $E =$ chemical composition, chemical name, and mineral name; $F =$ star in the upper right corner which indicates that the card contains data of high reliability; an open circle indicates low reliability; no indication means average reliability; G, H, and I columns tabulate all the lines in their correct order as they appear on the diffractogram, G giving the d spacing, H, the relative intensity, and I, the Miller index which has been assigned by the author. The rest of the card contains experimental, mineralogical, and crystallographical data and references to the original publications that served as the data source.

6. Final confirmation that the unknown is actually the mineral described on the card is provided by a detailed comparison of the lines tabulated from the diffractogram (step 1) with columns G and H on the card (Fig. 6.15). The example chosen here correlates perfectly and fluorite is confirmed. It must of course also be confirmed by optical work in oil immersion and/or thin section.

7. Because individual species of the common mineral groups, such as feldspars, olivines, and pyroxenes, exhibit a considerable range of X-ray powder patterns, depending on their composition and structural state, the procedure for their identification and determination of composition and/or structural state requires more experience and precise measurement of selected parts of each diffractogram. These methods are described in Chapter 7.

8. Simple mixtures of, say, not more than three mineral components may be identified by powder diffractograms or powder films. This becomes an art rather than a science and requires the comparison of all the lines of the mixture with lines of each of the assumed separate phases. If the individual phases are known, then a comparison of the mixture pattern with patterns taken of individual separate phases helps a great deal. Guesses about the components may have to be made and confirmed or rejected by identifying all the lines of each component in the mixture.

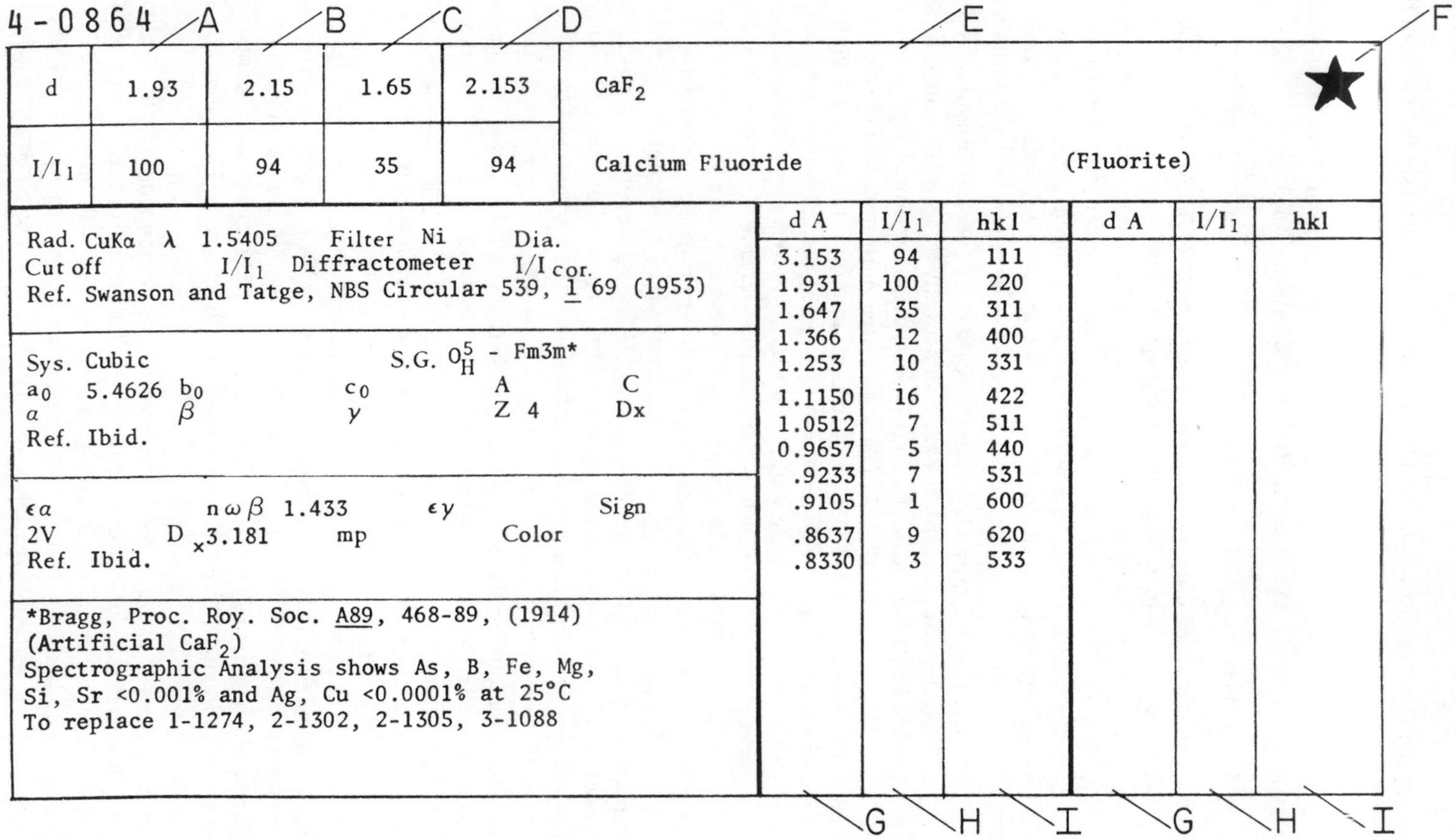

4-0864					
d	1.93	2.15	1.65	2.153	CaF$_2$
I/I$_1$	100	94	35	94	Calcium Fluoride

Rad. CuKα λ 1.5405   Filter Ni   Dia.
Cut off   I/I$_1$ Diffractometer   I/I cor. 69
Ref. Swanson and Tatge, NBS Circular 539, 1 69 (1953)

Sys. Cubic   S.G. O$_H^5$ - Fm3m*
a$_0$ 5.4626 b$_0$ c$_0$ A   C
α   β   γ   Z 4   Dx
Ref. Ibid.

εα   nωβ 1.433   εγ   Sign
2V   D$_x$3.181   mp   Color
Ref. Ibid.

*Bragg, Proc. Roy. Soc. A89, 468-89, (1914)
(Artificial CaF$_2$)
Spectrographic Analysis shows As, B, Fe, Mg,
Si, Sr <0.001% and Ag, Cu <0.0001% at 25°C
To replace 1-1274, 2-1302, 2-1305, 3-1088

(Fluorite)

d A	I/I$_1$	hkl	d A	I/I$_1$	hkl
3.153	94	111			
1.931	100	220			
1.647	35	311			
1.366	12	400			
1.253	10	331			
1.1150	16	422			
1.0512	7	511			
0.9657	5	440			
.9233	7	531			
.9105	1	600			
.8637	9	620			
.8330	3	533			

Fig. 6.15 An example of a card, 4-0864 FLUORITE, of the X-ray powder data file. See text of exercise 6.20 for description. By courtesy of Joint Committee on Powder Diffraction Standards, 1601 Park Lane, Swarthmore, Pennsylvania 19081.

Indexing the Powder Diffraction Lines

The function $Q = 1/d^2$ is frequently used in indexing X-ray diffraction lines. A table for conversion of d values to Q values from $d = 1.000$ to 10.000 can be found on pages 274 to 319 of Azároff and Buerger (1958). The relation between $Q = 1/d_{hkl}^2$ for each of the crystal systems is

System	$Q = \dfrac{1}{d_{hkl}^2} =$

CUBIC

$$\frac{1}{a^2}(h^2 + k^2 + l^2),$$

TETRAGONAL

$$\frac{h^2 + k^2}{a^2} + \frac{l^2}{c^2},$$

ORTHORHOMBIC

$$\frac{h^2}{a^2} + \frac{k^2}{b^2} + \frac{l^2}{c^2},$$

HEXAGONAL

(hexagonal indices)

$$\frac{4}{3a^2}(h^2 + hk + k^2) + \frac{l^2}{c^2}$$

(rhombohedral indices)

$$\frac{1}{a^2} \cdot \frac{(h^2 + k^2 + l^2)\sin^2\alpha + 2(hk + kl + lh)(\cos^2\alpha - \cos\alpha)}{1 - 2\cos^3\alpha + 3\cos^2\alpha},$$

MONOCLINIC

(first setting)

$$\frac{h^2/a^2 + k^2/b^2 - (2hk\cos\gamma)/ab}{\sin^2\gamma} + \frac{l^2}{c^2},$$

(second setting)

$$\frac{h^2/a^2 + l^2/c^2 - 2hl\cos\beta/ac}{\sin^2\beta} + \frac{k^2}{b^2}.$$

TRICLINIC

$$\frac{h^2}{a^2}\sin^2\alpha + \frac{k^2}{b^2}\sin^2\beta + \frac{l^2}{c^2}\sin^2\gamma$$

$$+ \frac{2hk}{ab}(\cos\alpha\cos\beta - \cos\gamma)$$

$$+ \frac{2kl}{bc}(\cos\beta\cos\gamma - \cos\alpha)$$

$$+ \frac{2lh}{ca}(\cos\gamma\cos\alpha - \cos\beta).$$

$$\overline{1 - \cos^2\alpha - \cos^2\beta - \cos^2\gamma + 2\cos\alpha\cos\beta\cos\gamma}$$

(Azároff and Buerger, 1958);

a, b, and c and α, β, and γ are the lattice constants and hkl are the Miller indices of the lattice planes which give rise to the X-ray powder reflections. A FORTRAN IV program to compute and tabulate d spacings, $Q(=1/d^2)$ values, 2θ and 4θ for any pre-scribed lattice, with a data input of cell parameters and X ray wavelength, may be obtained from Dr I. S. Kerr, Physical Chemistry Laboratories, Imperial College, London S.W.7, England. The program is suitable for an IBM 7090/94 or an IBM 360 machine (Cole and Villiger, 1969).

Indexing of powder patterns for minerals with low symmetry (orthorhombic, monoclinic, and triclinic) is difficult. Viswanathan (1968) has offered a systematic approach to the indexing of such powder patterns, using the computed Q values.

For cubic, tetragonal, and hexagonal crystals the graphical methods of Azároff and Buerger (1958, 61–77) should be used.

Nevertheless, for the vast majority of workers any mineral that is not cubic is considered too difficult to index with powder diffraction data alone. Hence, if it is known that the mineral is not cubic, it is generally recommended that the reader follow the indexing given on the appropriate J.C.P.D.S. powder data card.

Nevertheless, if a, b, and c and α, β, and γ are known from single crystal data for any particular mineral, the mineral's diffraction lines may be indexed by applying the appropriate equation given above for the various systems or the computer program of Cole and Villinger (1969) may be used.

Indexing of a cubic mineral is straightforward and a diffractogram of a cubic mineral can allow us to quote a in Å with a high degree of accuracy, especially if high angle lines are used in which the $K\alpha_1$ and $K\alpha_2$ doublets are resolved.

The recommended method for cubic mineral indexing is as follows:

1. To index the powder pattern of a cubic mineral first determine the interplanar spacings d_1, d_2, d_3, d_4, ..., d_n Å, where d_1 is the largest spacing (first line on a dif-fractogram) observed, and d_2, d_3, d_4, etc., are the successively smaller spacings in the correct order as they occur.

2. Next compute

$$\frac{d_1{}^2}{d_1{}^2}, \frac{d_1{}^2}{d_2{}^2}, \frac{d_1{}^2}{d_3{}^2}, \frac{d_1{}^2}{d_4{}^2}, \ldots, \frac{d_1{}^2}{d_n{}^2}.$$

3. This set of squared quotients will be discovered in one of the columns in Table 6.1. Of course not all planes in any column will be present because of possible extinc-tion (see exercise 6.22). The column thus identified then permits an indexing of the lines of the diffraction record by inspection.

(Method after Bloss, 1969.)

DISCUSSION

Graphical indexing of powder diffractograms for cubic minerals becomes less satis-factory for hkl reflections in which $N = (h^2 + k^2 + l^2)$ is large in value. This method indexes the high angle reflections as easily as it does those of lower 2θ angle.

Any method of indexing will result in ambiguities if the observed d values contain large experimental errors and h, k, l represent a reflection for which $N_1 = (h_1{}^2 + k_1{}^2 + l_1{}^2)$ is relatively large in value. As an example of the use of Table 6.1, fluorite (J.C.P.D.S. card 4-0864) has the following lines in increasing 2θ and decreasing d, Å order: $d_1{}^2$, which $= 9.9414$, is divided in turn by each d^2 to give column 2. A search

Table 6.1 Sets of d_1^2/d^2 Quotients for Indexing Powder Diffraction Patterns of Isometric Crystals. (after Bloss, 1969).

hkl for d	Lattice Type*	True reflection index corresponding to d_1													
		100 (1)	110 (2)	111 (3)	200 (4)	210 (5)	211 (6)	220 (7)	300, 221 (8)	310 (9)	311 (10)	222 (11)	320 (12)	321 (13)	400 (14)
100		1													
110	1	2	1												
111	F	3	1.5	1											
200	1, F	4	2	1.33	1										
210		5	2.5	1.67	1.25	1									
211	1	6	3	2	1.5	1.2	1								
220	1,F	8	4	2.67	2	1.6	1.33	1							
300, 221		9	4.5	3	2.25	1.8	1.5	1.12	1						
310	1	10	5	3.33	2.5	2	1.67	1.25	1.11	1					
311	F	11	5.5	3.67	2.75	2.2	1.83	1.37	1.22	1.10	1				
222	1, F	12	6	4	3	2.4	2	1.5	1.33	1.20	1.09	1			
320		13	6.5	4.33	3.25	2.6	2.17	1.62	1.44	1.30	1.18	1.08	1		
321	1	14	7	4.67	3.5	2.8	2.33	1.75	1.56	1.40	1.27	1.17	1.08	1	
400	1, F	16	8	5.33	4	3.2	2.67	2	1.78	1.60	1.45	1.33	1.23	1.14	1
410, 322		17	8.5	5.67	4.25	3.4	2.83	2.12	1.89	1.70	1.55	1.42	1.31	1.21	1.06
411, 330	1	18	9	6	4.5	3.6	3	2.25	2.00	1.80	1.64	1.50	1.38	1.29	1.13
331	F	19	9.5	6.33	4.75	3.8	3.17	2.37	2.11	1.90	1.73	1.58	1.46	1.36	1.18
420	1, F	20	10	6.67	5	4	3.33	2.5	2.22	2.00	1.82	1.67	1.54	1.43	1.25
421	1	21	10.5	7	5.25	4.2	3.5	2.62	2.33	2.10	1.91	1.75	1.62	1.50	1.31
322	1	22	11	7.33	5.5	4.4	3.67	2.75	2.44	2.20	2.00	1.83	1.69	1.57	1.38
422	1, F	24	12	8	6	4.8	4	3	2.67	2.40	2.18	2.00	1.85	1.71	1.50
500, 430		25	12.5	8.33	6.25	5	4.17	3.12	2.78	2.50	2.27	2.08	1.92	1.79	1.56
510, 431	1	26	13	8.67	6.5	5.2	4.33	3.25	2.89	2.60	2.36	2.17	2.00	1.86	1.63
511, 333	F	27	13.5	9	6.75	5.4	4.5	3.37	3.00	2.70	2.45	2.25	2.08	1.93	1.69
520, 432		29	14.5	9.67	7.25	5.8	4.83	3.62	3.22	2.90	2.64	2.42	2.23	2.07	1.81

* In this column, 1 indicates that the reflection is possible for a body-centered lattice; F for a face-centered lattice as well as a primitive lattice,

(continued)

Table 6.1 (Continued)

True reflection index corresponding to d_1

hkl for d	Lattice Type	100 (1)	110 (2)	111 (3)	200 (4)	210 (5)	211 (6)	220 (7)	300, 221 (8)	310 (9)	311 (10)	222 (11)	320 (12)	321 (13)	400 √(14)
521	1	30	15	10	7.5	6	5	3.75	3.33	3.00	2.73	2.50	2.31	2.14	1.88
440	1, F	32	16	10.67	8	6.4	5.33	4	3.56	3.20	2.91	2.67	2.46	2.29	2.00
522, 441	1	33	16.5	11	8.25	6.6	5.5	4.12	3.67	3.30	3.00	2.75	2.54	2.36	2.06
530, 433	1	34	17	11.33	8.5	6.8	5.67	4.25	3.78	3.40	3.09	2.82	2.62	2.43	2.13
531	F	35	17.5	11.67	8.75	7	5.83	4.37	3.89	3.50	3.18	2.92	2.69	2.50	2.19
600, 442	1, F	36	18	12	9	7.2	6	4.5	4.00	3.60	3.27	3.00	2.76	2.57	2.25
610	1	37	18.5	12.33	9.25	7.4	6.17	4.62	4.11	3.70	3.36	3.08	2.85	2.64	2.31
611, 532	1	38	19	12.67	9.5	7.6	6.33	4.75	4.22	3.80	3.45	3.17	2.92	2.71	2.38
620	1, F	40	20	13.33	10	8	6.67	5	4.44	4.00	3.64	3.33	3.08	2.86	2.50
443, 621, 540	1	41	20.5	13.67	10.25	8.2	6.83	5.12	4.56	4.10	3.73	3.42	3.15	2.93	2.56
541	1	42	21	14	10.5	8.4	7	5.25	4.67	4.20	3.82	3.50	3.23	3.00	2.63
533	F	43	21.5	14.33	10.75	8.6	7.17	5.37	4.78	4.30	3.91	3.58	3.31	3.07	2.69
622	1, F	44	22	14.67	11	8.8	7.33	5.5	4.89	4.40	4.00	3.67	3.38	3.14	2.75
630, 542	1	45	22.5	15	11.25	9	7.5	5.62	5.00	4.50	4.09	3.75	3.46	3.21	2.81
631	1	46	23	15.33	11.50	9.2	7.67	5.75	5.11	4.60	4.18	3.83	3.54	3.29	2.88
444	1, F	48	24	16	12	9.6	8	6	5.33	4.80	4.36	4.00	3.69	3.43	3.00
700, 632	1	49	24.5	16.33	12.25	9.8	8.17	6.12	5.44	4.90	4.45	4.08	3.77	3.50	3.06
710, 550, 543	1	50	25	16.67	12.50	10	8.33	6.25	5.56	5.00	4.55	4.17	3.85	3.57	3.13
711, 551	F	51	25.5	17	12.75	10.2	8.5	6.37	5.67	5.10	4.64	4.25	3.92	3.64	3.19
640	1, F	52	26	17.33	13	10.4	8.67	6.5	5.78	5.20	4.73	4.33	4.00	3.71	3.25
720, 641	1	53	26.5	17.67	13.25	10.6	8.83	6.62	5.89	5.30	4.82	4.42	4.08	3.79	3.31
721, 633, 552	1	54	27	18	13.50	10.8	9	6.75	6.00	5.40	4.91	4.50	4.15	3.86	3.38
642	1, F	56	28	18.67	14	11.2	9.33	7	6.22	5.60	5.09	4.67	4.31	4.00	3.50
722, 544	1	57	28.5	19	14.25	11.4	9.5	7.12	6.33	5.70	5.18	4.75	4.38	4.07	3.56
730	1	58	29	19.33	14.5	11.6	9.67	7.25	6.44	5.80	5.27	4.83	4.46	4.14	3.63

(*continued*)

Table 6.1 (Continued)

		True reflection index corresponding to d_1							300,						
hkl for d	Lattice Type	100 (1)	110 (2)	111 (3)	200 (4)	210 (5)	211 (6)	220 (7)	221 (8)	310 (9)	311 (10)	222 (11)	320 (12)	321 (13)	440 (14)
731, 553	F	59	29.5	19.67	14.75	11.8	9.83	7.37	6.56	5.90	5.36	4.92	4.54	4.21	3.69
650, 643		61	30.5	20.33	15.25	12.2	10.17	7.62	6.78	6.10	5.55	5.08	4.69	4.36	3.81
732, 651	1	62	31	20.67	15.5	12.4	10.33	7.75	6.89	6.20	5.64	5.17	4.77	4.43	3.88
800	1, F	64	32	21.33	16	12.8	10.67	8	7.11	6.40	5.82	5.33	4.92	4.57	4.00
810, 740, 652	1	65	32.5	21.67	16.25	13	10.83	8.12	7.22	6.50	5.91	5.42	5.00	4.64	4.06
811, 741, 554		66	33	22	16.50	13.2	11	8.25	7.33	6.60	6.00	5.50	5.08	4.71	4.13
733		67	33.5	22.33	16.75	13.4	11.17	8.37	7.44	6.70	6.09	5.58	5.15	4.79	4.19
820, 644	1, F	68	34	22.67	17	13.6	11.33	8.5	7.56	6.80	6.18	5.67	5.23	4.86	4.25
821, 742		69	34.5	23	17.25	13.8	11.5	8.62	7.67	6.90	6.27	5.75	5.31	4.93	4.31
653	1	70	35	23.33	17.5	14	11.67	8.75	7.78	7.00	6.36	5.83	5.38	5.00	4.38
822, 660	1, F	72	36	24	18	14.4	12	9	8.00	7.20	6.55	6.00	5.54	5.14	4.50
830, 661	1	73	36.5	24.33	18.25	14.6	12.17	9.12	3.11	7.30	6.64	6.08	5.62	5.21	4.56
831, 750, 743	1	74	37	24.67	18.5	14.8	12.33	9.25	8.22	7.40	6.73	6.17	5.69	5.29	4.63
751, 555	F	75	37.5	25	18.75	15	12.5	9.37	8.33	7.50	6.82	6.25	5.77	5.36	4.69
662	1, F	76	38	25.33	19	15.2	12.67	9.5	8.44	7.60	6.91	6.33	5.85	5.43	4.75
832, 654		77	38.5	25.67	19.25	15.4	12.83	9.62	8.56	7.70	7.00	6.42	5.92	5.50	4.81
752	1	78	39	26	19.5	15.6	13	9.75	8.67	7.80	7.09	6.50	6.00	5.57	4.88
840	1,F	80	40	26.67	20	16	13.33	10	8.89	8.00	7.27	6.67	6.15	5.71	5.00
900, 841, 744, 663		81	40.5	27	20.25	16.2	13.5	10.12	9.00	8.10	7.36	6.75	6.23	5.79	5.06
910, 833	1	82	41	27.33	20.5	16.4	13.67	10.25	9.11	8.20	7.45	6.83	6.31	5.86	5.13
911, 753	F	83	41.5	27.67	20.75	16.6	13.83	10.37	9.22	8.30	7.55	6.92	6.38	5.93	5.19
842	1, F	84	42	28	21	16.8	14	10.5	9.33	8.40	7.64	7.00	6.46	6.00	5.25
920, 760		85	42.5	28.33	21.25	17	14.17	10.62	9.44	8.50	7.73	7.08	6.54	6.07	5.31
921, 761, 655	1	83	43	28.67	21.50	17.2	14.33	10.75	9.56	8.60	7.82	7.17	6.62	6.14	5.38
664	1, F	88	44	29.33	22	17.6	14.67	11	9.78	8.80	8.00	7.33	6.77	6.29	5.50

d,Å	$\dfrac{d_1{}^2}{d^a}$	Column (3) hkl	$h^2 + k^2 + l^2$ N	a, Å $= d\sqrt{N}$
3.153	1.00	111	3	5.461
1.931	2.67	220	8	5.462
1.647	3.67	311	11	5.462
1.366	5.33	400	16	5.464
1.253	6.33	331	19	5.462
1.1150	8.00	422	24	5.462
1.0512	9.00	511, 333	27	5.462
0.9657	10.66	440	32	5.463
0.9233	11.66	531	35	5.462
0.9105	11.99	600, 442	36	5.463
0.8637	13.33	620	40	5.462
0.8330	14.33	533	43	5.462

for the values of column 2 above shows that they occur only in column 3 of Table 6.1 Right away the appropriate Miller symbols can be given and it can also be seen that the lines present could indicate a face-centered lattice. As a final check on the accuracy of the indexing the Miller symbol hkl is converted to $N = (h^2 + k^2 + l^2)$ and, since the mineral is cubic, the unit cell edge a in Å is calculated for each line from the relationship $a = d\sqrt{N}$. It can be seen from the fluorite example that all the Miller symbol allocations give rise to an a which is within ± 0.001 of a mean value 5.462 Å, thereby indicating how good this indexing method is. Any anomalously calculated value for a would indicate a wrong allocation of Miller symbol, provided of course that the d value was known to be accurate.

EXERCISE 6.22

Deductions Regarding the Lattice Type

Not all possible lines are present in an X-ray diffractogram. Possible reflections are systematically absent in certain minerals and these systematic absences can be used to tell something about the type of lattice that characterizes the mineral.

The following *systematic absences* indicate the *lattice type*:

Absence	Lattice Type
hkl reflections absent when $(k + l)$ is odd	— A centered
hkl reflections absent when $(h + l)$ is odd	— B centered
hkl reflections absent when $(h + k)$ is odd	— C centered lattice
hkl reflections absent when $(h + k + l)$ is odd	— I centerd lattice
hkl reflections absent when indices are mixed odd and even	— F centered lattice

If a rhombohedral lattice is indexed by using Miller-Bravais axes with indices of the type $(hkil)$, then only reflections with $(-h + k + l) = 3n$ or $(h - k + l) = 3n$ will be present (n is an integer).

$hk0$ absent when h is odd   — glide of $\dfrac{a}{2} \perp c$

$hk0$ absent when k is odd   — glide of $\dfrac{b}{2} \perp c$

$hk0$ absent when $(h + k)$ is odd — glide of $\dfrac{a + b}{2} \perp c$

and similar rules applying to 0kl and h0l reflections. Also

$h00$ absent when h is odd — screw axes parallel to a
$0k0$ absent when k is odd — screw axes parallel to b
$00l$ absent when l is odd — screw axes parallel to c.

(Zussman, 1967b)

The powder method may allow one, by carefully noting these systematic absences, to tell something about the lattice symmetry. However such a study is usually reserved for single crystal diffraction, which is not included in this book.

EXERCISE 6.23

Special Powder Diffraction Camera Techniques

Two camera techniques deserve special mention:

1. The Guinier-De Wolff Quadruple Focusing Camera (N. V. Verenigde Instrumenten Fabrieken, Enraf-Nonius, Roentgenweg 1, P.O. Box 483, Delft, Holland). The camera can simultaneoulsy record the powder diffraction patterns from $2\theta = 3$ to $90°$ of four separate samples on a single film. A crystal monochromator is incorporated to give extremely high angular precision. The dispersion is equivalent to a Debye-Scherrer camera of 229.2 mm diameter. Bragg angles are obtained with high precision by measuring the distance from the incident beam line and angular reflection. This distance is 4 mm/2θ. Because of the high angular precision the camera has found wide application for clay minerals. The simultaneous record of four patterns makes the camera ideal for comparison of specimens, which may have been glycolated or heat treated, with selected standards. The measurement of the films is facilitated by using a Guinier Viewer (Enraf-Nonius, Delft).

2. The High Temperature Guinier Lenne Camera (Enraf-Nonius, Delft, Holland). This camera is capable of producing an X-ray powder pattern of a mineral in the range 2θ 3 to $90°$ while the specimen is being continuously heated from room temperature to a maximum of $1200°C$. The developed film shows continuous changes in the X-ray pattern with temperature by a bidimensional film recording device. The temperature at which structural changes occur can therefore be monitored and compared with, for example, differential thermal data. The crystalline phases that are stable within certain temperature ranges may thus be identified by their powder patterns.

APPLICATION OF X-RAY POWDER DATA TO SPECIFIC MINERAL GROUPS

7

X-ray determinative methods have been successfully applied to a number of mineral groups. These methods are extremely valuable and are widely used to give information that cannot be obtained by any other method, or can be had much more quickly by the X-ray powder method. The details vary with mineral group and the characteristic under determination. Chemical composition can generally be determined in an isomorphous series from the measurement of one or more easily identifiable X-ray reflections. Structural data, hence some indication of the crystallization temperature and cooling history of a mineral, can be determined in some cases.

A. FELDSPARS

Many granites or gneisses are sufficiently coarse grained for small fragments of fairly pure alkali-feldspar to be handpicked from a rock specimen that has been crushed coarsely with a hammer. The occurrence of the alkali-feldspar as phenocrysts in granite or augen in gneiss and as large crystals in pegmatite further assists the easy separation of small pure alkali-feldspar fragments.

With hardly any exception, alkali-feldspars in plutonic and metamorphic rocks are perthitic on one of several scales—*perthitic* when the potassium- and sodium-rich phases are visible as discrete lamellae or patches to the naked eye, *microperthitic* when the exsolution lamellae are visible only under the microscope, and *cryptoperthitic* when the existence of two crystalline phases in an alkali-feldspar crystal is detected only by X-ray methods. These perthitic alkali-feldspars have exsolved from a homogeneous single-phase sanidine, which existed at high temperatures in a plutonic rock and the exsolution occurred in the solid state during cooling. During metamorphism a homogeneous sanidine phase would never have been attained except in metamorphism of extremely high temperature and the majority of metamorphic alkali-feldspars would normally be perthitic on growth.

Extensive use has been made of X-ray methods in the study of feldspars, especially in granitic and metamorphic rocks. Tuttle and Bowen (1958) were among the first workers to make frequent use of X-ray methods to determine the composition of alkali-feldspars.

Figure 7.1 (after Kuellmer, 1959) shows that if we take an X-ray powder diffractogram of a perthite (micro- or cryptoperthite) we will obtain two superimposed patterns—one from the microcline (or orthoclase) (potassic component) and the other from the albite (sodic component). We confine our interest to the range of 2θ 21

180

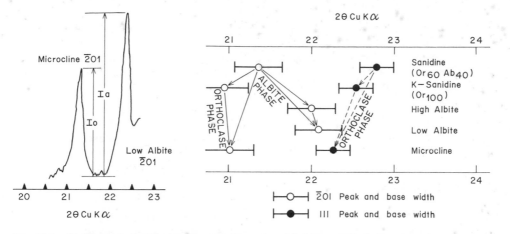

Fig. 7.1 (*Left*) Typical diffractogram of the region 2θ 21° to 23° CuKα of a microcline-low albite perthite (After Kuellmer, 1960.) for bulk composition Or$_{31}$Ab$_{69}$% by weight. The background intensity is measured at the mid-point between the two peaks. (*Right*) Variation of $\bar{2}01$ reflections of alkali-feldspars as a result of unmixing and symmetry change (after Kuellmer, 1959). *Peak* refers to the 2θ position measured on a powder goniometer, and *base width* to the angular range of the reflection at the background level. Homogeneous sanidine has only one $\bar{2}01$ reflection.

to 23° for CuKα radiation and are especially concerned with the $\bar{2}01$ reflections. Compared with the perthite, homogeneous (nonperthitic) sanidine will have only the one $\bar{2}01$ reflection. Figure 7.1 reveals that powder diffraction studies of $\bar{2}01$ spacings are not capable of distinguishing two kinds of potassium-rich feldspar phases or two or more sodium feldspar phases under most experimental conditions; the peaks of these individual reflections are so close together that in most cases they would not be resolved. The position of the $\bar{2}01$ line is proportional to the composition for a homogeneous feldspar (Tuttle and Bowen, 1958). Unmixing of the originally homogeneous alkali-feldspar should, if complete, produce two phases with compositions of nearly pure albite and nearly pure orthoclase (or microcline) (Tuttle and Bowen, 1958). The $\bar{2}01$ spacings of two such nearly pure alkali-feldspars would differ by about 0.20 Å or 1.04° 2θ CuKα radiation (Fig. 7.1). When the two phases in a perthite differ by this maximum possible amount, it follows that the composition of the individual perthite lamellae should be close to that of the end members albite and microcline. Accordingly, when a perthite has only two $\bar{2}01$ X-ray reflections and the difference between them is 0.19 Å (1.0° 2θ CuKα) or greater, the perthite behaves as if it were a mixture of two separate end-member components. When a perthite has only two $\bar{2}01$ reflections and they are closer together than 0.19 Å (1.0° 2θ CuKα), it may still be regarded as a two-component mixture, but of some intermediate components, not of the pure end members.

EXERCISE 7.1

Determination of Intensity Ratio

Extract an alkali-feldspar from a granite or a pegmatite. The hand specimen is coarsely crushed with a hammer. The alkali-feldspar may be handpicked. Careful handpicking and chipping with a chisel, with further separation under a binocular microscope, constitute the best method. If there is doubt about the distinction between alkali-feldspar and plagioclase, sodium cobaltinitrite staining may be used (see Chapter 2). Note that careless staining may allow the plagioclase to take up the yellow color in addition to the alkali-feldspar: alkali-feldspar generally takes on a smooth canary-yellow coating, whereas plagioclase, if it should stain, shows an irregular discontinuous color distribution because the color tends to occur along cracks and altered areas. If staining has been used to assist the visual separation of alkali-feldspar from plagioclase, the stain will have to be removed from the sorted material by immersing the alkali-feldspar grains in 1 N hydrochloric acid for 10 minutes at room temperature before washing and drying.

The most careful separation cannot avoid the inclusion of small grains of plagioclase and quartz in the alkali-feldspar separation. These grains may have occurred as crystal inclusions in the larger alkali-feldspar phenocrysts. Care should be taken to discard any alkali-feldspar that is known to have inclusions, but a small proportion of the inclusions will give such weak diffraction peaks that they will not interfere with the experiment.

The alkali-feldspar is crushed to the appropriate grain size for powder diffraction, usually to less than 325 mesh. Slurry the powder with acetone or with an acetone-durofix mixture, spread evenly over a glass slide, and mount on the goniometer.

Alternatively, a better mount with less preferred orientation of the grains is obtained by tamping the powder once into a metallic sample holder.

Scan at approximately $\frac{1}{8}°$ 2θ per minute, using 200 counts per full chart scale (or other sensitivity based on experience). The time constant is 4 seconds; there are $1°$ slits and 0.003 in. receiving slit. The chart speed is 15 in. or 300 mm/hr. These settings are only a guide and can be varied with experience.

The $\bar{2}01$ peaks which occur between 20.95 and 22.4° 2θ CuKα should be traversed in both directions and the angular positions averaged. The peak positions are determined accurately by locating the midpoint of the X-ray peaks at the $\frac{2}{3}$ peak height on either side to an accuracy of 0.01 in. on the chart.

The goniometer is now centered accurately on one peak at a time and the $\bar{2}01$ peak intensities measured by using a fixed time of 10 seconds with a 2-second time constant at the 2θ position of the peak. Count each peak three times and average the counts, ignoring any that is obviously significantly different from the other two on the same peak. In this way we can exclude a spurious count caused by interference in the amplifier: never rely on a single count in isolation. After the intensities of the two peaks have been counted, the goniometer is set at the midpoint between the two peak counting positions of the orthoclase- and the albite-rich $\bar{2}01$ peaks.

From the data

A = counts/10 seconds on microcline $\bar{2}01$,

B = counts/10 seconds on low albite $\bar{2}01$,

C = counts/10 seconds on the background (midpoint 2θ between the two peaks). The intensity ratio I_0/I_a (see Fig. 7.1) is calculated as $(A - C)/(B - C)$. Keep this value of I_0/I_a and proceed to exercise 7.2 for determination of the alkali feldspar bulk composition. With both values proceed to exercise 7.4, in which the temperature-structural state of the feldspar may be obtained.

EXERCISE 7.2

Determination of Bulk Composition of Alkali Feldspar

Crush the separated alkali-feldspar in a mortar with pestle and sieve. Discard the material retained on the 100 mesh but regrind it if it is still rich in alkali-feldspar. If any muscovite impurity were present, it would normally be retained in the + 100 mesh fraction: it can be discarded. Collect and use the material on the 200 mesh sieve. If the alkali-feldspar is considered free from muscovite impurity, grains of 2 to 3 mm size are suitable for heating. Put a small amount of the powder, enough for several powder diffraction photographs, into a platinum crucible. Place a lid on the crucible and maintain at 1050° C in a muffle furnace for 48 hours. The furnace must have a good temperature controller to avoid overheating and fusion. Since most natural alkali-feldspars are perthitic, this treatment is necessary to cause complete homogenization as a sanidine monoclinic phase. The lid on the crucible is considered necessary to prevent potassium contamination from the walls of the muffle furnace. Some furnaces become so badly contaminated that the furnace lining may eventually have to be replaced.

Cool the crucible by placing in the air, then grind enough of the treated powder in a mortar to a grain size suitable for powder photography. Add an approximately equal volume of potassium bromate ($KBrO_3$) as an angular standard. Mix thoroughly, and prepare a specimen for the X-ray goniometer.

Copper $K\alpha$ (nickel-filtered) radiation is used. The goniometer is run at a suggested scanning rate of $1/8°$ per minute, with a source aperture of $1°$ and a detector aperture of 0.2 mm. Scan from 19 to 23° 2θ Cu$K\alpha$.

Indexing of the alkali-feldspar diffractogram may be facilitated by reference to the calculated diffractogram patterns of Borg and Smith (1969b). Identify the $\bar{2}01$ reflection of the feldspar and the 101 reflection of the $KBrO_3$. The difference $\Delta 2\theta$ ($\bar{2}01$ feldspar $-$ 101 $KBrO_3$) for Cu$K\alpha$ radiation is plotted on the diagram (Fig. 7.2) (after Orville,

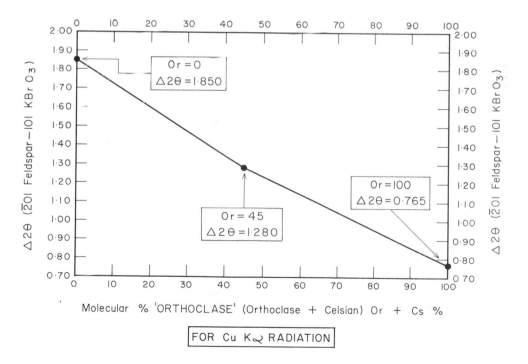

Fig. 7.2 Homogenized synthetic microcline-low albite X-ray determinative curve (Orville, 1967). The difference between $2\theta^o_{\bar{2}01}$ alkali-feldspar and $2\theta_{101}KBrO_3$ for Cu$K\alpha$ radiation is plotted against molecular percent Or + Cs. This value will be $\approx$Or%, since Cs is normally small. For conversion to absolute 2θ values of feldspar $\bar{2}01$ lines for Cu$K\alpha$ add 20.212° to each $\Delta 2\theta$ value. The graph is of two straight-line components $Or_{0 \text{ to } 45}$ and $Or_{45 \text{ to } 100}$.

1967) and the molecular percentage of the orthoclase molecule in the alkali-feldspar can be read off the graph.

If this experiment is done for the first time, the following are the recommended preliminary steps:

a. Make a powder X-ray film or diffractogram of pure $KBrO_3$ with $CuK\alpha$ radiation and keep it as a reference. Index the lines, using J.C.P.D.S. card 7-242, and note particularly the position of the first line 101 at $d = 4.39$ Å ($2\theta = 20.23°$ $CuK\alpha$); $KBrO_3$ is chosen as a useful standard because this line does not interfere with the $\bar{2}01$ line of the feldspar and there are no further lines until 012, $d = 3.21$ ($2\theta = 27.79°$ $CuK\alpha$).

b. Make a powder X-ray film or diffractogram of the homogenized feldspar and

Table 7.1 List of the First Six X-ray reflections of Potassium Bromate and the first 16 Lines of a Monoclinic Homogeneous Sanidine, for comparison of the 2θ $CuK\alpha$ Positions, relative Intensity, and Indexing of the Lines[a]

Potassium Bromate $KBrO_3$ J.C.P.D.S. Card 7-242 All measurements in degrees 2θ for $CuK\alpha$ radiation						Monoclinic Sanidine J.C.P.D.S. Card 13-456 47.6% $KAlSi_3O_8$: 42.4% $NaAlSi_3O_8$: 10.2% $CaAl_2Si_2O_8$			
$2\theta°$ (α_1)	α	(α_2)	d Å	I/I_1	hkl	$2\theta°$ α	d Å	I/I_1	hkl
	20.23 or 20.22		4.39	60	101	13.64	6.49	16	110, 020
						15.88	5.798	8	$\bar{1}11$
						21.40	4.152	45	$\bar{2}01$
						22.78	3.904	10	111
						23.68	3.758	30	200
						23.73	3.750	70	130
						24.73	3.600	12	$\bar{1}31$
						25.48	3.496	6	$\bar{2}21$
						25.79	3.455	40	$\bar{1}12$
						27.37	3.259	100	202, 040
(27.76)	27.79	(27.83)	3.21	100	012				
						27.71	3.219	90	002
						29.00	3.079	6	not indexed
(29.67)	29.70	(29.75)	3.008	70	110	30.02	2.977	35	131
						30.81	2.902	25	$\bar{2}22$, 041, +
						32.39	2.764	20	$\bar{1}32$
(32.92)	32.95	(33.01)	2.718	10	003	35.02	2.562	30	$\bar{3}12$, $\bar{2}41$, +
(36.16)	36.19	(36.25)	2.482	2	021				
(41.07)	41.10	(41.17)	2.196	49	202	Numerous other weak lines			
	Numerous other weak lines								

[a] The positions of the $KBrO_3$ lines form a good internal standard for alkali-feldspar because of their invariability of position. On the other hand, the positions of the alkali-feldspar lines will vary with composition, but this table should allow identification of any diffractogram by comparison of the relative positions and intensities. Please note, however, that line intensities can never be absolutely reproduced and may vary by up to $\pm 40\%$ relative.

keep it as a reference for future work. Index it after careful comparison with a monoclinic alkali-feldspar (sanidine) in the J.C.P.D.S. powder data file (card 13-456).

By keeping these two records as standards the 2θ range required for scanning on the goniometer for subsequent experiments can be readily predicted.

It would also be a useful first exercise to have taken a pattern of the mixed homogenized feldspar + $KBrO_3$ and to index all the lines by comparison with the records for the unmixed feldspar and $KBrO_3$ separately. The lines arising from the feldspar could be colored differently from those from the $KBrO_3$ to make future differentiation easy. In this way we can become confident about identifying the lines of the mixture without making any allocation mistakes.

Table 7.1 lists some important data regarding $KBrO_3$ and sanidine, which will be found useful in such a study; 2θ CuKα for 101 $KBrO_3$ has been given different values by various authors, namely, 20.23, 20.22, 20.212, and 20.19. Because of the lack of agreement, the reader would be well advised to measure his own $KBrO_3$ values, using silicon as a reliable standard, and not to rely implicitly on published values.

DISCUSSION

The determinative graph in Fig. 7.2 (Orville, 1967, p. 75) is for common plutonic or metamorphic microclines (microcline-low albite solid solution). It is based entirely on synthetic microclines. Jones et al. (1969) have carried out a similar study with natural

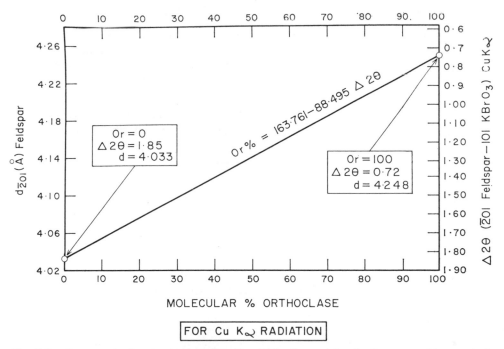

Fig. 7.3 Homogenized natural microcline-low albite X-ray determinative curve (Jones et al. 1969). Either $d_{\bar{2}01}$(Å) feldspar or the difference between $\bar{2}01$ feldspar and 101 $KBrO_3$ in degrees 2θ CuKα is plotted against molecular % Or. In this angular calibration, 2θ CuKα $KBrO_3$ 101 is taken as 20.19 ($d_{101} = 4.3985$ Å) after Jones et al. (1969). The straight-line graph can be drawn from the values Or % = 0 at $d_{\bar{2}01}$ 4.033 Å, $\Delta2\theta$ CuKα 1.85; Or % = 100 at $d_{\bar{2}01}$ 4.248 Å $\Delta2\theta$ CuKα 0.72.

microclines, homogenized for 48 hours at 1050°C and scanned at $\frac{1}{4}°$ per minute with a 0.1 mm receiving slit and 40 mm per degree of chart drive; KBrO$_3$ was used as an internal standard. Their results differ slightly but significantly from those of Orville (1967), but because they agree fairly closely with those of Wright (1968), also based on natural microclines, they are shown here on Fig. 7.3 as what might be considered a good determinative curve for homogenized low-temperature (plutonic and meta-morphic) alkali-feldspars. Jones et al. (1969) suggest that barium may be the cause of the difference between the curves of Figs. 7.3 and 7.2. Hence they suggest that when read from the curve of Fig. 7.2 the composition should be quoted as ORTHOCLASE + CELSIAN (Or + Cs)% instead of simply Or%. Figure 7.3 plots Or% against $d_{\bar{2}01}$ of the homogenized alkali-feldspar determined directly on a well-calibrated goniometer.

From the direct $d_{\bar{2}01}$, $Or(\omega t\%) = 465.5 \times d_{\bar{2}01}$ (Å) $- 1877.5$. The K$_2$O weight percentage is also shown. Equivalent percentage Or $= \%K_2O \times 5.91$. Incomplete homogenization of perthites will cause the $\bar{2}01$ spacing method for estimating bulk composition of perthites to give results eroneously high in albite (Parsons, 1968).

To facilitate the use of this diagram I have recalculated the values of Jones et al. (1969) and recast them as a plot of Or% against $\Delta 2\theta$ ($\bar{2}01$ feldspar $- 101$ KBrO$_3$) for CuKα radiation. This conversion is based on a d_{101} of KBrO$_3$ of 4.3985 Å (Jones et al., 1969). In this case the regression equation is Or% $= 163.716 - 88.495 \times \Delta 2\theta$.

EXERCISE 7.3

Determination of Composition of Sanidine from Volcanic Rocks

When determining the composition of sanidine from volcanic rocks (e.g., a trachyte), from high-level granite that is known to contain high-temperature monoclinic homogeneous alkali-feldspar (ascertained from optical studies), or from the high-temperature contact sanidinite facies of metamorphism, the procedure is identical to that outlined in exercise 7.2.

Use should be made of the determinative graph in Fig. 7.4 (after Orville, 1963). This graph has been constructed from synthetic sanidine-high albite alkali-feldspars and as far as I am aware no complete study has yet been made on the determination of natural high-temperature alkali-feldspars by the X-ray method. Until such time Fig. 7.4 will have to suffice. There is little difference between this graph (Fig. 7.4) and that of Fig. 7.2 (Orville, 1967), but the difference is more pronounced at high values of Or% in the range 60 to 100%; Or% determinations should be accurate within 2 mol% by this method with careful X-ray measurement.

EXERCISE 7.4

Determination of Temperature-Structural State of Alkali-Feldspar

In exercise 7.1 we determined the intensity ratio I_0/I_a for the $\bar{2}01$ peaks of the perthitic feldspar. In Exercise 7.2 we determined the bulk composition in Or% of the same feldspar after homogenization. Now plot I_0/I_a against Or% on a diagram such as Fig. 7.5 (after Kuellmer, 1959).

In general it is found (Kuellmer, 1960) that for a rock which crystallized under placid conditions (no regional deformation) for any bulk composition Or%, the lower the temperature-structural state of the feldspar, the lower the $\bar{2}01$ I_0/I_a intensity ratio.

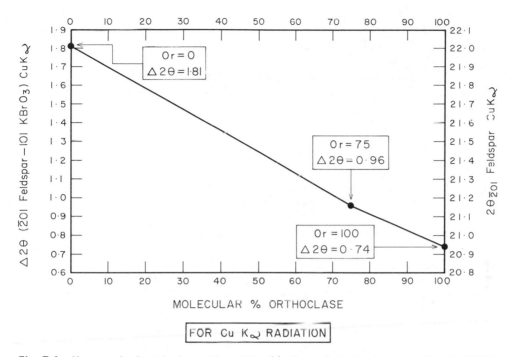

Fig. 7.4 Homogenized synthetic sanidine-high albite X-ray determinative curve (Orville, 1963). The difference in 2θ between $\bar{2}01$ for alkali-feldspar and 101 for $KBrO_3$ for $CuK\alpha$ is plotted against the molecular Or %. For conversion to absolute values of $2\theta_{\bar{2}01}^{\circ}$ feldspar for $CuK\alpha$ add 20.212° to each 2θ value. The graph is of two straight-line components, $Or_{0\,to\,75}$ and $Or_{75\,to\,100}$.

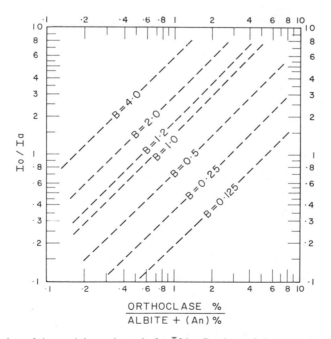

Fig. 7.5 Variation of the peak intensity ratio for $\bar{2}01$ reflections of the potassic and sodic phases in a perthite (I_o/I_a) with the bulk composition Or % for numerous values of the reflection broadening ratio B (dashed lines). (After Kuellmer, 1959.)

From the plots of Or% against I_0/I_a on Fig. 7.5 a broadening ratio (Kuellmer, 1959) can be read off. X-ray reflection broadening indicates relative grain-size distortion or structural mistakes in the two phases of the perthite. It is expressed by B (Fig. 7.5).

The line-broadening ratio B is an indication of the structural state of the perthite. With the same composition Or%, the smaller the broadening ratio B, the lower the temperature-structural state.

DISCUSSION

The relationship between the Or% and the I_0/I_a ratio for a suite of feldspar specimens collected from different parts of a single pluton might be expressed as a perthite unmixing law.

Thus, if all results from X-ray analyses of perthites from an intrusive body are plotted on this kind of diagram (Fig. 7.5), we might be able to construct isograd-like zones. Parts of a pluton might be compared, one with another, or one single pluton might be compared with another of a different tectonic setting. Differences in the temperature-structural state of the perthites will show up in the distribution of points in a diagram like Fig. 7.5. The Kuellmer parameter, however, is a poor substitute for direct measurement of perthite member compositions before homogenization and comparison with experimental solvi or the Wright-Stewart method (exercise 7.6) of accurately characterizing the composition and structural state of alkali-feldspar.

EXERCISE 7.5

Determination of the Triclinicity of Alkali-Feldspar

Monoclinic sanidine has a single 131 reflection. In triclinic orthoclase or microcline the 131 peak is replaced with a doublet 131 and $1\bar{3}1$. With increase in Al/Si disorder in the alkali-feldspar the spacings of hkl and $h\bar{k}l$ approach each other and eventually coalesce as a single hkl reflection when the mineral becomes monoclinic.

Goldsmith and Laves (1954) observed that the maximum value obtained for $d_{131} - d_{1\bar{3}1}$ is slightly less than 0.08 Å. Therefore for convenience they proposed a scale as follows:

TRICLINICITY

$$\Delta = 12.5 \, (d_{131} - d_{1\bar{3}1}).$$

Triclinicity expressed in this manner would mean that $\Delta = 1$ fully ordered (maximum microcline), that is, the most triclinic microcline possible. $\Delta = 0$ fully disordered; the alkali-feldspar is monoclinic. Make a diffractogram from $CuK\alpha$ 29 to 31° at $\frac{1}{8}$° 2θ per minute. Identify the peaks by reference to Tables 7.2, 7.3, and 7.4.

Carefully measure the 2θ values for 131 and $1\bar{3}1$ of an alkali-feldspar and convert the values of d Å by using a set of conversion tables. Then calculate the triclinicity with the equation given above. There is no need for an internal angular standard, for the difference between two closely neighboring peaks is used in the calculation.

DISCUSSION

It will be expected that alkali-feldspars with triclinicity values near unity will have crystallized slowly under plutonic conditions in deep-seated granites and in most metamorphic rocks.

Alkali-feldspars with low triclinicity values will be expected to have cooled quickly and their presence would indicate a volcanic or a high-level subvolcanic environment. Border zones of a pluton may contain feldspar of lower triclinicity than interior zones of the same pluton because of more rapid cooling.

EXERCISE 7.6

Determination of Structural State of Alkali-Feldspar

The following procedure is recommended for measurement of unknown feldspar X-ray patterns by Wright and Stewart (1968) and by Wright (1968):

1. A standard diffractogram should be run on a powder smear of untreated alkali-feldspar from 2θ 13 to $52°$ Cu$K\alpha$.

2. The position and shape of the $\bar{2}01$ reflection(s) should be checked to determine whether the feldspar consists of a potassic phase, a sodic phase, or both (see Fig. 7.1) and whether the phase(s) has a limited or broad range of composition (i.e., whether the peaks are sharp or broad). The $\bar{2}01$ reflection will be sharp for samples in which the Or% range is 5% or less. Samples with greater Or% variation will show broadening of the top of the $\bar{2}01$ reflection.

3. If a potassic phase is present, the 131 reflection should be checked. If it is a single sharp reflection, the potassic phase may be indexed in monoclinic symmetry with the help of Table 7.3 (after Wright and Stewart, 1968). If 131 is split into two peaks, 131 and $1\bar{3}1$, the feldspar is triclinic, and if both reflections are sharp they may be indexed with the help of Table 7.2 (after Wright and Stewart, 1968). Sometimes both monoclinic and triclinic phases may appear in the sample or the 131 reflections will be broad and indistinct. This indicates that the structural state is highly variable in the sample, in which case this method cannot be used.

4. Albite in a perthite intergrowth may be indexed by reference to Table 7.4 (after Wright and Stewart, 1968). In most cases albite will not yield enough unique reflections for the determination of its structural state; however, the structural state of the albite in a perthite may be inferred by determining the structural state of the perthite's potassic phase.

5. Natural anorthoclases, including those with an appreciable anorthite content, may be indexed by referring to Table 7.3 (after Wright and Stewart, 1968).

6. Patterns for measurement should be run three times with an internal standard (to check the accuracy of the 2θ readings), from high to low values of 2θ with a chart speed that yields approximately $1° 2\theta = 1$ in. (or 2 cm) on the chart paper. The wavelength Cu$K\alpha_1$ with $\lambda = 1.5405$ Å should be used. Measurement of each peak should be as near the top of the peak as practicable. The indexing of the peaks should be done with the help of Tables 7.2 to 7.4, as appropriate.

7. The five reflections $\bar{2}01$, 002, $\bar{1}13$ (if present), 060, and $\bar{2}04$ should be identified and measured carefully. Table 7.5 will assist in this indexing (after Wright, 1968). The $\bar{2}01$ and the 060 reflections may be indexed unambiguously in all natural feldspars. The $\bar{2}04$ reflection is strong and single in most orthoclase, microcline, anorthoclase, and albite samples. In sanidine or orthoclase with anomalous cell dimensions $\bar{2}04$ is commonly either overstepped by or joined by one or more additional reflections.

8. If reflections in the vicinity of $\bar{2}04$ are sharp, 2θ values of $\bar{1}13$ and 002 reflections may be used to make an unambiguous identification of the reflection whose index is $\bar{2}04$ by the method in step 9.

Table 7.2   CuKα2θ Guide to the Indexing of Microcline Powder Photographs

	Maximum Microcline			Intermediate Microcline	
	Observed $2\theta°$ (range)	Approximate Intensity	Miller Index	Observed $2\theta°$ (range)	Approximate Intensity (range + average)
	About 13.1	8	110		
	About 13.6	10	$\bar{1}10$		
			001	Approximately similar to	
			020	maximum microcline in this	
	About 14.9	8	$\bar{1}\bar{1}1$	range; $\bar{2}01$ is identified as the	
	About 15.2	8	$\bar{1}11$	first strong line	
	About 19.25	5	021		
			$0\bar{2}1$		
****	20.96–21.04	35	$\bar{2}01$	21.00–21.05	30–65 (45)
	22.29–22.38	20	111	22.46–22.56	15–30 (20)
	22.61–22.69	10	$1\bar{1}1$	About 22.55	10
	23.18–23.24	25	130	23.44–23.51	20–40 (30)
	23.97–24.02	30	$\bar{1}30$	23.67–23.73	20–40 (30)
	24.26–24.37	10	$\bar{1}\bar{3}1$	About 24.5	10
	24.70–24.77	10	$\bar{2}\bar{2}1$		
	About 24.9	10	$\bar{1}31$	About 24.8	15
	25.53–25.56	40	$\bar{1}\bar{1}2$	25.62–25.70	40–100 (65)
	About 25.6	10	$\bar{2}21$		
	About 25.7	<10	112		
	26.40–26.47	40	220	26.71–26.80	40–70 (50)
	27.04–27.10	40	$\bar{2}02$	27.07–27.11	60–100 (80)
	About 27.05	10	$\bar{2}20$		
****	27.44–27.52	>100	002	27.50–27.59	>100
**	29.42–29.55	30	131	29.72–29.82	25–45 (40)
**	30.12–30.24	30	$1\bar{3}1$	29.92–30.01	15–45 (35)
	30.72–30.82	30	041	30.81–30.87	25–65 (40)
	About 30.75	<10	$\bar{2}22$		
			$0\bar{2}2$		
			$0\bar{4}1$		
	About 32.1	10	$\bar{1}32$	32.25–32.50	15
	About 32.3	10	$\bar{3}11$		
	32.44–32.47	15	$\bar{1}\bar{3}2$	32.37	15
	34.19–34.27	15	$\bar{2}\bar{4}1$		
	About 34.3	10	221		
	About 34.6	15	$\bar{3}12$		
			$\bar{3}\bar{1}2$	About 34.5	15

Continued

190

Table 7.2 Concluded

	Maximum Microcline			Intermediate Microcline	
	Observed $2\theta°$ (range)	Approximate Intensity	Miller Index	Observed $2\theta°$ (range)	Approximate Intensity (range + average)
			$\bar{2}41$	About 34.65	15
	34.88–35.03	10	112	About 35.1	25
			$\bar{2}41$		
	35.1	10	$2\bar{2}1$		
	35.48–35.57	15	$\bar{2}41$		
	About 35.9	6	$\bar{3}10$		
	About 36.8	10	$\bar{2}40$		
	36.98–37.00	10	$\bar{1}50$		
****	about 38.6	10	$\bar{1}13$	38.61–38.76	10–20 (15)
****	41.78–41.85	30	060	41.72–41.79	15–45 (30)
			241	About 42.7	10
			$2\bar{4}1$	About 42.8	10
			311	About 43.7	10
			$3\bar{3}1$		
	About 45.5	8	222		
			$\bar{4}22$		
	About 45.6	8	$\bar{4}21$	About 44.2	10
	About 46.2	8	$2\bar{2}2$		
	About 48.7	8	113	48.96–49.10	15
****	50.51–50.66	25	$\bar{2}04$	50.58–50.78	20–40 (30)
			$0\bar{4}3$		
	About 50.9	10	$0\bar{6}2$		

**** Lines used in exercise 7.6.
**   Lines used in exercise 7.5.
Figure in parenthesis under intensity is an average value.

9. To identify $\bar{2}04$ when more than one reflection occurs between 50° and 51° 2θ $CuK\alpha$

$$2\theta_{\bar{2}04} = 1.1780\,(2\theta_{\bar{1}13}) + 5.1048 \text{ to within } \pm 0.0202\,2\theta_{\bar{2}04}.$$

Also, as a double check,

$$2\theta_{\bar{2}04} = 1.6886\,(2\theta_{002}) + 4.1690 \text{ to within } \pm 0.0317\,2\theta_{\bar{2}04}\,CuK\alpha$$

(Wright, 1968). Thus $\bar{2}04$ can be unambiguously identified from the positions of $\bar{1}13$ and 002. In rare cases in which reflections near to and including $\bar{2}04$ are fuzzy this method cannot be applied.

Table 7.3 Cu$K\alpha$ 2θ Guide to the Indexing of Orthoclase, Sanidine, and Anorthoclase in Powder Diffractometry

	Orthoclase, Sanidine, and Homogeneous alkali-Feldspar			Anorthoclase	
	Observed $2\theta°$ (range)	Approximate Intensity (average)	Miller Index	Calculated $2\theta°$ (range)	Average Intensity
	13.5–13.7	5–15 (10)	110		
			020	13.64–13.73	10
			001	13.79–13.88	5
	15.1	5	$\bar{1}11$	15.3–15.6	<5
			$0\bar{2}1$	19.0–19.3	<5
****	20.87–21.13	15–100 (35)	$\bar{2}01$	21.65–21.88	65
			$1\bar{1}1$	22.85	25
	22.52–22.64	10–40 (20)	111	23.11–23.56	35
	23.06–23.14	5–10 (8)	200		
			$\bar{1}30$	23.66	45
	23.54–23.64	25–100 (65)	130	23.90–24.30	40
			$\bar{1}31$	24.5–24.6	10
	24.48–24.69	10–25 (15)	$\bar{1}31$	24.85–25.3	5
	25.12–25.22	5–15 (10)	$\bar{2}21$		
			$\bar{1}\bar{1}2$	25.6–25.7	15
	25.65–25.82	25–100 (60)	$\bar{1}12$	25.9–26.3	20
	26.83–27.04	30–100 (70)	220		
	27.04–27.22	25–100 (55)	$\bar{2}02$		
	About 27.5	Less than (002)	040	27.48–27.66	>100
****	27.51–27.74	>100	002	27.78–27.96	>100
			220	27.75–28.3	40
**	29.82–29.97	25–100 (50)	131		
**			$1\bar{3}1$	29.64–29.91	25
			$0\bar{4}1$	30.32–30.60	20
			$0\bar{2}2$	30.54–30.81	25
**			131	30.50–31.28	25
	30.46–30.52	10–25 (15)	$\bar{2}22$		
	30.79–30.99	15–60 (25)	041	31.1–31.8	<5
	About 30.85	Usually 10	022		
			$\bar{1}32$	31.80–32.16	10
	32.31–32.46	10–40 (20)	$\bar{1}32$	32.75–33.48	5
	34.39–34.51	10–30 (15)	$\bar{3}12$		
			$\bar{2}41$	35.2–35.5	25
	34.80–34.91	10–60 (30)	$\bar{2}41$	35.25–35.6	25

Continued

Table 7.3 Concluded

	Orthoclase, Sanidine, and Homogeneous alkali-Feldspar			Anorthoclase	
	Observed 2θ° (range)	Approximate Intensity (average)	Miller Index	Calculated 2θ° (range)	Average Intensity
	35.10–35.22	10–45 (15)	112		
	35.58–35.74	5–15 (10)	310	37.0–37.5	<5
	36.14–36.22	5–10 (8)	240	36.9–37.6	<5
			$\bar{1}$51	About 37.1	5
	37.16–37.25	5–20 (10)	$\bar{1}$51	37.4–38.0	5
	37.64–37.78	5–15 (10)	$\bar{3}$31	38.7–39.0	5
****	38.60–39.00	5–15 (8)	$\bar{1}$13		
			$\bar{3}$31	38.7–39.2	<5
			1$\bar{5}$1	40.5–40.9	<5
	About 41.0	<10	151		
****	41.60–42.00	10–60 (30)	060	41.74–42.03	15
			$\bar{1}$52	42.2–42.7	<5
			151	41.6–42.7	5
	42.46–42.67	10–20 (12)	241		
			2$\bar{4}$1	42.61–42.86	10
	About 42.8	About 10	$\bar{4}$02		
			$\bar{4}$01		
	About 43.8	<10	202		
			311		
	44.04–44.14	5–15 (8)	061	44.5–45.3	<5
	45.06–45.29	5–15 (10)	$\bar{4}$22		
	About 45.1	<10	$\bar{4}$21		
	45.96–46.11	10–25 (15)	222	47.2–48.2	5
	About 47.3	<10	400		
	47.50–47.58	5–15 (10)	$\bar{4}$03	48.6–49.0	5
			400	48.9–49.6	5
	About 48.2	<10	260	48.9–49.7	<5
			1$\bar{1}$3	49.6–49.8	<5
			0$\bar{6}$2	49.7–50.4	5
	49.01–49.14	5–25 (10)	113	49.92–50.62	10
	About 50.5	10	$\bar{2}$62		
	About 50.6	10	170		
			$\bar{3}$50	About 50.6	<5
	50.63–50.75	10–40 (20)	062		
	50.78–50.88	10–20 (15)	043		
****	50.50–51.10	15–55 (30)	$\bar{2}$04	51.02–51.31	25

**** Lines used in exercise 7.6.
** Lines used in exercise 7.5.
Values in parenthesis under intensity indicate average values.

193

Table 7.4   CuKα 2θ Guide to the Indexing of Albite in Perthite

Miller Index	ALBITE (perthite bulk more sodic than Or₅₀Ab₅₀)		ALBITE (perthite bulk more potassic than Or₅₀Ab₅₀)	
	Observed $2\theta°$	Average Intensity	Observed $2\theta°$	Remarks
001	13.84–13.88	20		
020	About 13.85	1		
$\bar{2}01$	22.04–22.23	15	21.90–22.10	
$1\bar{1}1$	23.07–23.10	7	23.05–23.35	
111	23.54–23.57	25	23.50–23.60	
$\bar{1}31$	About 24.25	20		
$\bar{1}30$	24.26–24.31	15	24.05–24.35	May conflict with K phase $\bar{1}31$
$\bar{1}\bar{1}2$	About 25.45	10		
$\bar{2}\bar{2}1$	About 25.57	1		
112	26.42–26.44	7		
002	27.92–27.95	100	27.90–28.10	May conflict with 002 of highly distorted K-phases
040	About 28.1	50		
220	About 28.35	9		
$\bar{1}31$	30.12–30.18	9	30.10–30.20	Conflicts with $1\bar{3}1$ and $\bar{2}\bar{2}2$ of triclinic K phases
$0\bar{2}2$	About 30.51	15		
$0\bar{4}1$	About 30.52	16		
$\bar{1}31$	31.22–31.49	7	31.05–31.30	
$1\bar{3}2$	About 31.5	1		
041	About 32.2	1		
$\bar{1}32$	33.92–33.99	5	33.90–34.00	
$\bar{2}\bar{4}1$	35.00–35.06	7		
$\bar{3}12$	About 36.7	1		
$\bar{2}41$	About 36.8	3		
$\bar{3}\bar{3}1$	About 38.84	3		
$\bar{2}42$	About 41.3	3		
$1\bar{5}1$	About 41.3	4		
060	42.48–42.54	7	42.45–42.55	
151	About 42.8	5		
$2\bar{4}1$	43.56–43.60	1		
061	About 45.9	3		
222	48.13–48.19	7		
$0\bar{6}2$	50.00–50.10	5		
113	50.56–50.66	7		
$\bar{2}04$	51.12–51.19	10		

Table 7.5   Position of the Alkali-Feldspar Reflections Used in the Three-Peak Method of Exercise 7.6

Miller Index	Approximate Intensity	Approximate Range $2\theta°$ CuKα		
		Potassic phase	Anortho-clase	Albite
$\bar{2}01$	40	20.8–21.2	21.6–21.9	21.9–22.1
002	>100	27.4–27.8	27.8–28.0	27.9–28.1
$\bar{1}13$	8	38.6–39.0	Not present	Not present
060	25	41.6–42.0	41.7–42.0	42.2–42.6
$\bar{2}04$	30	50.5–51.1	51.1–51.3	51.2–51.5

10.   The 2θ CuKα values for 060 and $\bar{2}04$ are plotted one against the other in Fig. 7.6 (after Wright, 1968). From the point plotted read off the approximate 2θ value for $\bar{2}01$ from the cross contours of the diagram. If this value agrees within 0.1° 2θ with the actual measured value obtained directly from the scan, the feldspar may be assumed to have normal unit cell dimensions. If the $\bar{2}01$ 2θ value obtained from Fig. 7.6 exceeds the measured value by 0.1° 2θ or more, the feldspar may be termed anomalous.

11.   For normal feldspar the structural state is obtained directly from Fig. 7.6 in terms of the closest plotted feldspar of known structural state. The following names may be used (refer to Fig. 7.6):

HIGH SANIDINE: MONOCLINIC. The structural state as defined by the axial ratio $b:c$ corresponds to the high sanidine-high albite series.

LOW SANIDINE: MONOCLINIC. The $b:c$ ratio lies between the orthoclase and the high-sanidine-high albite series.

ORTHOCLASE: MONOCLINIC. The $b:c$ ratios fall along the line shown on Fig. 7.6.

INTERMEDIATE MICROCLINE: TRICLINIC, α, γ, b, and c are distinct from the maximum microcline-low albite series and points fall on Fig. 7.6 intermediate between the maximum microcline and orthoclase series.

For the soda end-members:

HIGH ALBITE. Unit cell parameters b and c correspond to the end-member of the high sanidine-high albite series.

INTERMEDIATE ALBITE. Parameters b and c fall between the high sanidine-high albite and the orthoclase series.

LOW ALBITE. Parameters b and c correspond to the sodic end-member of the maximum microcline-low albite series.

ANORTHOCLASE: TRICLINIC. The bulk composition must contain between 5 and 40% Or.

12.   Composition of the normal feldspars can be obtained once the structural state has been determined from Fig. 7.6 by using the following equations (Wright, 1968).

HIGH SANIDINE-HIGH ALBITE SERIES:

$$Or\% = 2030.05 - 92.18 \times 2\theta_{\bar{2}01}.$$

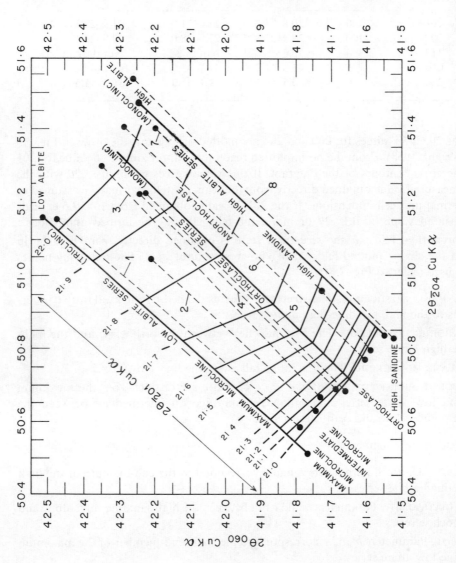

Fig. 7.6 Corresponding values of $2\theta_{\bar{2}04}$, $2\theta_{060}$, and $2\theta_{\bar{2}01}$ for alkali-feldspars using $CuK\alpha_1$ radiation (after Wright, 1968). The three major series—maximum microcline-low albite, orthoclase, and high sanidine-high albite—are shown in solid lines. Dashed lines are for specific alkali-feldspars which have been well studied. In order of increasing structural state they are (lowest) 1, Spencer U; 2, Spencer B; 3, low albite 111; 4, SH 1070; 5, Benson; 6, S62-34; 7, Puye; and 8, S63-30 (highest). (From Wright, 1968.)

196

ORTHOCLASE SERIES:

$$\text{Or} \% = 1930.77 - 87.69 \times 2\theta_{\bar{2}01}.$$

MAXIMUM MICROCLINE-LOW ALBITE SERIES:

$$\text{Or} \% = 2031.77 - 92.19 \times 2\theta_{\bar{2}01}.$$

13. For anomalous feldspars, the apparent structural state may be derived as it is for a normal feldspar; it is not yet known how accurate this determination will be. At present it can be only a guess, in which case the composition *cannot* be determined from $2\theta_{\bar{2}01}$.

14. As examples, a specimen with $2\theta_{\bar{2}01}$ 21.180, $2\theta_{060}$ 41.610, and $2\theta_{\bar{2}04}$ 50.840 from Fig. 7.6 can be shown to be a sanidine with a structural state between that of high sanidine and Puye equivalents. The composition is estimated from $\bar{2}01$ as Or 76%. A specimen with $2\theta_{\bar{2}01}$ 21.026, $2\theta_{060}$ 41.741, and $2\theta_{\bar{2}04}$ 50.675 is a slightly anomalous feldspar with an approximate structural state intermediate between equivalents SH 1070 and Spencer B. Composition cannot be accurately determined from $\bar{2}01$ and any value obtained would be little better than a guess. A specimen with $2\theta_{\bar{2}01}$ 20.940, $2\theta_{060}$ 41.891, and $2\theta_{\bar{2}04}$ 50.775 is highly anomalous. The apparent structural state is close to that of Spencer B. Its composition cannot be estimated.

DISCUSSION

The three-peak method outlined above will identify and describe alkali-feldspars encountered routinely in petrological studies. Wright (1968) and Wright and Stewart (1968) give further details for computing the unit cell dimensions from data derived from this method; for most petrological work, however, this refinement will not be necessary.

EXERCISE 7.7

Determination of Plagioclase Structural State

Separate some plagioclase from a rock and prepare a diffraction smear on a glass slide. Use $KBrO_3$ or silicon to calibrate the goniometer 2θ readings properly. Scan with $CuK\alpha$ radiation by using divergent and scatter slits of $1°$ and a receiving slit of 0.006 in., with a scanning rate of $\frac{1}{4}°$ per minute and chart speed of 2 in. (or 2 cm) or equivalent per degree. The appropriate 2θ region may be scanned two or three times and the values averaged arithmetically. Readings are made to the nearest $0.01° 2\theta$. The scanning range is 27 to $37° 2\theta$ $CuK\alpha$. The peaks to be identified are $1\bar{3}1$, 131, $24\bar{1}$, and $\bar{2}41$. These peaks can be readily identified by reference to Fig. 7.7 (after Smith and Yoder, 1956) or to a similar diagram in Bambauer et al. (1967, p. 337).

All the peaks of interest lie on the high-angle side of the 100% intensity composite plagioclase peak 002 to 040, which is the most prominent feature of a plagioclase diffractogram (shown in Fig. 7.7 going off scale). These peaks are chosen as suitable in that (a) they are clearly resolved from neighboring reflections, (b) their 2θ values are sensitively related to composition, and (c) their reflections are not too far apart (hence chart-scale errors are minimized).

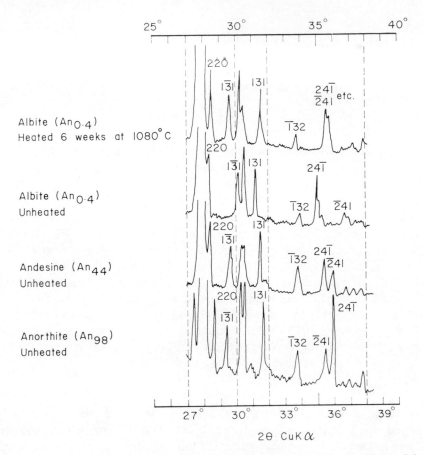

Fig. 7.7 Powder diffraction patterns for plagioclases showing the relative positions of the $1\bar{3}1$, 131, $24\bar{1}$, and $\bar{2}41$ peaks for high-temperature albite, low-temperature albite, andesine, and anorthite.

The following are the positions of the significant reflections:

Index	$2\theta°\mathrm{Cu}K\alpha$ Low-Temperature Albite	High-Temperature Albite	Anorthite	Remarks
$1\bar{3}1$	30.1	29.6	29.4	May or may not be resolved in perthitic (in homogeneous) low-temperature plagioclase in range An_6 to An_{17}
131	31.2	31.6	31.6	Resolved from An_0 to An_{100}.
$24\bar{1}$	34.9	35.7	35.9	Resolved in all but high-temperature albite and in the region An_{75} to An_{85}; $24\bar{1}$ is at a lower angle than $\bar{2}41$ except for anorthite, where $24\bar{1}$ is at a higher angle than $\bar{2}41$
$\bar{2}41$	36.7	35.8	35.5	

NOTE. If the J.C.P.D.S. cards are used in the indexing of plagioclase, line $24\bar{1}$ may equally well appear indexed as $\bar{2}41$ on the cards. Furthermore, in anorthite (J.C.P.D.S. card 12-301) line $1\bar{3}1 = 1\bar{3}2$, $131 = 132$, $\bar{2}41 = \bar{2}42$, and $24\bar{1} = \bar{2}4\bar{2}$.

No internal standard is required for the measurement of $\Delta(\theta)_1 = 2\theta_{131} - 2\theta_{1\bar{3}1}$.

Determine from the diffractogram $\Delta(\theta)_1 = 2\theta_{131} - 2\theta_{1\bar{3}1}$ and $\Delta(\theta)_2 = 2\theta_{\bar{2}41} - 2\theta_{24\bar{1}}$ for $\mathrm{Cu}K\alpha_1$. Note that $\Delta(\theta)_2$ values pass through zero somewhere between labradorite and bytownite; the algebraic difference must therefore be used. Albite to labradorite has a $+ve$ value for $\Delta(\theta)_2$ and bytownite and anorthite have $-ve$ values. These values are related to composition and the structural state of the plagioclase.

Bambauer et al. (1967) used the following convention for structural state: D-plagioclase = plagioclase with disordered Al, Si distribution. O-plagioclase = plagioclase with ordered Si, Al distribution, and O/D-plagioclase = plagioclase of intermediate state. High-temperature plagioclase possesses the highest possible stable state of Al, Si disorder. It is summarized as *plagioclase* (*high*). For a given chemical composition it possesses a maximum value for $\Delta(\theta)_1$ and a minimum value for $\Delta(\theta)_2$.

Low-temperature plagioclase constitutes the feldspars with stable states of highest possible Al, Si ordering. They are summarized as *plagioclase* (*low*) and possess a minimum $\Delta(\theta)_1$ and a maximum $\Delta(\theta)_2$ value for a given chemical composition.

Figure 7.8 shows a plot of $\Delta(\theta)_1 = 2\theta_{131} - 2\theta_{1\bar{3}1}$ for $\mathrm{Cu}K\alpha_1$ against chemical composition of the plagioclase. The top scale is the Si/Al ratio in the plagioclase and has been plotted as a linear scale, for the lattice geometry bears a linear relation to Si/Al rather than to An content (Bambauer et al., 1967). The bottom scale shows its equivalence to An (mol)%.

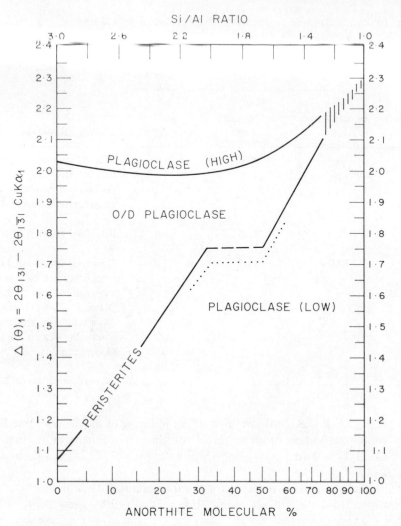

Fig. 7.8 Relationship between $\Delta(\theta)_1$ and plagioclase composition. Solid lines are for Or% < 0.5 to 0.8. Dotted lines are for Or ≃ 4% (molecular) in the plagioclase. (After Bambauer et al., 1967.) $\Delta(\theta)_1 = 2\theta_{131} - 2\theta_{1\bar{3}1}$ for copper $K\alpha_1$ radiation.

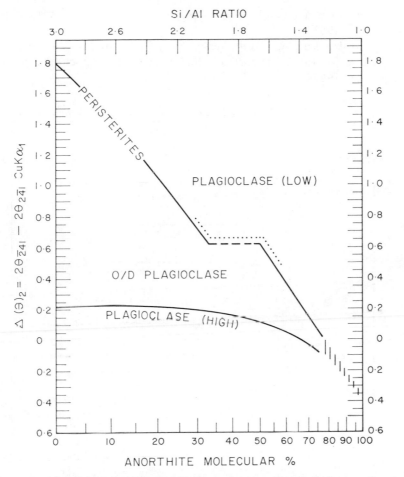

Fig. 7.9 Relationship between $\Delta(\theta)_2$ and plagioclase composition. Solid lines are for plagioclase with Or% less than 0.5 to 0.8 mole%. Dotted lines are for Or mole% $\simeq$ 4. (After Bambauer et al, 1967.) $\Delta(\theta)_2 = 2\theta_{\bar{2}41} - 2\theta_{24\bar{1}}$ for copper $K\alpha_1$ radiation.

Figure 7.9 shows a plot of $\Delta(\theta)_2 = 2\theta_{\bar{2}41} - 2\theta_{24\bar{1}}$ CuKα_1 versus composition (Bambauer et al., 1967).

Figure 7.10 shows a plot of $\Delta(\theta)_1$ versus $\Delta(\theta)_2$ CuKα_1 and the anorthite contents along this curve for the O-plagioclases, that is, the low-temperature plagioclases (Bambauer et al., 1967).

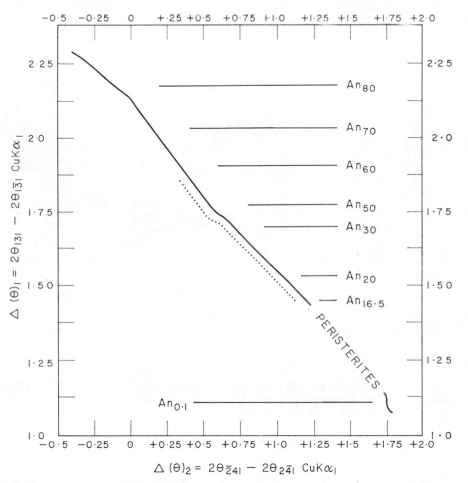

Fig. 7.10   Relationship between $\Delta(\theta)_1$, $\Delta(\theta)_2$, and plagioclase composition. The solid lines are for O-plagioclases with Or < 0.5 to 0.8 mol%. The dotted line is for Or ≃ 4%. (After Bambauer et al, 1967.) 2θ(Cu Kα_1) for 131, 1$\bar{3}$1, 24$\bar{1}$, and $\bar{2}$41 are used.

DISCUSSION

X-ray data cannot be used to determine the exact composition of a plagioclase. If, however, the composition is known by chemical analysis, universal stage measurements, or refractive index measurements, the X-ray method may supply some information regarding the structural state of the plagioclase.

The composition and structural state of albite and oligoclase may be determined from the optical orientation, with confirmation, if desired, from the refractive indices and optic axial angle. It should be noted that if the plagioclase is high the 131 parameter offers no significant compositional information.

For andesine and labradorite the An content may be estimated from refractive indices or from optic orientation; the structural state can be estimated only in a crude manner from the optic orientation.

For bytownite and anorthite the An content may be estimated accurately from refractive indices or from optical properties, but there is little hope of deducing the structural state from optical properties.

The variation of X-ray properties with An content shows that (a), if the An content is known by other methods, as, for example, by exercise 7.8, the structural state may be determined from $\Delta(\theta)_1$ and $\Delta(\theta)_2$, and (b), if the structural state is known by petrographic information, X-ray methods may be used to estimate the composition.

In Figs. 7.8 and 7.9 the curve plagioclase (high) is for high-temperature plagioclase from volcanic or subvolcanic rocks or for natural low-temperature plagioclases which have been heated for considerable time. The plagioclase (low) curve is for low structural state plagioclase that occurs in granite, pegmatite, and plutonic igneous rocks in which cooling has generally been slow. Usually the plagioclase from a volcanic rock has an intermediate structural state (O/D plagioclase) which shows that it has partly changed in structural state from its original disordered condition. Hence we should not expect any particular plagioclase to lie exactly on either of the (high) or (low) curves in Figs. 7.8 and 7.9; hence these curves cannot be used to estimate chemical composition.

Although the composition of the plagioclase generally must be known first before the graphs can be used, it will nevertheless be seen that a $\Delta(\theta)_1$ value less than 1.6 and a $\Delta(\theta)_2$ value greater than 1.0 almost conclusively prove a low structural state without knowledge of the chemical composition.

The following features characterize the curves in Figs. 7.8 and 7.9 (Bambauer et al., 1967):

1. The boundary curve for plagioclase (high) is continuous at least to Si/Al 1.25 (An_{36}). At about An_{76-80} there may be a discontinuity.

2. The boundary curve for plagioclase (low) shows the following sequence:

a. Si/Al 3.0 to 2.97 ($An_{0 \text{ to } 1.3}$); linear increase in $\Delta(\theta)_1$ and decrease in $\Delta(\theta)_2$. This represents the stability region of *Low albite*.

b. Si/Al 2.97 to 2.45 ($An_{1.3 \text{ to } 16}$); the peristerite gap. It is commonly believed that no homogeneous members of the plagioclase (low) series occur in this region. Plagioclase is present as peristerite containing lamellae of two different plagioclase compositions. This is the region of the moonstones, which show schiller light interference in hand specimen.
Peristerites. It is usually impossible to examine both components of peristerites by means of powder methods and often even the identification of a peristerite is difficult. For a well-resolved peristerite we find two patterns representing albite (low) and oligoclase (low) in the diffractogram. In most cases, however, the lines of one or both phases are broad and diffuse. Because of this, a persisterite may fall outside the plagioclase field as shown in Fig. 7.8.

c. Si/Al 2.45 to 2.02 ($An_{16 \text{ to } 33}$); linear increase in $\Delta(\theta)_1$ and decrease in $\Delta(\theta)_2$.

d. Si/Al 2.02 to 1.67 ($An_{33 \text{ to } 50}$); the continuation is uncertain; it is probably horizontal.

e. Si/Al 1.67 to at least 1.25 ($An_{50 \text{ to } 76}$); linear increase in $\Delta(\theta)_1$ and decrease in $\Delta(\theta)_2$.

f.   Si/Al 1.25 to 1.0 ($An_{76\ to\ 100}$); there is a discontinuity around $An_{70\ to\ 80}$. An unequivocal distinction of plagioclase (high) and plagioclase (low) is not really possible in this region with $\Delta(\theta)$ values. In addition to the Si, Al order/disorder process, the "districtive transformation" occurs here.

The curves also include dotted lines for plagioclase containing up to 4 mol% Or, and, as can be seen, the K content of a plagioclase has a significant effect on the position of the curve. It should be clear that not only the An content but also the isomorphously incorporated Or content of the plagioclase must be known in order to analyze it structurally on the basis of $\Delta(\theta)$ values. For this purpose $\Delta(\theta)_1$ is more sensitive than $\Delta(\theta)_2$.

If it could be known with certainty that the structural state of a plagioclase is low, that is, if its point fell on or very near the plagioclase (low) boundary curve of Figs. 7.8 and 7.9, then Fig. 7.10 could be used to estimate its composition. The structural state of any plagioclase, however, cannot be assumed. Intrusions that have cooled slowly over millions of years may contain O/D-plagioclase of various degrees of ordering. Hence there is no possibility of meaningful composition estimation by X-ray methods. In Fig. 7.10 $\Delta(\theta)_1$ is plotted against $\Delta(\theta)_2$ for plagioclase (low) with Or content less than 0.5 to 0.8 mol %. This graph may be used for the majority of plutonic plagioclases.

The following specimen types do not lie on this curve:

1.   Those with a higher Or content. The values lie to the left of the curve for Or contents less than Or 0.5 to 0.8. The curve for Or approximately equal to 4 mol % is shown dotted. Thus it is possible to estimate the Or content of a plagioclase by the amount of its deviation from the curve.

2.   Some O/D-plagioclases lie to the right of the curve.

In summary, all plagioclases (unless authigenic) are metastable at room temperature. High plagioclase is the least stable (most metastable). Plagioclases falling on the boundary curves are either ORDERED (plagioclase low) or DISORDERED (plagioclase high). Those falling within the boundary lines are called O/D-plagioclases; that is, they are of intermediate order/disorder structural state. A quantitative subdivision into grades of order or disorder should be avoided because no quantitative relation has yet been found between $\Delta(\theta)$ valued and the actual Al, Si distribution with the exception of albite. Therefore use only D-plagioclase if the value lies on or near the plagioclase (high) curve, O-plagioclase if it lies on or near the plagioclase (low curve), and O/D-plagioclase for all intermediate plagioclases.

Or contents of more than 1 mol% in plagioclase should not be ignored because of their significant effect on the $\Delta(\theta)$ values.

Provided these analytical and theoretical limitations are observed, relative statements on the structural state of different plagioclases may be made with the help of the diagrams in Figs. 7.7 to 7.10 and in some favorable cases the plagioclase composition in terms of An% and Or% may be estimated.

EXERCISE 7.8

Determination of Plagioclase Composition

A rather elegant method of studying plagioclases in which the plagioclase separated from a rock is converted to its potassium equivalent by cation exchange has been introduced by Viswanathan (1971). The method is relatively simple and consists of the following steps:

1. The plagioclase must be crushed and passed through a 325 U.S. standard sieve (grains of $< 44\mu$).

2. About 30 mg of the sample to be analyzed is mixed with approximately 900 mg of KCl in a platinum crucible. The lid is put on and the crucible is held in a muffle furnace at 850°C for 16 hours.

3. After cooling, the powder is washed with water to remove the excess KCl and dried.

4. A diffraction smear is made of the powder (ground further if necessary) to which about 1 mg of silicon has been added as internal angular standard.

5. Determine first whether any K-plagioclase has been formed. An examination of the 2θ CuKα_1 region 26.30 to 28.6 will allow the strongest alkali feldspar peaks, 220, $\bar{2}02$, $\bar{2}20$, 040, and 002, to be identified by reference to Tables 7.2 and 7.3.

6. If K-plagioclase has been formed, identify the $\bar{2}01$ peak (again by reference to Tables 7.2 and 7.3); 2θ CuK$\alpha_1\bar{2}01$, corrected by the internal standard, can be plotted at the top of Fig. 7.11 to give the composition of the original plagioclase to ±2 or 3% An.

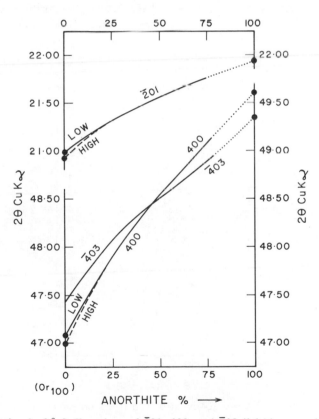

Fig. 7.11 Variation in 2θ CuKα values of $\bar{2}01$, 400, and $\bar{4}03$ K-feldspar peaks with anorthite content of the cation exchanged plagioclase. (After Viswanathan, 1971.)

7. With the An$\%$ value obtained from $\bar{2}01$ use Fig. 7.11 to find the approximate position of K-feldspar peaks 400 and $\bar{4}03$, which lie in the 2θ region 47° to 49.6 CuKα_1. With this knowledge, and with the help of Tables 7.2 and 7.3 locate the exact position of these two almost equally intense peaks on the diffractogram. With the precise 2θ

values of 400 and $\bar{4}03$, the An% of the plagioclase can be determined to an accuracy of $\pm 1\%$ An. The two lines, 400 and $\bar{4}03$, overlap in the compositional range 35 to 55% An and form a broad line which reduces the accuracy. At composition An 46%, however, they coincide exactly in a sharp single peak with accompanying high accuracy.

DISCUSSION

If no K-feldspar peaks showed in the diffractogram, the K cation exchange must have been far from complete. Incomplete exchange results from coarse grain size of the plagioclase and high anorthite content greater than An_{50}, in which case the powdered and sieved plagioclase should be ground further in an agate mortar before heating with KCl. Provided the maximum grain size does not exceed about 1 μ, complete cation exchange should be obtained within 16 hours of heating with KCl. Nevertheless, it is not necessary to cause complete exchange, for a partial exchange will produce the same K-feldspar. Our only requirement is that enough of the plagioclase be exchanged to allow identification of the K-feldspar peaks.

A knowledge of the approximate structural state is necessary if compositions in the range 0 to 15% An have to be determined accurately. The positions of $\bar{2}01$ and 400 are influenced by the structural state. Plot the 2θ values of 060 and $\bar{2}04$ in Fig. 7.12

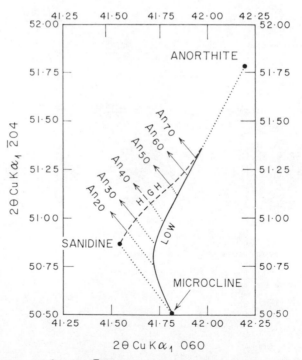

Fig. 7.12 Correlation of 2θ CuKα $\bar{2}04$, with 060 for K cation exchanged plagioclases. (After Viswanathan, 1971.)

and from them the structural state of a sodic plagioclase can be approximately estimated. The 060 line is much stronger than adjacent lines (see Tables 7.2 and 7.3), hence can be identified easily. The $\bar{2}04$ line can generally be distinguished from the neighboring less intense reflections $0\bar{4}3$ and $0\bar{6}2$ except near the approximate composition Or_{87}/An_{13}, where they all overlap.

If there is difficulty in identifying $\bar{2}04$, its approximate value can be obtained from Fig. 7.12 by plotting 2θ 060 and reading the $\bar{2}04$ value corresponding to the anorthite content determined previously with $\bar{2}01$ or 400 in Fig. 7.11.

B. GARNET GROUP

Determination of Garnet Composition

Because garnet cannot be specifically identified as any particular species in thin section, the need for further petrographic work is always necessary, more so than for any other mineral group. Parameters that should be determined are refractive index (N), which is relatively easy because the mineral is cubic, density (G) which should be found on a small pure fragment with the help of a Berman density balance, and unit cell edge (a) in Å by using a powder camera or a diffraction goniometer.

Since many garnets contain appreciable concentrations of iron, it is recommended that cobalt radiation generally be used for garnet X-ray analysis to prevent the high background that results from the fluorescence radiation of iron by copper radiation.

EXERCISE 7.9

Measurement of the Unit Cell Edge a in Å of a Garnet

Extract a fragment of garnet from a rock. Crush and make a cylindrical preparation for a camera or a smear with acetone for a goniometer. If the goniometer is used, add an almost equal quantity of potassium bromate as a 2θ standard. Scan carefully at a slow speed for $CoK\alpha_1$ 2θ 34 to 42°. Identify the 110 peak of the potassium bromate ($d_{110} = 3.008$, 2θ $CoK\alpha_1 = 34.60$). The garnet 420 peak is readily and unambiguously identified because it is the first and most intense. It is always given the intensity 100% on garnet diffractograms. Table 7.6 shows that this peak occurs at a 2θ $CoK\alpha_1$ angle in the region 38.41 to 40.86, depending on the composition of the garnet. Measure from the diffractogram the angular 2θ distance between 110 $KBrO_3$ and the garnet$_{420}$. Add 34.60 to this value and convert the total 2θ $CoK\alpha_1$ to a d value for garnet$_{420}$.

If the diffraction goniometer is well aligned, the $2\theta_{420}$ garnet line may be directly converted to d_{420} by using cobalt radiation tables without making use of the $KBrO_3$ standard.

Because garnet is cubic, $a^2 = d^2_{hkl}$ $(h^2 + k^2 + l^2)$. Therefore $a^2 = 20\ d^2_{420}$ and $a, \text{Å} = 4.472\ d_{420}$.

If a powder camera is used, it is recommended that a high-angle garnet line, usually 10 4 0 be used. This line appears at 2θ angles varying from 104.80 to 114.43 $CoK\alpha_1$. At these high angles the peak will appear as a well-resolved doublet; the $K\alpha_1$ line should be identified from the $K\alpha_2$ line and the proper conversion table used to find d, Å.

$$a^2 = 116\ d^2_{10\ 4\ 0}.$$

Therefore $a = 10.77\ d_{10\ 4\ 0}$. For identification of the garnet 10 4 0 line compare your powder photograph with J.C.P.D.S. cards 2–1008, 3–0801, 10–288, and 10–367. Guessing the correct line is not permissible. The line must be unambiguously found.

Table 7.6   Refractive Index, Specific Gravity, and Powder Diffraction Data on the Garnet End Members[a]

(all angles are for COBALT $CoK\alpha_1$ radiation, wavelength $\lambda = 1.7889$ Å)

Species[a]	N[b]	D	a, Å	d 420	2θ 420[c]	d 1040	2θ 1040[d]	Composition
Pyrope	1.714	3.582	11.459	2.5623	40.86	1.0639	114.43	$Mg_3Al_2Si_3O_{12}$
Almandine	1.830	4.318	11.526	2.5773	40.62	1.0701	113.47	$Fe_3Al_2Si_3O_{12}$
Spessartite	1.800	4.190	11.621	2.5985	40.27	1.0789	112.00	$Mn_3Al_2Si_3O_{12}$
Grossular	1.734	3.594	11.851	2.6499	39.46	1.1003	108.77	$Ca_3Al_2Si_3O_{12}$
Andradite	1.887	3.859	12.048	2.6940	38.78	1.1186	106.20	$Ca_3(Fe, Ti)_2Si_3O_{12}$
Uvarovite	1.86	3.848	12.00	2.6833	38.94	1.1141	106.80	$Ca_3Cr_2Si_3O_{12}$
Hydrogrossular	1.734—	3.594—	11.85—	2.6497—	39.46—	1.1002—	108.77—	$Ca_3Al_2Si_2O_8($
	1.675	3.13	12.16	2.7190	38.41	1.1290	104.80	$SiO_4)_{1-m}(OH)_{4m}$.

[a] Values generally after Skinner (1956). Uvarovite from J.C.P.D.S. card 11–696.

[b] Grossular, spessartite, andradite, and uvarovite may be slightly birefringent, in which case the refractive index given is only an average value.

[c] The 420 line is the first very dense line and cannot be mistaken; it is usually assigned a 100% intensity.

[d] For the position of the 1040 line see the example given in J.C.P.D.S. card 10-288.

Identification of the Garnet

To identify the garnet specifically it is necessary to determine, in addition to the a, Å, the refractive index N and the specific gravity G.

A garnet corresponding to one of the end members listed in Table 7.6 is rare, except in the case of hydrogrossular, which may be considered a separate species. It is readily identified from all other garnets by its low specific gravity and low refractive index. Any garnet is usually named after the most dominant end member in that particular crystal. Garnets are divided into two main series:

PYRALSPITE, between pyrope, almandine, and spessartite.
UGRANDITE, between uvarovite, grossular, and andradite.
HYDROGROSSULAR, a separate species.

It should be noted, however, that most garnets do not fit entirely into either of the two main series and it is not uncommon for the composition of a garnet to be described in terms of five end members, three from one series and two from the other.

Having determined a, N, and G, we can make use of the determinative diagrams of Winchell (1958) which have been redrawn and expanded here, following the values of synthetic end members given by Skinner (1956).

The a and the N values are the most reliable for garnet, for G is normally less exact because of the common occurrence of small inclusions. Care must be taken to exclude impurities and to determine G as accurately as possible. In Figs. 7.13, 7.14, and 7.15 more emphasis is put on the a and N values because of their superior accuracy.

Figure 7.13 shows pyrope (pyr), almandine (alm), grossular (gro), and andradite (and) with straight-line connections in pairs to represent chemical variations in the corresponding binary series.

Ticks along the lines are at 10 % intervals (compare with the triangular diagram in Fig. 13.1). Figure 7.14 also includes spessartite (sp). The composition point for uvarovite (uv) is shown in both figures.

The use of the diagrams is best described by taking an actual example. If a garnet has the determined parameters $a = 11.550$ Å, $N = 1.770$, and $G = 3.880$, its composition can be estimated by reference to Fig. 7.13. The intersection of the a and N values shows that the garnet lies within the triangle pyr-alm-gro. The ratio pyr:alm is found from the intersection of a line through the garnet point to the gro apex (point 1). This line cuts the base at pyr:alm = 46:54. The join from the alm apex through the composition point intersects the pyr/gro base at (point 2) pyr:gro = 72:28; the join from the pyrope apex indicates (point 3) the ratio alm:gro = 75:25.

From the example worked in Chapter 13 (Fig. 13.1) we get the garnet composition

$$\text{PYROPE } \% = \frac{100}{1 + 54/46 + 28/72} = \frac{100}{2.563} = 39.0\%,$$

$$\text{ALMANDINE } \% = \frac{100}{1 + 46/54 + 25/75} = \frac{100}{2.185} = 45.8\%,$$

$$\text{GROSSULARITE } \% = \frac{100}{1 + 72/28 + 75/25} = \frac{100}{6.571} = 15.2\%.$$

Therefore the garnet composition is $\text{pyr}_{39}\text{alm}_{46}\text{gro}_{15}$.

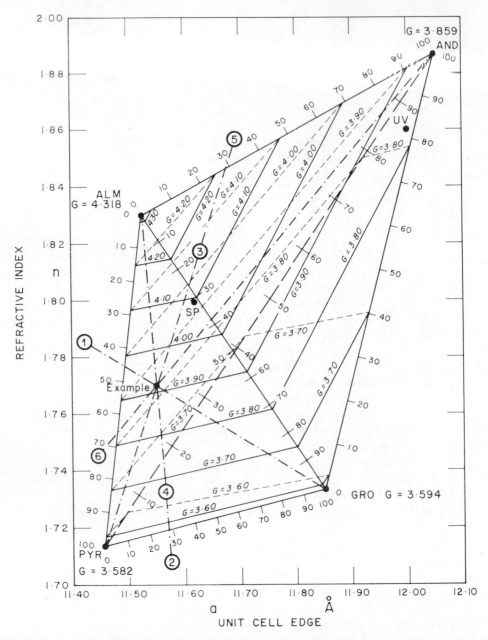

Fig. 7.13 Relationships of refractive index (*N*), unit cell edge (*a*) and specific gravity (*G*) for the garnet ternary series between alm (almandine)-pyr (pyrope)-gro (grossular), alm-gro-and (andradite), alm-pyr-and, and pyr-and-gro. The positions of sp (spessartite) and uv (uvarovite) are shown. See text for use of the diagram and for the composition determination of the example. (Data after Skinner, 1956.)

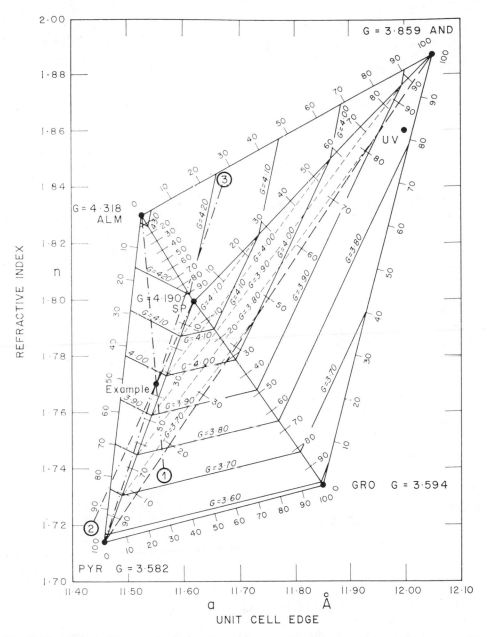

Fig. 7.14 Relationship of refractive index (*N*), unit cell edge (*a*), and specific gravity (*G*) for the garnet ternary series between alm (almandine)-pyr (pyrope)-sp (spessartite), pyr-sp-gro (grossular), gro-sp-and (andradite), alm-sp-and, pyr-sp-and. Winchell (1958) shows also the series between alm-sp-gro, but this has been omitted because in terms of *N* and *a* the three end members alm, sp, and gro lie nearly on a straight line. See text for determination of the example. (Data after Skinner, 1956.)

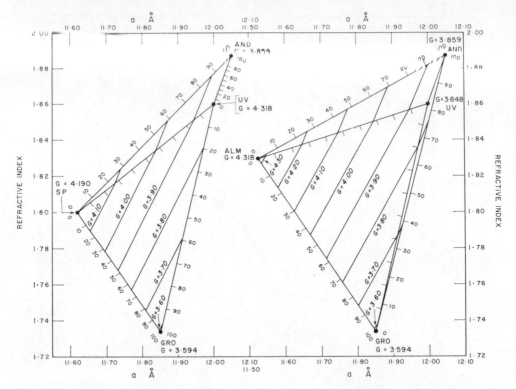

Fig. 7.15 Relationships between refractive index (N), unit cell edge (a), and specific gravity (G) for the ternary garnet series: sp (spessartite)-and (andradite)-uv (uvarovite), sp-uv-gro (grossular), alm (almandine)-and-uv, alm-uv-gro, and gro-uv-and. (Data after Skinner, 1956.)

DISCUSSION

The point established for $a = 11.550$ and $N = 1.770$ has been shown to be consistent with a $pyr_{39} alm_{46} gro_{15}$ composition. From the density contour lines in Fig. 7.13 such a garnet would have a density of $G = 3.916$, which is close to the measured value of 3.880. We may then assume that this discrepancy is due to some inclusions in the garnet so that the actual measurement was low. So we may accept the $pyr_{39} alm_{46} gro_{15}$ composition. To get a reasonably precise estimation of the garnet composition, however, we must be able to rely on the density determination. Let us now assume that we determined the density on a nice clean garnet fragment and that the determination $G = 3.880$ is accurate.

Similarly, in Fig. 7.13 the garnet point could equally well be taken within the triangle pyr-alm-and. By joining up the point already found at $a = 11.550$, $N = 1.770$ to the three spices the garnet composition can be determined as

$$\text{PYROPE } \% = \frac{100}{1 + 34/66 + 17/83} = \frac{100}{1.720} = 58.1 \% \text{ (points 6 and 4),}$$

$$\text{ALMANDINE } \% = \frac{100}{1 + 66/34 + 29/71} = \frac{100}{3.350} = 29.9 \% \text{ (points 6 and 5),}$$

$$\text{ANDRADITE } \% = \frac{100}{1 + 71/29 + 83/17} = \frac{100}{8.330} = 12.0 \% \text{ (points 5 and 4),}$$

giving $pyr_{58}alm_{30}and_{12}$, and for this composition the specific gravity would be 3.839 from the G lines on that triangle. This value likewise may be correct if the specific gravity determination of 3.880 was a poor one on a garnet with numerous inclusions that could not be extracted.

Now go to Fig. 7.14. The same garnet point ($a = 11.550$, $N = 1.770$) lies within the pyr-alm-sp triangle and its composition may be determined as

$$\text{PYROPE } \% = \frac{100}{1 + 19/81 + 58/42} = \frac{100}{2.615} = 38.2\% \text{ (points 2 and 1)},$$

$$\text{ALMANDINE } \% = \frac{100}{1 + 81/19 + 86/14} = \frac{100}{11.406} = 8.8\% \text{ (points 2 and 3)},$$

$$\text{SPESSARTITE } \% = \frac{100}{1 + 14/86 + 42/58} = \frac{100}{1.887} = 53.0\% \text{ (points 1 and 3)},$$

which gives the composition as $pyr_{38}alm_9sp_{53}$ with a specific gravity of $G = 3.971$. This value is quite far removed from the measured value of 3.880; hence it is the least likely composition to be correct.

We may now list our computed possibilities:

	Computed Composition	Computed Density	Actual Density
1.	$Pyr_{39}alm_{46}gro_{15}$	3.916	
2.	$Pyr_{58}alm_{30}and_{12}$	3.839	3.880
3.	$Pyr_{38}alm_9sp_{53}$	3.971	

We may proceed as follows if we are confident that the actual specific gravity determination is accurate at 3.880, that is, if we are sure that the small garnet fragment weighed on the Berman density balance had no impurities: 3.880 lies between 3.839 and 3.916, 0.041 from 3.839 and 0.036 from 3.916. The total interval between compositions 1 and 2 is 0.077. Hence the measured G lies $0.036/0.077 = 0.47$ of the way from composition 1 and $0.041/0.077 = 0.53$ of the way from composition 2. A composition that would satisfy a, N, and G for the determined garnet would be found by adding 0.53 of $pyr_{39}alm_{46}gro_{15} = pyr_{21}alm_{24}gro_8$ to 0.47 of $pyr_{58}alm_{30}and_{12} = pyr_{27}alm_{14}and_6$ to give a total of $pyr_{48}alm_{38}gro_8and_6$.

Also our given density $G = 3.880$ lies between composition 2 and 3 at an interval 0.041 from 2 and 0.091 from 3, that is, at $0.041/0.132 = 0.31$ from 2 and 0.69 from 3. Therefore we get the final composition by adding 0.69 of $pyr_{58}alm_{30}and_{12}$ $= pyr_{40}alm_{21}and_8$ to 0.31 of $pyr_{38}alm_9sp_{53} = pyr_{12}alm_3sp_{17}$ for a total of $pyr_{52} alm_{23}and_8sp_{17}$. Both values $pyr_{48}alm_{38}gro_8and_6$ and $pyr_{52}alm_{23}and_8sp_{17}$ are consistent with the given parameters of the garnet, $a = 11.550$, $N = 1.770$, and $G = 3.880$. In this, as in every case, some chemical information is necessary to determine which of the possible deduced compositions is the correct one. If we ignore uvarovite, there are five end members, that require four parameters for unambiguous determination. A chemical parameter is therefore *always* needed in addition to the three physical parameters. Levin (1950) accordingly supplemented his determined physical parameters with partial chemical analyses for total manganese and for ferrous iron and found that the composition of a garnet deduced on the basis of its physical properties, together with a limited chemical analysis, agrees well with the composition derived from a complete chemical analysis. For the example worked above a determination of MnO and CaO in the specimen would indicate which of the alternative compositions is correct, because the presence of grossularite requires CaO and the presence of spessartite requires MnO in the analysis. This confirmation could be made rapidly on an X-ray spectrometer or a simple optical spectrometer.

In the absence of chemical data mineral associations within the rock would give useful indications of the best choice of composition. The chief value of the foregoing physical methods technique lies in showing the relative variation among garnets from similar metamorphic terranes, for example, across or between isograds. These variations will show up on diagrams plotting a, Å against RI without resort to chemical analysis, either complete or partial.

For garnets with a chromium content Fig. 7.15 may be used for the ugrandite series and certain end members of the pyralspite series.

The use of the diagrams in Figs. 7.13, 7.14, and 7.15 is based on the assumption that the physical properties represented are additive functions of the molecular proportions of the end members and that components other than the six common end members (alm, and, gro, pyr, sp, and uv) are relatively insignificant. A linear relationship, which illustrates Vegard's law, between cell edge and composition and between refractive index and composition has been demonstrated for the almandine-pyrope and grossularite-pyrope series and a similar relationship has been demonstrated for specific gravity in the spessartite-almandine and almandine-pyrope series. Thus it is probable that the additive relationship applies throughout the whole garnet group.

Hydrogrossular can be determined separately, since the H_4^{4+} group replaces Si^{4+}, resulting in lower specific gravities and refractive indices but in an increase in the cell edge. The artificial end-member $3CaO \cdot Al_2O_3 - 6H_2O$ has $N = 1.605$, $a = 12.56$ Å, and $G = 2.52$ (Deer, Howie, and Zussman, 1966 p. 27).

Perhaps the most satisfactory determinative method for garnets would be to analyze with X-ray fluorescence spectrometry or atomic absorption spectrophotometry for Ca, Mg, Mn, Ti, Fe (total), Al, and perhaps Si as well, and then make use of the refractive index N and the unit cell edge a_0 to deduce Fe^{2+} versus Fe^{3+}.

C. CARBONATES

EXERCISE 7.11

Determination of Percentage of Dolomite in a Carbonate Rock

The analysis of limestone specimens by this method can quickly indicate highly dolomitized zones with locations favorable to mineral deposition.

Make a diffractometer smear of a representative sample of the limestone, marble, or dolomite; $CuK\alpha$ radiation is used. The region 26 to 34° is scanned first to check for interference from minerals other than calcite and dolomite in the specimen and to select a proper place for background determination.

The lines used in this method are the 104 calcite and 104 dolomite at 29.43 and 30.98° 2θ $CuK\alpha$, respectively. Table 7.7 will facilitate the identification of the calcite and dolomite lines. Counts are made for 10 seconds on the two peaks as well as on a suitable

Table 7.7 Diffraction Lines of Calcite and Dolomite for $CuK\alpha_1$

	Calcite (J.C.P.D.S. card 5-0586)				Dolomite (J.C.P.D.S. card 11-78)		
	2θ	d,Å	I/I_1	hkl	d,Å	I/I_1	2θ
				101	4.03	3	22.05
	23.04	3.86	12	102	3.96	5	24.11
***	29.43	3.035	100	104	2.886	100	30.98
	31.44	2.845	3	006	2.670	10	33.56
				015	2.540	8	35.33
	36.00	2.496	14	110	2.405	10	37.39
	39.43	2.285	18	113	2.191	30	41.18
				021	2.066	5	43.82
	43.18	2.095	18	202	2.015	15	44.98
	47.16	1.927	5	204	1.848	5	49.31
	47.53	1.913	17	108	1.804	20	50.60
	48.55	1.875	17	116	1.786	30	51.14
	Many other lines but all less than 8% relative intensity				Many other lines but all less than 10% relative intensity		

background position at a 2θ value slightly (about 1 to $1\frac{1}{2}°$ 2θ) higher than that for the dolomite line and at a 2θ value (about 1 to $1\frac{1}{2}°$ 2θ) just below the calcite line. The ratio

$$\frac{\text{counts 10 sec calcite}_{104}\text{ peak} - \text{counts 10 sec background}}{\text{counts 10 sec dolomite}_{104}\text{ peak} - \text{counts 10 sec background}}$$

is plotted in Fig. 7.16 (taken from Tennant and Berger, 1957) and the percentage of dolomite in the sample is read off the graph.

DISCUSSION

The peaks of calcite at 29.43° 2θ $CuK\alpha_1$ and dolomite at 30.98° 2θ $CuK\alpha_1$ are chosen because they have the greatest intensity for calcite and dolomite, respectively, and are relatively free of interference from peaks of other minerals that may occur in small amounts in carbonate rocks; they are also close together, thereby facilitating accurate measurement. The peaks may be quickly measured for height (intensity) directly on a diffractometer or diffractogram tracing and measurement of the background intensity determined by extrapolating the background line underneath the respective peaks from 2θ 20 to 35°. Royse et al. (1971) confirm that the Tennant and Berger (1957) curve (Fig. 7.16) can be used to estimate a composition within ±6% dolomite with 95% confidence.

Gulbrandsen (1960) gave the following relationship:

$$x = \frac{y - 0.3921}{0.7885},$$

where

$$x = \log_{100}, \frac{\%\text{ calcite}}{\%\text{ dolomite}}$$

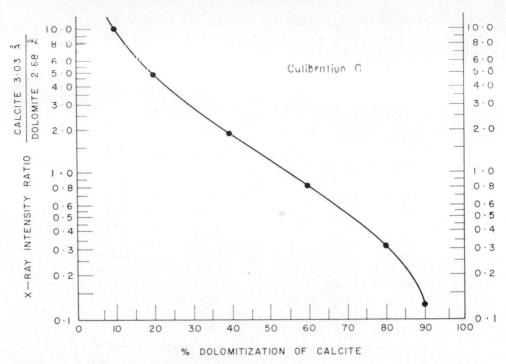

Fig. 7.16 Calibration curve of % dolomitization versus the intensity ratio of the calcite and dolomite peaks. (After Tennant and Berger, 1957.)

and

$$y = \log_{100} \frac{\text{peak height calcite} - \text{background calcite}}{\text{peak height dolomite} - \text{background dolomite}}.$$

He used a plot of logarithms of the ratios so that peak-height ratio plotted against the percentage would be a linear function. Royse et al. (1971) suggest, however, that the Gulbrandsen (1960) relationship cannot be recommended and the graph in Fig. 7.16 is preferable.

Royse et al. (1971) recommend the use of their relationship

$$\frac{\text{dolomite}}{\text{dolomite} + \text{calcite}} = 0.010 \text{ wt \% dolomite} - 0.023$$

as being more accurate. They conclude that the X-ray peak height (fixed counting time) provides the most rapid and precise procedure for determination of calcite and dolomite. These intensities are entered at the left in the above equation.

EXERCISE 7.12

Determination of the Amount of Ca Replaced by Mg in Calcite

Goldsmith et al. (1955) gave determinative curves for the molecular percentage of $MgCO_3$ in calcite in terms of the d spacing of the most intense carbonate peak 104 (see Table 7.7). The method they used is by comparison of the 2θ position of this peak in any given calcite with the position of the same peak in specpure (analar) $CaCO_3$. The method is not easily applied in practice because in most X-ray apparatus it is impossible to record simultaneously two diffractograms, one of the unknown, the other of the specpure $CaCO_3$ standard. I have modified the Goldsmith et al. (1955) technique and, using their data, have devised the following method:

Grind a sample of calcite, with silicon powder as a standard, and prepare a goniometer smear on a glass slide. Scan at the lowest possible speed and maximum 2θ sensitivity from 28 to 31° 2θ $CuK\alpha_1(\lambda = 1.5418$ Å). Identify the most intense silicon$_{111}$ peak, which occurs at 2θ 28.44. Identify the most intense calcite$_{104}$ peak, which occurs at approximately 29.42 2θ for pure $CaCO_3$ and at 30.98 for pure $CaMg(CO_3)_2$ (i.e., calcite containing 50 mol% $MgCO_3$). Both the silicon$_{111}$ and the carbonate $_{104}$ peaks are intense and unmistakable. Determine their 2θ values to the nearest 0.01°. Then determine calcite$_{104}$ − silicon$_{111}$ 2θ $CuK\alpha$.

The following regression equation can be computed from the 33 analyzed specimens of Goldsmith et al. (1955):

2θ calcite$_{104}$ − 2θ silicon$_{111}$($CuK\alpha$, $\lambda = 1.5418$) = 0.9616 + 0.0296   ($MgCO_3$ mol%) within the range mol % $MgCO_3$ 0 to 22; or mol % $MgCO_3$ in the calcite = 33.784 (2θ calcite$_{104}$ − 2θ silicon$_{111}$) − 32.486.

DISCUSSION

This regression line can apparently be projected all the way from calcite through dolomite to magnesite (i.e., over the whole range $MgCO_3$ mol% 0 to 100) and gives only slight errors. The determined data and the determined regression equation, however apply strictly to the range 0 to 22%.

Silicon$_{111}$ was chosen as a standard reference line because it lies on the low-angle side of calcite$_{104}$ and only about 1° from it, so that errors of measurement are minimized.

Potassium bromate is unsuitable as a standard for calcite work because its 110 peak would coincide with the carbonate 104 peak somewhere between calcite and dolomite composition. Silicon is an ideal standard because it has no interfering peak in the calcite 104 range.

Waite (1963) described a method for determining small changes in the position of the calcite peak relative to CdO as a standard and made use of diffracted intensities integrated over each 0.2° 2θ angle. The method is complicated, hence I cannot recommend it for general use, though it is claimed to have high sensitivity. Integration of the peak intensities over 0.2° 2θ ranges will not be possible for the majority of X-ray diffractometers.

EXERCISE 7.13

Determination of Carbonate Minerals and Quartz in Rocks

The following method is proposed for carbonate mineral quantitative determination in sediments. It can be applied to other rocks equally well.

1. The bulk samples are washed, air dried, and placed in an oven at 115°C for 2 hours.

2. A 100-g split of the bulk sample is crushed in a pulveriser to coarse sand size.

3. An 8-g split of this sample is hand ground in a mortar until it passes through a 250 mesh sieve. *Note.* Aragonite inverts to calcite on prolonged grinding. Care must be taken to keep grinding to a minimum time.

4. The powdered sample is back-packed into an aluminum sample holder and analyzed on an X-ray diffractometer with Cu nickel filtered radiation, 40 kV, 20 mA, slit system 1°, 0.1 mm, 1°, scanning at 1/8° per minute, rate meter 16, time constant 1 second, and range 2θ 25.5 to 32.0°.

5. Pure calcite, aragonite, and quartz specimens are ground separately to 250 mesh, mixed in known proportions, and homogenized in a mixing machine for 60 minutes. Enough of this spike mixture is prepared for a full number of analyses (0.5 g per sample). The preparations of calcite, aragonite, and quartz are determined to be as close as possible to that of the unknown.

6. The previously analyzed sample (step 4) is mixed and homogenized for 30 minutes with the spike mixture (step 5) in a ratio 1:1. The spiked sample is analyzed in the same X-ray diffractometer under the same conditions as the unspiked sample in step 4.

7. *Calculation.* The following peak areas are measured on both the spiked and unspiked sample diffractograms by using a polar planimeter, 111 of aragonite at 26.22° 2θ, 10$\bar{1}$1 of quartz at 26.64° 2θ, and 10$\bar{1}$4 of the (low-magnesian calcite + high magnesian calcite) peak at 29.4 to 30° 2θ, depending on the magnesium content. Sometimes, at a very slow scanning rate of 1/8° per minute, the calcite peak may show two distinct modes, attributed to low magnesian and high magnesian components. The total calcite peak area is calculated, and the difference between twice the half area of the major mode and that of the total curve is considered to be that of high magnesian calcite. The first mode at or greater than 29.60° 2θ ($\pm$0.02° 2θ) is that due to high magnesian calcite.

8. Using the peak areas so calculated, we obtain the weight percentage of the different minerals from the equation

$$\frac{I_a}{I_b} \times \frac{I_b'}{I_a'} = \frac{[W_a(W_b + S_b)]}{[W_b(W_a + S_a)]},$$

where I_a = peak intensity for mineral a,
    W_a = weight percentage of a in the mixture,
S_a, S_b = percentage of a and b in the spike mixture (the spike ratio must be
        1:1),
I_a, I_b = the unknowns,
I_a', I_b' = the spiked mixtures,

If one of the four minerals is omitted from the spike mixture (i.e., $S_b = 0$), the other minerals can be readily determined by using I_b and I_b' as a reference peak. In a sample containing little or no high magnesian calcite it is necessary to choose another component for omission from the spike, preferably one that is a major constituent of the sample.

(Method after Gunatilaka and Till, 1971.)

DISCUSSION

The weight percentage of high magnesian calcite cannot be determined from the equation. The total carbonate content of the samples may be determined by digesting 1.5 g powdered sample in 3% acetic acid and weighing the insoluble residue (essentially quartz + clay minerals). The high magnesian calcite is then found by the difference. Thus all carbonate minerals and quartz can be determined. The results by this method are precise and the accuracy is good.

D. OLIVINE

Determination of the Composition of Common Olivine

Make an accurate powder diffractogram of olivine extracted from an ultramafic rock (dunite, peridotite, or olivine gabbro). The separation need not be pure. Use Table 7.8 to identify the lines on the diffractogram. Carefully identify line 130 and convert $2\theta_{130}$ to d_{130}, using standard conversion tables for the wavelength. Forsterite can be

Table 7.8   X-Ray Powder Lines of the Two End Members of the Common Olivine Series[a]

	Forsterite				Fayalite		
2θ	d,Å	I/I_1	hkl	I/I_1	d,Å	2θ	
20.20	5.10	50	020	20	5.25	19.61	
24.01	4.30	10	110				
26.63	3.883	70	021	20	3.972	26.03	
27.80	3.723	10	101	10	3.784	27.34	
29.65	3.496	10	111	30	3.558	29.12	
29.80	3.478	20	120				
34.61	3.007	10	121	5	3.075	33.82	
34.79	2.992	10	002	10	3.047	34.14	
**** 37.70	2.768	60	130	100	2.831	36.84	
			022	20	2.634	39.70	
			040	20	2.621	39.91	
41.72	2.512	70	131	50	2.566	40.80	
42.68	2.458	100	112	70	2.501	41.91	
44.09	2.383	5	200	5	2.416	43.46	
44.80	2.347	20	041	20	2.408	43.61	
45.44	2.316	10	210	5	2.350	44.74	
46.43	2.269	40	122	20	2.313	45.50	
46.85	2.250	30	140	30	2.303	45.71	
48.90	2.161	10	220, 211	10	2.194	48.12	
52.23	2.032	5	132	5	2.071	51.17	
56.95	1.876	20	150				
			113	5	1.842	58.10	
60.15	1.785	5	151	10	1.834	58.38	
61.48	1.750	40	222	30	1.778	60.40	

The remaining lines are all weaker than 20% relative intensity

Forsterite: J.C.P.D.S. card 7-74 Mg_2SiO_4 96%; Fe_2SiO_4 4%
Fayalite: J.C.P.D.S. card 7-164 Mg_2SiO_4 5.9%; Fe_2SiO_4 94.1%

[a] 2θ values are for cobalt $K\alpha$ radiation of wavelength 1.7889 Å.

run with copper radiation even without pulse height analysis, but fayalite should be run with cobalt or iron radiation, if a pulse height analyzer is not available, to prevent fluorescence of the iron in the specimen. However, with a well-set pulse height analyzer to discriminate against the iron fluorescence fayalitic olivine will produce a good diffractogram with Cu radiation.

The olivine $_{130}$ line is used because (a) it is sufficiently intense to appear in rock matrices containing as little as 10% olivine, (b) the 130 peak is not overlapped by peaks of other phases in the system Mg-Fe-Si-O-OH, and (c) d_{130} varies enough to permit reasonably precise estimates of composition.

Yoder and Sahama (1957) give the following data for common olivine:

$$\text{for Fo} = 100 \text{ (pure forsterite)} \quad d_{130} = 2.7659 \text{ Å};$$
$$\text{for Fo} = 0 \quad \text{(pure fayalite)} \quad d_{130} = 2.8328 \text{ Å}.$$

They also give the following relationship which can be used to determine the olivine composition:

$$\text{Fo (mol\%)} = 4233.91 - 1494.59 \ d_{130}.$$

The error in the determination of Fo% will be up to 3 or 4 mole %, depending on the composition.

Agterberg (1964), however, recalculated these data and gave the following refined relationship, which is to be preferred in calculating the composition of a common olivine.

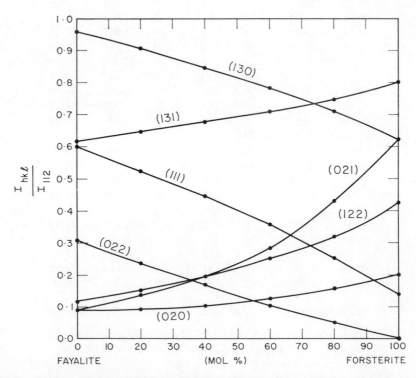

Fig. 7.17 Variation of the calculated intensity ratios I_{hkl}/I_{112} with composition for several reflections of olivine. (After Jahanbagloo, 1969.)

For Fo greater than 30% (i.e., for most natural olivines, excluding fayalite) $\%Fo = 4088.89 - 1442.44 \, d_{130}$. The olivine need not be separated from the rock if the rock contains more than 10% olivine. It has been found possible to obtain a suitable X-ray pattern from an uncovered thin section of an olivine-bearing rock. The only requirement is that the intense olivine 130 peak be definitely identified and measured.

Alternative Methods. Jambor and Smith (1964) used the high-angle reflection 174. Since the back reflection was employed, this method is suited to a small-diameter (57.3 mm) camera instead of a diffractometer. The identification of the 174 line is facilitated by reference to their paper in *Mineralogical Magazine*, **33**, 1964, 734, Fig. 1. With FeKα radiation the 2θ range was 136.57 for Fo$_0$ to 143.61 for Fo$_{100}$.

Fo $\%$ (molecular) $= 4151.46 - 3976.45 \, d_{174}$ for olivine with Fo greater than 30% is given. For olivines within the range Fo$_0$ to Fo$_{30}$ (i.e., nearly pure fayalite) the determinative curve shows a sharp break from this straight-line relationship.

Jahanbagloo (1969) has described a method for Fo$\%$ determination for the complete Fo–Fa series in which the intensity of the olivine lines 130, 131, 111, 021, 122, 022, and 020 (see Table 7.8), in each case divided by the intensity of the strongest line 112 on the diffractogram, is related to mol $\%$ forsterite. His graph is shown here as Fig. 7.17.

DISCUSSION

The linear variation of cell dimensions with composition in a solid-solution series is known as Vegard's law, which states that the cell dimension in a cubic mineral a_0 varies linearly with composition. The forsterite-fayalite olivine series has been shown to be thermodynamically ideal within the range Fo$_{100}$ to Fo$_{30}$ and to offer a good illustration of Vegard's law.

Values for synthetic olivines differ from natural olivines, however. Fisher and Medaris (1969) found the relationship Mg_2SiO_4 mol$\%$ (Fo mol$\%$) for synthetic olivine $= 15.8113\sqrt{3.0358 - d_{130}} - 7.2250$ with an accuracy within ± 2 mol$\%$ forsterite, but they clearly state that this relationship cannot be used for natural olivines.

Smith (1966) has shown that the divergence between the natural and the synthetic olivines is accounted for by the presence in natural iron-rich olivines of minor amounts of Ca^{2+} and Mn^{2+}. After correction for these impurities the d_{130} values for natural olivines fitted well the relationship given for synthetic olivines.

E.   CORDIERITE

Determination of the Structural State of Cordierite

Indialite is a hexagonal mineral polymorphic with cordierite, which is pseudohexagonal. X-ray diffractograms for indialite and cordierite show general similarities (see Table 7.9). The pattern in the range 2θ CuKα 29 to $30°$ is the most distinctive (Miyashiro, 1957). At least three peaks (possibly as many as four or five) appear in cordierite, whereas all unite as a single peak in indialite (Fig. 7.18).

Table 7.9   Diffraction Data on Cordierite: All Angles Are for CuKα₁

	Low Cordierite				High Cordierite (Indialite)			
	$2\theta°$	d,Å	I/I_1	hkl	$2\theta°$	d,Å	I/I_1	hkl
	10.36	8.54	80	020				
	10.47	8.45	80	110	10.43	8.43	100	100
	14.12	6.27	2	111				
	18.07	4.91	30	130	18.14	4.89	30	110
	18.25	4.86	12	200				
	19.00	4.67	12	002	18.96	4.679	16	002
	20.41	4.35	4	131				
	20.80	4.27	6	040				
	21.00	4.23	2	220				
	21.72	4.09	75	112	21.71	4.094	50	102
	26.28	3.39	70	132				
	26.44	3.37	60	202	26.38	3.379	55	112
	27.88	3.20	4	240				
	28.33	3.15	45	042				
***	28.50	3.13	100	222	*** 28.44	3.138	65	202
***	29.29	3.046	65	151				
***	29.42	3.033	55	241				
***	29.61	3.014	60	311	*** 29.51	3.027	85	211
	31.40	2.846	2	060				
	31.74	2.817	2	330				
					33.95	2.640	25	212
					36.82	2.441	6	220
	All remaining lines are less than 25% relative intensity				All remaining lines are less than 10% relative intensity			

Figure 7.18 illustrates diffractograms of both minerals (Miyashiro, 1957). The 2θ spaces between 511, 420, and 131 increase with the degree of distortion of the cordierite structure. In indialite all these spaces are zero. The following value (Miyashiro, 1957) has been proposed to indicate an INDEX of the degree of distortion in the cordierite structure:

$$\text{DISTORTION INDEX } \Delta = 2\theta_{131} - \frac{2\theta_{511} + 2\theta_{421}}{2} \text{ for } CuK\alpha_1.$$

Cordierite showing the maximum degree of distortion is called perdistortional cordierite, with $\Delta \geq 0.29$; subdistortional cordierite has Δ between 0.29 and 0, and indialite has $\Delta = 0$. The distortion is probably due to some order-disorder change in the Si, Al atoms in the cordierite-indialite Si_5AlO_{18} rings. Indialite ($\Delta = 0$) is stable at very high temperatures. It occurs in fused sediments at the Bokaro coalfield, India. High perdistortional cordierite ($\Delta = 0.29$ to 0.31) is stable at medium temperatures. It has been found in andesite in xenoliths at Asama and also occurs in furnace linings.

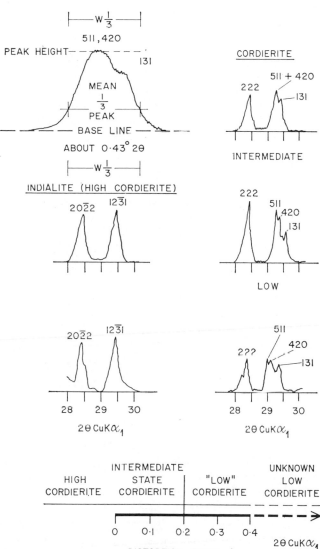

Fig. 7.18 Several illustrations of typical partial diffractograms of cordierite and indialite in the region $CuK\alpha_1$ 2θ 28 to 30°. See text for description. The classification in terms of distortion index and the meaning of the width index at one-third peak height are both illustrated.

High subdistortional cordierite ($0.29 > \Delta > 0$) has been found in andesite from Kaso-to and also in fused sediments at Bakaro.

Low subdistortional cordierite ($0.29 > \Delta > 0$): most cordierites, if not all in ordinary metamorphic rocks, pegmatites, and quartz veins, belong to this class (Miyashiro, 1957).

The lower diagram on Fig. 7.18 illustrates the classification of cordierites after Langer and Schreyer (1969) in terms of the distortion index. They propose that a better way of defining early stages in the structural transition from high to low cordierite is to measure the WIDTH INDEX of the peaks occurring in the range 29 to

$30°2\theta$ $CuK\alpha_1$. This index is taken as the width of the complex peak at one-third its peak intensity and is briefly referred to as $W_{1/3}$. It was determined with the following instrumental settings: the critical region from 2θ $CuK\alpha_1$ 29 to 30° is run by using 50 kV, 30 mA, Ni filtered radiation, slits 1/6°, 0.1, 1/6°, scanning rate 1/8°/min, ratemeter adjustment 400/0.5/0, and chart speed 1200 mm/hr.

Langer and Schreyer conclude that the Δ index cannot be used as a measure of the degree of Si/Al order/disorder, even in specimens of the same bulk composition. Zeck (1969) warned that care must be taken to measure the $K\alpha_1$ reflection and not to mix $K\alpha_1$ and $K\alpha_2$ reflections in the range 29 to 30°. The α_1 and α_2 reflections are separated by 0.075° 2θ $CuK\alpha$ radiation.

The distortion index Δ is in common use; the width index is not, because most natural cordierites have well-defined 511, 420, and 131 peaks. As an example of the application of the Δ index, Harwood and Larson (1969) have applied it in Maine and have shown that the index for cordierite increases from 0.23 to 0.29 inward toward the Cupsuptic pluton. In all measured rocks the changes in Δ index could be related to the thermal environment during crystallization of cordierite.

F. PYROXENE

EXERCISE 7.16

Determination of Pyroxene Composition in the Range Enstatite to Hypersthene

METHOD A

Goniometer smears of orthopyroxene separations, mixed with at least 10% by volume lithium fluoride as standard, should be scanned from 34.5 to 40.2° 2θ $CuK\alpha$ with a 1° divergent slit, a 4° preslit, and a 1° receiving slit at 1/4° per minute and chart scale 2 in. per degree 2θ (only suggested for the goniometer; may be changed to suit your own instrument).

The lithium fluoride 111 peak is identified by reference to J.C.P.D.S. card 4-0857 at 2θ $CuK\alpha$ 38.75° (d,Å = 2.325). It has an intensity of approximately 95% and is the first LiF peak.

The orthopyroxene 131 peak should be carefully identified by reference to the following diffraction patterns:

1. *J.C.P.D.S. card 7-216 (Enstatite) hkl (311).* This peak is actually 131, for the a and b axes have been reversed. It occurs at 2θ $CuK\alpha_1$ 35.39° (d, Å = 2.534) and has a relative intensity of 43%.

2. *Geol. Soc. America Memoir 122 (page 275) (Hypersthene) hkl (131)* occurs at d, Å 2.560 (2θ $CuK\alpha_1$ 35.02°) with an intensity of 57%.

The 131 orthopyroxene peak is therefore to be expected in the range 2θ $CuK\alpha_1$ 35.02 to 35.39°.

By subtraction, determine $\Delta 2\theta = 2\theta$ $CuK\alpha_1$ LiF_{111} − orthopyroxene$_{131}$. Himmelberg and Jackson (1967) gave the following relationship: $\Delta 2\theta = 4.112 - 0.009132$ Mg, where Mg is the ionic percent of Mg in the orthopyroxene. Mg = (Mg × 100) divided by Mg + Fe^{+2} + Mn + Ni + Co + Cu + Al^{+6} + Fe^{+3} + Cr + V + Sc + Ti

+ Zr + Mo + Sn; 2 sigma for this regression equation is $\pm 0.012°$ 2θ, corresponding to about $\pm 1.3\%$ Mg.

Expressed in another way, which we can use directly,

$$\text{Mg ionic } \% = \left(\frac{\text{Mg} \times 100}{\text{total octahedral cations}}\right) \% = \frac{4.112 - \Delta 2\theta \text{Cu}K\alpha_1}{0.009132}.$$

DISCUSSION

This regression equation is for orthopyroxene from the Stillwater Complex, Montana, only. Because X-ray parameters are affected by aluminum and calcium contents this determinative equation should therefore be used with caution when comparing or estimating orthopyroxenes from a different chemical, thermal, or pressure environment.

METHOD B

In place of the 131 orthopyroxene peak the 060 peak may be used. This peak occurs at 2θ $\text{Cu}K\alpha_1$ $63.20°$ (d, Å $= 1.470$) in enstatite (En% about 94) (see J.C.P.D.S. card 7-216) and has a recorded intensity of 22%. It occurs at 2θ $\text{Cu}K\alpha_1$ $62.36°$ (d, Å $= 1.4878$) in hypersthene (En% about 52) [see Geol. Soc. America Memoir 122 (1969), 275)] and has an intensity of 16%.

The position of the orthopyroxene 060 peak should be carefully determined in comparison with the position of LiF_{220}, which occurs at $65.56°$ 2θ $\text{Cu}K\alpha$ (d Å $= 1.424$) (see J.C.P.D.S. card 4-0857). If this LiF peak does not occur exactly at this 2θ position, the $+ve$ or $-ve$ error should be algebraically added to the 2θ value of orthopyroxene 060.

Desborough and Rose (1968) gave the following relationship for orthopyroxenes in the range En_{87} to En_{76}:

$$\text{mol}\% \text{ En} = \frac{2\theta\text{Cu}K\alpha_1 \text{ for orthopyroxene}_{060} - 61.31}{0.02015},$$

with a standard error of ± 0.3 mol% En.

DISCUSSION

As in method A, this regression equation is strictly for orthopyroxene within the range En_{76} to En_{87} and is only for pyroxene from the Bushveld Complex, South Africa. It is not known yet how accurately this relationship will hold for orthopyroxenes from other provinces.

G. CLAY MINERALS

The most comprehensive guide to the identification and measurement of clay minerals is edited by Brown (1961). A more condensed guide to their X-ray identification (Carroll, 1970) contains useful tabulated data.

X-ray Diffraction Preparation of Clays

1.   Start with the raw unconsolidated clay sample of less than 100 g.

2.   Disaggregate it in water using (a) an ultrasonic vibrator or (b) a mechanical shaker, in which case a peptizing agent (NH_4OH, $NaOH$, Calgon or sodium hexametaphosphate) has been added to the water.

3.   Allow to stand in a cylinder containing 100 ml water long enough to sediment the sand fraction ($>62\mu$).

4.   Siphon off the clay and silt fraction which is still in suspension.

5.   Remove the silt from the clay by centrifuging.

6.   Allow the slit to dry at room temperature and weigh and store it in a bottle for separate analysis.

7.   Remove the water from the clay by filter candle under vacuum until about 200 ml of slurry remains.

8.   Remove about 10 ml of the clay slurry with an eye dropper and place evenly on a glass slide or semiporous ceramic tile to dry at room temperature.

9.   Place a sample of the clay slurry in a snap-cap vial for reference and storage.

10.   Allow the remainder of the slurry to dry at room temperature in a large evaporating dish. When dry, scrape off and store the cake in a bottle.

Oriented Mounts

11.   Use heat resisting VYCOR glass. It is preferable, however, to use $48 \times 48 \times 6$ mm semiporous unglazed tiles (Coors porcelain) for drying the crystal precipitates. Cut in half, they fit easily into the Philips X-ray diffractometer (obtainable from Lordon Ceramics).

12.   The clay slurry is easily dried at room temperature because water drains down through the tile (especially if it is set on a styrofoam pad) and is also lost by evaporation. It is good practice to prepare slides in batches of six at a time.

Unoriented Mounts

13.   Dry clay sample ($<2\mu$) is packed into an aluminum sample holder. To increase random orientation of the clay minerals the sample should be sieved into the holder and gently cut down into the holder hole with a small spatula blade held with its thin edge at right angle to the holder. When the holder is full, the powder should be lightly pressed down with the wide blade to obtain the necessary flat surface that will coincide with the flat surface of the aluminum specimen holder.

(Method after Carroll, 1970.)

DISCUSSION

Quantitative analysis of individual clay minerals in a mixture is not directly proportional to peak intensity because of different mass absorption coefficients of the different minerals. The modal analysis method of Norrish and Taylor (1962) outlined in exercise 3.10 is recommended for clays.

Routine X-ray Diffractometer Procedure

1.   Make an oriented mount (as in exercise 7.17, step 11) on a tile or VYCOR glass slide from the dispersion of the original sample ($<2\mu$).

2.   Allow to dry at room temperature (4 to 8 hours).

3.   Place in a dessicator at constant humidity until the clay appears dry. This is especially necessary for montmorillonite which reacts to a humid climate.

4.   Make a diffractogram with CuKα radiation (Co radiation if the clay is iron-bearing) from 2θ 2 to 37°. This scan will allow a preliminary identification and will include most common nonclay minerals. The suggested speed should be 1° 2θ per minute. An untreated ordered mount will give the most information. The goniometer should have been well aligned. When complete, leave the chart on the recorder.

5.   Identify and label the lines according to their d spacings. This is facilitated by making a comparison with previously run diffractograms of standard clays and by comparison with the various J.C.P.D.S. powder data cards.

6.   Place the mounted sample in a small dessicator containing about ½ pint of ethylene glycol. When the dessicator contains 6 to 12 mounted samples, place it in an oven at 60°C for at least 1 hour. Several hours are better. Leave the samples in the dessicator until ready for use.

7.   Make a diffractogram on each of the glycolated samples on the same chart used for the same sample before it was glycolated but choose a different color chart ink and set the chart zero back to the identical value, as in step 4. It is always convenient to run clay samples in batches of six, first all unglycolated, then all glycolated, in the same order.

8.   Inspect and label the d spacings, particularly any in the region 14 to 17° 2θ CuKα.

9.   Make a preliminary identification of the minerals present after comparison of the untreated and glycolated results. The results of glycolation may be summarized as follows:

NO CHANGE in the following minerals—kaolinite ordered, kaolinite disordered, halloysite, mica 2M, illite 1Md, vermiculite, chlorite Mg-form, chlorite Fe-form, mixed clay minerals: regular (no change unless an expandable component is present), mixed clay minerals: random (expands only if montmorillonite is present as a component), attapulgite (palygorskite), sepiolite, and allophane.

Expansion of *d* spacings. In the montmorillonite group the normal 15 Å (001) and its integral series of basal spacings expand to 17 Å together with its rational sequence of basal spacings.

10.   The specimens are placed in a cold muffle furnace and allowed to heat up to a controlled maximum temperature, chosen to give the characteristic heat reaction of the various minerals according to the following scheme. If the clay contains organic matter, it should be removed before X-ray preparation by hydrogen peroxide.

Likewise iron oxide coatings should be removed by sodium dithionite.

Once the maximum characteristic temperature has been reached, the furnace is allowed to cool to room temperature, and only then is the specimen removed.

NOTE.   A specimen placed in an already hot furnace is likely to break. The slow cooling within the furnace prevents the clay aggregate from flaking off the glass slide. Withdrawal of a specimen from the hot furnace will almost certainly result in the clay flaking off the glass slide.

Mineral	Temperature °C	Effect
Kaolinite (well crystallized)	575 to 625	No diffraction pattern
Kaolinite (disordered)	550 to 562	No diffraction pattern
Dickite	665 to 700	No diffraction pattern
Nacrite	625 to 680	No diffraction pattern
Halloysite	125 to 160	Water is removed
	560 to 605	No diffraction pattern
Meta halloysite	125 to 150	Water is removed
	560 to 590	No diffraction pattern
Allophane	140 to 180	Water is removed; no diffraction pattern ever
Mica, well crystallized (muscovite)	700	(001) spacings remain even up to 1000°C
Illite and clay micas	125 to 250	Loss of water
	350 to 550	Reverts to mica structure
	700	(001) spacings of mica remain
Glauconite	530 to 650	Loss of water; reverts to mica structure
Celadonite	500 to 600	Reverts to mica structure
Biotite	700	Phlogopite is similar to muscovite; biotite shows breakdown 700 to 1000°C
Vermiculite	About 300°C	Water is removed in stages, with 14, 13.8, 11.6, and 9 Å; the initial (001) spacing is controlled by the humidity
Montmorillonite group	300	The original 15 Å spacing disappears; 9 Å develops
Chlorite group	600 to 800	No change in pattern
Mg-chlorite	650	14 Å is intensified; (004) at 3.54 Å is not affected.
Fe-chlorite	500	14 Å less intense and becomes broad and diffuse
Mixed layer clays	<600	Various effects; depends on the kinds of minerals present
Sepiolite	<200	Rapid dehydration
	>200	Spacing at 12 Å above 350°C becomes weak and diffuse; A spacing at 9.8 Å appears and the spacing at 7.6 Å becomes more intense; recrystallization occurs at 800°C
Palygorskite	400 to 440	Rapid dehydration but no structural change <400°C; above 400°C 10.5 Å peak becomes broad and diffuse; near 800°C the structure is destroyed

11. After heating the specimens for 1 hour and cooling them in the muffle furnace make a diffractogram on the same chart, using another color recorder ink with the identical zero settings. If blue ink is selected for the untreated sample, green for the glycolated sample, and red for the heated sample, all on the same chart and the same 2θ scale, interpretation of the diffractograms and their filing will be simplified.

(Method after Carroll, 1970.)

DISCUSSION

Clay minerals cannot be identified by X-ray diffraction alone. Use should also be made of differential thermal analysis (Chapter 14) to supplement the data.

Oriented mounts give the most information. They give (001) reflections only, and these are the reflections that vary most with hydration and glycolation. Unoriented mounts give (hkl) reflections; these spacings do not vary appreciably with the various treatments.

Iron-bearing minerals have a high mass absorption coefficient for $CuK\alpha$ radiation. Thus, when a clay contains appreciable iron, Co radiation should be used; otherwise the peaks from the Fe-bearing components will be unfairly suppressed.

Clay fractions ($<2\mu$) give the highest X-ray diffraction intensities. Silt gives lower intensities. Hence it is necessary to separate the slit from the clay during the initial preparation. As soon as the grain size exceeds 10μ, X-ray intensity falls off appreciably.

Kaolinite occurs in two states: ordered and disordered. This has led to a definition of crystallinity index for the kaolonite minerals. It is illustrated and defined in Fig. 7.19.

The following features are diagnostic in identifying the common clay minerals:

Kaolinite, ordered. A ~ 7 Å (001) and a 3.57 Å to 3.58 Å (002) reflections in ordered mounts. A (060) reflection at ~ 1.49 to 1.50 Å. Collapse of structure at 550°C. For distinction between kaolinite and chlorite see under chlorite.

Kaolinite, disordered. Ordered mounts show broad (001) and (002) reflections at ~ 7 and 3.57 Å. The broad peaks give rise to a crystallinity index that differs from the ordered kaolinite (Fig. 7.19).

Dickite. Similar to kaolinite, but peaks are more intense.

Nacrite. Similar to kaolinite.

Halloysite. This mineral has less sharp basal reflections than kaolinite. The 10.1 Å spacing collapses to about 7.2 Å on dehydration.

Allophane. Allophane is similar to amorphous clay and is not identified by X-ray diffraction alone.

Muscovite. Muscovite is easily recognized by its intense reflection at ~ 10 Å and a small reflection at ~ 5 Å. The 1 *Md* polytype has basal spacings only—(060) at ~ 1.50, and (020) at about 4.4 Å. The 1 *M* polytype has basal spacings and hkl's for $11\bar{1}$, 021, $1\bar{1}\bar{2}$, 022, 024, 112, $11\bar{3}$, 130, $13\bar{1}$, 200, etc. The 2 *M* polytype has more numerous hkl reflections than the other polytypes.

Illite. Similar to the 1 *Md* muscovite. There may be some glycol expansion if the mineral is interlayered with montmorillonite.

Glauconite. Newly formed glauconite has 1 *Md* reflections. Other glauconite has 1 *M* reflections.

Celadonite. This mineral is similar to 1 *M* muscovite and cannot be distinguished from it except chemically.

Biotite. Biotite is similar to muscovite, but spacings differ; d (060) is 1.52 to 1.53 Å, whereas muscovite is 1.50 Å. The intensity of the basal reflections decreases with Fe content and is always less than muscovite if $CuK\alpha$ is used.

Vermiculite. In an ordered mount the lowest reflection (001) is at 14 Å; (060) occurs at 1.50 to 1.53 Å but may be at 1.48 to 1.49 Å.

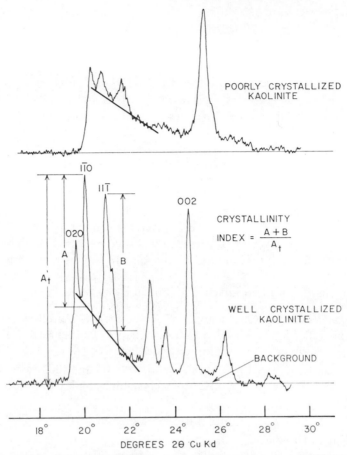

Fig. 7.19 Crystallinity index for kaolinite as determined on an unoriented diffractometer mount. (After Carroll, 1970, p. 9.) 2θ positions are for Cu $K\alpha$ radiation.

Montmorillonite. This mineral is identified by its swelling on glycolation from 15 to 17 to 18 Å.

Chlorite. Ordered mounts show a series of basal spacings from (001) 14 Å. Unordered mounts show many *hkl* lines. Most Mg and low Fe chlorites give a clear sequence of the first basal reflections, based on $d(001) = 14$ Å. Fe-rich chlorites give weak first and third basal reflections but strong second and fourth reflections; $d(001)$ does not expand on glycolation. With warm 1 N HCl treatment well-crystallized chlorite dissolved only slightly, but less ordered chlorite will dissolve. Kaolinite does not dissolve at all. Heating to 600°C for 1 hour does not destroy the 14 Å peak which intensifies. On the other hand, kaolinite is destroyed. Kaolinite and chlorite can be routinely differentiated in clays by resolving, respectively, the kaolinite (002) and the chlorite (004) peaks at 3.5 Å. However, the resolution of the 3.5 Å peaks should be used in conjunction with the usual diffraction scan to detect other minerals or chlorites of different composition which may contribute other peaks in the 3.5 Å region and obscure or preclude the kaolinite-chlorite distinction by the method (Biscaye, 1964).

Swelling chlorite. Some chlorites are intermediate between chlorite and montmorillonite or vermiculite. They have a *d* around 28 Å, which increases to 32 Å on glycolation and collapses to about 13.8 Å on heating to 450°C.

Dioctahedral chlorite. This mineral has $d(060)$ 1.496 Å, whereas normal trioctahedral chlorite has $d(060)$ at 1.53 to 1.54 Å.

Chamosite. Only basal reflections are present and they are diffuse. A trio of medium-intensity lines at 2.40, 2.15, and 2.67 Å is diagnostic and a band at 4.5 Å corresponding to (002) of disordered kaolinite is common. The (060) spacing is at 1.56 Å for ferrous chamosite and at 1.514 Å for ferric chamosite. Kaolinite-type chamosite resembles kaolinite; chlorite-type chamosite resembles chlorite.

Mixed-layer clay minerals. Regularly stratified clays are identified by their 001 reflections. Various treatments are necessary to identify the individual components. Randomly interstratified clays are difficult to interpret. Mixed layering of mica is indicated by a series of reflections on the low-angle side of the 10 Å spacing and results in broad lines with humps and shoulders. Mixed layering of montmorillonite, vermiculite, and chlorite is indicated by reflections on the high-angle side of the 14 Å spacing. Glycolation and heating will identify the component that causes the humps and shoulders of the peaks. In the preliminary analysis of a two-component system no large errors result if it is assumed that the d spacings of the peaks move linearly from the position of the one pure component (10.0, 12.4, 14.0 Å) toward the nearest position of the other component as the proportion of the components change. If the distances from the neighboring A and B positions are x and y, respectively, the proportion of A is deduced as $y/(x + y)$ (Carroll, 1970).

Palygorskite. Reflections occur at 10.5, 4.5, 3.23, and 2.62 Å. Above 400°C the 10.5 Å peak is replaced by a broad band and near 800°C the structure is destroyed.

Sepiolite. Sepiolite has a wide variety of crystallinity. A number of peaks occur between 12.07 and 2.07 Å, and poorly crystalline material may show only a few poor peaks at about 12.6, 4.31, and 2.62 Å. Heating above 300°C causes the 12 Å reflection to become weak and diffuse, and a new peak occurs at 9.8. The 7.6 Å peak increases in intensity, and heating to 700°C causes the mineral to become amorphous.

H.   SUMMARY OF APPLICATIONS TO OTHER MINERALS

Aluminum silicates. Johnson and Andrews (1952) have used powder photographs for the quantitative estimation of mullite, kyanite, andalusite, and sillimanite in mixtures.

Amphiboles. Cell parameters of amphiboles have been determined from powder and single-crystal data and have been correlated with composition:

> Whittaker (1960) for amphiboles in general;
> Borley and Frost (1963) for ferrohastingsites;
> Frost (1963) for arfvedsonites and ferrohastingsites;

Klein and Waldbaum (1967) for the cummingtonite-grunerite series. Refined lattice parameters and 2θ FeKα values for selected reflections are used to determine the mol% of $Fe_7Si_8O_{22}(OH)_2$. Composition can be determined within ± 2 to 3 mol %.

Viswanathan and Ghose (1965) have shown that for cummingtonite $d_{061} = 2.65019$ Å $- 0.000615$ Å (mg) where mg $= (100 \times Mg^{2+})/(Mg^{2+} + Fe^{2+})$.

Analcite. Wilkinson (1963) has related the unit cell edge and $2\theta_{639}$ to the composition of analcite in the range nepheline-albite.

Apatite. Skinner (1968) has used X-ray diffraction techniques to detect compositional fluctuations in the apatite group.

Aragonite. Davies and Hooper (1963) have used X-ray diffraction to determine the ratio of aragonite to calcite in fossil shells.

Arsenopyrite. Morimoto and Clark (1961) determined the arsenic:sulfur ratio from the cell constants.

Carbonates. Morelli (1967) used the X-ray diffraction intensities for some typical reflections for several isomorphous compounds of the $MgCO_3$-$FeCO_3$ system as well as for some ternary compounds $MgCO_3$—$FeCO_3$—$CaCO_3$. The method can be used to determine composition, provided preferred orientation can be avoided in the powder preparations.

Chlorite. The basal spacing d_{001} appears to be influenced principally by substitution of aluminum for silicon and the d_{060} value is a useful guide to iron content (see Hey, 1954; Brindley and Gillery, 1956). The intensity relationships among basal reflections have been used determinatively (see Schoen, 1962; Petruk, 1964).

Chromite (see *Spinel*).

Feldspars (*Plagioclase*). Borg and Smith (1968) have published a guide to the indexing of complex plagioclase X-ray powder patterns which includes low and high albite (An_0), oligoclase (An_{29}), bytownite (An_{80}), and anorthite (An_{100}). These patterns will be helpful in the more detailed work of exercise 7.8 and will supplement Fig. 7.7. Hamilton and Edgar (1969) have shown the variation of the $\bar{2}01$ reflection of plagioclase with composition.

Ilmenite. Cervelle (1967) has described the application of X-ray diffraction to the study of composition in the series ilmenite-geikielite.

Loellingite. The sulfur and cobalt contents of loellingite can be determined by X-ray diffraction by the use of 101, 120, 111, and 210 lines (Clark, 1962).

Micas. The powder patterns of di- and trioctahedral micas can generally be distinguished with the 060 reflection, which for dioctahedral micas is near 1.50 Å and for trioctahedral is between 1.53 and 1.55 Å. Also, with the exception of lepidolite and glauconite, the basal reflection at about 5 Å is strong for most dioctahedral and weak for most trioctahedral micas.

Sodium and calcium micas can be distinguished from potassium micas by their smaller basal spacings.

Basal spacings of coexisting muscovite and paragonite in pelitic schists fit the regression equation

$$d_{(002)_{2M}}(\text{paragonite}) = 12.250 - 0.2634\, d_{(002)_{2M}}\,(\text{muscovite}) \pm 0.006 \text{ Å}$$

(Zen and Albee, 1964.)

The powder patterns of $1M$, $2M$, and $3T$ polymorphs of dioctahedral micas are in general distinguishable from one another (see Yoder and Eugster, 1955), but for the trioctahedral micas the powder patterns of $1M$ and $3T$ polymorphs are identical. The part of the powder pattern that is most useful for distinguishing polymorphs is the spacings between 4.4 and 2.6 Å; $1M$ and $1Md$ (disordered mica) are distinguishable by the absence of hkl reflections from the latter.

Franzini and Schiaffino (1965) show a graph relating the intensity ratio

$$(I_{004} + I_{006})/I_{005}$$

as a function of octahedral occupancy of Fe^N and Mg^N atoms for the phlogopite-annite series.

Hormann and Morteani (1969) show, however, that this curve is not applicable to all biotites.

Nepheline. The replacement $Na \rightleftarrows K$ in nephelines affects the cell parameters and causes variation in 2θ values of appropriate reflections (see Smith and Sahama, 1954). At the kalsilite end of the Na–K range the separation of reflections $10\bar{1}2$ and $10\bar{1}1$ can be used to estimate composition (Sahama and Yoder, 1956).

Pyroxene (Clinopyroxene). The variation of cell parameters over the field diopside-hedenbergite-ferrosilite-clinoenstatite has been given by Viswanathan (1966).

The cell parameters can be obtained from indexed powder patterns: $a \sin \beta$ can be deduced more directly than a and $\sin \beta$ separately and can be a useful parameter for determination purposes. Linear regression graphs of cell parameters against composition have been provided by Winchell and Tilling (1960).

Rutstein and Yund (1969) have shown that in calc-silicate skarns, if one is sure that elements other than Ca, Fe, Mg, Si and O are known to be present only in very low concentrations, the composition of the clinopyroxene (diopside-hedenbergite series) coexisting with wollastonite and garnet can be estimated within ± 5 mol % from

$$d_{220} = 3.2329 + 0.000416 \ X,$$
$$d_{310} = 2.9505 + 0.000439 \ X,$$
$$d_{13\bar{1}} = 2.5640 + 0.000206 \ X,$$
$$d_{150} = 1.7538 + 0.000194 \ X,$$

where X is the mol % of hedenbergite, accurate to within ± 5 mole % hedenbergite.

Pyrite. Riley (1968) related d_{511}, Å of the series pyrite (FeS_2) – Cattierite (CoS_2) to weight percent cobalt and found a good straight-line relationship.

Pyrrhotite. The value of x in the pyrrhotite formula $Fe_{1-x}S$ can be correlated with the lattice spacing d_{102} (Arnold and Reichen, 1962). X-ray methods can also be used to determine quantitatively the occurrence of hexagonal and monoclinic pyrrhotite in a specimen (Graham, 1969).

Scapolite. Burley et al. (1961) have shown a variation of $2\theta_{400} - 2\theta_{112}$ in scapolite in terms of composition.

Serpentine. Whittaker and Zussman (1956) have shown how a serpentine mineral can usually be identified as chrysotile, lizardite, or antigorite by X-ray diffraction.

Sphalerite. Skinner et al. (1959) have shown how the unit cell edge of sphalerite varies with the substitution of iron for zinc.

Short and Steward (1959) have described a method of determining the proportions of cubic and hexagonal (wurtzite) structures in sphalerite.

Spinel. The two parameters a, Å and $I(220)/I(111)$ can be used in conjunction to determine the composition of magnesium spinel in the solid solution series between $MgAl_2O_4 - MgCr_2O_4 - MgFe_2O_4$ (Allen, 1966). Deer et al. (1962, 5, 61) have shown how a combined knowledge of the unit cell edge, refractive index, and specific gravity may be used to define a spinel. Stevens (1944) has related the unit cell edge a, Å to the Cr_2O_3 content of chromites. His data were recalculated and supplemented by Hutchison (1972) to form the relationship

$$\text{Al cations per unit cell} = 407.3 - 48.6 \times a \text{ (in Å)},$$
$$\text{Cr cations per unit cell} = 58.46 \times a \text{ (in Å)} - 473.88.$$

MacGregor and Smith (1963) gave the approximate relationships for chrome spinels as

$$\%Al_2O_3 = 45.4 - 104.8 \ (a, \ \text{Å} - 8.00),$$
$$\%Cr_2O_3 + Fe_2O_2 = 114.7 \ (a, \ \text{Å} - 8.00) + 22.2,$$
$$\%Cr_2O_3 = 96.0 \ (a, \ \text{Å} - 8.00) + 22.3,$$
$$\%FeO = 33.7 \ (a, \ \text{Å} - 8.00) + 11.6,$$
$$\%MgO = 18.2 - 34.3 \ (a, \ \text{Å} - 8.00).$$

They also gave the refractive index of chrome spinels as $n = 1.7008 \ a, \ldots - 2.0332$.

Topaz. Ribbe and Rosenberg (1971) have shown that the weight $\%$ fluorine in topaz is $155.6 - 35.7 \ (\Delta \ 021)$, where $\Delta \ 021 = 2\theta \ NaCl_{200} - 2\theta \ topaz_{021}$, using $CuK\alpha$ radiation. Reagent grade NaCl is the internal standard. A diffractometer trace from $2\theta \ CuK\alpha$ 26.5 to 32.5 at a slow speed is run, and the separation between $topaz_{021}$ at 27.9 and the $NaCl_{200}$ peak near 31.7 is measured to the nearest $0.001°$ to obtain $\Delta 021$. The regression equation gives $\pm 0.5\%$ F by weight error.

Tourmaline. Horn and Schulz (1968) have described an X-ray method of determination of the chemical composition of tourmaline.

Vesuvianite (Idocrase). X-ray powder data have been given for vesuvianite by Domanska and Nedoma (1969).

SPECIFIC GRAVITY DETERMINATION 8

The specific gravity or density of a crystalline substance is one of its fundamental properties and is characteristic of the substance. As such, it is of diagnostic value and should be carefully determined. An accurately calculated specific gravity can help in the identification of a mineral if its chemistry is more or less fixed and in its composition determination if the mineral species belongs to an isomorphous series. As an example, the density of olivine $= 4.4048 - 1.353\ X - 0.0435\ X^2$, where $X =$ the mole fraction forsterite in the Mg-Fe olivine ($=$ mol % forsterite/100) (Fisher and Medaris, 1969). Other examples may be found in mineralogy textbooks such as Berry and Mason (1959) and Mason and Berry (1968). To give some other familiar examples, the specific gravity of plagioclase varies linearly from 2.62 for pure albite to 2.76 for pure anorthite. The specific gravity of the garnet group has already been fully described as a diagnostic property in Chapter 7 (exercise 7.10).

The specific gravity of rock specimens also gives some indication of their composition, though this is by no means so useful a property in petrography as it is in mineralogy; nevertheless it may have definite application in the description of a suite of fine-grained to glassy rocks.

The accurate determination of specific gravity requires considerable care. The most serious errors to guard against are (a) in the method used, (b) arising from inhomogeneity of the sample, and (c) arising from the operator himself. It is important to select the method most suited to give accurate results with the material available.

Errors arising from specimen inhomogeneity require special mention. It is frequently difficult to obtain a large piece of homogeneous material, even if a large hand specimen has been obtained, because of the inclusion of foreign material that invariably is present. Generally, therefore, the best results will be obtained by using a small carefully selected part of the larger specimen. This small part should be carefully checked for purity and homogeneity under a binocular microscope before measurement. Great difficulty arises when fine-grained porous materials are dealt with, for the air trapped in the pores will lead to an erroneously low specific gravity determination. Care should always be taken to expel any air from porous specimens before weighing in the liquid. This can be done by placing the liquid container in a bell jar and using a vacuum pump to help draw out any air trapped in the pores of the submerged specimen before it is weighed. Errors due to the observer himself can be obviated only by a careful operating procedure.

Three basic methods are in common use for geological materials:

1. The weight is measured directly and the volume determined by the principle of Archimedes. Suitable balances are the Kraus-Jolly and the Berman, both of which are described on the following pages.

2. The weight is measured directly and the volume found from the weight of liquid displaced in a pycnometer. An accurate pycnometer method is described.

3. The specific gravity is determined by direct comparison with that of a heavy liquid of known specific gravity.

Each of these methods is described in the following exercises.

EXERCISE 8.1

Density Determination by the Archimedes' Principle

The volume is determined by measuring the apparent loss in weight when a weighed mineral fragment is immersed in a suitable liquid of known specific gravity. The fragment displaces an amount of liquid equal to its own volume and its weight is apparently diminished by the weight of the liquid displaced.

If W_1 is the weight of the fragment in air and W_2 is the weight of the fragment when immersed in a liquid of specific gravity L, the specific gravity of the fragment $G = (W_1 \times L)/(W_1 - W_2)$. Water (specific gravity $L = 1$) is frequently used as the immersing liquid, in which case $G = W_1/(W_1 - W_2)$. Water is not a good choice of liquid for accurate determinations, however. It should be used only when a rough result is all that is required. This is because water has a high surface tension and does not wet solids readily. Bubbles are tenaceously held on the solid and the result is an erroneously low specific gravity determination.

Organic liquids of known density and high purity such as toluene and carbon tetrachloride are much better in this respect because their surface tension is one-third to one-fourth that of water. Cabri (1969), however, recommends that carbon tetrachloride should be discontinued as an immersion liquid, for in addition to its toxicity it suffers from photochemical reactions. He recommends hexachloro-1, 3-butadiene (also called hexachlorobuta-1,3-diene), with formula CCl_2:$CCl \cdot CCl$: CCl_2 (obtainable at high purity from Hooker Chemical Co., Niagra Falls, New York, or B.D.H. Chemicals Ltd., Poole, Dorset BH 12 4NN, England). This chemical has a specific gravity at 25°C of 1.67067 g/cc with a density correction factor of 0.00149 g/cc(°C).

METHOD A

Use the 5000 *Kraus Jolly balance* (supplied by Eberbach Co. P.O. Box 1024, Ann Arbor, Michigan 48106, or by Ward's Natural Science Establishment Inc., P.O. Box 1712, Rochester New York 14603), illustrated in Fig. 8.1:

1. Check the two verniers and scales (A and B, Fig. 8.1) at the top of the instrument for correct zero. Adjusting screws with lock nuts may be turned for accurate adjustment.

2. Place a beaker of 100 to 250 ml (large enough for the specimen to be contained in it freely) filled about three-quarters with hexachlorobuta-1,3-diene, toluene, or carbon tetrachloride on the support shelf C (Fig. 8.1). Distilled water may be used if only a rough determination is required.

3. Determine the weight of the sample to the nearest gram by any other balance.

(a) Use the light spring supplied when the sample weighs between 1 and 10 g.

(b) For a sample weighing between 10 and 20 g trim the length of the light spring until its maximum elongation of 40 cm is reached when a load of 22 g is suspended from it.

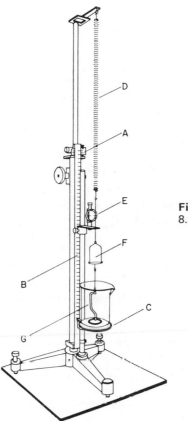

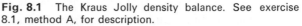

Fig. 8.1 The Kraus Jolly density balance. See exercise 8.1, method A, for description.

(c) Use the heavy spring supplied when the sample weight is between 20 and 50 g.

(d) For a sample weighing between 50 and 100 g trim the length of the heavy spring until the maximum elongation of 40 cm occurs when a load of approximately 100 g is suspended from it. *NOTE.* Greatest accuracy is attained when the spring operates near its point of maximum elongation.

4. The appropriate spring should be suspended from the overhanging support arm at the top of the instrument with its coils of smaller diameter at its upper end. Attach the index rod E with reading disc, then the metal pan F, and the glass pan G (Fig. 8.1).

5. Immerse the glass pan in the liquid (make sure that no air bubbles adhere) until the metal pan is about 3 to 4 cm above the liquid level.

6. Level the instrument by means of the screws in the tripod base by aligning the index rod concentrically with its adjustable guide and parallel with the pillar.

7. Further zeroing should be done at this point. Line up the index disc E with the line on the mirror. Coarse adjustment is accomplished by raising or lowering the mirror support on the support rod, fine adjustment by manipulation of the screw adjustment on top of the mirror and by moving the inner rod carrying the spring support.

8. Place the thoroughly cleaned and dried pure and homogeneous specimen on the metal pan in air. Using the large hand wheel on the left side of the instrument, move the scale up until the index disc is in line with the mirror. To do this the scale clamp on the lower right side of the instrument must be loose.

9.   The weight of the sample in air W_1 can be obtained by taking the reading on the left-hand side of the scale from the top down, employing the fixed vernier.

10.   Clamp the scale on the lower right side of the instrument.

11.   Move the sample by hand or by forceps to the glass pan in the liquid by lowering the beaker, changing the sample into the lower pan, and raising the beaker again until the metal pan is again approximately 3 to 4 cm above the liquid level. It is important that no air bubbles remain on the sample or pan.

12.   Turn the large hand wheel clockwise to lower the scale until the index disc is in line with the line on the mirror. The scale clamp is still locked as in step 10.

13.   The second reading required can now be made from the right-hand side of the scale reading down by employing the movable vernier. This reading represents the loss in weight $W_1 - W_2$ between the weight of the specimen in air and in the liquid.

14.   The specimen is removed and the specific gravity calculated from the equation

$$G = \frac{W_1 \times L}{W_1 - W_2}.$$

15.   The value of L, the specific gravity of the immersion liquid, is obtained by consulting Table 8.1 and using a thermometer to check the temperature of the immersion liquid at the end of step 13.

Table 8.1   Densities of the Three Recommended Immersion Liquids at Various Temperatures. Intermediate Values May Be Arithmetically Extrapolated

Temperature (°C)	Density (g/ml)		
	Hexachlorobuta -1,3-diene $CCl_2:CCl.CCl:CCl_2$	Carbon Tetra- chloride CCl_4	Toluene C_7H_8
17	1.68268	1.59986	0.86842
18	1.68119	1.59793	0.86750
19	1.67970	1.59599	0.86657
20	1.67821	1.59405	0.86564
21	1.67672	1.59211	0.86472
22	1.67523	1.59017	0.86379
23	1.67374	1.58823	0.86286
24	1.67225	1.58629	0.86193
25	1.67076	1.58434	0.86100

DISCUSSION

Distilled water should be used only when a rough determination is sought. It can be taken as having a specific gravity of 1. The Kraus-Jolly balance is recommended for larger specimens exceeding 1 g. More accurate results can be obtained by using smaller, purer fragments on the Berman density balance. The Kraus-Jolly balance should be kept clean and a light coat of vaseline kept on the pillar rack. Springs should never be left hanging on the balance when the instrument is not in use.

METHOD B

Use the Berman density torsion balance supplied by the Bethlehem Instrument Co. Inc., Bethlehem, Pennsylvania, and illustrated in Fig. 8.2.

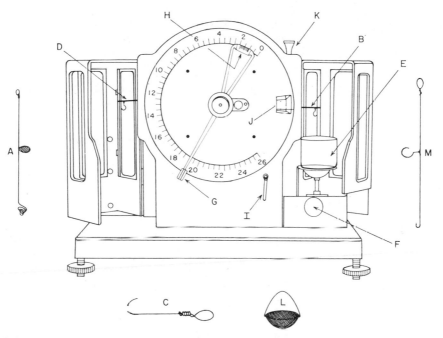

Fig. 8.2 Berman density torsion balance for accurate density determination of small fragments in the range 15 to 25 mg. See exercise 8.1, method B, for details.

1. Set up the balance on a solid bench. Adjust the leveling screws on the two front feet until the bubble in the spirit level is centered.

2. Hang the double weighing pan A on the right end of beam B (Fig. 8.2) and the counterweight C on the left end of beam D (Fig. 8.2).

3. Fill the liquid dish E with one of the liquids listed in Table 8.1 and turn knob F until the lower section of the double weighing pan A is completely immersed.

4. Adjust index level G until zero on the vernier H coincides exactly with zero on the dial. Release beam BD by rotating the beam clamp lever I until the lever points directly forward or back. Balance pointer J should now coincide with the zero line. If it does not, rotate the zero adjustment knob K until the beam is in balance and the needle on J is exactly zeroed. The balance pointer J is attached rigidly to the beam BD and swings in front of a small mirror. To eliminate errors due to parallax, the eye, the balance pointer, and the image of the eye in the mirror must all be in one line. Do not adjust K, thereafter.

5. The index lever G can be adjusted to different positions by loosening the central knurled knob so that the arm does not obscure the balance dial.

6. When the balance is brought to zero, lock the beam by rotating beam clamp lever I. The balance is now ready for use.

7. Select a pure homogeneous mineral fragment for measurement. A sample in the range 15 to 25 mg proves to be most satisfactory. Accuracy of measurement decreases rapidly below 10 mg. Samples greater than 25 mg should be avoided also because of loss of accuracy. Use a single grain and examine it under a binocular microscope to make certain it is free from cracks and inclusions. The sample should be a fracture or cleavage fragment. Wash the grain in the pure immersion liquid to remove dust and make sure that it is thoroughly dried.

8.   Slide the grain onto the top tray of the suspended double weighing pan A in a trough made of aluminum foil. Do not lift the grain with a forceps because small chips may be broken off. Close both balance doors to prevent drafts.

9.   Move the index lever G to a position on the scale heavier than that of the grain tested. This is to prevent the weighing pan from dropping abruptly into the immersion liquid.

10.   Release clamp I and move index lever G gradually toward the lighter readings until a balance is reached. Gentle tapping on the right side of the instrument will prevent a sudden relaxation of surface tension when the lower section of the weighing pan support is slowly released into the dish. Always approach the final weight from the heavier rather than from the lighter side to allow the weighing pan to move down into the liquid rather than up. When the beam is exactly balanced, as indicated on dial J (Fig. 8.2), the index vernier H gives the weight W_1 of the specimen in air.

11.   Weighing in air should be immediately followed by weighing in liquid to prevent inaccuracies caused by change in temperature if any delay is made. All readings should be taken with both doors closed to prevent air currents from introducing errors. To transfer the grain from the top tray to the bottom wire loop clamp the balance (lever I) and remove the weighing pan A with tweezers. Do not move knob F which would change the level of the liquid.

12.   Slide the grain from the top tray onto the foil trough and then slide it into the bottom wire loop of the weighing pan, which is secured by the forceps. The weighing pan should be held vertically with the loop just touching the table surface to provide support.

13.   Replace the weighing pan A on the balance hook B. Close the draft-excluding doors. Unlock the balance clamp lever I and weigh the specimen in liquid (steps 9 and 10). When the balance is exactly at zero on dial J the vernier reading H will give the weight of the specimen W_2 immersed in liquid. At the completion of the measurement check the zero point of the balance after removing the sample (step 8). It is suggested that at least three readings be taken of the specimen in air and three subsequently in liquid, using the average value for the specific gravity determination.

14.   While the specimen is being weighed in liquid a thermometer should be hung up in the right-hand draft excluder. The temperature should be measured and the specific gravity of the liquid L read from Table 8.1 for that temperature.

15.   The specific gravity of the mineral is now computed as

$$G = \frac{W_1 \times L}{W_1 - W_2}.$$

DISCUSSION

The balance may also be used for samples between 25 and 75 mg, although the accuracy will be less. With no weight on the left hook D, the balance records to full scale of 25 mg, as indicated on the dial. If as much as 50 mg is to be weighed, a 25-mg counterweight is placed on the left hook. The zero on the balance is then obtained when H points to 25 mg, and this amount must be added to any reading when the specimen is weighed. When weighing up to 75 mg, a counterweight of 50 mg is placed on the left hook and this amount is added to any reading on the dial. Weighings in excess of 75 mg should not be done because the balance will be damaged. The values of the various counterweights are marked on their respective containers. When using the double weighing pan A or the basket and wire hook combination L and M (Fig. 8.2), a compensating weight must always be hung on the left hook. When a counterweight and a compensating weight are in use simultaneously, one is suspended below the other on the left hook.

When a single grain is not available, a coarse powder may be used. Coarse powders are first weighed in a small wire-mesh basket L suspended on the upper hook M from arm B. None of the powder grains should be fine enough to escape through the wire mesh. Before introducing the powder, the hook M with the wire basket L must be counterbalanced by an appropriate weight hung on the left arm of the balance. The following measurements need to be made:

1. Weight of the basket in air (on upper hook).
2. Weight of the basket in the liquid (basket has been moved to the lower hook).
3. Weight of the basket + mineral in air (on upper hook).
4. Weight of the basket + mineral in toluene (on lower hook).

Readings $3 - 1$ give the weight of the mineral in air W_1. Readings $4 - 2$ give the weight of the mineral in the liquid W_2. When using the wire basket, care must be taken that all air bubbles, which may be trapped between the grains in the basket, are allowed to escape before the weighings.

METHOD C

The following method is particularly suited for top-pan balances, requires no special apparatus, and is time saving.

1. A rigid-sided container partly filled with a liquid of known density is placed on the pan of a top-pan balance.

2. The combined mass of liquid and container is measured. Let it equal a g.

3. The solid to be measured is suspended by a thin wire from a rigid beam (e.g., from a laboratory stand) into the liquid. The weight of the container plus liquid plus suspended solid is taken as b g.

4. The solid is detached from the rigid beam and allowed to drop into the liquid to the bottom of the container. The weight obtained is c g.

5. The density of the solid is given by $(c - a)/(b - a) \times d$ g/cm^2, where d is the density of the liquid.

(Method after Bensch and Brynard, 1972.)

EXERCISE 8.2

Density Determination with a Pycnometer

The pycnometer method requires good technique for accurate results. The design of the pycnometer stopper is critical. Figure 8.3 illustrates a superior design. It is made from heavy-walled capillary tubing with a bore of 2 mm. The base is formed by sealing one end with a hot torch and then pressing the softened glass on a graphite

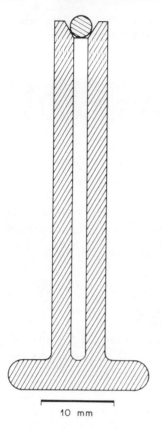

Fig. 8.3 Pycnometer design after May and Marinenko (1966). See exercise 8.2 for details of use in determining the specific gravity of powdered minerals.

10 mm

block. The open end is ground flat with a fine carborundum abrasive on a polishing lap and the outer edge gently beveled. The capillary is tapered by grinding with fine carborundum and a 60° tapered copper tool on a lathe and finished with polishing abrasive.

The metal ball used for closure is a $\frac{1}{8}$-in. precision monel ball bearing. The capillary height is 45 mm, giving a volume of 0.144 ml. With the ball closure a precision of ± 0.0001 ml can be obtained for volume measurement. (Design after May and Marinenko, 1966.)

1. Weigh the pycnometer empty (with ball bearing in place). Weight W_1.

2. Weigh the pycnometer containing the mineral sample (usually a powdered mineral of grain size -100 to $+200$ mesh of weight about 10 mg) with ball bearing in place. Weight W_2.

3. Place the pycnometer in a small bell jar fitted with a Teflon stopcock. Evacuate the jar with a vacuum pump and deliver enough tetrachloroethylene or toluene through the stopcock to cover the sample. The vacuum helps to remove any trapped air from the pores between the sample grains.

4. Remove the pycnometer and place it on a filter paper. Fill it completely with the same liquid. Use metal-tipped tweezers to seat the ball on the tube. Wipe any overflow and remove most of the excess liquid on the tip of the pyncometer by gently dabbing with hardened filter paper.

5. Place the pycnometer on the balance pan and observe the liquid-ball interface through a magnifier against a white background. The weight is recorded at the very instant that a small air bubble begins to appear just below the contact between the ball and the glass. Weight W_3.

6. Empty the pycnometer and wash out all the grains with the same liquid.

7. Fill the pycnometer with the same liquid—tetrachloroethylene or toluene. Repeat the exact procedure from step 4 to obtain the weight. Weight W_4.

8. Calculation:

$$\text{weight of sample} = W_2 - W_1,$$

$$\text{volume of pycnometer} = \frac{W_4 - W_1}{x},$$

where x is the specific gravity of the liquid; x may be determined from Table 8.1 for the appropriate temperature.

$$\text{volume of sample} = \frac{W_4 - W_1}{x} - \frac{W_3 - W_2}{x} = \frac{W_2 - W_1 + W_4 - W_3}{x},$$

$$\text{specific gravity of sample} = \frac{x(W_2 - W_1)}{W_2 - W_1 + W_4 - W_3}.$$

(Method after May and Marinenko, 1966.)

DISCUSSION

The specific gravity of the sample can be obtained by using the specific gravity of the tetrachloroethylene (or toluene) at the appropriate temperature. If a table is not available, a calibration curve can be prepared by measuring the specific gravity of the liquid at various temperatures, using a 10-ml specific gravity bottle.

NOTE.   The pycnometer is best placed off center on the balance pan so that the air bubble will tend to form on the front side. Unless this is done, it will not be possible to select the precise moment for weighing, when the bubble just begins to form. If the bubble forms behind the ball, it will not be seen until it grows too big.

This method gives very satisfactory results for fine mineral powders of small amount. Such powders cannot be measured satisfactorily by any other method.

EXERCISE 8.3

Density Determination with Heavy Liquids

1. Equipment required:

A glass cylinder or beaker

A burette

A complete set of density markers: 21 glasses whose specific gravities range from 2.28 to 4.08 (obtainable from Gemmological Instruments Ltd., Saint Dunstan's House, Carey Lane, London E.C.2, England).

A set of Sink-Float accurately calibrated and marked glass density capsules of

2 cm length covering the range 0.25 to 7.50 is available from R. P. Cargille Laboratories Inc., Cedar Grove, New Jersey 07009.

A suitable heavy liquid (see list in exercise 5.5).

A suitable diluent (see list in exercise 5.5).

2. Carry out a preliminary test to determine that one of the selected glasses of the set of density markers is denser than the mineral.

3. A suitable quantity of the pure heavy liquid is placed in the glass tube or beaker, together with the mineral fragment whose density is to be measured and the density marker that is known to be denser than the mineral. The heavy liquid should be denser than the solids so that both will float.

4. The appropriate diluent is added from the burette, with thorough stirring. A magnetic stirrer placed under the beaker and activated intermittently is ideal. Burette volume readings are taken just before, during, and after the sinking of the density marker. Stirring should be slowed and should cease when the solid sinks.

5. The process is continued. The next less dense marker of the series is added, and the diluent is added gradually until it sinks. The total burette reading is then taken (step 4).

6. The process goes on until the mineral fragment sinks. Burette readings are likewise taken for its sinking. It is essential to proceed beyond this stage to observe the sinking of at least one or two density markers less dense than the mineral fragment.

7. A plot is made of burette volume readings versus density and the density of the unknown fragment obtained by interpolation on the smooth curve drawn between the points.

(Method after Embrey, 1969.)

DISCUSSION

For a very small crystal fragment it may be necessary to centrifuge the tube at the critical time to see whether the fragment is sinking or floating. For larger grains centrifuging will be unnecessary.

This method requires neither an accurately known initial volume of heavy liquid nor an accurately calibrated burette. Any drift in temperature during the procedure is compensated for by the markers and merely alters the slope of the curve.

After completion the heavy liquid may be recovered (exercise 5.6).

EXERCISE 8.4

Density Determination with a Pycnometer and Heavy Liquid

1. Make and assemble the following items of equipment:

   a. A Pyrex pycnometer of dimensions shown in Fig. 8.4, heated in a furnace for 18 hours at 600°C to remove all internal stresses and cooled slowly.

   b. A water bath, with means for thermostatically or manually controlling the temperature, assembled as shown in Fig. 8.4.

2. Clean the pycnometer. Place the crystal or mineral fragment in it and weigh it, complete with stopper.

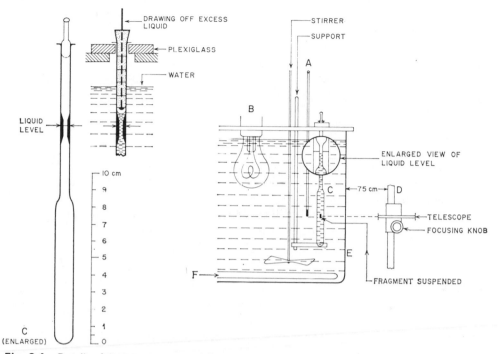

Fig. 8.4   Details of the pycnometer and the water bath for density determination by the suspension method. A = thermometer; B = heating element; C = pycnometer just before removing it from the bath; D = telescope or magnifying glass; E = glass jar; F = cooling coil (optional). (After Straumanis, 1953.)

3.   Fill the clean pycnometer containing the mineral with triple distilled and de-aerated water and hold it immersed in the water bath at each of the temperatures 25, 30, 35, and 40°C on subsequent occasions. When the temperature is stabilized, remove or add water to the pycnometer to bring it exactly to the mark. Pour the water into an accurate measuring cylinder and prepare a calibration chart of liquid volume against temperature.

4.   Place the mineral fragment in a test tube with a heavy liquid of approximately the same density as the substance. The fragment will either rise or sink slowly. The heavy liquid is prepared by mixing it with the appropriate diluent.

5.   To obtain the suspension of the substance in the liquid, when the two densities are exactly matched, transfer both liquid and substance to the pycnometer (Fig. 8.4), which is immersed in the water bath. Put the stopper on.

6.   Raise or lower the temperature of the bath until the substance becomes suspended in the liquid half way up the pycnometer. The density of the liquid is taken to be equal to that of the substance when the latter remains suspended in the middle for about one-half hour at constant temperature. Read the temperature.

7.   At this suspension temperature, the excess liquid is removed from the pycnometer (down to the mark) by means of a capillary pipette and by needle-like filter paper strips. The adjustment of the liquid level is checked by means of a telescope or magnifying glass. The pycnometer is now closed with a stopper.

8.   Remove the pycnometer, dry with chamois, place in the balance case for 10 minutes, and weigh.

9.   The density of the mineral is calculated as follows. a is the weight of pycnometer + mineral fragment (step 2), b is the weight of the pycnometer + mineral fragment + liquid (step 8), V is the volume of the liquid held in the pycnometer at the temperature of suspension (temperature, step 6); it is then obtained from the graph in step 3. The density of the liquid, hence of the mineral, is $(b - a)/V$.

(Method after Straumanis, 1953.)

DISCUSSION

The accuracy of the method is at least 3×10^{-4} g cm^{-3}. About 3 hours are necessary to perform one determination. If slightly less accuracy can be tolerated, step 3 can be shortened by simply determining the volume of the liquid in the pycnometer at room temperature and assuming that it will be constant over the temperature range. In fact, this assumption introduces little error into the density determination.

REFRACTIVE INDEX DETERMINATION 9

Minerals of more or less fixed chemical composition may usually be readily identified on the basis of their refractive indices (RI). On the other hand, a mineral species belonging to a group that shows a wide range of chemical isomorphism may have its composition indirectly determined by an exact RI determination. A large number of tables relating RI with composition for the majority of mineral groups has been prepared but they are too numerous even to attempt to detail here. The reader is referred to standard systematic and optical mineralogy textbooks for details, especially to Tröger et al. (1971), Deer et al., (1962a, b, c; 1963a, b; 1966); Heinrich (1965); Kerr (1959), and Winchell and Winchell (1951).

Precise RI determination is therefore a most valuable tool in mineralogy. It may be done rapidly within rough limits for routine identification or very precisely, if necessary, with more care and time. The degree of precision will naturally depend on the purpose. It is always combined with the simultaneous collection of related optical data, for a polarizing microscope is used, and it is convenient to collect other data, such as $2V$ and sign, at the same time.

EXERCISE 9.1

Routine Method for Refractive Index Determination

1. Crush a small fragment of the mineral in a steel pistontype percussion mortar with a close-fitting steel cylindrical pestle of 28-mm diameter (Griffin and George Ltd., Ealing Road, Alperton, Wembley, Middlesex, England, or Gallenkamp, P.O. Box 290, Technico House, Christopher Street, London E.C.2). Use a percussion action by hand on the pestle rather than a grinding action.

2. Sieve briefly and sort out the grains that pass the 100 and collect on the 170 mesh sieve to obtain the best grain size for immersion study.

3. With a small spatula blade scatter a *few* grains on the central area of a clean glass slide and cover with a circular coverglass.

4. Every laboratory should have a complete stock of bottled RI oils. They may be obtained from R. P. Cargille Laboratories Inc., Cedar Grove, New Jersey 07009, Buehler Ltd. and A. I. Buehler, Inc., 2120 Greenwood Street, Evanston, Illinois, 60204, or Buehler-Met AG, Postfach 4000, Basel 23, Switzerland. These oils are supplied in small conveniently sized bottles with screw caps that incorporate a glass rod for easy application. The bottles are labeled with all the necessary data. Oils of RI 1.300 to 2.11 are available in 207 different indices. The temperature coefficients and dispersions are marked on each label. The oils are also supplied in sets covering specific ranges of intervals .002, .004, .005, .01 and .02. Details may be obtained from R. P. Cargille.

5. From a bottle containing an oil of refractive index close to that of the mineral apply with the glass rod a minimum of liquid to the contact between the cover glass and the slide. The oil will soak in by capillary attraction. Apply only enough oil to fill the space. Wipe away any excess with a filter paper. Mark the value of the oil on the slide. At this stage it is not essential that the oil be matched to the RI of the mineral, and really any oil of approximately equivalent RI is suitable.

6. Place the preparation on the microscope and observe under crossed polars. A thorough examination should be performed to determine whether the mineral is ISOTROPIC, in which case all grains remain extinguished under the microscope and the solid may be cubic or glass; or ANISOTROPIC, in which case the grains will alternatively extinguish and be illuminated as the stage is rotated and will show birefringence. It is now necessary to use a higher power objective ($\times 40$) and to look for interference figures. Most information is obtained from those grains that are nearly isotropic, for they give nearly centered optic axis figures. Since the grains are small, the best figures may be obtained not by using the Bertrand lens but by removing the eyepiece ocular. The eye can be carefully adjusted in position above the microscope tube to hold it centered on the interference figure. Observe whether the mineral is UNIAXIAL or BIAXIAL. The sign should also be determined; if biaxial, the curvature of the isogyre in the 45° position should be estimated to give an approximate value of $2V$. The procedure from now on will depend absolutely on whether the mineral is isotropic, uniaxial, or biaxial and on the sign of the biaxial mineral.

7. *For isotropic (cubic or glass) solids only.* Such solids have only a single refractive index N for any wavelength; hence they require the simplest procedure. The RI of the grains is compared directly with that of the oil by using white light and a $\times 10$ objective and either plane polarized or unpolarized light. Any grain in any orientation will give the same result, for we have only one value N to determine. The methods of comparison of solid and liquid for RI are outlined in exercise 9.2.

8. *For uniaxial minerals only.* Every grain, even those in random orientation, will allow the determination of the ordinary RI $N\omega$. Only specially oriented grains, however, will allow the correct value of the extraordinary RI $N\varepsilon$. Because both are needed, it is best to make determinations only on those grains that are appropriately oriented. First immerse the grains in an immersion oil that is approximately equal in RI to the average of the mineral grains. Under crossed polars and $\times 40$ objective search all grains for one that remains isotropic. Check that it gives a CENTERED OPTIC AXIS figure. This grain is held centered on the cross wires, the polars are uncrossed and a $\times 10$ objective is used. It permits the comparison of $N\omega$ with the oil for any stage position. After $N\omega$ has been compared with the oil search the grains again with crossed polars and high power for a grain that will give a centered flash figure. This grain will be a prismatic section, readily identified because it will have the highest birefringence. Make sure that the flash figure is as perfectly centered as any grain in the preparation can allow. We assume that the polarizer vibration direction of the microscope in use is N–S (easily checked with a granite slide, in which case under plane-polarized light all prismatic sections of biotite are in their dark pleochroic position when the cleavage traces lie N–S).

For a grain that gives a centered flash figure both $N\omega$ and $N\varepsilon$ may be determined at each of its extinction positions. Observe the interference figure at its cross position (extinction). By rotating one or two degrees of the stage either way from extinction note the direction along which the isogyre breaks up and disappears from the field of view. If it is N–S, the optic axis will lie horizontally N–S, a direction that is equivalent to $N\varepsilon$, whereas $N\omega$ will lie E–W and vice-versa if it is E–W.

Still with crossed polars, but using a × 10 objective, adjust the section exactly to its extinction position in which the optic axis was found to lie N–S, that is, parallel to the microscope polarizer vibration direction. Now uncross the polars and, using plane polarized N–S light, we are in a position to compare $N\varepsilon$ of the mineral with the refractive index of the immersion oil. Rotate the section exactly through 90° and we are in a position to determine $N\omega$ in the same way. The numerical difference between the determined $N\omega$ and $N\varepsilon$ gives the birefringence of the mineral. If $N\omega$ is found to be less than $N\varepsilon$, the mineral is optically $+ve$, but of course this information will already have been found in step 6. In step 8, therefore, it is not essential to know whether we are determining $N\omega$ or $N\varepsilon$ in a prismatic section, for the optical sign already determined will confirm which is which.

9.   *For biaxial minerals only.* In biaxial minerals three principal refractive indices may be determined: $N\alpha$, the smallest, N_β, and $N\gamma$, the largest. Of course it is not necessary to determine each of these, for quite often the value of $2V$ may have been determined by a rotation method and the relationships are the following.

For optically $+ve$ minerals, for which $2V$ is the acute optic angle,

$$\cos^2 V\gamma = \frac{\alpha^2(\gamma^2 - \beta^2)}{\beta^2(\gamma^2 - \alpha^2)},$$

and for optically $-ve$ minerals, for which $2V$ is the acute angle,

$$\cos^2 V\alpha = \frac{\gamma^2(\beta^2 - \alpha^2)}{\beta^2(\gamma^2 - \alpha^2)}.$$

These calculations, however, should be attempted only if two of the refractive indices are known with high accuracy; otherwise large errors may occur in the computation.

Furthermore, most determinative tables for mineral groups show the relationship between the three principal refractive indices and the chemical composition of the members of the series. Accordingly, it is necessary to determine only one of these indices to make accurate use of the tables.

Randomly oriented grains are not suitable for these determinations; one has to search in the slide for centered interference figures and then perform the measurements only on those specially oriented sections.

The most useful orientation is that which gives a centered optic axis figure. Such grains are the most easily discovered in a slide, for they remain isotropic while the stage is rotated. Select only those grains that are isotropic or show the lowest interference colors. Check that they give a centered optic axis figure. When a suitable grain is found, uncross the polars and, using a × 10 objective, N_β of the mineral may be compared with the immersion oil for any position of the stage. Optic axis figures are always the most informative, for not only are they most readily found because of their isotropism but in addition to giving β they allow $2V$ and the sign to be estimated.

The next most easily located grains are those that will give an optic normal figure (flash figure) in which both X and Z lie on the plane of the section. They are identified first because of all the grains they show the highest interference colors. Select such highly birefringent grains, then eliminate all but those that give a centered flash figure.

Using a × 10 objective and plane polarized light, rotate the grain to one of its extinction positions; uncross polars and compare the refractive index (which will be $N\alpha$ or $N\gamma$) with the immersion oil. Now rotate the grain exactly through 90° on the microscope stage and the other refractive index ($N\gamma$ or $N\alpha$) may be determined.

Other centered interference figures may be utilized, although there are no good rules for finding these suitably oriented grains; this must be done simply by trial and error.

A centered *BXa* figure of an optically +*ve* mineral will allow $N\alpha$ and N_β to be compared with the oil at each of the extinction positions and for an optically −*ve* mineral $N\gamma$ and N_β. The direction in which either X or Z lies may be determined by noting along which line the crossed isogyre breaks up when the grain is rotated on either side of its extinction position.

A centered *BXo* figure of an optically +*ve* mineral will allow $N\gamma$ and N_β to be compared with the oil at each of its extinction positions and for an optically −*ve* mineral $N\alpha$ and N_β. Care must always be taken to ensure the correct identification of the type of interference figure obtained, for it is well known that as $2V$ increases the distinction between a *BXa* and a *BXo* figure becomes progressively difficult. Thus it may be advisable to specialize on optic axis figures that are unambiguous and supplement this information with optic normal or flash figure grains.

EXERCISE 9.2

Special Method of Orientating Grains for Refractive Index Measurements

The method of exercise 9.1 depends on one or more grains in a random preparation being in suitable orientation for a centered interference figure to be obtained. Many grains may not allow such a conveniently centered figure and many preparations may not give a suitable grain at all. Hence it may become necessary to use a modified method. Several are available and some examples are listed here:

METHOD A

A mineral grain may be attached to the axis of a spindle stage and rotated into any orientation that will give the necessary centered interference figure. Setting up the grain and the method have already been described in exercises 4.3 and 4.4 and need no repetition here.

Alternatively, the universal stage may be used with a grain immersion mount to align selected grains to give centered interference figures. The methods of orientation are described in exercises 4.7 and 4.8. Once the figures are centered the appropriate refractive indices of the mineral may be compared with that of the immersion oil. The universal stage, however, has the disadvantage that oils are less easy to change than in the spindle stage, and that the latter is preferable for RI determination.

METHOD B (*for uniaxial minerals only when an accuracy of better than* ±0.0005 *is required*)

1. Crushed fragments are sieved and the +0.3-, −0.7-mm fraction is retained. Four to six grains are put into the crystal grinder (supplied by Scientific Instruments Division, Enraf-Nonius P.O. Box 483, Rontgenweg 1, Delft, Holland), which is operated by compressed air, The resulting spheres usually range in diameter from 0.1 to 0.3 mm. If grinding is continued up to 10 minutes, excellent spheres of 0.02 to 0.03

mm are obtained. For optical work the spheres must have a fine finish. Hence grinding must end up with a very fine abrasive.

2. Two or three spheres of different sizes are mounted on the five-axis universal stage. A suitable immersion oil with an RI approximately midway between $N\omega$ and $N\varepsilon$ has to be selected. A very thin camel's hair brush, slightly moistened with the immersion oil, is used to pick up the spheres and put them on a glass slide (Corning No. 2950 is suitable). The glass slide must be approximately 55 to 60 mm long. This length allows a slight protrusion from the universal stage mount which is useful for moving and rolling the spheres into subsequent suitable orientation.

3. The diameter of each sphere is measured to an accuracy of at least 0.001 mm. This may be done with the Leitz or Zeiss screw micrometer eyepiece, both of which can give the necessary accuracy. The image-splitting measuring eyepiece (supplied by Vickers Instruments) is preferable, however, to the simple graticule and filar-type measuring eyepieces of Leitz and Zeiss for this accurate measurement. The eyepiece should be calibrated accurately against a stage micrometer. An image-splitting measuring eyepiece can have an accuracy as high as 0.0001 mm. The spheres must be measured dry. They must not be surrounded by immersion oil at this stage, for the fluid will induce a serious error.

4. The immersion medium and the upper coverglass are placed on the spheres, more medium is put on top of the coverglass, and the upper hemisphere of the universal stage is attached. The sphere to be studied must be centered on the cross wires. It is essential that it remain centered throughout the procedure and that the universal stage be perfectly aligned. One problem is that the largest diameter sphere takes all the weight of the upper hemisphere, thus causing the sphere and the overlying coverslip to be strained. Accordingly, the largest sphere should not be used for determinations.

5. Manipulate the universal stage to place the optic axis of the centered sphere horizontal and N–S. One of the problems of universal stage work is the achievement of precise extinction position, which can be facilitated by using a Nakamura half-shadow plate or a Mace de Lepinay half-shadow wedge (E. Leitz, Wetzlar, Germany). These accessories allow almost perfect extinction to be obtained.

6. The principal birefringence $N\varepsilon - N\omega$ is now determined. The crystal is rotated 45° on the microscope stage so that the optic axis is parallel to the accessory slot of the microscope. The path difference between ε and ω is measured with a Wright combination quartz wedge (obtainable from E. Leitz). This wedge operates in a slot in a special Wright eyepiece (E. Leitz). The standard Wright wedge has zero to four orders. Standard variant compensators have an upper compensation limit of four orders. Accordingly only minerals of low to medium birefringence, and only of small diameter spheres, can be studied. To solve this problem four- and eight-order quartz plate compensators of known path difference may be inserted below the Wright wedge, on the accessory slot of the microscope tube, in the substage assembly, or in the objective. This increases the thickness of the section without affecting the gentle slope of the Wright wedge and extends the upper compensation limit to 8, 12, or 16 orders without impairing the accuracy. However, Carl Zeiss can supply a range of Eringhaus compensators suitable for large path-difference determinations up to and beyond seven orders so that a single compensation plate will suffice. The principal birefringence is determined from

$$(\varepsilon - \omega) = \frac{\Delta}{T},$$

where $(\varepsilon - \omega)$ = the principal birefringence,

Δ = the path difference in nanometers (millimicrons),

T – the thickness of the section (diameter of the sphere) in millimeters.

7. The optical orientation is varied until the RI of the sphere equals that of the immersion fluid. First the optic axis is returned exactly to N–S horizontal, in which position, for a N–S polarizing microscope, the RI of the crystal equals $N\varepsilon$. From this position rotation on the outer E–W axis of the universal stage will bring the optic axis progressively upward from a horizontal toward a vertical N–S direction. Progressively, the RI will change from ε toward ω. A critical point will be reached when the crystal RI equals that of the immersion oil, thus giving a value to ε'. The Becke line method is used for matching RI (oblique illumination is found to be less accurate than central illumination). The angle of rotation on the outer E–W axis from ε to ε' is the complement of the angle ϕ from the microscope tube to ε; ϕ must be corrected for the difference in RI between the glass hemispheres and the immersion liquid. The standard tilt angle correction tables are used. The temperature of the immersion medium in the mount is read so that its RI can be corrected. Thus the true crystallographic $\varepsilon - \varepsilon'$ angle is computed.

8. With the axes of the universal stage locked in position such that ε' = RI of the immersion liquid the microscope is rotated to the 45° position and the path difference between ε' and ω measured by the Wright combination wedge. The partial birefringence $(\varepsilon' - \omega) = \Delta/T$. The refractive index of ω is determined by algebraically adding the partial birefringence $(\varepsilon' - \omega)$ to the refractive index ε'.

9. The index of refraction ε is determined by algebraically adding the principal birefringence to the refractive index of ω.

10. As an example of the procedure, the mineral quartz has been determined.

   a. Orient the quartz sphere with optic axis N–S horizontal.
   b. Principal birefringence.
      Compensation (path difference) ε to ω = 2224.9 nm $(\varepsilon - \omega)$:

$$\text{Principal birefringence} = \frac{\Delta \text{ (path difference)}}{T \text{ (diameter of sphere)}}$$

$$= \frac{2224.9 \text{ nm}}{0.2445 \text{ nm}} = 0.0091.$$

   c. Determine ε' and ω.
      ϕ corrected (degrees from ω to ε') = 33.7°.
      Temperature of mount = 18.7°.
      ε' corrected for temperature = 1.5469 at 20°C.
   d. Determine RI ω.
      Compensation (path difference) ε' to ω = 639.9 nm. $(\varepsilon' - \omega)$.
      Partial birefringence = Δ/T = 639.9 nm/0.2445 nm = 0.0026.
      $\omega = \varepsilon' \pm (\varepsilon' - \omega) = 1.5469 - 0.0026 = 1.5443$.
      [Quartz is $+ve$; therefore $(\varepsilon' - \omega)$ must be subtracted from ε'. If the crystal under study were $-ve$, $(\varepsilon' - \omega)$ would be added to ε'.]
   e. $\varepsilon = \omega \pm (\varepsilon - \omega) = 1.5443 + 0.0091 = 1.5534$

(Method after Sunderman, 1970)

METHOD C

Special techniques have been designed to simplify the determination of certain mineral groups. Details are to be found in mineralogy textbooks; for example, determinative graphs of plagioclase composition are available which relate composition to the refractive indices of glasses formed by melting a small amount of the plagioclase, quenching it to a glass, and determining its single refractive index. The refractive index of an isotropic glass is more easily determined than the RI of oriented fragments of the biaxial crystals, and for plagioclase the rate of change in refractive index of the glass with composition is approximately twice as great as the indices for the crystals. This method can also provide an average composition for zoned or exsolved crystals (Deer et al., 1966, p. 329).

Because crushed grains of plagioclase tend to lie preferentially on cleavage directions (001) and (010), determinative curves of composition versus the refractive indices found on these oriented grains are available in most mineralogy textbooks (Tröger et al., 1971). The grains must be identified by their parallel edges as cleavage fragments, and the two refractive indices of each of the two types of grain (010) and (001) determined at their extinction positions. Morse (1968) gives plagioclase determinative graphs based on method D in exercise 9.3.

EXERCISE 9.3

Matching Refractive Index of Solid and Immersion Oil

Having oriented the grain into its proper position for comparison of its selected RI with that of the immersion medium by the methods in exercise 9.1 or 9.2, we are now ready to make a direct comparison of the refractive indices of the solid and the immersion oil. Eventually we hope to match the two exactly. By knowing that of the oil we deduce the RI of the solid. Various properties and techniques are in use for making the comparison and the exact matching, and it is common procedure to employ more than one of the following procedures:

Relief. If a colorless isotropic substance like glass is immersed in a colorless immersion oil of the same refractive index, the glass will be invisible in the liquid because of the absence of refraction and reflection effects at the junction of solid and liquid. If the refractive indices are different, the RELIEF, or strength, of the border phenomena between the solid and liquid depends on the magnitude of the difference between the refractive indices. High relief means that the solid can be seen in the liquid. No relief means that it is difficult to see, implying equality of the refractive indices. The property of relief therefore is not absolute. A glass of RI 1.54 will have no relief if immersed in an oil of RI 1.54 but will have equally high relief if immersed in an oil of 1.34 or 1.74. Its relief simply indicates the difference between the refractive index and that of the liquid in which it is presently immersed. Relief, however, does not tell us if the refractive index of the solid is greater or less than that of the oil; it gives us the scale of the difference but not the sign.

Because the object of the exercise is to immerse the grain in an oil of identical RI, we must keep on changing the immersion oil, all the time lowering the relief, until the final preparation contains an oil that causes no relief. If white light is used and relief disappears by matching the RI of the solid and liquid, the outlines of the grain will be bordered by color fringes, usually green and red. The appearance of such fringes confirms that the two refractive indices are approximately equal. The fringes are there because the refraction phenomena at the grain-liquid boundary act differently on the different wavelengths of the white light.

Because relief tells us only of the similarity of or difference between the two refractive indices and does not tell us whether the solid is greater or less in RI, we need to use an additional technique, such as one of the following, to indicate whether the RI of the solid is greater or less than the liquid so that we will know which oil to use next to gain a subsequent exact match.

METHOD A   BECKE LINE

1.   Focus the microscope (best using the × 10 objective) on the junction between the grain and the immersion liquid. White plane polarized light is employed. At this stage it is better that the lighting not be too strong. Closing the substage diaphragm and even slightly lowering the substage condenser will benefit the observations.

2.   Now slowly *increase* the distance between the objective and the microscope stage. Some microscopes lower the stage, others raise the tube. The result is the same. The junction between the grain and the liquid will slowly move out of focus. As the grain is moving out of focus, the border region at the contact between grain and liquid will become accentuated, whereas under focusing conditions it was evenly illuminated (Fig. 9.1). A border zone of greater brightness, parallel to the outline of the grain,

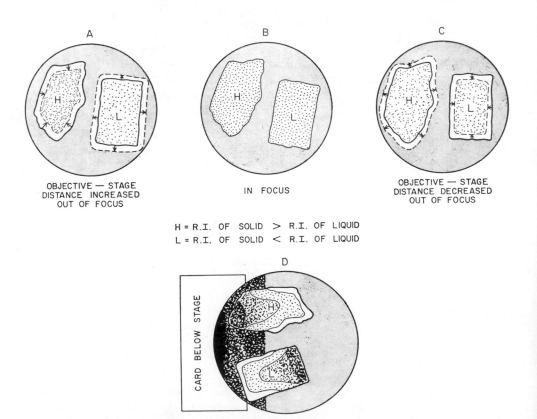

Fig. 9.1   Methods of comparing the RI of grains with that of an immersion oil: A, B, and C, Becke Line method; D, oblique illumination method. See exercise 9.3 for details.

called the Becke Line, will move in or out of the border as the microscope becomes more and more out of focus (Fig. 9.1). The rule is as follows: as the distance between stage and objective is INCREASING, the Becke Line (zone of brightness) moves into the substance of the *higher* RI. Thus in Fig. 9.1 the middle diagram B shows two grains in focus. The borders are not accentuated. Diagram A shows the results when two grains are slowly put out of focus by increasing the objective to stage distance. A border zone of brightness moves into the grain which has a higher RI than the immersion oil (H), whereas the border zone of brightness moves out into the immersion oil away from the grain whose RI is less than the oil (L). If the grains are slowly put out of focus by reducing the distance between the objective and oil (C in Fig. 9.1), the opposite effects are observed.

The Becke line is not to be regarded as a thin line of brightness. It is a zone of variable thickness that moves in or out of the grain away from the junction. It is always brighter than the surrounding region. As the focus is being removed, the Becke line becomes broader and fainter.

WARNING.   Some grains have two zones of brightness that move in opposite directions. One of these is the true Becke line; the other is caused by reflection from the cleavage faces at the mineral grain junction. The false line may be caused by using a power objective that is too high; never use one higher than X10. The false line may be eliminated by closing down the substage diaphragm to permit only minimum light to illuminate the grain. Lighting that is too bright is often the cause of the false line. If there still remains doubt about the identity of the true Becke Line, method B or C may be used instead of the Becke Line test.

3.   If the grain is found to have a higher RI than the immersion liquid, a new preparation is made with a liquid of higher RI value and the procedure is repeated. When, eventually, the grain is immersed in an oil that has approximately the same RI, the relief of the grain will have reduced almost to zero and the white Becke line will be replaced by color fringes. The appearance of color fringes in place of the white Becke line indicates that for a specific wavelength in the white spectrum the RI of the grain is equal to that of the oil. We will not, however, be able to determine which.

4.   At this stage it is essential to replace the white light of the microscope with a monochromatic light source. The refractive indices of minerals, as tabulated in mineralogy textbooks, are determined at a standard wavelength of 589 nm unless stated otherwise. This is the wavelength of a sodium vapor lamp (color yellow-orange). The white built-in lamp of the microscope should be replaced with a sodium vapor lamp, which can be put directly into the microscope housing in place of the white light or obtained in a separate lamp housing which makes use of a mirror to direct the light into the microscope. Sodium lamps are available from both Carl Zeiss and Ernst Leitz and many other microscope suppliers. A convenient alternative to the sodium lamp is to fit a continuous interference filter (Carl Zeiss or Vickers Instruments) on top of the microscope light source. Such a filter is continuously variable from 400 to 700 nm and may be set at 589 nm for the monochromatic determination. Alternatively a narrow band-pass interference filter of 589 nm (Carl Zeiss) is an ideal attachment to the light supply beneath the microscope condenser.

5.   With the monochromator attached the Becke line will appear as a yellow-to-orange band with no color fringes. The oils are changed until an immersion mount has been made with a selected oil in which the grains have zero relief and the Becke line is no longer seen, neither moving in nor out of the grain as the objective to microscope stage distance is increased. In this situation the RI of the grain exactly matches that of

the oil for a wavelength of 589 nm. To gain this exact match it may be necessary to mix two of the available immersion oils; for example, with the grains immersed in an oil of R.I 1.540) we may have found that the RI of the grains is greater than that of the oil. We remake the mount with the next available oil of RI 1.542 (R. P. Cargille's most complete set). In this case we may find that the RI of the grains is less than that of the oil, but there is no oil available of intermediate value in the set. We can now say that the mineral RI under observation is >1.540, <1.542. This precision is adequate for most work. If we wished to be more precise, we could mix together on a clean glass slide one drop each of the two oils to give an intermediate value before making the immersion mount; for example, a mixture of two similar drops of oil of 1.540 and 1.542 should result in an oil of approximately 1.541. Different proportions of adjacent value oils may be mixed to gain intermediate values. The precise RI of the mixed oil may be determined subsequently by means of a refractometer once the match has been made in the immersion mount.

NOTE.   When mixing oils, never touch the glass rod of one bottle with that of another. If a rod from one bottle is contaminated by any other liquid, it is essential that it be wiped clean before being returned to its correct bottle. Contamination of the liquids must be avoided at all cost.

METHOD B   OBLIQUE ILLUMINATION

1.   The substage diaphragm should be opened and a low power (maximum $\times 10$) objective used. Oblique illumination is obtained by placing a card or a finger over half the light supply or by shadowing the field with an accessory plate or the analyser frame (Fig. 9.1D).

2.   Under these conditions it will be seen that when a substance has a *higher* RI than the immersion oil a shadow will appear on the *same side* of the grain as that on which the light was cut off; the opposite border of the grain will be more strongly illuminated. This is illustrated in Figure 9.1D. The field of view is darker on the side on which the card has been placed. The grain with the higher RI (H) is darker on the same side and lighter on the illuminated side of the field of view. On the other hand, a grain whose RI is less than that of the oil (L) gives the opposite effect (Fig. 9.1D).

WARNING.   Care must be taken to place the card always in the same position with respect to the condenser, because the effect can be reversed by placing the cutoff card above or below the focus of the condenser. This method of oblique illumination should be used to confirm the result obtained from the Becke Line. When there is doubt, one may be used to check the result from the other.

METHOD C   USE OF PHASE CONTRAST EQUIPMENT

Phase contrast equipment is used to accentuate any slight difference in refractive index between the solid grain and the immersion liquid. The difference is shown up by different intensity of illumination of the grain and the liquid. At the exact matching of the two the grain and liquid illumination becomes identical. Matching is achieved by changing oils, temperature, or wavelength until the grains become invisible by having identical illumination. Monochromatic light is used as in the final stage of methods A and B. To convert a microscope for phase contrast study a special condenser must be fitted (the Heine condenser, E. Leitz, or the 11 Z, POL phase contrast condenser, Carl Zeiss). These instruments must be used in conjuction with the Pv phase contrast objectives (E. Leitz) or the Ph phase contrast objectives (Carl Zeiss). In addition, the Jamin-Lebedeff transmitted light interference equipment (Carl Zeiss) or the Zernike condenser in conjunction with the Phaco phase contrast objectives (F. Leitz) make suitable combinations and readily convert most microscopes for a phase contrast study. Unfortunately, most phase contrast condensers are suitable only for isotropic substances. The Jamin-Lebedeff transmitted light interference equipment (Carl Zeiss) may be used in conjunction with polarized light; hence it can be applied to anisotropic as well as isotropic minerals.

The exact match of the RI of mineral grain and immersion oil can be judged visually for equal illumination or a more refined match may be attained by the following procedure:

1. Each immersion oil supplied by, for example, R. P. Cargille laboratories will have full information marked on its label. As an example,

$$N25°C\ 5893\ \text{Å} = 1.560 \text{ temperature coefficient } dN/dt$$
$$\text{at } 15 \text{ to } 25° = 0.0004/°C;$$
$$\text{at } 25 \text{ to } 35° = 0.00039/°C.$$

This means that for monochromatic light of 589 nm the RI of the oil in this bottle is exactly 1.560 at 25°C. However, all refractive index oils lower their values with increasing temperature and vice versa. The rate of change is dN/dt, which is given on the label. Hence one can deduce that this oil has the following RI values for the given temperatures:

refractive index	1.5612	1.5608	1.5604	1.560	1.5596	1.5592	1.5588
temperature °C	22	23	24	25	26	27	28

and so on; the complete range of RI values may be deduced from the information on the label. Figure 9.2A shows a plot of the data given on this R. P. Cargille refractive index bottle. Similar plots may be made for adjacent oils.

2. Use is made of the change in refractive index with temperature to gain an exact match between solid grain and oil by varying the temperature until the match is perfect and noting the temperature at which this happens. This is a more satisfactory procedure than mixing oils at constant temperature.

3. The preparation is best heated on an electrically heated support glass with built-in thermistor. The temperature can be observed on the microammeter of the thermistor thermometer or directly on a thermometer integrated into the stage. Suitable heating stages are the Kofler micro cold/hot stage (supplied by C. Reichert, optische Werke AG, Wein XVII, Hernalser Haupstrasse 219, A 1171 Vienna, Austria), which has a temperature range of −50 to +80°C, a heating and cooling stage 47 80 15 (Carl Zeiss), which can be used for temperatures of −30 to +70°C with an accuracy of 1/40°C, and a cold/hot stage (Leitz), which has a similar range of −20 to +80°C and high accuracy.

4. We start with a mineral grain whose RI is just less than that of the oil. Monochromatic light is used.

5. The specimen is gradually heated, causing the RI of the oil to reduce progressively. At a certain temperature the RI of the mineral will match that of the oil exactly. This temperature must be read and the RI of the oil deduced. The exact moment of match is found when the illumination of the fragment is exactly equal to that of the enveloping oil. This is elegantly done by attaching to the top ocular of the microscope a photometer used for reflectivity measurement in conjunction with galvanometer or digital readout (exercise 2.12 and Fig. 2.3).

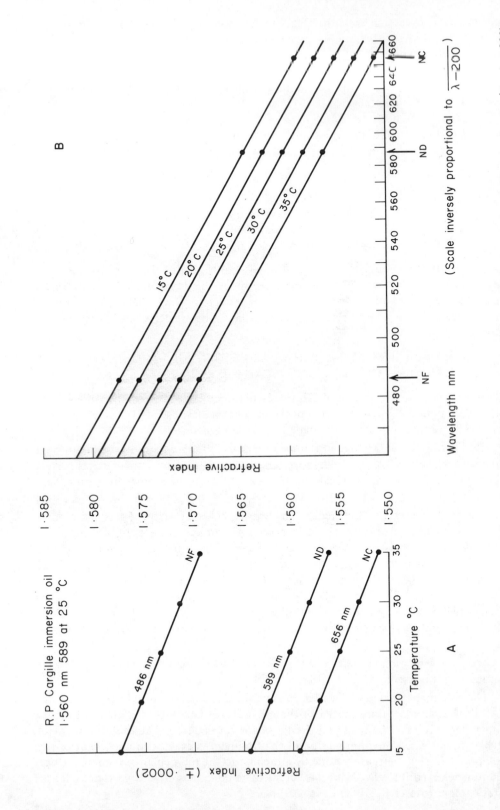

Fig. 9.2 Variation of RI of an immersion oil : A, temperature on a linear scale ; B, wavelength on a Hartmann scale (inversely proportional to $\lambda(nm) - 200$).

6. The tube of the ocular should have the smallest pinhole diaphragm fitted.

7. With the crystal centered on the cross wires, the light intensity passing through the crystal will give a light readout.

8. The objective can be deliberately put off center, so that rotation of the microscope stage by, say, 10° will swing the centered crystal off the cross-wire position and allow light to pass through the immersion oil into the photometer. This will give a light readout.

9. Now quickly and repeatedly move the microscope stage back and forth from the centered crystal grain position to the position at which the crystal is out of the measuring line; take light readings at each position.

10. As the temperature rises, the light readings on the grain will gradually fall, but the readings on the oil will increase steadily at a faster rate. At one specific moment (at which the temperature must be noted) the light readings on the crystal will equal that on the oil exactly, thus indicating exact matching of the two refractive indices.

(Method after Chromy, 1969.)

DISCUSSION

This photometer method allows a much more precise matching temperature than by visible observation. When glasses or isotropic substances are determined, the accuracy can be as good as $+0.00005$ for the RI.

METHOD D   USE OF WAVELENGTH DISPERSION

Microscopic immersion measurement of RI at room temperature by wavelength dispersion is notably rapid, convenient, and accurate and deserves much wider utilization.

1. Prepare a chart to plot RI for the slide temperature on a linear scale against wavelength on a Hartmann scale (scale inversely proportional to wavelength in nm − 200) for each of the neighboring immersion oils in an R.P. Cargille set. A plot of one oil is shown in Fig. 9.2B. By using the Hartmann scale the variation of RI with wavelength is linear. The liquids used must be of similar dispersion, and this will be ensured if neighboring oils of the R. P. Cargille sets are used.

2. At room temperature a grain preparation is made with each of at least two neighboring immersion oils. Temperature is not varied, but wavelength is varied by using a continuously variable interference filter adjustable between 400 and 700 nm (Carl Zeiss) until an exact matching of RI between solid and liquid is obtained for each of the neighboring oils. The wavelength at the point of matching is noted in each case. The Becke line method is adequate for judging the matching point, but oblique illumination or phase contrast may also be used.

3. There are two procedures for plotting the results (Fig. 9.3).

   a) The wavelength of matching of each immersion liquid is plotted on the liquid dispersion curve (open circles in Fig. 9.3). If two or three different liquids have been used, and matched, the three intersections should lie on a straight line. The intersection of this line with the value of N_D, gives the refractive index of the mineral for a wavelength of 589 nm (N_D). Figure 9.3 shows the example of $N\omega$ for the mineral xenotime.

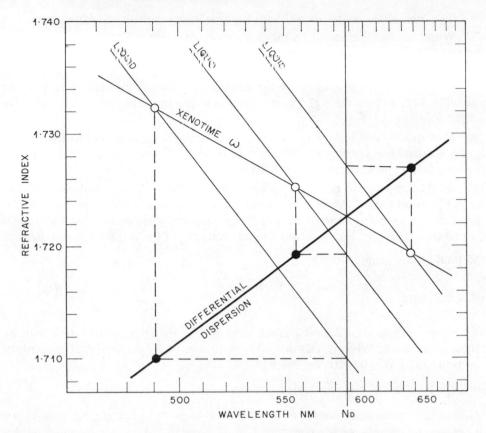

Fig. 9.3 Plot of optical dispersion lines for three liquids of identical dispersion of neighboring RI values. The RI scale is for ambient temperature and is linear. The wavelength scale is inversely proportional to wavelength (nm) — 200. The dispersion and differential dispersion curves of xenotine are shown as an example. They intersect in the xenotine N_D refractive index value. (After McAndrew, 1972.)

b) N_D of each liquid may be plotted against the exact wavelength of matching of solid and liquid (solid points on Fig. 9.3). The line through these points represents the differential dispersion of the solid and liquids and intersects the standard D wavelength at N_D of the solid at the same value as the dispersion curve through the open circles.

(Method after McAndrew, 1972.)

DISCUSSION

The differential dispersion curve is sufficiently close to a straight line to determine N_D within 0.001 by linear extrapolation. Only N_D of the liquids need be measured to determine this differential dispersion curve, thereby avoiding the refractometry of the liquids at varying wavelengths. All liquids used should have the same dispersion $(N_F - N_C)$ to give parallel dispersion lines, as in Fig. 9.3. The ambient slide temperature should be determined to 0.1°C while the RI matching is being made. See Morse (1968) for this method applied to low-temperature plagioclase.

To Check the Refractive Index of an Immersion Liquid

When two bottled oils have been mixed before a RI matching or if an unmixed bottle requires checking, some means of directly measuring the RI of the oils is required.

METHOD A

The simplest instrument is the Leitz-Jelley refractometer (E. Leitz, Wetzlar, Germany), which is supplied in two models. One has a range of 1.116 to 2.35 and the other of 1.33 to 1.92. It is illustrated in Fig. 9.4.

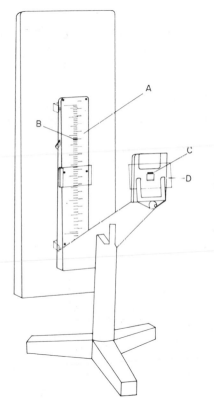

Fig. 9.4   The Leitz-Jelley refractometer. A = reading scale with sliding wire for centering on the refracted beam; B = aperture for incident sodium light source; C = viewing slit; D = glass plate with hollow prism in which a drop of oil is mounted. See exercise 9.4 for description.

1.   Set up the instrument with a sodium lamp shining through the aperture B behind the reading scale. Place the correct glass plate with cemented prism on the stage clip in front of the viewing slit C. If the reading scale A shows 1.116 to 2.35, this range must be engraved on the glass plate D. A scale of 1.33 to 1.92 must accompany a similarly engraved glass plate.

2.   Place a drop of the immersion liquid on the prism space of the glass plate. Look through the prism of liquid directly toward the incident monochromatic beam. Two images will be seen on the screen—that of the direct beam from B and that of the extrapolated beam which has been refracted by the immersion oil. Slide the wire on the reading scale until it coincides exactly with the refracted beam. A direct beam that is too bright gives rise to a broad line. Aperture B may be narrowed for greater sensitivity.

3. Now read directly on the scale the value under the wire. This is the RI of the liquid. The scale is calibrated to the second decimal place and the third decimal place can be estimated.

4. To replace the immersion oil by another the glass plate and prism must be thoroughly washed in acetone and dried before beginning with a new oil.

DISCUSSION

By using white light in place of the sodium lamp the image of the slit is drawn out into a colored band. The readings at the two extremes of the band will give a rough estimate of the dispersion of the liquid.

METHOD B

More precise determination of RI is made possible by using the Abbe refractometer (Carl Zeiss, 7082 Oberkochen/Wuertt, West Germany; Griffin and George Ltd.). This instrument has a measuring range of 1.3 to 1.7 and special prisms of 1.17 to 1.56 and 1.45 to 1.85. The accuracy of a determination may be as good as the fourth decimal place. The Abbe refractometer is illustrated in Fig. 9.5.

1. Bend the instrument over on its stand until the prism faces are nearly horizontal (Fig. 9.5A). Open the hinge while releasing the spring clamp lever.

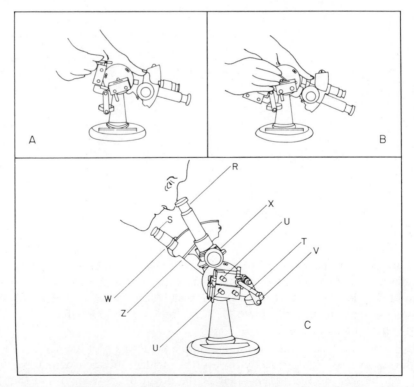

Fig. 9.5 Abbe refractometer. A, position for opening the prism; B, application of liquid to the prism face; C, reading position. R = telescope; S = index arm; W = knob for moving index arm; Z = adjustment for RI calibration; X = compensator knob; T = thermometer; V = mirror; U = rubber hose attachments for supplying heating water.

2. Take a drop of the selected liquid and transfer it to the clean face of the prism, as in Fig. 9.5B. Close the prism and make sure that the clamp is firmly locked.

3. Rotate the instrument upright on its stand to the position shown in Fig. 9.3C.

4. Move the index arm S (Fig. 9.5C) until a colored patch, either red or blue, is seen in the microscope tube R.

5. The mirror V should be set to reflect white light from a window along the axis of the telescope.

6. Having secured good illumination in the field of the telescope, focus the eyepiece on the cross wire and the reader S on the scale. Adjust the Abbe compensator by the milled head X until the colored patch in the telescope field is replaced by a clearly defined, nearly colorless edge to the dark shadow in the field. Move the index arm S by means of knob W until the edge of the dark shadow rests exactly on the cross line. The RI of the liquid is read from the scale through eyepiece S.

DISCUSSION

White light may be replaced by a sodium lamp for determining the RI at 589 nm. To obtain accurate readings over the entire range of refractive indices it is necessary only to ascertain that the reading at any one point is correct. This is conveniently done by using the glass test piece supplied with the instrument, the RI of which is engraved on it. It is provided with two perpendicular polished surfaces that meet in a sharp line. The larger of these faces is placed in contact with the upper prism; light enters through the other. The light used must be diffused. A drop of liquid of higher RI than the glass makes an optical contact between glass and prism. No surplus liquid should lie in front of the test piece. The dividing line of the field should be achromatized by knob X. The edge on the intersection is set on the field cross wire by adjusting the index arm and the reading is noted. If the reading does not agree exactly with that engraved on the glass, an adjustment is made by turning a small screw Z (Fig. 9.5) so that when the field dividing edge is exactly centered the reading through S is exactly as engraved on the test piece.

A thermometer T may be fitted to the refractometer and, if necessary, water may be circulated through the prisms to adjust the refractometer to any desired temperature to determine the RI values at different selected temperatures.

METHOD C

The microscope stage refractometer 47 37 61 (Carl Zeiss) is attached directly to the microscope stage. It is used to determine the RI of small liquid volumes with accuracy to the third decimal place. The microscope is focused on the measuring body which has a volume of 33 mm^3 and the RI is read in the eyepiece. Light of 589 nm is obtained with a narrow band-pass interference filter.

METHOD D

Every set of Jamin-Lebedeff interference equipment (Carl Zeiss) is supplied with micro-interference refractometers, which are specimen slides of accurately known RI with a spherical segment ground into their centers. In conjunction with the interference equipment, they allow the RI of liquids to be determined. If the refractometer is used as a specimen slide, the mounting method makes it possible to determine not only the RI of the immersion medium but also that of the solid immersed grain. In such a case the refractometer is part of the specimen, guaranteeing equality of temperature, which is important when RI equality of grain and oil has been achieved.

X-RAY FLUORESCENCE SPECTROMETRY 10

Oversimplified general introductory descriptions of X-ray spectrometry usually make the false claim that this is a rapid nondestructive technique. Such a claim is undoubtedly valid if the method is to be used purely qualitatively. For precise and accurate quantitative analysis, however, the method is certainly not nondestructive and specimen preparation may not always be so rapid as many reviews claim. Nevertheless, the method is rapid when compared with wet chemical analysis. For major element quantitative analysis much effort is required in specimen preparation to attain accuracy. For trace element analysis it may be possible to gain high accuracy with a less exacting preparation technique.

Because of their quite distinct requirement of technique the applications to qualitative analysis, quantitative major element analysis, and quantitative minor and trace element analysis are dealt with separately and in that order in this chapter. Basic procedure is described in section A, so that if the operator's main interest is in section B he will need to refer to section A initially for the general methods of operation before proceeding to section B for the specialized techniques.

The theory and general outline of X-ray fluorescence spectrometry are not fully described here, and some background knowledge will be assumed. Those readers lacking this background are referred to Norrish and Chappel (1967), Jenkins and De Vries (1970a), Bertin (1970), and Müller (1972). A useful companion volume to Jenkins and DeVries (1970a) for those who wish to practice the various calculations in X-ray spectrometry is Jenkins and De Vries (1970b).

A wide variety of X-ray fluorescence spectrometers is available commerically. It is impossible to refer to the variations of each. I have therefore decided to describe only the Philips universal vacuum PW 1540 spectrometer, which is perhaps the most popular model in mineralogy laboratories. All quoted code numbers are for Philips accessories which fit this spectrometer. The dimensions of specimens in this chapter are specifically for this spectrometer. The operator is advised to check that the size quoted here is appropriate for his particular machine and to make any necessary adjustment. The methods otherwise are directly applicable to any X-ray spectrometer of whatever make.

A. QUALITATIVE ANALYSIS

Because the analysis is only qualitative, specimen preparation should be kept as simple as possible. Any solid or liquid specimen that can fit into the specimen holder is suitable. It may have to be trimmed to size if it is too large. The standard specimen holder (PW-1527/20) accommodates a specimen no bigger than 32 × 29 mm height. These holders should have a circular 28 mm diameter aluminum window frame fitted at the base. If the specimen is small, irregularly shaped, or an unconsolidated powder, it can be supported by a thin (6 to 7 μ) sheet of polyester film (PW 1526/00) fitted to the base of the specimen holder. Such supporting windows are readily fitted by screwing off the holder base, pressing the circular window frame over the polyester against the holder, trimming off excess film with a razor blade, and screwing the base on again tightly. With a film base the same specimen holder may be used for liquids or the acid-resistant PVC holder (PW 1527/30) may be preferred. Contaminated polyester windows may be cleaned or replaced between use.

Because the polyester window intervenes between the specimen and the X-rays, thereby causing a reduction of both incident and fluorescent radiation intensity, which is especially serious for long wavelengths (elements lighter than Ca), it is preferable, wherever possible, to make specimens into self-supporting discs which can then be used in the same specimen holder from which the polyester window has been removed.

The following method describes how to make these discs:

EXERCISE 10.1

To Make Pressed-Powder Self-Supporting Specimen Discs of 30.4-mm Diameter

METHOD A

1. The various metal tools shown in Fig. 10.1 should be machined. They are a cylindrical tool steel base A for snugly fitting into cylinder C. A steel ring B for fitting the groove around the base of cylinder C. A steel cylinder C of tool steel. It fits snugly on A. A hollow aluminum tube D which fits loosely into cylinder C. A tool steel plunger E which fits snugly into C. A perspex plunger F which fits loosely into the aluminum tube D. Its ends should be polished. The dimensions of all these parts for making discs of diameter 30.4 mm are given in Fig. 10.1. The internal diameters may be adjusted if different diameter discs are required. A similar die for pelletizing samples has been described by Fabbi (1970).

2. Place ring B in the groove G at the base of cylinder C. Then fit cylinder C onto base A so that it seats properly. Make sure that all the inside surfaces, especially S of the steel base, have been perfectly cleaned.

3. Place the aluminum tube D into the well of C.

4. The rock or mineral powder, which has been ground in an agate mortar or in a Tema Mill for 10 to 12 minutes, is poured into the aluminum tube. A suitable amount of powder is 4 to 5 cc or 2.5 to 3 g. This may be conveniently measured in a small plastic vial.

5. Hold the aluminum tube firmly down in contact with surface S of the steel base A. Now introduce the perspex plunger F into the aluminum tube and press the powder firmly against surface S to compact it. The perfection of the compaction can be viewed through the perspex. When the powder is well compacted, slowly and carefully withdraw the perspex plunger, looking through it all the time to make sure that the powder disc does not break on withdrawal of the plunger. While the perspex plunger is being slowly withdrawn, it will be necessary to withdraw the aluminum tube simultaneously. Experience shows that when the aluminum tube is withdrawn, it will take the powder disc with it. Therefore it will be necessary to drop the perspex plunger

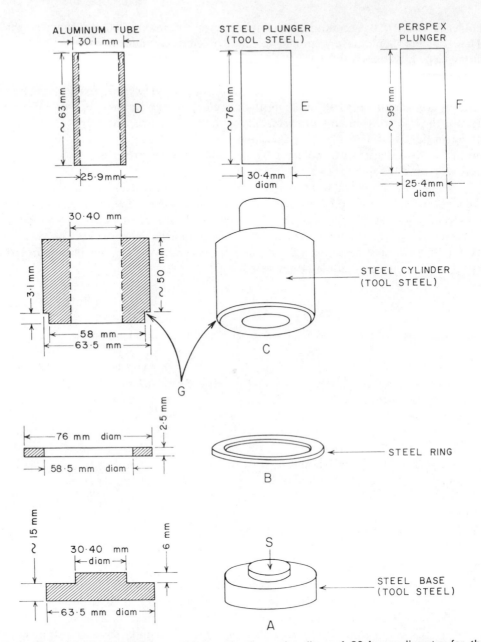

Fig. 10.1 Machined tools for making pressed powder discs of 30.4-mm diameter for the Philips X-ray spectrometer. Design modified after Norrish and Chappell (1967). Parts A, C, and E must be of hard tool steel; Part B can be of ordinary steel; Part D is aluminum, F is perspex, both circular ends of which should be polished so that one can clearly see through its length. See exercise 10.1 for details.

once again, slowly and carefully, to prevent the powder disc from being pulled up with the aluminum tube. By simultaneous withdrawal and reintroduction of the perspex plunger and aluminum tube it should be possible eventually to withdraw both perspex and tube and to leave behind a perfect pancakelike powder disc sitting on the central area of surface S of the steel base A. The pancake should be separated from the walls of the steel cylinder by the space that was formerly filled by the aluminum tube. If the pancake collapses, repeat the procedure from step 2.

6. Carefully pour one or two teaspoons of boric acid or methyl cellulose first to fill the annular space around the powder disc, then to cover it completely.

7. Steel plunger E is now slowly pushed into the hole of cylinder C. Because it is a snug fit, it will enter reluctantly. The base A should be firmly held onto the cylinder C while the plunger is being pushed in; otherwise air pressure will push off the base and expel the powder.

8. The whole assembly is now placed between the jaws of a hydraulic press. The press is put under pressure and plunger E is driven home, expelling all air from the cylinder C (Fig. 10.2).

9. When the plunger is fully in, increase the hydraulic pressure to 4000 psi and hold for a minute. Then release the pressure. Fabbi (1970) recommends a pressure of 30,000 psi.

10. The whole assembly is removed from the jaws of the press and inverted as shown in Fig. 10.3. A steel ring made to the dimensions of E (Fig. 10.3) is placed over the top

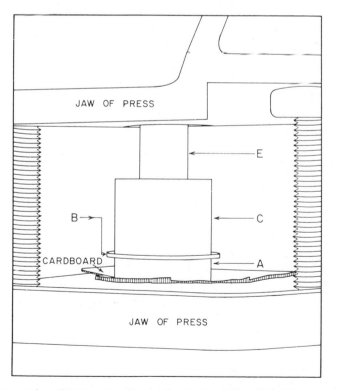

Fig. 10.2 Compressing the powder disc in the tools of Fig. 10.1. Pressure used is usually 4000 psi.

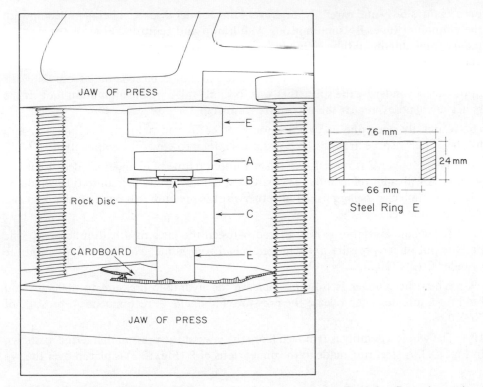

Fig. 10.3 Extrusion of the rock/boric acid disc from the cylinder. Steel ring E can be made of soft steel. It is useful to place a card between steel ring E and the upper metal jaw of the press during compression to prevent base A from sticking onto the upper jaw and falling down onto and breaking the rock disc.

of A to rest against the ring B. The whole assembly is now compressed under the press. This causes the piston E to push the rock disc and boric acid backing out of the cylinder C with the base A ahead of it. Pressure may be released and A removed from the top of the rock disc. If the rock disc has not been fully extruded, ring E is replaced and more hydraulic pressure is applied until the rock disc is completely extruded.

11. The rock disc is completely surrounded on its sides and back by solid boric acid. The front rock surface must not be touched. The specimen number may be marked directly on the boric acid back with a marking pen and the whole disc stored, rock face up, in a plastic ointment or pill vial of suitable size.

12. Boric-acid rimmed and backed discs made in this way may be mounted directly into a specimen holder, type PW 1527/20, from which the polyester window has been removed. There is then no film to cut down the energy of the X-rays to and from the rock or mineral specimen.

13. After use all the internal parts of the disc-making equipment should be thoroughly cleaned with tissue paper soaked in acetone. It is most important to clean surface S of the steel base A (Fig. 10.1) completely, for it is this surface that comes in direct contact with the next specimen.

METHOD B

To produce discs of 40-mm diameter suitable for specimen holder PW 1527/70 the following equipment is required:

1. A 12-cc capacity measure for the powder sample.
    A 2-cc capacity measure for the liquid sample.
    Several evaporation dishes.
    A stainless steel spatula.
    One infrared lamp for drying the sample.
    One porcelain mortar and pestle.
    A supply of aluminum sample cups (Philips PW 1527/81).
    One pressing mold with piston (Philips PW 1527/82) (this item is similar to items A, C, and E on Fig. 10.1, but of different dimensions).

2. Dissolve 25 g of polybutyl metacrylate polymer (Du Pont trade name Lucite 44 Acrylic Resin) in 100 cc acetone and store in a reagent bottle.

3. One 12-cc measure of the rock powder is put in an evaporation dish.

4. One 2-cc measure of the metacrylate solution is added and thoroughly mixed with the powder with the spatula.

5. Allow the mixture to dry under an infrared lamp for approximately 2 minutes.

6. The mixture is now lightly ground in a porcelain mortar.

7. The powder is placed in an aluminum sample cup (PW 1527/81) and firmly packed with the spatula until the cup is filled to the brim.

8. The cup is then placed on the base plate of the mold (PW 1527/82) and surrounded by the cylinder of the mold. The metal piston is introduced into the cylinder as in Method A. The mold is now pressed hydraulically at 2000 kg/cm^2 (1 psi = 0.0703 kg/cm^2). Pressure can be released at once.

9. To extract the specimen from the mold invert the mold and place a ring of sufficient height on top, as in method A, step 10. The application of slight pressure to the press will extrude the base plate and the specimen from the mold cylinder.

10. The compressed powder has a strong, smooth surface; rubbing it will not remove the powder. The specimen number may be marked on the aluminum backing. Such a specimen disc is so strong that it can withstand dropping. It is ready for loading in the PW 1527/70 specimen holder.

EXERCISE 10.2

Alignment of the Spectrometer

These instructions relate specifically to the Philips PW 1540 spectrometer and will need modification for other makes.

The spectrometer will operate smoothly only if the goniometer is properly adjusted so that its axis is in line with the axis of the crystal holder shaft of the vacuum drum. The procedure is as follows:

1. Screw G should initially be loosened (Fig. 10.4).

2. Turn the three leveling screws A (Fig. 10.4) in the goniometer base plate and move the locating bar B with screws C and D in the positioning strip E until the hollow shaft of the goniometer is roughly parallel to the shaft in the vacuum drum of the spectrometer. The coincidence and parallelism of the centers of the goniometer and vacuum drum shafts can be determined by looking through the empty hollow goniometer shaft F and inserting the auxiliary adjustment shaft (metal rod supplied

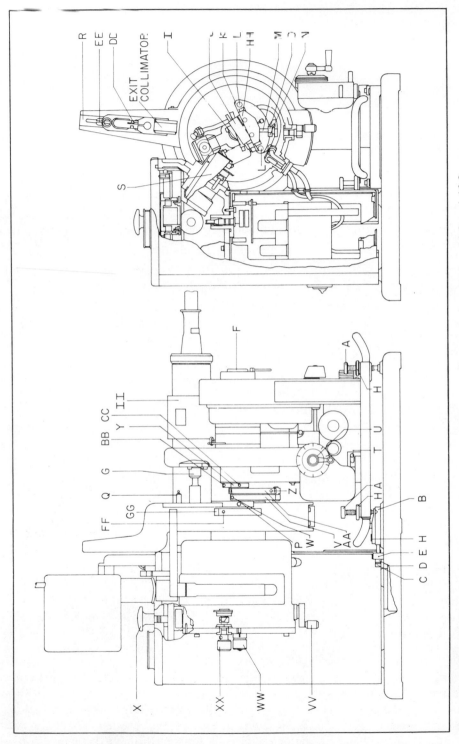

Fig. 10.4 Philips PW 1540 vacuum spectrometer. Letters refer to adjustments in exercises 10.2 and 10.3.

with the goniometer) into the center hole of the crystal holder shaft. A good adjustment is obtained when the auxiliary shaft can be slid smoothly into the hole of the crystal holder shaft over the whole length of the hole (approx. 5 mm).

3. Screw G should be tightened.

4. The position of the goniometer is now fixed by tightening the three locking nuts H of the leveling screws and by fixing the locating bar on the base plate with screws H (Fig. 10.4). The auxiliary shaft should not be left in but fully removed from the hollow F and kept by the goniometer for future checking of the alignment.

EXERCISE 10.3

Adjustment of the Crystal Holders

1. Slide the dispersing crystals into the crystal holders (Philips, PW 1546/00). Care must be taken not to damage the crystals during insertion. In the following description of the adjustment of counters and crystal holders it is supposed that the crystal with the smallest d spacing for measuring the heavier elements (e.g., LiF) is mounted in the inner fixed crystal holder I (Fig. 10.4) and the crystal with the largest d spacing (e.g., PET) for lighter elements in the separately adjustable holder J. The adjustable crystal holder is mounted on the left-hand (front) side of sledge K, as shown in Fig. 10.4, so that the crystal plane of this holder is in the advanced angular 2θ position of $15°$ with respect to the plane of the rear fixed crystal holder (Fig. 10.4). Several crystal holder sledges K will be needed if a range of crystals is in use. Each will require adjustment according to the following steps.

2. Slide the crystal holder sledge K on the two bars L of the crystal holder guide. The pin M with the ball bearing can be sunk and lifted after the clamping screw N in the arm O of the crystal holder positioning handle assembly has been loosened.

3. Turn the flow counter to its advanced angular position (30° on the 2θ range) by turning disc P at the right-hand side of the counter arm to position 2 (number 2 is visible on the disc).

4. Turning the flow counter is possible only after turning the small locking lever Q on the counter arm counterclockwise. The position of the flow counter is then locked by turning lever Q clockwise.

5. Turn the scintillation counter arm R to a position in which it is in line with the long entrance collimator S. Looking over the edge of a straight ruler will help to determine this position accurately. Block the goniometer by means of the clutch lever T and adjust the degree indication to 0.00° after loosening the three screws U on the indicator dial. Place a card between the circular dial and the goniometer body to form a free movement space between dial and goniometer and retighten the screws. Remove the card after the screws are tight.

6. Turn the crystal holder positioning handle VV to position 1 (the number 1 is visible from the front side on the arm). The crystal for the heavier elements is now in the proper position for diffraction. The crystal holder shaft should be turned until the plane of the crystal (the one on the right-hand side) is also in the zero line. In this position of the shaft the crank V should hang down vertically, for which it may be necessary to loosen the crank clamping screw W for a moment.

7. Remove the exit collimator from the front of the scintillation counter and set the counter so that it is well in line with the arm.

8. A specimen, say of copper metal or copper oxide, is inserted into the specimen holder and placed in the irradiating position by turning knob X through 180°.

9. The X ray unit is switched on and set to a suitable setting for CuKα (e.g., 13 kV, 6 mA), using W radiation. The scintillation counter is set at about 800 V (HT potentiometer approx. 175). Timer and scaler are at infinity, window at threshold, attenuation Z at 2, lower level at zero, and time constant at 1 second. The recorder is switched on to a low speed. The ratemeter should be adjusted so that any pen deflection stays on scale but makes full use of the recorder chart.

10. Turn the goniometer to the 2θ value of the Kα radiation from the specimen for that particular dispersing crystal (e.g., CuKα from an LiF crystal 2θ = 45.00°). The angle should always be approached from the low-angle side.

11. Turn the crystal holder shaft V by hand with the crank and carrier arm (the carrier arm shaft is not yet fixed to the goniometer shaft by inserting clamping screws in the clamp Y) until the recorder shows a ratemeter indication and pen deflection. We may now be seeing the Kα or the Kβ line of the copper. As soon as we get a pen deflection, it will be easy to fix on Kα, which is usually seven times more intense than Kβ.

12. A fine adjustment should be made by turning the adjustment screw Z in the carrier arm AA after having fixed the carrier shaft in the goniometer (screws BB and CC in clamp Y) until maximum recorder indication is obtained. Whenever the adjustment screw Z in the carrier arm is turn counterclockwise, the whole carrier arm AA should be slightly pulled forward to have the play in the 2 : 1 gearing of the goniometer on the correct side for scanning toward higher 2θ values.

13. Set the goniometer to angle values (in steps of 0.1°) which are higher and lower than the theoretical table value for the specimen and crystal concerned. Try to find a higher recorder indication in each goniometer position by turning the screw Z in the carrier arm. The position in which the maximum intensity is found is the correct position of the scintillation counter arm and of the analyzing crystal. The angle indication of the goniometer should be set to the 2θ value for the specimen and the analyzing crystal concerned, after locking the clutch lever T.

14. Attach the exit collimator to the scintillation counter arm and determine its correct position by turning the adjusting screw DD in the arm until maximum recorder indication is obtained. Turn the scintillation counter back and forth over a small angle to find the maximum recorder deflection (loosen screws EE for a moment). The scintillation counter is now properly set up in respect to the first crystal.

15. Turn the flow counter in line with the scintillation counter (position 1 of the disc P; turning is possible only after turning lever Q counterclockwise. Once disc P is turned lever Q is again locked by turning it clockwise).

16. Adjust the gas flow so that the floating metal cone in the gas density stabilizer (PW 1548) is centered on the indicator line. The gas pressure going into the density stabilizer should be about 1 kp/cm^2 or slightly less.

17. Switch on the flow counter, giving it a high tension of about 1600 V (HV reading about 450) with attenuation Z = 2.

18. The goniometer should be set at the same angle used for setting up the scintillation counter in step 13. (For CuKα and an LiF crystal 2θ = 45.00°.) If the goniometer has to be adjusted, the 2θ value should be approached from the low-angle side.

19. Set the flow counter arm with its collimator to its correct angular position with the adjustment screw FF in the cam of the arm at the front side, after loosening the smaller locking screw GG in the upper side of this cam. The adjustment screw FF should be turn until maximum intensity is shown by the recorder and screw GG is then tightened.

20. Turn the flow counter to the advanced angular position (position 2 of disc P) and turn the crystal holder positioning handle VV to position 2(large d-spacing crystal).

21. Look up the 2θ angle for the specimen and the crystal now concerned (e.g., if it is still Cu$K\alpha$ and the crystal is PET, the 2θ angle is 20.32) and adjust the goniometer to an angle value equal to this table value minus 30° (every 2θ value must be minus 30° from the table value when position 2 is used with the counter in the advanced position). The goniometer reading for Cu$K\alpha$ would therefore be 990.32° (i.e., 9.68° below zero, so that when 30° is added to it, the flow counter is actually at $30 - 9.68 = 20.32°$). Alternatively, one could look for the second-order CuK line at 41.32° 2θ by setting the goniometer at 41.32 minus 30 = 11.32°.

22. Adjust the position of the adjustable crystal holder by turning the adjustment screw HH with an Allen key until maximum recorder indication is obtained.

23. Adjust the position of the flow counter arm with the adjustment screw in the cam at the rear side of the arm (similar to screws FF and GG, but at the rear side of the arm), as shown in Fig. 10.4. The small locking screw (equivalent to GG on the other side of the arm) has to be loosened. Turn the adjustment screw to give maximum recorder indication.

24. Repeat the adjustments in steps 22 and 23 because they depend slightly on one another.

25. The locking screw (on the opposite side of the cam from GG) is tightened. Both crystals are now properly aligned in the spectrometer in relation to 2θ and to the counter arms.

DISCUSSION

There is no need to make a very refined adjustment to the 2θ angle at which any wavelength is diffracted, for it will fluctuate with the ambient operating temperature and the d-spacing of dispersing crystals is temperature dependent. Rather, a large number of simple composition specimen discs should be prepared by exercise 10.1 of, for example, CuO, Mg metal, Al_2O_3, and FeO, which can be put into the spectrometer and the goniometer angle turned until the $K\alpha$ line from the major element of the specimen is accurately placed in its diffraction position. The 2θ on the goniometer need not coincide exactly with the theoretical table value. The important thing is that one is confident of looking at $K\alpha$ radiation from that particular element. Such specimen discs can also be used for the determination of the best counter settings and pulse height analyzer settings as well as for setting 2θ precisely.

It is wise at some stage to determine whether the X-ray tube is set to give maximum intensity. At any time when a peak is being observed on the recorder loosen clamp II (Fig. 10.4) and slightly rotate the X-ray tube in its housing. Then clamp it in the position that gives maximum peak intensity.

Selection of Optimum X-ray Tube and Its Operating Conditions

For Light Element Detection. For light elements from fluorine (atomic number 9), which is the lightest that can be detected in most spectrometers, through sodium (11), magnesium (12), aluminum (13), silicon (14), phorphorus (15), sulfur (16), chlorine (17), potassium (19) to calcium (20) the ideal tube for exciting X-ray fluorescence is a chromium anode tube (Philips PW 2168/21) which has a 2000-W rating. The use of any other anode will give less fluorescence yield. The spectrum so produced will include Cr lines from the tube.

The $K\alpha$ wavelengths of each of these elements are long and easily absorbed by air. It is essential therefore that a vacuum path be used in the spectrometer when detecting elements from fluorine (wavelength 18.307 Å) to calcium (3.360 Å). The spectrometer lid needs to be fixed, knob WW (Fig. 10.4) tightened, and the vacuum pump activated.

Because the fluorescence yield from light elements is considerably less than from heavier elements, when detecting elements in the range F (9) through to Ca (20) the coarse primary collimator of the spectrometer must be utilized to allow maximum radiation through [by turning knob XX (Fig. 10.4) *fully* counterclockwise toward letter C].

Detection of fluorescence radiation from light elements from fluorine to calcium must be performed with a gas flow proportional counter that uses a flow of argon/ methane gas. For highest accuracy it is essential to have a gas density compensator (Philips, PW 1548) to achieve high stability.

Optimum voltage for the chromium tube is 44 kV. It should not be operated at higher values because the background would increase more than the peaks to be measured; 44 kV gives the best peak to background ratio. The choice of tube current is not critical, for both peak and background increase approximately linearly. The choice therefore will depend on the intensity required. A normal working current would be mA 20, but it may be increased to get a stronger intensity that may be necessary for the lighter elements in view of their poor fluorescence yield. Of course, the product of kV × mA must not exceed both the rating of the generator and the rating of the tube. If the generator rating is 1600 W, and the tube, 2000 W, then kV × mA must not exceed about 90% of the generator rating. Also, the kilovolt rating of the shielding of the spectrometer must not be exceeded. The Philips PW 1540 spectrometer, for example, has a clear warning "Do not exceed 60 kV."

For Heavy Element Detection. For elements from titanium (atomic number 22), vanadium (23), chromium (24), manganese (25), iron (26), and all others, to the very heaviest, the ideal exciting X-ray source is a Tungsten anode X-ray tube (Philips, PW 2164/10) which has a 2000-W rating. Of course, if a tungsten tube is used, the X-ray spectrum will, of necessity, include the tungsten lines from the tube spectrum. Hence, even if the specimen contains no tungsten, the diffracted spectrum will include it. Therefore to analyze a specimen for tungsten it is essential to select an X-ray tube of another anode material. In such a case a gold tube (Philips, PW 2161/10) may be used in place of the tungsten. If the laboratory possesses only chromium and tungsten tubes, the chromium tube could be used when tungsten is being measured. However, it is not ideal for exciting tungsten radiation. Radiation from an element in the specimen is always best excited by a tube whose anode has a heavier atomic number than the element being analyzed.

Optimum voltage for operating the tungsten tube varies with the atomic number of the element being observed. For Cr (atomic number 24) optimum voltage is 50 kV; and optimum kV increases with atomic number until at about atomic number 50 it is around 100. Of course, many spectrometers do not allow such high kV settings. Accordingly, it is normal practice to set a tungsten tube at 50 kV as regular practice, but if the spectrometer allows, kV may be increased up to 100 for higher atomic number elements. As with the light elements, mA is adjusted to suit the peak intensity required. Average running conditions would be 50 kV and 20 mA, but for a high concentration mA may be reduced, or for trace elements increased, to 25. Again kV × mA must not exceed the generator and tube rating.

An air path is used for all elements and the fine primary collimator is employed throughout (rotate knob XX, Fig. 10.4, *fully* clockwise). The ideal detector is the scintillation counter, although the gas flow counter may be preferable for titanium, vanadium, chromium, manganese, and iron. For all elements heavier than iron the scintillation counter must be used.

Relationship Between Kilovolts and the Production of $K\alpha$ Radiation

K **Spectrum.** Because the $K\alpha$ radiation from a specimen is much more intense than the *L* spectrum lines, we hope to use the $K\alpha$ peaks whenever possible. *K* radiation, however, can be produced on most spectrometers only from elements lighter than atomic number 68 or thereabouts. For elements heavier than this we have no choice but to produce only the *L* lines. To know whether the energy produced by an X-ray tube is sufficient to excite the *K* spectrum from any element in the sample, we should first note the excitation potential for the *K* wavelengths of the various elements. This information may be found in Berman and Ergun (1969), for example, and a summary is given in Table 10.1. The variation of excitation potential with atomic number is displayed in Fig. 10.5.

To excite the *K* wavelengths from any element in a specimen efficiently the kilovolts applied to the X-ray tube should exceed the excitation potential of that element by at least 10% and preferably 20. A glance at Table 10.1 will show, for example, that if we wish to obtain tin $K\alpha$ (Z = 50) lines the high tension on the X-ray tube should be about 20% greater than 29, or of the order of 34 kV. If we were to operate the tube at only 20 kV we should not be able to excite tin $K\alpha$ radiation from our specimen. If a spectrometer is normally run at 50 kV, then Table 10.1 shows that the heaviest common element to have its *K* radiation excited will be cerium (atomic number 58). All heavier elements produce no *K* radiation.

A tungsten tube run at 50 kV will produce no tungsten *K* lines. This is to our advantage, for we will not usually be interested in having any more lines than necessary to complicate our specimen spectrum.

On the other hand, since the *K* lines are more intense than the *L* lines from any element, a spectrometer of a higher rating can be used to advantage to measure the heavier elements. A spectrometer that allows a tube to run at 100 kV can be used to excite *K* lines from elements up to lead or bismuth (atomic numbers 82 and 83), but for thorium and uranium the *K* lines will never be obtained.

In the analysis of a specimen use may be made of the excitation potential to avoid deliberately exciting the *K* spectrum of a particular element in the specimen, especially if it is a major element and its $K\alpha$ peak obscures the peak of another element in which we are more interested. As an example, a specimen rich in tin may be run in the spectrometer at kV 24 and no tin $K\alpha$ lines will be excited. A barium-rich specimen run at kV 30 will produce no K spectrum for barium, and so on.

In summary, Table 10.1 and Fig. 10.5 should be consulted to ensure that the kV setting on the X-ray tube exceeds by 10 to 20% the excitation potential of the element we are measuring. If the rating of the generator will not permit this, we have no choice but to use an L wavelength for analysis.

L **Spectrum.** Provided the potential applied to the X-ray tube exceeds 26 kV, the complete spectrum of L lines will be excited from all elements in a specimen, including uranium. Table 10.2 lists the wavelengths for the most intense L lines, the excitation

Table 10.1 K Spectrum Wavelength, Excitation Potential, and Absorption Edge for the Commonly Occurring Elements

Atomic Number Z	Element Name	Wavelength(Å)[a]		Excitation[b] Potential for K Spectrum (keV)	Absorption Edge (Å)
		$K\beta$	$K\alpha$		
9	Fluorine		18.32	0.68	
11	Sodium	11.575	11.91	1.07	
12	Magnesium	9.52	9.89	1.30	9.512
13	Aluminum	7.96	8.34	1.56	7.951
14	Silicon	6.753	7.126	1.84	6.744
15	Phosphorus	5.796	6.158	2.14	5.787
16	Sulfur	5.032	5.373	2.47	5.018
17	Chlorine	4.403	4.729	2.81	4.397
18	Argon	3.886	4.193	3.19	3.871
19	Potassium	3.454	3.742	3.60	3.437
20	Calcium	3.089	3.359	4.03	3.070
21	Scandium	2.780	3.032	4.49	2.758
22	Titanium	2.514	2.750	4.96	2.497
23	Vanadium	2.284	2.505	5.46	2.269
24	Chromium	2.085	2.291	5.99	2.070
25	Manganese	1.910	2.103	6.53	1.897
26	Iron	1.757	1.937	7.11	1.744
27	Cobalt	1.621	1.790	7.71	1.608
28	Nickel	1.500	1.657	8.33	1.488
29	Copper	1.393	1.542	8.98	1.381
30	Zinc	1.295	1.437	9.66	1.281
31	Gallium	1.207	1.341	10.37	1.195
32	Germanium	1.129	1.256	11.10	1.116
33	Arsenic	1.057	1.177	11.86	1.045
34	Selenium	0.992	1.106	12.65	0.980
35	Bromine	0.933	1.041	13.47	0.920
36	Krypton	0.879	0.981	14.33	0.866
37	Rubidium	0.829	0.927	15.20	0.816

(continued)

Table 10.1 (continued)

Atomic Number Z	Element Name	Wavelength (Å)[a]		Excitation[b] Potential for K Spectrum (keV)	Absorption Edge (Å)
		$K\beta$	$K\alpha$		
38	Strontium	0.783	0.876	16.10	0.770
39	Yttrium	0.741	0.830	17.04	0.727
40	Zirconium	0.702	0.788	17.99	0.688
41	Niobium	0.665	0.748	18.98	0.653
42	Molybdenum	0.632	0.710	20.00	0.620
45	Rhodium	0.545	0.614	23.22	0.534
46	Palladium	0.521	0.587	24.34	0.509
47	Silver	0.497	0.561	25.51	0.486
48	Cadmium	0.475	0.536	26.64	0.464
49	Indium	0.455	0.514	27.94	0.444
50	Tin	0.435	0.492	29.19	0.425
51	Antimony	0.417	0.472	30.49	0.407
52	Tellurium	0.400	0.453	31.81	0.390
55	Cesium	0.355	0.402	35.82	0.345
56	Barium	0.341	0.387	37.43	0.332
57	Lanthanum	0.328	0.373	38.91	0.319
58	Cerium	0.316	0.359	40.43	0.307
73	Tantalum	0.190	0.217	67.37	0.184
74	Tungsten	0.184	0.211	69.48	0.178
78	Platinum	0.163	0.187	78.34	0.158
79	Gold	0.159	0.182	80.67	0.153
80	Mercury	0.154	0.176	83.04	0.149
81	Thallium	0.150	0.172	85.45	0.144
82	Lead	0.146	0.167	88.05	0.141
83	Bismuth	0.142	0.162	90.44	0.137
90	Thorium	0.117	0.135	109.50	0.113
92	Uranium	0.111	0.128	115.39	0.107

[a] The value $K\alpha = \frac{1}{3}(2K\alpha_1 + K\alpha_2)$. The value of $K\beta$ is $K\beta_1$.
[b] The excitation potential is that for the shortest wavelength of the K spectrum. Values after Berman and Ergun (1969) and Jenkins and De Vries (1970a).

potential for the L spectrum, and the absorption edges for common elements heavier than atomic number 50. Table 10.2 and Fig. 10.5 show that a tungsten anode tube operating at any value in excess of about 14 kV will produce a complete L spectrum.

For analysis of the heaviest elements in a specimen, for which the $K\alpha$ line could not be excited, use is made of the $L\alpha_1$ line, which is the most intense of the L spectrum. Any spectrometer operating at a standard 50 kV setting will excite this line from all elements in the specimen.

M **Spectrum.** *M* spectrum lines are easily excited at low kilovolt ratings; for example, the excitation potential for the uranium *M* spectrum is only 5.50 keV. However, the intensities of *M* spectrum lines are usually so weak that they may be completely ignored.

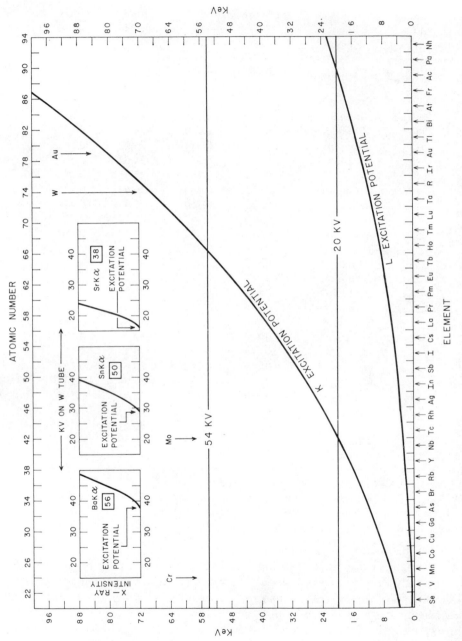

Fig. 10.5 Variation of the *K* and *L* excitation potential with atomic number and three examples of X-ray intensities of *Kα* lines at various kV settings.

Table 10.2 L Spectrum Wavelengths, Excitation Potential, and Absorption Edges for the Commonly Occurring Heavier Elements

Atomic Number Z	Element Name	Wavelength[a]				Excitation[b] Potential for L Spectrum (keV)	Absorption Edges (Å)		
		$L\alpha_1$	$L\beta_1$	$L\beta_2$	$L\gamma_1$		L_I	L_{II}	L_{III}
50	Tin	3.600	3.385	3.175	3.001	4.46	2.778	2.982	3.156
51	Antimony	3.439	3.226	3.023	2.851	4.69	2.639	2.830	3.000
52	Tellurium	3.289	3.077	2.882	2.712	4.93	2.510	2.687	2.856
55	Cesium	2.892	2.683	2.511	2.348	5.70	2.167	2.314	2.474
56	Barium	2.775	2.567	2.404	2.241	5.97	2.068	2.204	2.363
57	Lanthanum	2.665	2.458	2.303	2.142	6.07	1.973	2.103	2.259
58	Cerium	2.561	2.356	2.209	2.048	6.53	1.890		2.164
73	Tantalum	1.522	1.327	1.284	1.138	11.67	1.061	1.113	1.256
74	Tungsten	1.476	1.282	1.245	1.099	12.09	1.024	1.074	1.215
78	Platinum	1.313	1.120	1.102	0.958	13.87	0.893	0.934	1.072
79	Gold	1.276	1.084	1.070	0.926	14.34	0.864	0.903	1.040
80	Mercury	1.241	1.049	1.040	0.896	14.84	0.836	0.872	1.009
81	Thallium	1.207	1.015	1.010	0.868	15.33	0.808	0.844	0.979
82	Lead	1.175	0.982	0.983	0.840	15.84	0.782	0.815	0.950
83	Bismuth	1.144	0.952	0.955	0.813	16.38	0.757	0.789	0.924
90	Thorium	0.956	0.765	0.794	0.653	20.42	0.606	0.630	0.761
92	Uranium	0.911	0.720	0.755	0.615	21.72	0.569	0.592	0.722

[a] There are numerous additional lines. The four tabulated are the most intense of the spectrum.
[b] The excitation potential is that for the shortest wavelength of the L spectrum. Values after Berman and Ergun (1969) and Jenkins and De Vries (1970).

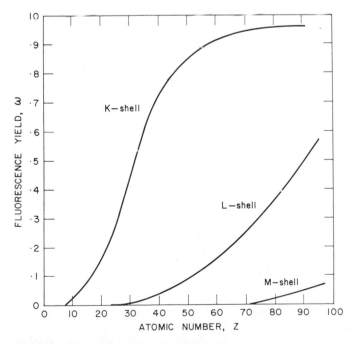

Fig. 10.6 Fluorescence yield for the K, L, and M spectra as a function of atomic number. (After Berman and Ergun, 1969).

Fluorescence Yield. Figure 10.6 illustrates the fluorescence yields for K, L, and M spectrum lines as a function of atomic number. This figure shows that the most intense lines will always be the K lines; hence they are to be preferred for analysis whenever possible. If a K line cannot be obtained, the next best will be an L line. The M lines are always so weak that they are never used for analysis. Figure 10.6 shows that the intensity of any X-ray line will increase with atomic number and light elements give rather low intensities. The intensity of any particular line, for example, $K\alpha$ is therefore proportional to the concentration of the element in a sample and to the atomic number of the element. By way of example, a specimen composed of 100% tin ($Z = 50$) will give approximately a $K\alpha$ line twice as intense as a sample made of 100% zinc ($Z = 30$). Hence in qualitative analysis one must correct for the fluorescence yield by using Figure 10.6 if the results are to be given a semiquantitative estimate.

EXERCISE 10.5

Angular 2θ Limits of the Spectrometer

Before operating a spectrometer one should determine precisely the permissible 2θ angular range of adjustment possible and make sure that this range is NEVER exceeded. With the spectrometer housing open the goniometer can be manually turned to its very lowest and its very highest 2θ limits. These limits will be reached when some internal component just begins to touch some other component; for example, a collimator housing may be in danger of hitting the diffracting crystal, in which case it is important to select a limit of 2θ movement that stops short of such a collision. This limit must never be exceeded in practice; otherwise permanent damage will be done to the crystal or to the goniometer.

The limits for the Philips PW 1540 spectrometer have been precisely determined as follows:

Minimum 2θ Angular Limit

1. With the gas flow counter in its normal No. 2 ($+30°$) advanced position limit $=$ 983.50° (i.e., 16.50° below zero).

2. With the gas flow counter in its parallel No. 1 position limit $= 15.50°$.

Maximum 2θ Angular Limit

1. With gas flow counter in advanced No. 2 position limit $= 117.20°$

NOTE. Between 114.5 and 117.2° 2θ on no account should the primary collimator be changed from fine to coarse, or vice versa, because it will hit the analyzing crystal during the change. It must be selected as fine or coarse when 2θ is less than 114.50°.

2. With the gas flow counter in its parallel position No. 1 limit $= 133°$ 2θ.

It may be good practice to set the automatic angular stops on the goniometer permanently to a lower 2θ of 15.50° and a higher 117.20° and to release these stops temporarily only when necessary for special settings. The employment of the automatic stops will prevent an inexperienced operator from damaging the spectrometer.

Choice of Dispersing Crystal

The following are factors that control the choice of dispersing crystal in the spectrometer:

1. The solution to the Bragg Equation $n\lambda = 2d \sin \theta$, where n is an integer 1 (first-order), 2 (second-order), etc., λ is the wavelength in Å of the radiation being dispersed, d is the lattice spacing of the planes of the dispersing crystal being used, and θ is the angle of reflection from the dispersing crystal.

2. The angular 2θ limits of the spectrometer.

3. The dispersing and reflecting efficiency of the crystal.

4. The temperature and vacuum stability of the crystal.

The best dispersing crystals available are suitable for shorter wavelengths (radiation from the heavier elements). For dispersing the longer wavelengths (from lighter elements) we have no choice but to use somewhat inferior crystals, but fortunately newer crystals which have a more efficient dispersion of the longer wavelengths are being developed. Recommended crystals are discussed next.

Basic Combination of Three Crystals. The following three crystals together are a minimum basic requirement.

Name	Abbreviation	Reflecting Plane	$2d$ (Å)	Reflection Efficiency
1. Lithium fluoride	LiF	200	4.028	Intense
2. Penta erythritol	PET or PE	002	8.742	High
3. Potassium acid phthalate	KAP	010	26.632	Average

NOTES. (a) LiF (200) is used for all elements, from the heaviest down to and including calcium ($Z = 20$). For calcium $K\alpha$ radiation the Bragg equation gives for first-order $K\alpha$ ($n = 1$) $\sin \theta = 3.359/4.028$, from which $2\theta = 113.20°$. This angle is possible on the spectrometer in which the upper 2θ limits are $117.20°$. The LiF crystal, however cannot be used for the next lighter element potassium, for which first-order $K\alpha$ would occur at a 2θ of $136.50°$ ($\sin \theta = 3.742/4.028$) because this angle is beyond the spectrometer limit. The LiF crystal, because of its superior nature, is used to its maximum, which is down to and including calcium.

(b) The next best crystal is the PET. It is used for elements in the range of potassium ($Z = 19$) down to and including aluminum ($Z = 13$), but no lighter than aluminum.

(c)   The KAP crystal, which is not good because it fluoresces during irradiation, is reserved for magnesium, sodium, and fluorine. At the present time X-ray spectrometry cannot be extended to elements lighter than fluorine.

Additional Crystals for Specific Elements. The following crystals should be utilized for superior dispersion of specific elements. They supplement the above list.

Name	Abbreviation	Reflecting Plane	$2d$ (Å)	Reflection Efficiency
1. Rubidium acid phthalate	RAP or RbAP	001	26.12	Average
2. Ammonium di hydrogen sulfate	ADP	110	10.65	Low
3. Germanium	Ge	111	6.53	Average
4. Polycrystalline graphite	C (Ho graphite)	002	6.71	High
5. Sorbitol hexa acetate	SHA	110	13.98	Average
6. Lithium fluoride	LiF	420	1.802	Intense

NOTES. (a)   The RAP crystal is preferable to KAP for analysis of fluorine, sodium, and magnesium because it produces less fluorescence and has a higher peak to background ratio (Jenkins, 1972).

(b)   The ADP crystal is preferred by many workers for magnesium determination.

(c)   The Ge crystal is used by many workers for phosphorus and sulfur determination.

(d)   Polycrystalline graphite (Ho graphite) gives superior results for phosphorus, sulfur, chlorine, and potassium. It is one of the most expensive crystals, however.

(e)   SHA theoretically should be ideal for magnesium, sodium, and fluorine analysis, but present experience shows that a high-quality untwinned crystal has not yet been produced commerically. At present the RAP is still superior to SHA (Jenkins, 1972), but it is hoped that better technology will produce a good SHA crystal in the near future.

(f)   LiF (420) can replace LiF (200) to advantage for wavelengths shorter than 0.4 Å, that is, for the K spectrum of elements heavier than atomic number 52.

(g)   All the crystals described above may be obtained from the suppliers of X-ray fluorescence spectrographic equipment, for example, from Philips in Holland or in the United States.

EXERCISE 10.7

Routine Procedure for X-ray Spectroscopy Scanning

Scanning over a range of 2θ values is used only for qualitative or at the best semi-quantitative analysis, to determine which elements are present in a specimen and to gain an approximate estimate of their concentration.

1. Prepare the specimen as in exercise 10.1.

2. Place it in a sample holder and position it in the spectrometer irradiation position.

3. Select the proper combination of operating conditions according to the following outline:

	For Elements from Calcium ($Z = 20$) to the Heaviest Uranium ($Z = 92$)	For Elements from Aluminum ($Z = 13$) to Potassium ($Z = 19$)
Exciting radiation	TUNGSTEN	CHROMIUM
Kilovolts	50	44
Milliamperes		
(major elements)	6	20
(trace elements)	24	24
Path in spectrometer	AIR	VACUUM
Primary collimator	FINE	COARSE
Detector	SCINTILLATION	GAS FLOW PROPORTIONAL
Detector voltage	HV 160 ($= 740$ V)	HV 440 ($= 1550$ V)
Attenuation	$Z = 2$	$Z = 2$
Pulse height analyser	THRESHOLD	THRESHOLD
Lower level	ZERO	ZERO
Dispersing crystal	LiF ($2d = 4.028$)	PET ($2d = 8.742$)
Scanning 2θ range	9 to 114°	46 to 146°*
Scanning speed	1°/min	1°/min
		$\frac{1}{2}$°/min
Time constant	1 or 2 seconds	1 or 2 seconds
Chart speed	600 mm/hr	600 mm/hr

* See *SPECIAL NOTE below.*

DISCUSSION

The HV settings given above for the detectors are average settings which will be applicable to most elements, but are not strictly ideal for every element in the range of the scan. Ideal HV settings can be obtained for each element by following the advice of exercise 10.9, in the following section on quantitative analysis. The ratemeter must be adjusted for any particular scan to just keep the most intense line on the graph paper and to make full use of the width of the chart paper.

SPECIAL NOTE.   In the above table the upper scanning 2θ angle of 146° would appear to be beyond the angular range of the spectrometer (see exercise 10.3). The Philips spectrometer, however, uses the PET crystal in combination with the gas flow counter in an advanced $+30°$ position, in which case 30° must be added to the goniometer scale readings for the actual 2θ value and a 2θ upper limit of 146° will actually read 116° on the goniometer scale. This is still possible within the angular limit of 117.20° for the Philips spectrometer. It is not recommended that scanning be done for elements lighter than aluminum. Although Mg, Na, and fluorine can be detected with a KAP or an RAP dispersing crystal, a simple scan will often fail to show peaks, even for major element concentrations in the sample, because precise settings of the pulse height analyser and HV of the gas flow counter are necessary for each element to give maximum peak to background ratio and to reduce the fluorescence from the crystal to discriminate in favor of the characteristic radiation from the element we wish to observe. Detection of light elements is therefore delayed until a later section on quantitative analysis.

4. Having obtained a spectrogram, the 2θ values of each peak should be converted to a wavelength by using the Bragg equation $n\lambda = 2d \sin \theta$ and the wavelength listed in Tables 10.1 and 10.2 so that the element causing the peak may be identified. Tables converting 2θ to wavelengths and to characteristic lines for the range of elements for all common dispersing crystals are available from the manufacturers of the crystals or the following may be obtained:

Powers (1960), conversion tables for topaz, LiF, NaCl, EDDT, and ADP crystals; Berman and Ergun (1969) for topaz, LiF, NaCl, quartz, PET, EDDT, ADP, gypsum, and KAP crystals. Tables for specific crystals may be obtained from N. V. Philips, Eindhoven, Holland or M.E.L. Equipment Co., Manor Royal, Crawley, Sussex, England. For the most complete conversion tables, see White and Johnson (1970).

The interpretation of a spectrogram is best illustrated by an example. Figure 10.7 is a spectrogram of a powdered sample made up as a boric-acid-backed disc. We wish to analyse it qualitatively for zinc, copper, and nickel. The spectrogram was run with an LiF (2d 4.028) dispersing crystal and tungsten radiation at 50 kV and 6 mA. Other operating conditions are those in step 3. Powers (1960) gives the following 2θ positions for the elements we are interested in. Of course, we could calculate these ourselves

Atomic Number	Element	Symbol	Order	$2\theta°$ $K\alpha_1$	$K\alpha_2$	$K\beta_1$
28	Nickel	Ni	1	48.61	48.71	43.73
			2	110.82	111.12	96.28
29	Copper	Cu	1	44.96	45.08	40.43
			2	99.75	100.11	87.44
30	Zinc	Zn	1	41.74	41.86	37.54
			2	90.88	91.20	80.11

from the Bragg equation and the wavelengths in Table 10.1. In addition, we should be aware that the L lines from the tungsten tube will be present

			Order	$L\alpha_1$	$L\beta_1$
74	tungsten	W	1	42.99	37.12
			2	94.26	79.07

There will be other weak tungsten lines but too weak for detection normally.

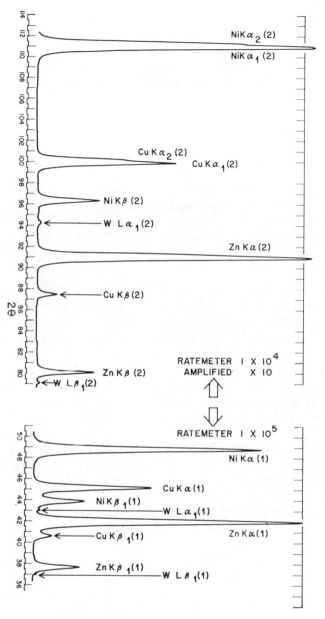

Fig. 10.7 X-Ray spectrogram showing first- and second-order K lines from nickel, copper, and zinc in the specimen and tungsten lines from the tube obtained from an LiF ($2d$ 4.028 Å) dispersing crystal.

Figure 10.7 shows the spectrum obtained for the first-order lines ($n = 1$) from 2θ 37 to 50° and the second-order lines from 79 to 112°. The second-order lines are only about one-tenth as intense as the first-order. They are shown amplified × 10.

The definite identification of any element in the specimen is confirmed by finding its complete spectrum or at least its $K\beta$ and $K\alpha$ lines at the appropriate 2θ position for that particular dispersing crystal. The peak heights on such a spectrogram cannot be directly related to element concentration because they have to be corrected for fluorescence yield (Fig. 10.6), which varies with atomic number, and to interelement absorption and other matrix effects such as particle size effect. Such corrections can be carried out only if a precise specimen preparation technique has been applied, as in the sections B and C. A scan such as that shown in Figure 10.7 should therefore be used qualitatively or at the best semiquantitatively.

B.   QUANTITATIVE MAJOR ELEMENT ANALYSIS

Specimen Preparation

Need for Specimen Preparation. When I is the intensity of monochromatic X-radiation that emerges from a specimen, I_0 is the intensity of the incident radiation, μ/ρ is the mass absorption coefficient of the specimen for the monochromatic radiation, Px is the mass per unit area of the absorbing element in the specimen, then

$$I = I_0\, e^{-\mu/\rho \cdot Px}.$$

For any given element the mass absorption coefficient increases with increasing wavelength, except for important discrepancies at particular wavelengths known as absorption edges. These absorption edges are listed in Tables 10.1 and 10.2. Mass absorption coefficients have a fixed value for a given element for a particular wavelength of X-rays. The most reliable values are to be found in Heinrich (1966), and a more complete tabulation appears in Clark (1967) and in Jenkins and De Vries (1970a). See also White and Johnson (1970). Discrepancies will be found in these listings, for there is uncertainty about the best values for mass absorption coefficients. Table 10.3 lists the most reliable values for the major rock and mineral-forming elements.

As an example, a specimen of diopside $CaMgSi_2O_6$ is being analysed for chromium using the $CrK\alpha$ ($\lambda = 2.291$ Å) wavelength. The absorption coefficient of diopside for this wavelength is found from the following tabulation

Element	Weight %	μ/ρ for Wavelength 2.291 Å	
O	44.32	39	
Mg	11.23	119	
Si	25.94	184	
Ca	18.51	469	(Table 10.3)

μ/ρ diopside $= 0.443 \times 39 + 0.112 \times 119 + 0.259 \times 184 + 0.185 \times 469 = 165$ (after Norrish and Chappell 1967).

In the same way it can be found that diopside has a different mass absorption coefficient for other wavelengths; for example, for Mg $K\alpha$, Si $K\alpha$, and so on. Therefore, if the chemical composition of the specimen is known, it is possible to predict the amount of absorption caused by the specimen to the various analytical wavelengths. Here is the very crux of the matter; in any kind of analytical technique we do not know the chemical composition of the specimen until we have analyzed it.

The problem is overcome by always comparing the X-ray response of an unknown with a standard specimen of similar bulk chemistry (hence of similar mass absorption coefficient). The chemical composition of the standard is usually well authenticated by accurate wet chemical analyses. Thus for the analysis of an unknown granite a granite standard is used, a gabbro for a gabbro unknown, a peridotite for a peridotite, a hornblende for a hornblende, and so on.

Availability of Standards.   Rock and mineral international analytical standards are available, usually free of charge, from a variety of sources. Details and addresses may be obtained by reference to Flanagan (1970), Flanagan (1969), Fleischer (1969), and Sine et al. (1969). A wide range of well-known standards may be had on request from Dr. F. J. Flanagan, U.S. Geological Survey, Washington, D.C. 20242.

Outline of Method.   Any major difference between a standard and an unknown may be further reduced by diluting both by a standard method outlined in the following descriptions, by adding a fixed amount of heavy absorbing element, or both. Dilution and addition of a heavy absorber make the mass absorption coefficients of standard and unknown closely similar. If the standard and unknown are similar initially, then dilution and addition make the two almost perfectly identical, and so the comparison of their X-ray intensities will become directly proportional to the concentrations of their elements.

If we can obtain the following data,

	Oxide (%) or Cation (ppm)	Peak Intensity $K\alpha$ or $L\alpha_1$ (cps)	Back- ground (cps)
STANDARD	X	P	B
Unknown	x	p	b

then

$$x = \frac{X(p - b)}{P - B}.$$

Although the comparison is always for element by element, it is normal to express the concentration in an oxide form, for example, $MgO\%$. This presents no complication except in analyses for iron, which will have been reported in the standard wet chemical analysis as both FeO and Fe_2O_3. Thus X for the standard is in two forms: $X_1 = FeO$ and $X_2 = Fe_2O_3$; X for the standard must be expressed EITHER as total iron in the form of FeO, in which case $X = X_1 + 0.8998 \times X_2$, OR as total iron in the form of Fe_2O_3, in which case $X = 1.111 \times X_1 + X_2$. If X has been expressed as FeO, the unknown x will be recorded as total iron FeO, and if X has been expressed as Fe_2O_3, then x will also be expressed in the same form.

It is customary to express total iron in the form of FeO because most rocks and minerals have more ferrous than ferric iron. Fe_2O_3, however, may be preferred for materials richer in ferric than ferrous iron.

Table 10.3 Mass Absorption Coefficients for the Common Rock-Forming Elements [a]

$K\alpha$ Wavelengths

Emitter Wavelength / Absorber	Na 11.909	Mg 9.889	Al 8.337	Si 7.126	P 6.155	S 5.373	Cl 4.728	K 3.742	Ca 3.359	Ti 2.748	Cr 2.291	Mn 2.103	Fe 1.937
1 H	22	12.5	7.85	5.05	3.50	2.30	1.65	1.00	0.90	0.66	0.54	0.50	0.43
3 Li	169	99	60.7	38.6	25.3	17.1	11.8	6.00	4.4	2.5	1.5	1.1	0.9
4 Be	418	245	150	96	63.3	42.9	29.8	15.2	11.2	6.3	3.7	2.9	2.2
6 C	1534	904	557	356	235	160	111	57.3	42.1	23.8	14.2	11.1	8.8
8 O	4109	2432	1503	965	638	435	303	157	116	65.7	39.4	30.9	24.5
9 F	5169	3066	1897	1220	809	552	385	199	147	83.9	50.3	39.6	31.4
11 Na		5409	3359	2168	1440	986	690	359	266	151.9	91.4	72.0	57.2
12 Mg	770	463	4376	2824	1877	1284	899	468	346	198	119	93.8	74.6
13 Al	1021	614	385	3493	2324	1593	1116	582	431	247	149	117	93.4
14 Si	1332	802	503	327	2840	1949	1367	715	530	304	183	145	115
15 P	1695	1021	640	417	279	2370	1663	870	645	370	223	176	140
16 S	2102	1265	794	517	346	239	1965	1030	765	439	266	210	167

															L_I L_II L_III

														K
17 Cl	2578	1552	974	634	425	293	207	1210	898	516	312	246	196	Cl 17
19 K	3729	2245	1408	917	615	424	299	158	1189	685	415	328	261	K 19
20 Ca	4412	2656	1667	1086	728	502	354	187	139	772	469	371	296	Ca 20
22 Ti	6056	3646	2288	1490	999	689	486	256	191	110.6	597	472	377	Ti 22
24 Cr	7943	4782	3000	1954	1310	904	637	336	250	145	88.2	69.9	474	Cr 24
25 Mn	9041	5443	3415	2225	1491	1029	726	383	285	165	100.5	79.5	63.5	Mn 25
26 Fe	10166	6120	3840	2502	1677	1157	816	431	321	185	113	89.4	71.4	Fe 26
27 Co	11464	6902	4330	2821	1891	1305	920	486	362	209	127	100.8	80.6	Co 27
28 Ni	12805	7709	4837	3151	2112	1458	1028	543	404	233	142	112	90.0	Ni 28
29 Cu	12165	8569	5376	3503	2348	1620	1143	603	449	259	158	125	100	Cu 29
30 Zn	9690	9506	5965	3886	2605	1797	1268	669	498	288	175	139	111	Zn 30
33 As	2683	1655	6782	5163	3461	2388	1684	889	662	383	233	184	147	As 33
37 Rb	3824	2359	1513	1006	4111	3339	2355	1243	926	535	326	258	206	Rb 37
38 Sr	4150	2560	1642	1092	4454	3595	2536	1339	997	576	351	278	222	Sr 38
39 Y	4501	2776	1781	1184	809	3300	2730	1442	1073	621	378	299	239	Y 39
40 Zr	4878	3008	1930	1283	876	3561	2939	1552	1155	668	405	322	257	Zr 40

L

[a] After Heinrich. 1966. Stepped lines are absorption edges.

Another matrix problem is that of particle size effect, which may be overcome by fusing both standard and unknown to a glass. The following methods have been evolved to overcome matrix problems and to produce good specimens of similar mass absorption coefficients for direct analytical comparison

EXERCISE 10.8

Specimen preparation

METHOD A [*most commonly used*; *suitable for Mg to Fe (inclusive) analysis*]

The following fusion technique achieves uniform samples, removes the particle size effect, and greatly reduces matrix effects.

1. Prepare the following high purity chemicals:

    (a) Ignite lithium tetraborate (anhydrous) in a large platinum or palau crucible in a muffle furnace at approx. 500°C.
    (b) Ignite lithium carbonate at approx. 500°C
    (c) Ignite lanthanum oxide at 900°C

2. After ignition store in a dessicator until required.

NOTE. Platinum or palau crucibles may be obtained from Engelhard Industries Inc., 113 Astor Street, Newark, New Jersey 07114, or from their Baker Platinum Division, St. Nicholas House, St. Nicholas Road, Sutton, Surrey, England. Whenever possible, a plastic reshaper should be used to reshape the crucible when it is distorted out of shape.

High purity lanthanum oxide and lithium tetraborate may be obtained from K and K Laboratories Inc., 121 Express Street, Engineers Hill, Plainview, N.Y. 11803, or Alfa Inorganics Inc., P.O. Box 159, Beverley, Mass. 01915. High purity lithium carbonate may be obtained from E. Merck AG, Darmstadt, West Germany.

3. Weigh accurately and separately

    38.0 g lithium tetraborate (ignited),
    29.6 g lithium carbonate (ignited),
    13.2 g lanthanum oxide (ignited).

Mix them together with a spatula in a suitable crucible, preferably a 250-ml platinum crucible with lid (Engelhard) or, if funds are lacking, a 250-ml graphite crucible with lid ("E.K. 41" 82.5 diameter × 70 mm high with lid 82.3 mm diameter × 13 mm high; from Ringsdorff-werke GMBH, Bad Godesberg-Mehlem, Germany).

NOTE. If possible, the graphite crucible should be avoided because an uncontrolled amount of graphite pours out with the melt. This makes comparison of one batch with the next impossible. Use of the more expensive platinum crucible is more than justified because each batch is then reproducible and contains no contamination.

4. Place the crucible with lid in the muffle furnace and hold at 1000°C for 10 minutes until the mixture has completely melted. The melt is then poured onto a thick sheet of polished aluminum. To prevent the aluminum from melting and the glass from fusing onto it the melt should be poured over an area as wide as possible and the aluminum

sheet should be at least 3 mm thick. A short-handled crucible tongs and asbestos gloves should be used for the pouring.

Thus we obtain approximately 63 g of glass (actually it does not remain glass but crystallizes as a white opaque mass). This is suitable for making 28 rock or mineral specimens. It is preferable to make several batches, one after the other, so that enough of the glass will be available for all the rock and mineral preparations to be analyzed in any particular study.

5. Grind the glass to a coarse powder in a mortar and pestle and store it in a labeled airtight container. Grinding that is too fine should be avoided because it promotes the absorption of water. Periodically, before use, it should be heated to approximately 450°C to dry before weighing.

6. Specimens are prepared by accurately weighing into a 15 to 20 cc (90% Pt, 10% Au) platinum-gold alloy crucible or a palau (80% Au, 20% Pd) crucible:

2.25 g borate glass,
0.03 g sodium nitrate,
0.42 g powdered rock or mineral sample.

Norrish and Hutton (1969) use only two-thirds these amounts. I find, however, that it is preferable to use the amounts quoted, for they allow for some loss of glass adhering to the crucible.

NOTES. Pure platinum crucibles are inconvenient because the melt wets the crucible and a considerable amount adheres to the crucible after pouring. This has subsequently to be removed by time-consuming sodium carbonate fusion. Gold-platinum alloy crucibles give much less trouble. They may be obtained from Engelhard Industries Inc. A suitable size is shown in Fig. 10.8.

7. An electric muffle furnace or a Meker burner is used to heat the crucible to 1000°C. The temperature may be regulated by noting when sodium fluoride just melts (980°C).

8. A graphite disc and an aluminum plunger are kept on a hot plate at approximately 220°C. A brass ring is put over the disc and the melt quickly poured into its central depression. *Immediately* the plunger is brought down *gently* on top of the melt to mold and quench it. The plunger is *immediately* withdrawn and the brass ring removed.

This operation is greatly facilitated by firmly attaching the aluminum plunger to the rod of a simple hand-press whose base is a flat metal plate which is set on the hot plate. A shallow (1 mm) well in the center of the base-plate keeps the graphite disc in place. In its rest position the plunger is in perfect contact with the top of the graphite disc so that both will be maintained hot. When ready to pour the melt, the handle of the hand-press is raised about 15 cm, the melt poured, and the handle immediately dropped to press and quench the glass, and immediately raised again. With such a hand-press, the brass ring (Fig. 10.8) is unnecessary.

NOTES. Figure 10.8 shows the dimensions of the equipment needed. They may be made in any workshop. The aluminum plunger has a depression on its base with a sharp edge. Care must be taken not to destroy this edge. The graphite discs are machined from pure graphite rod or may be purchased from Ringsdorff-Werke GMBH, 5320 Bad Godesberg, Mehlem, P.O. Box 9087, West Germany, quoting their drawing 21.52.061.0.41 dated 29.5.68.

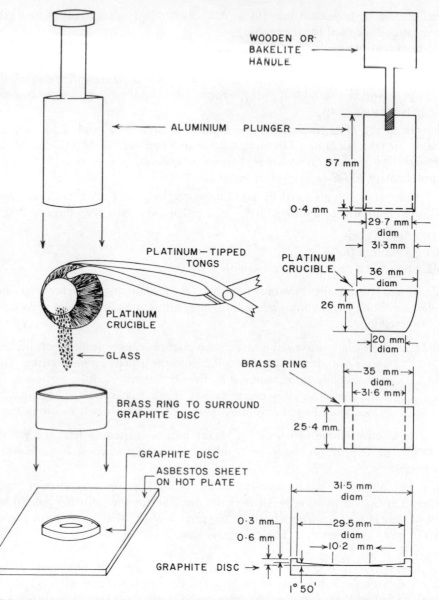

Fig. 10.8 Equipment needed for making the fusion discs. The brass ring is placed on the asbestos sheet to surround the graphite disc. The melt is quickly poured onto the graphite disc and the aluminum plunger quickly brought onto it and quickly removed. The dimensions of the components are from Norrish and Chappell, 1967.

9. The glass disc is now put between two clean asbestos mats on another hot plate kept at approximately 200°C. Alternatively the glass disc may be left on its graphite disc and placed together between the asbestos mats. After a few minutes the mats may be removed from the hot plate with the glass disc (and graphite disc) sandwiched between them and left to cool on the bench.

NOTE.   Do not remove the top mat for a considerable time and prevent any cold drafts from blowing over the glass disc which will cause it to splinter.

10. After cooling, projecting edges may be filed off so that the discs will fit perfectly in the X-ray spectrometer holder.

11. Sample discs should be stored flat, face (the face that was in contact with the aluminum plunger) upward in plastic ointment or in pillboxes.

(Method after Norrish and Hutton, 1969.)

DISCUSSION

If a disc breaks, it may be remelted and remade. If the break is clean, in two pieces, the disc may be taped on the back side (the side that was in contact with the graphite disc). A piece of tape on that side may also be used as a label for the specimen number.

The flat surface, which will be used for analysis (the surface that was in contact with the aluminum plunger), is always kept clean. Do not finger it. It may be cleaned with an acetone-soaked tissue paper.

The temperature of fusion must be kept above 950°C. The melt must be poured *quickly* on removal of the crucible from the source of heat; otherwise crystallization will take place. Observation of the final disc under a polarizing microscope will show abundant small crystals in the glass if the technique is poor. A few evenly distributed small crystals can be tolerated. Strain shadows in the glass disc are normal and do not affect the results.

Most normal silicate rocks and minerals are suitable for making discs by this method. Sulfur may be partly lost from the sample during fusion. Chlorine in micas and amphiboles is partly lost.

Ores give poor results. Cassiterite does not dissolve. Copper ores must not be used because the copper passes into and contaminates the platinum of the crucible. Ferrous iron also alloys with platinum.

I obtain good results with this method, but many authors report that a certain expertise is required to produce good discs and they have been unsuccessful in their own laboratories (e.g., Fabbi, 1972). The presence of lanthanum interferes with magnesium determination when the magnesium is present in the samples in low concentrations. A general discussion of the use of lithium metaborate flux in specimen preparation can be found in Ingamells (1970).

METHOD B [*a simpler method requiring less skill, suitable for Mg to Fe, inclusive, analysis*]

A sample: flux: binder ratio of 1 : 14 : 3 is used.

1. $LiBO_2$ (Southwestern Analytical Chemicals, P.O. Box 485, Austin, Texas, 78767) is dried in batches at 650°C for 30 minutes in a one-third to one-half full platinum evaporating dish covered with a radial ribbed watch glass. When ordering, specify "high purity anhydrous $LiBO_2$."

2. Dried $LiBO_2$ is sieved through a 16-mesh screen and several dried batches are homogenized.

3. 0.1000 g of powdered rock or mineral sample and 1.4000 g of dried and screened $LiBO_2$ are placed in a 30-ml porcelain crucible, mixed with a spatula, and carefully transferred to a preignited (950°C at 20 minutes) crucible (graphite crucible A2726, grade UF 4S, full radius inside bottom, from Ultra Carbon Co. P.O. Box 747, Bay City, Michigan, 48709.

4. The mixture is fused in a furnace for 15 minutes at 900°C.

5. The fused bead is cooled and weighed; 0.3 g of chromatographic cellulose (Whatman CF-11, catalog No. 10864-10, from Matheson Scientific, 1600 Howard Street, Detroit, Michigan 48216), plus the equivalent weight loss of H_2O, CO_2, etc., is added to bring the weight of the bead and the cellulose to a total of 1.8000 g.

6. The bead is crushed in a hardened tool steel mortar, transferred with grinding ball and cellulose into a grinding vial (polystyrene with tungsten carbide ball and end caps), and ground in a mixer mill (Spex) for 10 minutes. The ground powder is further ground and mixed by hand in a boron carbide mortar to ensure complete comminution and homogeneity.

7. This powder is compressed into a self-supporting pellet at 30,000 psi (exercise 10.1).

(Method after Fabbi, 1972.)

DISCUSSION

The pellets may be backed with methyl cellulose (Matheson Scientific, MX 850). Backing with methyl cellulose produces homogeneous, durable, and high-quality discs. Boric acid is unsatisfactory as a binder and as a backing material for quantitative work (Fabbi, 1972). Under the vacuum of the spectrograph the boric acid gives off water and causes severe errors in the X-ray intensities of the lines under observation.

Because $LiBO_2$, when finely subdivided, can take up water, pellets should be stored in a dessicator cabinet before analysis.

METHOD C [*suitable for Na to Fe (inclusive) analysis*]

It is claimed that the following method produces standard and unknown discs which give linear calibration curves.

1. Crush the rock in a jaw crusher.

2. Grind in a pulveriser to approximately −30 mesh.

3. Split the sample of approximately 40 g to a lesser volume suitable for a ball or Tema mill.

4. Fine grind for 4 minutes. Split in half, with each half carried through independently to the final steps.

5. Dry in a vacuum oven at > 20-in. vacuum and 100°C for 10 to 15 minutes. Cool in a vacuum dessicator.

6. Weigh on an analytical balance in porcelain crucibles.

7. Heat in a muffle furnace for 10 minutes at 850°C.

8. Cool in a vacuum dessicator.

9. Reweigh and determine loss on fusion (or gain).

10. Using a top-loading balance, tare the graphite crucible (Yu-40 grade graphite, crucible A-2726, supplied by Ultra Carbon Corporation). Add 5.2 g $Li_2B_4O_7$ (X-ray spectrographic grade, supplied by Allied Chemical Corporation). Add 2.8 g specimen powder from the fine-grained split.

11.   Stir thoroughly to ensure complete mixing.

12.   Fuse for 20 minutes at approximately 1050°C. Ten crucibles are set on a graphite carrier (supplied by Ultra Carbon Corporation P.O. Box 747, Bay City, Michigan 48709) and placed in a furnace preheated to 1100°C. Fusion time-count is begun when the furnace temperature begins to rise after insertion of the cold crucibles. Final temperature is about 1100°C used throughout the fusion period.

13.   The fusion beads are air cooled, and reweighed. Weight loss should be 8 g less fusion loss. If significantly different, an original weighing error is probable and the specimen should be remade.

14.   The fusion bead is crushed or broken into $\frac{1}{2}$- to 1-cm chips.

15.   The fusion bead chips are fine ground for 4 minutes.

16.   The final powder is pressed into a coherent specimen at 30,000 psi with bakelite or methyl cellulose backing and rim (exercise 10.1) and no internal binder. The specimens are then ready for X-ray spectrometry analysis.

(Method after Welday et al., 1964.)

DISCUSSION

This method is designed for analysis of batches of 30 or more specimens. For maximum efficiency at least two persons should prepare the specimens and a third person operate the spectrograph. A balance with relatively low sensitivity (0.01 g), such as the Mettler K, is ideal. A computer program is used to determine the least-squares calibration curve for several standards (prepared in the same manner and with the same dilution as the unknowns). Boric acid should not be used to back the specimens because it loses water under vacuum and this affects the X-ray intensities.

METHOD D [*suitable for Na to Fe (inclusive) analysis*]

1.   Weigh 3.5 g of rock or mineral powder, previously ground for chemical analysis, and 7.0 g Spec-pure lithium tetraborate into a polystyrene mixing phial and mix for 10 minutes with lucite balls on a ball mill.

2.   Pour the mixture into a graphite crucible and place in a muffle furnace at 1000°C for 15 minutes. The recommended graphite crucibles are the same as those in method C. They may be obtained with a graphite carrier, which holds 10 at a time, from Ultra Carbon Corporation, or from Heyden and Son Ltd., 64 Vivian Avenue, Hendon, London. N.W.4. Boxes of 15 crucibles (A-2726) may be ordered with carriers for 20 crucibles. The carrier may be put in the furnace as a single unit. The total weight of the sample and flux and the internal diameter of the graphite crucibles is critical. The final beads are designed to fit the standard Philips specimen holders of 32-mm disc diameter.

3.   After cooling, the glass beads drop out of the crucibles and their flat surfaces are ground on a carborundum wheel and polished briefly with diamond paste. Total grinding and polishing time is about 15 minutes per bead. The beads are ready for analysis. They are permanent, and if necessary their surfaces may be cleaned from time to time with carbon tetrachloride.

(Method after Hooper and Atkins, 1969.)

DISCUSSION

The beads are homogeneous. Volatiles are lost during fusion, which causes an increase in the concentration of the other ions of the rock or mineral powder to the extent of one-third of the volatile if the 2 : 1 dilution is used. Thus the summation of the analysis will be $100 + \frac{1}{3}$ volatile percentage and the oxides therefore can be corrected accordingly.

Matrix absorption effects by this method are still particularly noticeable for silicon and aluminum, but particle size effects are removed. Sodium can be analyzed satisfactorily.

NOTE.   It has been observed that lithium tetraborate glass discs, which have been irradiated for 10 hours or more, show significant decrease in contrate for Na and Mg and increase for Al, Si, and P. This effect is thought to be due to surface diffusion in the glass (Le Maitre and Haukka, 1973). Although the discs are usually never irradiated for such long times during analysis, it is recommended that at least one master standard or reference disc, which is rarely irradiated, be kept for periodic checks against a working standard disc whose irradiation is more frequent.

EXERCISE 10.9

Determination of Optimum Detector Voltage for All Elements Excluding Sodium

A detector is ideally set up when its response remains constant over a range of voltage fluctuation. To determine the best operating voltage the following procedure is recommended: settings should be determined by the operator himself for each element to be analyzed quantitatively by X-ray fluorescence. It is bad practice to use any casually recommended value which a friend or colleague may have been using for years because it may have been badly selected for no logical reason and may not suit your equipment.

1.   Select the exciting radiation that will be used in the final analysis, usually W for determining Cr $(Z = 24)$ and all heavier elements and Cr for V $(Z = 23)$ and all lighter elements.

2.   Make up self-supporting sample discs (one each, exercise 10.1), backed and rimmed with boric acid, of compounds containing only one major cation each (some boric acid may actually be mixed with the compound if it is not self-supporting); for example, for Ca use $CaCO_3$, for Sr, $SrCO_3$, for Al, Al_2O_3, and so on. These discs should be stored in a cabinet near the spectrometer. They are used to set the goniometer accurately on the correct 2θ for that element for any particular dispersing crystal and for determining the ideal detector voltage and pulse height selector settings. They should be stored face up in plastic capsules. The surfaces to be exposed to the X-rays must not be coated with any material—mylar, shellac, or grease—and must be kept clean. The specimens must be used in a specimen holder from which the polypropylene window has been removed because any material placed between the specimen and the X-rays will shift the energy spectrum. In all quantitative analysis nothing should be interposed between the X-rays and the specimen.

3. The scintillation counter will be used for all elements from Co ($Z = 27$) and heavier, and the gas flow proportional counter fitted with a 1-μm polypropylene window for elements from Fe ($Z = 26$) and lighter.

NOTE. The thin window has a limited life, which varies from several weeks to months. Once it has been found that the gas flow counter intensities have dropped dramatically, the window should be renewed. A supply of metal-supported windows ready for fitting may be obtained from Philips. The gas flow counter works only efficiently when a vacuum is applied to the spectrometer. The high tension should be applied to the detector only after the gas flow has been established.

4. The following examples apply only to Philips equipment and are not necessarily the same for other makes. By way of example we take Fe$K\alpha$ radiation. Put an FeSO$_4$ specimen disc in the spectrometer holder and turn it into the irradiating position by rotating knob X (Fig. 10.4) through 180°. Use Cr radiation. Adjust the flow counter to a position parallel with the scintillation counter (position 1 of disc P, Fig. 10.4) and 2θ at 57.5° for the LiF ($2d$ 4.028 Å). Use the fine collimator. Set time and scalar to infinity, attenuation Z to 2, HV = 450 (voltage approx. 1580). Raise the kV to exceed the excitation potential of Fe by at least 20% (see Table 10.1). Keep mA as low as possible, with the strip chart recorder running at 600 mm/h, lower level at 0, window at threshold, and adjust the 2θ on the goniometer to get maximum deflection of the recorder pen from the Fe$K\alpha$ peak. Lock 2θ at this value. Adjust the ratemeter so that the pen deflection is somewhere in the middle of the chart.

5. Set the timer to 10 seconds, the time constant to 1 second. With HV of the flow counter at 300 (voltage approx. 1145) count for 10 seconds. Immediately the count has stopped increase HV by 10 divisions and repeat the 10-second count. Continue increasing HV by 10 and counting for 10 seconds. Proceed until the recorder output begins to increase rapidly as shown in Figure 10.9. The operating plateau will have been drawn out by the pen.

6. Select an HV setting equivalent to a position on the plateau but toward the low-voltage end of the midpoint. Do not operate from the midpoint up because the higher voltage thus selected will lead to more noise in the circuit and poor operating conditions. This HV setting is ideal for Fe$K\alpha$ detection at attenuation 2. Figure 10.9 illustrates several gas flow counter plateaus and shows clearly that the beginning of the plateau moves to higher voltages with increasing attenuation (Z number) for any particular element. The arrows on the figure indicate suitable HV (voltage) operation values.

DISCUSSION

The plateau also changes position with increasing atomic number and certain elements (e.g., Ti) have an extremely narrow operating plateau; hence the HV must be carefully selected. Most elements can be detected on the flow counter within the HV range 450 to 500 (1580 to 1725 V) but 450 should be preferred. At 450 elements from Ni ($Z = 28$) down to Si ($Z = 14$) may be operated on attenuation 2, but Mg and Na may have to be operated on attenuation 1. Nickel, iron, and titanium can equally well be

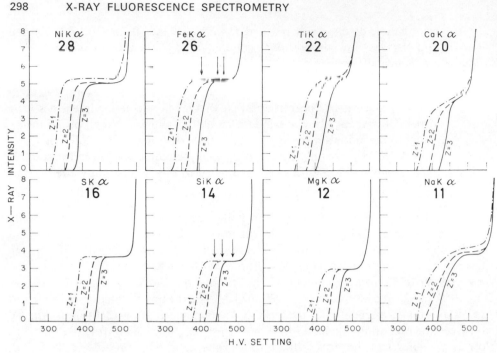

COUNTER VOLTAGE = 275 + 2.9 X READING ON H.V. CONTROL

Fig. 10.9 Gas flow counter response curves for the lighter elements $K\alpha$ lines for attenuation 1, 2, and 3. The arrows indicate suitable operating voltages. Note that the voltage varies with attenuation and atomic number of the element. The HV settings shown here are for Philip's equipment in my laboratory and need not be identical to those of the reader.

operated on attenuation 3. In summary, the heavier the element, the more attenuation, and very light elements cannot be attenuated much if the voltage is to be kept at a reasonable value of HV 450 to 500.

For the scintillation counter the procedure is the same and a reasonable operating range is HV 150 to 200 (voltage approx. 713 to 863). It will be seen from Fig. 10.10 that for elements from Cr ($Z = 24$) and lighter the scintillation counter plateau is narrow and the response of the counter becomes unsatisfactory. Hence for these elements the flow counter is preferable. From iron ($Z = 26$) and heavier the plateau is broad and the scintillation counter becomes ideal. Iron has a complex plateau. All elements can be determined on attenuation 2 between HV 150 and 200, but the heavier elements, for example, Sn and Ba, would be better determined at $Z = 3$, and the lighter elements, for example, Fe, Co, and Ni, would be better determined at HV 150 on $Z = 1$. If $Z = 2$ is used, the HV should be raised to at least 175.

Depending on the kV rating of the spectrometer, the K spectrum of the heavier elements will not be excited (see Table 10.1) and the L spectrum, especially the $L\alpha_1$ line, will have to be utilized. The scintillation counter gives a narrow plateau for $L\alpha_1$ for lanthanum ($Z = 57$) but this broadens with atomic number. A suitable operating HV is 150 to 175 on attenuation 2 (Fig. 10.10).

Figures 10.9 and 10.10 show the operating plateaus for a selected range of elements. The operator is advised to determine the best combination of HV and attenuation for every element (K or L spectrum, whichever is used) to be analyzed quantitatively. The optimum values should be noted in a reference notebook.

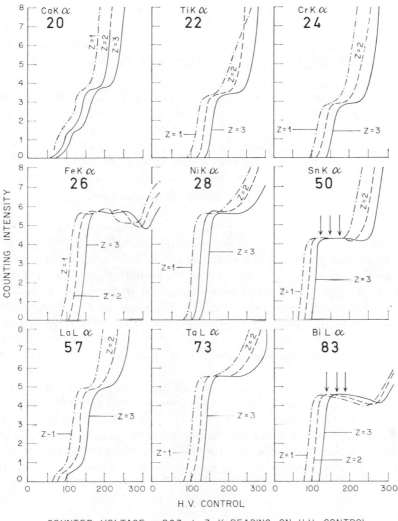

Fig. 10.10 Scintillation counter response curves for the heavier elements $K\alpha$ and some $L\alpha_1$ lines for attenuation 1, 2, and 3. The arrows indicate suitable operating voltages. Note that the voltage varies with attenuation and atomic number of the element. The HV settings shown here are for Philips' equipment in my laboratory and need not be identical to those of the reader.

EXERCISE 10.10

Determination of Pulse Height Analyzer Settings for all Elements Excluding Sodium

For precise quantitative analysis the pulse height analyzer, properly set for each element in turn, should be employed. This means that the setting of lower level, attenuation, and window must be determined for every element (whether $K\alpha$ or $L\alpha_1$ is used).

1. Take a sample containing an abundant concentration of the element to be analyzed (the same sample used in exercise 10.9, step 2). It must be self-supporting and the sample holder must be without a polypropylene window. Place it in the spectrometer irradiation position.

2. Set the goniometer to the correct 2θ position for the line to be detected from this sample. Select detector HV and Z as appropriate for that element as determined in exercise 10.9.

3. Proceed exactly as in exercise 6.18 from step 1 on, the only difference being that instead of CuKα radiation from a crystal diffraction we may be using the Kα (or Lα_1) radiation from a single element in the specimen disc. The procedure is the same whether we are using the scintillation detector or the gas flow detector. It is essential to set 2θ exactly on the value that gives maximum deflection on the recorder before locking it at this 2θ position. Proceed right through as far as step 13 of exercise 6.18.

DISCUSSION

When 2θ is set exactly on the peak, the count rate should not be higher than 16,000 cps. For heavy elements it may be too high, in which case put mA at its lowest possible value and kV at the lowest value that exceeds the excitation potential by 20% for that element (Table 10.1). For light elements there will be no problem, and normally 16,000 cps will certainly not be exceeded. For light elements kV will normally be put at 44 and mA can then be put as high as possible. Select a ratemeter setting for the peak deflection to fill at least 60% of the chart. If the count rate of 16,000 cps is exceeded, the electronic counting circuit will be saturated and this situation must be avoided.

If the energy distribution curve obtained is normal, as shown in Fig. 10.11, calculate the correct settings of lower level and window to be combined with detector HV and attenuation. Set window equal to $2\frac{1}{2} \times (V_2 - V_1)$ and lower level at $V_{max} - \frac{1}{2}$ window. In some cases, however, the energy distribution is not normal, for example, FeKα, and it is best not to calculate the settings of lower level and window but to obtain them just by observation of the energy distribution curve. The values of atten-

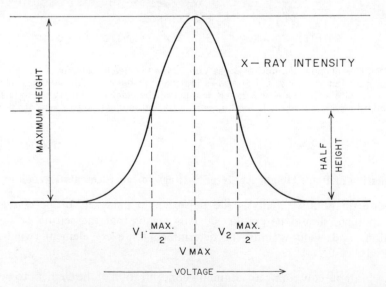

Fig. 10.11 Pulse height normal distribution curve. Normally the window is set at $2\frac{1}{2}(V_2 - V_1)$.

uation, lower level, window, and detector HV should be recorded in a notebook for every element that will be measured on the spectrometer.

By way of example, the settings for $SnK\alpha$ have been found to be the following:

Scintillation Counter HV 150, Z = 2.   Lower level 265 (1.325 V), window 240 (1.2 Volts) for accepting 98.2% of the total $K\alpha$ energy, and lower level 235, Window 300 for accepting 99.7% of the total $K\alpha$ energy.

Because a slight change in detector HV will affect the symmetrical setting of the pulse height distribution, the timing is now set to infinity, and with the lower level and window set properly adjust the ratemeter so that the pen deflection is between 50 and 100% of the chart. Now very slowly increase or decrease the detector HV. If the pen deflection decreases, it means that the pulse distribution is moving out of its ideal setting in the window. Now lower or increase the HV setting slowly. The pen deflection will slowly and steadily increase until for a particular setting of HV the pen shows maximum deflection. The HV setting that gives maximum deflection is the optimum setting and sets the energy spectrum symmetrically within the window. This kind of adjustment should be made frequently, at least every hour when the machine is being used and certainly any time the machine is being used afresh. The values of lower level, window, and Z can always be set as determined, but HV should be adjusted from its predetermined value to get maximum pen deflection.

Because the energy spectrum gradually drifts with time out of the symmetrical window setting, thereby causing the count rate to drop, it is desirable to add to the electronic panel a Philips Electronics Instruments Model 2100 PHA-scope. This instrument will simultaneously display on an oscilloscope screen the energy distribution curve of the wavelength being analyzed and the settings of lower level and window of the pulse height analyser. In this way one can continuously monitor that the energy spectrum is centered within the window. Any drift is immediately seen and corrected by a slight adjustment of the detector HV. (The Pha-Scope is obtainable from Philips Electronics Instruments, 750 So. Fulton Avenue, Mount Vernon, New York 10550.)

The pulse height for the gas flow counter depends, among other things, on the gas density, which in turn is influenced by the ambient temperature and the atmospheric pressure. Consequently it is important to check the flow counter HV more frequently than the scintillation counter—at least once an hour.

To minimize any drift it is recommended that the electronic equipment be switched on some hours in advance of measurement. A 24-hour switch clock can be utilized to turn on the apparatus every morning, say 2 hours before measurements begin.

It is also necessary to check the HV adjustment when a new bottle of argon-methane gas is opened. The use of a gas density stabilizer is highly recommended.

Because the mean pulse amplitude is proportional to energy, a different setting of the pulse height selector will be required for each wavelength to be measured. Automatic pulse height selection can be made available in some spectrometers, but it is not difficult to work out empirically, as shown in exercise 6.18, the value of lower level, window, Z, and HV for all elements to be measured. Then at each subsequent use only a slight adjustment of HV will obtain maximum count rate.

A poorly set pulse height analyzer is far worse than none at all. This instrument, once properly set, will give the spectrometer greater sensitivity. It discriminates against background and other unwanted wavelengths, but it must at all times be properly set, or it will discriminate against the very wavelength we wish to measure.

The ADP crystal is frequently used for $MgK\alpha$ detection. If the $MgK\alpha$ peak is set symmetrically in the pulse height analyzer window, the background will still be high, though crystal fluorescence is reduced. The background may be reduced further if the counter high tension (HV) is deliberately increased to put the energy spectrum off center in the window. The KAP crystal can be used for Mg detection, but an ADP crystal set up as above will give a better peak-to-background response.

EXERCISE 10.11

Setting of Counter HV and Pulse Height Analyser for Sodium

Because of weak $K\alpha$ X-ray intensity from elements lighter than magnesium and also because of severe fluorescence from the dispersing crystal (KAP) which obscures the $K\alpha$ sodium radiation, a special method is needed for setting both detector HV and the pulse height analyzer.

1. Determine the ideal HV and Z (attenuation) setting for $MgK\alpha$ (KAP 2θ 43.59°) by the method in exercise 10.9; for example, an ideal setting for a KAP dispersing crystal, using chromium radiation is $Z = 2$, HV (flow counter) 500 (approx 1740 V).

2. Determine the ideal pulse height analyzer settings for $MgK\alpha$ by setting the window at $2\frac{1}{2}$ times the half-peak-height width. Ideal settings are lower level 162 and window 312.

3. Using these settings, put a sodium-rich powder specimen (exercise 10.1), for example, sodium nitrate, in the spectrometer and adjust to the correct 2θ position for the dispersing crystal (KAP 2θ 53.12°) for $NaK\alpha$. In this way the sodium line will actually be seen above the background. Without the settings of the pulse height analyzer assumed from $MgK\alpha$, the $NaK\alpha$ radiation would not be detected at all. Adjust 2θ to get maximum counter response.

4. Now *slightly* increase HV to get maximum deflection of the recorder pen. This will be a very slight increase. The comparative settings are the following:

	For $NaK\alpha$	For $MgK\alpha$
Flow counter HV	505	500
Z (attenuation)	2	2
Lower level	162	162
Window	312	312

DISCUSSION

If the counter HV is increased or reduced too much from the value determined for Mg, the $NaK\alpha$ radiation will become lost in a high background. Figure 10.13 shows a scan over the Na peak for which the settings determined in this exercise were used.

It is recommended that sodium be determined on an undiluted whole rock or mineral powdered specimen made into a pressed disc (exercise 10.1) and backed and sided by boric acid, or better still by methyl cellulose, because the low intensity of sodium $K\alpha$ radiation will not permit dilution or addition of a heavy absorber in the fused disc methods. Count rates of approximately 15 cps above background per 1 % Na_2O should be obtainable on a pure undiluted rock disc.

Routine Instrumental Settings for Major Element Analysis

1. A complete batch of rock or mineral specimens should be prepared in the same way as the standard rock or mineral specimens according to a method in exercise 10.8.

2. Every specimen is then measured sequentially for one element in the spectrometer. Instrumental settings for the major elements are given in Table 10.4. Elements lighter than magnesium (Na and F) ideally cannot be diluted by one of the methods in exercise 10.8 and should be determined on whole-rock undiluted pressed samples. All elements from Mg to Fe may be determined on diluted samples made according to the methods in exercise 10.8. Manganese cannot be determined in low concentrations with a chromium tube because the CrK_β line overlaps the MnK_α line. Therefore for Mn determination a tungsten tube must be used to excite the radiation.

3. For each element in turn the spectrometer is set up accurately for 2θ, HV, Z, lower level, and window by using a pressed disc containing a major concentration of each of the elements. Once accurately set up, the spectrometer is fed consecutively with the standard and unknown specimen discs until every unknown has been measured for that element. The specimens are then repeated for every element until all analyses have been performed. Instrumental settings for major element analyses are listed in Table 10.4; 2θ is locked at a particular angle throughout the analysis of any particular element and K_α lines are used. The coarse 480-μm primary collimator is adjusted for all. Vacuum and flow counter are used for all elements. The pulse height analyzer is symmetrically set for all elements, except in the case of the ADP crystal for MgK_α, where the HV is set so that the energy spectrum is increased to 25% above its symmetrical setting in the window; In Table 10.4 % means percentage of the oxide in the specimen before dilution by method A in exercise 10.8.

4. The spectrometer has four specimen positions. The rock or mineral standard is always kept in specimen position 1. A blank specimen, prepared identically but without any rock or mineral added to the recipe, is left in position 2. Unknowns are placed in positions 3 and 4. When using a spectrometer for the first time, it is essential to check the equivalence of each of the spectrometer positions. Place a standard specimen in a sample holder. Observe its counting rate (for any particular element and line) when placed consecutively in each of the spectrometer positions 1, 2, 3, and 4. If the count rate varies significantly, *either* check in the instrument operating manual for an adjustment to the spectrometer positions *or* calculate a correction factor for each of the positions. Position 1 can be taken for convenience as needing no correction; for example, if the standard gives a count rate of x in position 1 and y in position 2, all subsequent readings in position 2 should be multiplied by the correction factor x/y. The correction factors for spectrometer positions 2, 3, and 4 will have to be utilized every time a comparison of a specimen is made with the standard housed in position 1.

Likewise a correction factor should be established for each of the specimen holders, which should be labeled A, B, C, D, and so on. This should be done more frequently because the specimen holders may be distorted by frequent use or by dropping on the floor or table. To establish the holder correction factor the same standard specimen should be put consecutively in each of the specimen holders and placed consecutively in spectrometer position 1. If the count rates differ significantly, a holder correction factor may be calculated. Specimen holder A will be taken as

Table 10.4 Instrumental Settings and Count Rates[a]

| | INSTRUMENTAL SETTINGS | | | | | | Performance | | | | |
Oxide	Dispersing Crystal	$2\theta°$ for $K\alpha$	X-Ray Tube Target	kV	mA	Routine Counting Time (seconds)	Counts per Second per % (approx)	Background as a %	Counting Time for Direct Reading (seconds)	Counts for 1%	Sensitivity (%)
							Undiluted Pure Rock /Mineral pressed Discs (Exercise 1C.1)				
F	KAP	86.85									
	RAP[b]	89.00	Cr	44	20	800	1.2	1	800	10^3	0.1
Na$_2$O	KAP	53.12									
	RAP[b]	54.26	Cr	44	20	100	22	0.1	45	10^3	0.03
							Samples Diluted by Fusion Technique in Exercise 10.8. Method A				
MgO	ADP[b]	136.43									
	RAP	44.50	Cr	44	20	100	8	0.36	120	10^3	0.06
	KAP	43.59									

Al$_2$O$_3$	PE	145.01	Cr	44	20	50	70	0.12	14.2	10^3	0.03
SiO$_2$	PE	109.20	Cr	44	20	50	70	0.24	14.2	10^3	0.05
P$_2$O$_5$	PE	89.51									
	Ge	141.00	Cr	44	20	50	200	0.06	50	10^4	0.007
	Graphite[b]	133.07									
SO$_3$	PE	75.84									
	Ge	110.73	Cr	44	20	50	182	0.08	55	10^4	0.008
	Graphite[b]	106.40									
K$_2$O	PE[b]	50.69	Cr	44	20	10	2600	0.05	3.8	10^4	0.007
	Graphite	67.84									
CaO	LiF (2d 4.028)	113.06	Cr	44	20	10	4000	0.06	2.5	10^4	0.007
TiO$_2$	LiF (4.028)	86.12	Cr	44	20	10	5900	0.17	1.7	10^4	0.012
MnO	LiF (4.028)	62.94	W	50	20	10	1250	0.16	8	10^4	0.012
Fe$_2$O$_3$	LiF (4.028)	57.49	W	50	20	10	1600	0.18	6.2	10^4	0.013

[a] All $K\alpha$ lines.
[b] Preferred crystal.

correct and all others referred to it; for example, if the count rate using A is x and that using B is y, the correction factor for B is x/y.

Holder A is subsequently always used to hold the standard in spectrometer position 1. Counting rates for all other specimens must then be multiplied by their specimen holders and spectrometer position correction factors.

5. The specimen rotation motor should be left running during analysis. Each specimen should be counted three times to ensure that any particular count is not spurious. A single count may be inaccurate, for at the time of counting the instrument may be susceptible to electrical interference. By collecting three counts on each specimen the similarity of the three will ensure that any spurious count can be rejected. If the three are similar, they may be averaged. Any count that greatly diverges from the other two may be rejected and recounted. The counting time for each element is given in Table 10.4. Heavy elements may be counted for 10 seconds, and lighter elements, because of their poor fluorescence yield and poor dispersing crystal performance, need more time; for example Al needs counting for approximately 50 seconds.

6. The counting sequence should be 1 (standard), 2 (blank), 3 (unknown), 4 (unknown), 1; 1, 2, 3, 4, 1; 1, 2, 3, 4, 1. As little time as possible should be allowed to lapse between counts. When a count is completed, the specimen chamber should be rotated to the next specimen, and the count initiated, until the whole sequence has been covered. The six counts on specimen 1 and the three on all the others are averaged, respectively.

7. Figure 10.12 illustrates a useful method of tabulating the results. The instrumental settings can be entered on the top part of the sheet. The averages of the counts should have the background average deducted, then used to calculate the oxide percentage according to the example given.

8. The preset time or count for any analysis should be selected with four-figure significance. This feature may be used to make the scaler give direct concentration readings; for example, if the standard is granite G 2 with a given SiO_2% of 69.21 and a rate of 48,447 counts in 10 seconds, it is equivalent to $4844.7/69.21 = 70$ cps/1% SiO_2. The timer therefore can be set to 14.3 seconds (1000/70) so that 10^3 counts $= 1\%$ SiO_2.

For most timers 14.3 seconds could not be done on one go. One would have to count consecutively for

$$
\begin{aligned}
1 \times 10^1 &= 10 \quad \text{sec} \\
4 \times 10^0 &= 4 \quad \text{sec} \\
2 \times 10^{-1} &= 0.2 \text{ sec} \\
1 \times 10^{-1} &= 0.1 \text{ sec} \\
\hline
\text{Total} &= 14.3 \text{ sec}
\end{aligned}
$$

and add the total counts obtained; 1000 counts $= 1\%$ SiO_2, and so standard G2, counting for 14.3 seconds, would give a scaler reading of 69.279, that is, 69.28% SiO_2.

9. Unknowns could be counted consecutively in 14.3 seconds and their SiO_2 concentrations given directly on the scaler. The concentrations read-out are uncorrected for matrix and background.

SPECIMEN Basalt & Dolerite	Standard BCR-1 TiC₂ 2.25% analysis		TUBE Cr	Sheet No. 1	

Line	Kα	kV 44	mA 24	S.C. / F.C.	Collimator fine / coarse	Path air / vacuum	Flow Co. position 1 / 2	HV 410 volt 1460
Element	Ti							

Crystal LiF	goniometer 2θ 86.12 reading 86.12	Pulse Height 2 3	Lower level 64	Window 62	counting time sec. 10

Position	Specimen number	Counts per second	average	background	c. p. s. - background	oxide %
1	standard	1423, 1438, 1429	1425	118	1307	2.25
2	blank	114, 120, 122	–	118	–	–
3	W-1	726, 744, 746	738	118	620	1.07
4	I-3 Dol	1600, 1627, 1604	1610	118	1492	2.57
1	standard	1417, 1437, 1410	–	–	–	–
1	standard					
2	blank					

$$Calculation \quad TiO_2 \text{ % } in\ W_1 = \frac{2.25 \times 620}{1307} = 1.07 \text{ %}$$

Fig. 10.12 Data sheet for X-ray fluorescence analysis. Instrumental settings are entered on the headings. SC means scintillation counter; FC means gas flow counter. When the flow counter is used in position 2, the actual goniometer reading is $2\vartheta + 30°$.

DISCUSSION

Apart from the convenience of the method of direct reading, the counting precision increases with concentration in a desirable way; for example, for SiO_2 the relative counting error is 0.3% for 100%, 1% for 10%, and 3% for 1% SiO_2. (Norrish and Hutton, 1969.)

EXERCISE 10.13

Background Determination

In the calculation of element concentration the peak height at the selected 2θ equals peak intensity due to the element under measurement plus the true background directly under the peak. We need to know this background accurately to subtract it from the peak so that the part due to the element may be determined. There are two possible methods of determining background but usually only one is applicable to any particular specimen.

METHOD A [useful for pure undiluted pressed powder specimens (exercise 10.1)]

1.   Scan over the peak by at least $\pm 5°$ 2θ on either side.

2.   Observe that extrapolation of the background line (which has a gradual slope) from either side would give a reasonably true background value underneath the peak (see example for sodium in Fig. 10.13). This will be the case if there are no interfering peaks near the main peak and the spectrogram goes to a reasonably flat background on either side of the peak.

3.   Select a suitable 2θ distance on either side of the peak (say $\pm 1.50°$) which will lie on this extrapolated background. The distance must be equal on both sides and the positions chosen must lie on the true flat background extrapolated line.

4.   Now set 2θ exactly on the peak 2θ and count for the required time.

5.   Set 2θ exactly on the $\pm$ background position in turn (e.g., for Na$K\alpha$, peak at 53.12; background at $53.12 + 1.50$ and at $53.12 - 1.50°$) and count for the same time. Average the two background counts, and subtract the average from the peak count to give the peak minus background value, which is used to calculate the element concentration.

METHOD B [necessary for diluted specimens that have absorber additions (exercise 10.8)]

When the specimens have been prepared with the addition of a heavy absorber, as in method A of exercise 10.8, the background determination needs modification because of the additional peaks to the specimen from the added absorber. This is best illustrated by the example of Mn$K\alpha$ determined on a fused borate disc to which lanthanum has been added (Fig. 10.13). Background for Mn$K\alpha$ cannot be determined on the same specimen because at 2θ position of peak ± 1 to $2°$ the 2θ scan does not go down to background values. Instead there are strong neighboring lanthanum peaks, which are present in the specimen. The required background determination procedure is as follows:

1.   Scan $\pm 5°$ on either side of the analytical peak.

2.   If there are no adjacent peaks, for example, in the case of TiO_2 (Fig. 10.13), background may be determined on the specimen at peak $2\theta \pm 1$ to $1.50°$ and averaged as in method A above.

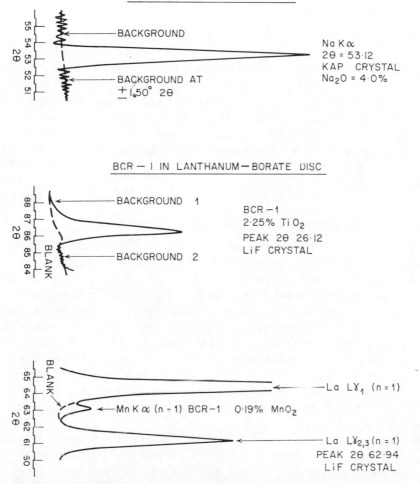

Fig. 10.13   X-Ray fluorescence spectrogram scans over certain $K\alpha$ peaks. Sodium is on a pure G-2 pressed powder specimen. Background may be determined on either side of the peak. Titanium and manganese are on BCR-1 specimens made into fused discs by method A of exercise 10.8. Although background could be determined for Ti by the same method, it is not possible for manganese because of neighboring lanthanum peaks. See exercise 10.13 for details.

3.   If, on the other hand, there are neighboring interfering peaks, background cannot be determined on the same specimen (case of $MnK\alpha$ in Fig. 10.13).

4.   In such a case background must be determined at the same 2θ position used for the peak determination, but on a specimen made by an identical recipe used for the specimen but to which no rock or mineral was added. Such a specimen is called a **BLANK**. In method A (exercise 10.8) a blank would be made from 2.25 g borate glass and 0.03 g sodium nitrate, to which no rock or mineral specimen is added. Figure 10.13 shows a scan for $MnK\alpha$ over the same 2θ range for a rock specimen and for a blank made by the same recipe (shown as dashed line). A count at 2θ 62.94 made on the specimen, then on the blank, will give peak plus background and background, respectively. The difference gives the peak height less background, which is used in the element concentration calculation.

DISCUSSION

A blank need not contain absolutely no specimen, but it *must* have all the diluents and heavy absorbers of the recipe; for example, a blank made with 2.25 g borate glass, 0.03 g sodium nitrate, 0.42 g pure SiO_2 (quartz) can be used as a suitable blank for determining all elements except silicon. A similar blank containing 0.42 g Al_2O_3 instead can be used for all elements except Al. Such pure element blanks are preferable to a blank containing no specimen because they are closer to the rock or mineral specimen in that they do include one major element of the specimen. A blank completely free of Si and Al is less ideal because the presence of Si and Al does contribute significantly to the overall background.

EXERCISE 10.14

Avoidance of High Count Rates and Dead-Time Determination

In operating the spectrometer care must be taken not to present the detector and amplifier combination with a count rate that is too high. If the equipment has a dead time of 5 μs, a counting rate in excess of 50,000 cps would become somewhat inaccurate and the scaler would show a smaller number of counts than the true value. Hence in counting *any* X-ray wavelength it is important that the count rate never be too large. Figure 10.14 illustrates this effect when the scintillation counter is used. The example given is for Sr$K\alpha$. The gas flow counter can be similarly affected.

The lower $K\alpha$ spectrum was determined with a very high 293,000 cps flux of Sr$K\alpha$. It can be seen that many of the incoming photons were entering the detector as a previously entered photon was still being registered; the result is a pulse of approximately twice the amplitude, occurring in the pulse height diagram at double the voltage of the main peak (Fig. 10.14, lower).

The upper $K\alpha$ spectrum was determined on the same specimen with an acceptable flux of 27,000 cps; it is evident that at this lower flux level the detector was able to accept all incoming pulses so that there was a complete absence of a doubled or sum peak and the whole energy was accepted as a normal pulse height distribution.

The correct lower level and window settings are shown on the upper spectrum. It is now warned that if a high flux of, say, 293,000 cps were given to the detector the pulse height analyzer would reject completely the "double or sum" peak and our recorded counting rate would be seriously in error. Our computed element concentration might be as much as 30 to 40% too low.

A counting rate of about 50,000 cps should therefore never be exceeded. When determining a major element in a sample, the counting rate may be reduced to an acceptable value by lowering the mA as far as possible. If this does not lower the rate sufficiently, kV may be lowered, but the operator must be sure that it always remains in excess of the excitation potential (Table 10.1) by 20% or more. If the count rate is still too high, a thin aluminum foil may be permanently placed in front of the detector for all specimen measurements, both standards and unknowns. This will require re-establishing the pulse height analyzer settings.

Determination of X-ray Detector Dead Time

1. Set up the spectrometer at the appropriate 2θ to detect $K\alpha$ radiation from a major element, for example Sn$K\alpha$ from a piece of tin plate.

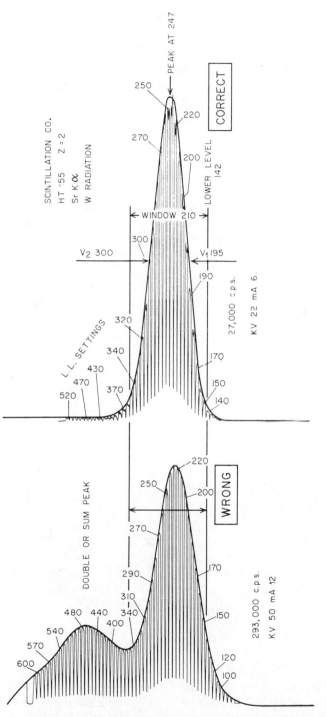

Fig. 10.14 Pulse height distribution for Sr$K\alpha$. (Lower) For a too high cps flux of 293,000. (Upper) For a proper low flux of only 27,000 cps. The high flux causes a double or sum peak which is not accepted by the scaler when the pulse height analyser is set at the lower level and window values indicated. The scaler value is then erroneously low by the amount of the sum peak.

2. The pulse height analyzer can be set symmetrically on the SnKα energy peak (optional).

3. Use W radiation, kV 50, and stabilize the equipment by allowing the generator, scintillation counter, and electronic panel to run continuously for 30 min.

4. Measure one counting rate in the normal way. Measure a second counting rate at the same mA setting while interposing an aluminum sheet of approximately 2 mm thickness as an absorber between the scintillation counter and its collimator.

5. Tabulate the readings as follows:

X-ray Tube Current	Counting Rate, (counts per second, no absorber)	Counting Rate (counts per second, through Al absorber)
6	$a(o)$	a
8	$b(o)$	b
10	$c(o)$	c
12	$d(o)$	d

6. *Calculation:* Where τ = dead time (in seconds)

$$\tau = \frac{\dfrac{a}{a(o)} - \dfrac{b}{b(o)}}{a\left(1 - \dfrac{b}{b(o)}\right) - b\left(1 - \dfrac{a}{a(o)}\right)}$$

(Method after Finger, 1971).

DISCUSSION

τ can be calculated using any pair of readings. The above equation is for pair a and b. Similar calculations can be performed for pairs a and c, a and d, and so on. The dead time for the scintillation counter is then taken as the average value of these calculations. A similar determination needs to be done for the gas-flow counter.

Whenever the counting rate exceeds 10^4 counts per second, the equation

$$N = \frac{N'}{1 - N'\tau}$$

is used to calculate the true from the observed counting rate, where N is the true counting rate (counts per second), N' is the observed counting rate (counts per second), and τ is the dead time of the detector system (seconds).

The foregoing equation is a first-order approximation of the more accurate $N = N'e^{N\tau}$, but for a count rate of 20,000 cps and a dead time of 5 μsec, the error introduced by using this first-order approximation is about 0.5%, which is not significant.

Matrix Corrections Only for Specimens Made by Method A in Exercise 10.8

If the standard is closely comparable to the unknowns, provided all specimens are made according to a method in exercise 10.8, especially by method A or B, matrix corrections should generally be unnecessary and, if performed, should make little difference to the results. Matrix corrections should be employed, however, if there is any significant divergence in major element concentration between standard and unknown. The following matrix correction method is recommended by Norrish and Hutton (1969) and is applicable ONLY to specimens prepared by method A in exercise 10.8.

All analyses depend on the intensity ratio of the unknown sample to the standard. All elements present in amounts greater than 1% of their oxide in the rock or mineral should be used in the matrix correction.

The method is outlined as follows:

1. The background must be determined for each element in turn by comparison with a standard. Both the standard and the specimen used for the background determination must be made identically by method A in exercise 10.8. The background must then be expressed as a percentage or as a weight fraction (%/100).

2. The specimen used for the standard may, for example, be made of the following: as an SiO_2 standard, 2.25 g borate glass + 0.03 g sodium nitrate + 0.42 g powdered pure quartz (SiO_2); this is said to be a 100% SiO_2 standard. Likewise, instead of containing the 0.42 g SiO_2, 0.21 g SiO_2 + 0.21 g Fe_2O_3 in a specimen could be used as a 50% SiO_2 and a 50% Fe_2O_3 standard. Quite a few standards can be made up in this way; for example, a 50% SiO_2 + 50% K_2O, a 50% SiO_2 + 50% MgO, a 100% CaO and so on. Also standards can be made up of the standard rocks, containing in each case exactly 0.42 g of the analyzed rock powder. All contents of standards should be carefully labeled.

3. In a similar way blanks should be made up for each element. Each blank will have no added element; for example, a blank for TiO_2 may contain 2.25 g borate glass + 0.03 g sodium nitrate + 0.42 g SiO_2; that is, the same specimen that serves as a 100% SiO_2 standard will also serve as a blank for all elements, excluding Si. It is wise to have a choice of blanks for each element; for example a similar specimen containing 0.42 g CaO will serve as a blank for all elements, excluding Ca. Standard rock specimens will not serve as blanks because they contain all the elements in varying degrees, and all blanks have to be specially made of single Analar chemicals or simple mixtures thereof. It is essential that all standards and blanks be made by the recipe in method A, exercise 10.8, with the only variation being in the 0.42 g of sample. Each specimen so made must be carefully labeled to show what has been added.

4. For each element in turn set the goniometer exactly on the precise 2θ for that element, and with proper HV, pulse height analyzer settings, etc., count the response from a standard; for example, if we use a 100% standard for SiO_2, let p = the counts per second at the SiO_2 peak position.

5. Now replace the specimen with an SiO_2 blank (containing no SiO_2). Let the counting rate measured be b cps with the same instrument settings.

6. The background reading can then be expressed as an apparent SiO_2 concentration (uncorrected) as background (uncorrected) = $100 \times b/(p - b)$ SiO_2%. It may also be expressed as a weight fraction $b/(p - b)$. It is wise to determine the background on several blanks of different composition, and in combination with several standards of different composition. Let us call this uncorrected background value B'; B' values must be determined for each element in turn.

7.   Each value of B' should be corrected for matrix effects by the equation

$$B = B'\left(M + \sum_1^n X_i E_i\right),$$

where B is the true background, B' is the background determined in step 6, M is the flux matrix correction listed in Table 10.5, X_i is the matrix correction for each element i in the sample, also given in Table 10.5, and E_i is the concentration of that element i in weight fraction.

As an example, if the blank used in step 6 contained 100% CaO and it was used as a blank for SiO_2, then B (for SiO_2) $= B'$ $(1.014 - 1 \times 0.042) = 0.972$ B'. As another example, if a blank containing 50% SiO_2 + 50% Fe_2O_3 were used as a blank for TiO_2, the value of B' determined in step 6 has to be corrected to a B (TiO_2): thus, B (for TiO_2) $= B'(0.851 + 0.5 \times 0.081 + 0.5 \times 0.110) = 0.947$ B'.

Another treatment may be illustrated by determining background for Fe, using a pure SiO_2 fusion as a blank and, say, 50% Fe_2O_3 and 50% SiO_2 for calibration.

First, ignoring background, the matrix correction for the standard for $FeK\alpha$ is $1.046 - 0.5 \times 0.027 - 0.5 \times 0.065 = 1.00$, so that its nominal Fe concentration is $50/1.00 = 50\%$. The background is measured on the SiO_2 sample, and its concentration of Fe determined in relation to this 50% sample. Because the background is only a fraction of a percent, the error in it, caused by not including background with the standard, will be slight.

However, this error can be removed by repeating the calculations, this time adding the background to the 50% standard. The calculations can be repeated, but should not be necessary. In step 6 several B' values are determined for each element by using different blanks and perhaps different standards. Each of these B' values for any element should be converted to a B by the method in step 7 and B for each element averaged.

8.   If the suite of unknowns is of granite, for example, a standard granite of known composition is kept in spectrometer position 1. The standard should always be similar to the unknowns; for example, peridotite used for peridotite and so on. Both standard and unknowns must be made identically by method A in Exercise 10.8.

9.   If, for example, we were to use granite G-2 as our standard, look up its published complete chemical analysis. The true composition is given as (Flanagan, 1967): $SiO_2 = 69.21$, $Al_2O_3 = 15.42$, $Fe_2O_3 = 1.02$, $FeO = 1.475$ (total iron computed as $Fe_2O_3 = 2.66$), $MgO = 0.75$, $CaO = 1.96$, $Na_2O = 4.09$, $K_2O = 4.45$, $H_2O + 0.46$, $H_2O - 0.18$, $TiO_2 = 0.47$, $P_2O_5 = 0.13$, $MnO = 0.04$, $CO_2 = 0.08$, $Cl = 0.01$, $F = 0.13$, $S = 0.01$, $BaO = 0.18$. Total $= 100.23$, less 0 of $0.07 = \overline{100.13}$. The loss on ignition of G-2 can be calculated as $0.46 + 0.18 + 0.08 + 0.01 + 0.13 = 0.86$. The general relationship for correction is $Si = Si'$ $\left(M + \sum_1^n X_i E_i'\right) - B$ and a similar equation for each element, for example, replacing Si and Si' with Ca and Ca', where $Si =$ the correct weight fraction of SiO_2 and $Si' =$ the apparent weight fraction of SiO_2 for X-R-F direct comparison purposes:

Table 10.5 Matrix Correction Coefficients Using Weight Fractions of the Oxides (wt %/100)

| M | X | | | | | | | | | | | |
Flux	Fe₂O₃	MnO	TiO₂	CaO	K₂O	SO₃	P₂O₅	SiO₂	Al₂O₃	MgO	Na₂O	Loss
Fe 1.046	−0.027	−0.031	0.146	0.134	0.126	−0.060	−0.060	−0.065	−0.074	−0.090	−0.110	−0.163
Mn 1.045	−0.044	−0.044	0.146	0.135	0.130	−0.037	−0.063	−0.063	−0.074	−0.078	−0.100	−0.163
Ti 0.851	0.081	0.077	0.179	0.647	0.644	0.194	0.181	0.110	0.078	0.069	0.051	−0.132
Ca 0.865	0.090	0.092	0.065	0.130	0.723	0.201	0.182	0.128	0.105	0.068	0.051	−0.134
K 0.897	0.098	0.086	0.017	000	0.069	0.182	0.179	0.119	0.101	0.080	0.057	−0.139
S 0.894	0.086	0.074	0.002	−0.023	−0.037	−0.053	0.167	0.131	0.112	0.087	0.063	−0.139
P 0.896	0.108	0.094	−0.020	−0.037	−0.047	−0.059	−0.063	0.127	0.110	0.094	0.046	−0.139
Si 1.014	0.082	0.086	−0.034	−0.042	−0.055	−0.057	−0.061	−0.061	0.122	0.093	0.063	−0.158
Al 1.056	0.112	0.116	−0.032	−0.037	−0.048	−0.056	−0.060	−0.088	−0.072	0.116	0.058	−0.164
Mg 1.050	0.136	0.126	0.010	−0.021	−0.043	−0.046	−0.016	−0.070	−0.078	−0.084	0.080	−0.163

M = the flux matrix correction (Table 10.5),

$\sum_i^n$ = sum of every element from 1 to n in the sample,

X_i the matrix correction for element i in the matrix (Table 10.5),

E_i' = the apparent weight fraction (%/100) of the oxide of element i as determined directly by X-R-F.

B = the background expressed as a weight fraction of SiO_2.

The calculation for G-2 is as follows (values of M and X are from Table 10.5):

$$0.6921 = Si' (1.014 + 0.082 \times 0.0266 + 0.086 \times 0.0004 - 0.034 \times 0.0047$$
$$- 0.042 \times 0.0196 - 0.055 \times 0.0445 - 0.057 \times 0.0001 - 0.061 \times 0.0013$$
$$- 0.061 \times 0.6921 + 0.122 \times 0.1542 + 0.093 \times 0.0075 + 0.063 \times 0.0409$$
$$- 0.158 \times 0.0086) - B$$
$$= Si' (0.991) - B.$$

B has been determined in step 7 on a specimen free of silicon. Suppose it gave a value of $B = 0.0024$ (weight fraction). Then $0.6921 = Si' (0.991) - 0.0024$ so that $Si' = 0.6945/0.991 = 0.7008$. This is the SiO_2 value for G-2 which is used in the standard G-2 borate discs for calibrating all subsequent granite unknowns. The height of the SiO_2 peak for standard G-2 is therefore taken as 70.08%. This is the gross peak height and no background correction is made, since it includes the background. The Si' values of all unknowns are determined by straight comparison with this value of 70.08 for G-2.

10. Other oxides are performed in the same way.

11. The unknowns are compared with the standard until a complete apparent composition is obtained for each unknown. Suppose, for example, that the counts per second on the G-2 standard at the SiO_2 2θ position was x and y on an unknown; then the apparent Si' % value of the unknown = $70.08 \times y/x$. Using the borate glass discs throughout, we take the background as constant so that determination involves one reading only, expressed as a percentage. This has already been done in step 7.

12. The total apparent composition of each unknown is converted to a real composition by using the values of M and X in Table 10.5. As an example, suppose a rock has the following apparent composition determined by X-R-F in comparison with the standard, as in step 11, uncorrected for matrix: $Fe_2O_3 = 13.00$, $MnO = 0.37$, $TiO_2 = 2.83$, $CaO = 14.09$, $K_2O = 1.48$, $P_2O_5 = 1.22$, $SiO_2 = 38.09$, $Al_2O_3 = 9.62$, $MgO = 13.62$, all determined on a fused borate disc by method A in exercise 10.8; $Na_2O = 3.24$ made on a pressed powdered undiluted sample (exercise 10.1); loss on ignition, 3.04. Using the matrix coefficients in Table 10.5, we compute the true concentration of SiO_2:

$$SiO_2 = 0.3809 (1.014 + 0.1300 \times 0.082 + 0.0037 \times 0.086 - 0.0283 \times 0.034$$
$$- 0.1409 \times 0.042 - 0.0148 \times 0.055 - 0.0122 \times 0.061 - 0.3809 \times 0.061$$
$$+ 0.0962 \times 0.122 + 0.1362 \times 0.093 + 0.0324 \times 0.063 - 0.0304 \times 0.158)$$
$$- background\ as\ a\ weight\ fraction\ of\ SiO_2;$$

B for SiO_2 has been determined on a specimen free of SiO_2 as 0.0024. SiO_2 (weight fraction) then = $1.015 \times 0.3809 - 0.0024 = 0.3842 = 38.42\%$. This is the true SiO_2 concentration in the sample, corrected both for matrix and for background.

A similar calculation must be made for each oxide in turn. In making the calculation by hand, the small products within the brackets may be neglected. It may be conveniently done by computer and may be iterated by using the new values obtained after each matrix correction (i.e., in the first calculation Si' is taken as 0.3809, 0.3842 in the next, and so on). It may also be iterated until each oxide concentration stabilizes. It is found, however, that if the standard is chosen close to the unknown the calculation need not proceed beyond the first calculation, and subsequent iteration produces insignificant change. Hence the whole calculation may be easily done on a desk calculator.

(Method after Norrish and Hutton, 1969.)

DISCUSSION

"Loss on ignition" at 1000°C is determined best on a separate portion of the rock or mineral powder.

Another way to perform this is by igniting all the specimens, including the standards, at 1000°C, in which case the "loss on ignition" becomes zero for all specimens and no matrix correction for loss is made either in standard or unknown.

For ordinary rocks and minerals the nominal concentrations, corrected for background, will be accurate within a few percent relative. Satisfactory corrections can be made by using only partial analyses. Low concentrations of certain oxides, for example, MnO and P_2O_5, may be ignored because they do not affect the correction significantly.

In the case of Fe_2O_3 the coefficients for SiO_2 and MgO are rather similar and the main matrix variation will arise from CaO and K_2O. Thus, if CaO and K_2O are measured approximately, reasonable Fe_2O_3 assays can be obtained from

$$Fe = Fe'[0.976 + 0.2(Ca' + K')] - B.$$

Similar relationships can be derived for other elements with appropriate rocks. The table of matrix coefficients X (Table 10.5) will allow us to predict when a correction is necessary. By this method, when the average oxide is greater than 1%, the coefficient of variation of the mean of several readings is less than 1%.

Using the sample preparation technique of method B (exercise 10.8) it is claimed that the high dilution overcomes most of the matrix effects and that no matrix correction is needed if the standards are chosen close to the unknowns in composition (Fabbi, 1972). Preferably, unknowns should be compared with more than one standard.

EXERCISE 10.16

Direct Measurement of Mass Absorption Coefficient

The direct measurement of the mass absorption coefficient A of an element for a chosen fluorescent wavelength can greatly facilitate the direct and reliable analysis of many minerals and rocks.

The mass absorption coefficient can be measured directly on most samples for wavelengths shorter than 3Å ; that is, for elements heavier than calcium.

Specimen Preparation

1. Samples must have a uniform density, thickness, and diameter. The most satisfactory method makes a 1 : 1 mixture of filter paper pulp ("Chromedia" CF11 from BDH Chemicals Ltd., or Whatman's ashless cellulose powder, standard grade) with the finely pulverized mineral or rock specimen. The filter paper pulp is added to strengthen the specimen into a self-supporting disc. The amount of filter paper pulp added is not critical, but standardization on 50% pulp 50% specimen is recommended.

2.   Weigh accurately about 0.2 to 0.3 g of specimen powder directly into a 2-in. long × $\frac{1}{2}$-in. diameter plastic vial. Add an exactly weighed equal amount of filter paper pulp. Add one $\frac{3}{8}$-in. diameter clear plexiglas pestle. Put on the plastic lid on the vial and set the vial on a WIG-L-BUG for about 10 minutes to effect a completely homogenized mixture (WIG-L-BUG model 3A, vials, caps, and plexiglas pestles are obtained from Crescent Manufacturing Co., 1837 South Pulaski Road, Chicago 23, Illinois).

3.   The homogenized mixture is pressed into a self-supporting disc with item B in Fig. 10.15. The disc so obtained has a diameter of 1.27 cm. A specially made piston-type die is used to press the disc (Fig. 10.15).

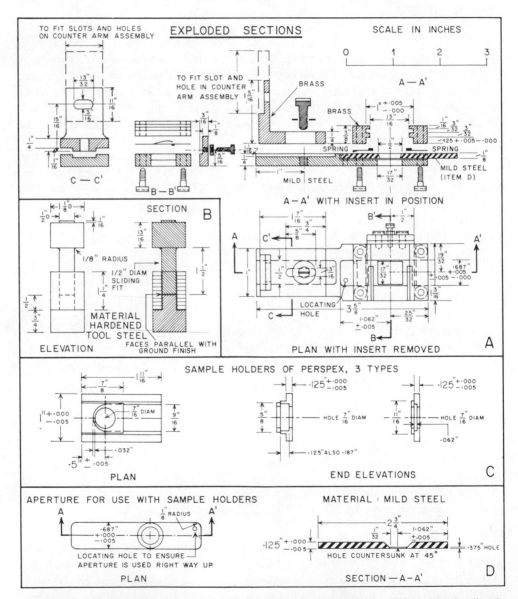

Fig. 10.15   Attachment for the vacuum spectrometer and accessories for making discs for directly determining the mass absorption coefficient for selected fluorescent wavelength. (By courtesy of Dr. K. Norrish, after a design by Dr. T. R. Sweatman. See exercise 10.16 for details.

4. The disc so made is then mounted in a perspex holder (C in Fig. 10.15). Three sizes of holder should be made to take different thicknesses of specimen disc.

5. A similar disc should be made of pure filter paper pulp to which no specimen has been added.

Measurement

6. The holder with the disc is then inserted on the spectrometer attachment (A in Fig. 10.15). The specimen disc now lies between the scintillation detector and its collimator. A standard aperture D is put into the slot to limit the X-ray beam to a selected width. Keep the same aperture in position for all measurements.

7. Set up the spectrograph to give a high intensity of the required wavelength; for example, a high intensity $SrK\alpha$ fluorescent radiation can be obtained if the specimen in the irradiation position of the spectrometer is pure $SrCO_3$, rimmed and backed by boric acid (method in exercise 10.1). This specimen is kept there permanently, the kV and mA is held stable, and 2θ is adjusted so that the $SrK\alpha$ line is dispersed by the crystal and detected by the counter. W or Mo primary radiation is used to excite the $SrK\alpha$. The pulse height analyzer is carefully set to accept only $SrK\alpha$ radiation.

8. The specimen (50% rock or mineral + 50% filter paper) is placed in the holder and the X-ray intensity I is measured.

9. The specimen disc is removed and the unattenuated intensity I_0 measured in the same way. The 100% filter paper disc is now put in the specimen holder and the attenuated intensity also measured. All intensities must be corrected for dead time (exercise 10.14) because the intensities measured may be quite high.

Calculation. By Beer's law

$$I = I_0 e^{-Apt},$$

Where I (cps) is the intensity attenuated by the specimen (corrected for dead time);
   I_0 (cps) is the primary unattenuated intensity (corrected for dead time);
   A is the mass absorption coefficient of the sample for the wavelength;
   p is the density of the sample (g/cc);
   t (cm) is the sample thickness.

The equation can be rewritten more conveniently as

$$A = \frac{1}{pt} \times \text{natural logarithm of } \frac{I_0}{I}$$

$$= \frac{\text{area of specimen in cm}^2}{\text{weight of specimen in g}} \ln \frac{I_0 \text{ cps}}{I \text{ cps}}.$$

By way of example, the following data are used to calculate the mass absorption A of the following samples for $SrK\alpha$:

	Standard W-1	Standard NBS-99
Area of sample (diameter 1.27 cm)	1.267 cm²	1.267 cm²
Weight of sample	0.2973 g	0.3427 g
I_0 (corrected for dead time)	105,600 cps	105,600 cps
I (corrected for dead time)	4,820 cps	18,240 cps
$A = \dfrac{\text{area}}{\text{weight}} \times \ln \dfrac{I_0}{I} =$	13.15	6.49

The calculation for W-1 is illustrated thus:

$$A = \frac{1.267}{0.2973} \ln \frac{105,600}{4,820} = 4.2616 \log_e 21.9087.$$

Natural log 21.9087 may be looked up, for example, in Abramowitz and Stegun (1965) as $\log_e 10 + \log_e 10 + \log_e 0.2190 = 2.3026 + 2.3026 - 1.5187$. A therefore = $4.2616 \times 3.0865 = 13.15$.

When the sample disc has been made (see steps 2 and 3) by adding a diluent to support the specimen, the determined absorption coefficient has to be corrected by the following relationship:

$$A_{sample} = \frac{A_{total} - (1 - P) A_{diluent}}{P},$$

where P is the proportion of the sample in the mixture. The values of I obtained for (the sample + diluent) mixture give A_{total}, and I obtained for a specimen made of 100% diluent give A for the diluent. A_{sample} is then computed from the above equation. For a standard procedure of making discs of 50% sample and 50% diluent, the relationship becomes

$$A_{sample} = \frac{A_{total} - 0.5 A_{diluent}}{0.5} = 2A_{total\ specimen} - A_{pure\ pulp}.$$

(Method after Norrish and Chappell, 1967; Moore, 1969.)

DISCUSSION

The usual diluent is filter paper pulp, but boric acid may also be used (Moore, 1969). The determination is for an X-ray spectrometer, but it may be done equally well on a diffraction goniometer using, for example, Co radiation.

Most measurements of the ratio I_0/I should give a ratio between 15 and 25. The thickness of the disc will have to be selected to give a ratio in this range. All count rates have to be adjusted for dead time, and pulse height analysis should be employed to eliminate possible higher order harmonics in the attenuated beam.

Ratios of I_0/I as high as 10^4 can be used, but when I becomes small a background correction becomes necessary.

If $I_0 \gg I$, the relative error in A due to counting errors equals the counting error $\%/\log_e(I_0/I)$. This shows that A can be measured accurately with relatively few counts if I_0/I is not much greater than unity.

EXERCISE 10.17

Composition Determination in Mineral Isomorphous Series by Mass Absorption Coefficient Determination

Mass absorption coefficient of a mineral for a given wavelength will vary progressively in an isomorphous series if there is sufficient contrast between the two end members and the series is essentially a replacement of one element by another.

The crystals to be analyzed are separated from a rock and treated by whatever means can give high purity. The sample that should be about 0.2 g is then made into a disc by the method in exercise 10.16 and its mass absorption coefficient for Sr$K\alpha$ is accurately determined.

The following results are given by Moore (1969):

PLAGIOCLASE (experimentally determined relationship).

$$An\%\{ = [Ca(+ Sr) \times 100]/[Ca(+ Sr) + Na + K]\} = 25.57\, A_{SrK\alpha} - 164.66$$

with a standard deviation of 1.61 % An.

ORTHOPYROXENE (calculated relationship).

$$Mg[= (Mg \times 100)/(Mg + Fe^{2+} + Fe^{3+} + Mn)] = 119.90 - 2.54\, A_{SrK\alpha} - 0.041\, A^2_{Sr\,K\alpha}$$

with a standard deviation of 1.38 % Mg.

SPHALERITE (calculated relationship).

$$In the series\ ZnS - (Fe, Mn)\, S,\ weight\ \%\ ZnS = 4.5\, A_{SrK\alpha} - 220.$$

(Method after Moore, 1969.)

C. QUANTITATIVE TRACE ELEMENT ANALYSIS

Because of the low concentrations of trace elements, the sample cannot be diluted nor can a heavy absorber be added as in major element analysis. If similar rocks and minerals are compared for trace elements, matrix effects should be minimal. The matrix effects will arise not because of the trace element but because of the major element variation between standard and unknowns.

Trace and Minor Heavy Element Analysis with Mass Absorption Coefficient and fluorescent Radiation Intensity

If the element being measured is the only major element with an absorption edge between the exciting and fluorescent radiation, we have

$$P = \frac{kCA}{K - C},$$

where k and K are constants that may be determined empirically on analyzed samples;

C is the total intensity per unit area from the fluorescent radiation;
P is the concentration of the element being considered;
A is its mass absorption coefficient for the fluorescent radiation.

In this equation K has a magnitude comparable to the fluorescent intensity observed from the pure element. When the fluorescent intensity from a sample is small compared with this magnitude (as in a trace element), the above equation becomes

$$P = KAC,\ where\ K\ is\ a\ constant.$$

METHOD

1. Determine the mass absorption coefficient A for the specimen by exercise 10.16 for the wavelength of the element to be analyzed; for example, if we want to determine the strontium content of a set of specimens, we use $SrCO_3$ in the spectrometer and determine $A_{SrK\alpha}$ for a standard and all the unknowns.

2. Determine the SrKα peak intensities in counts per second on pure 2.5 to 3 g powder pressed discs (made by the method in exercise 10.1) for the standard and the unknowns on boric acid rimmed and backed specimens. Use pulse height analysis and correct the intensities for background measured by the method in exercise 10.13 (method A). The intensities so determined give the C values in the above equation.

Calculation

3. As an example, rock W-1 was used as the standard. Its Sr content is given (Fleischer and Stevens, 1962) as 186 ppm. NBS-99 feldspar was used as the unknown. The values obtained are

	W-1	NBS-99
C (as in step 2)	4118 cps	5588 cps
A (as in exercise 10.16)	13.15	6.49

For W-1 we have

$$P = KAC \quad \text{or} \quad K = \frac{P}{AC}, \text{ where } K \text{ is a constant}$$

$$= \frac{186}{13.15 \times 411} = 0.00343 \,.$$

For NBS-99, we then have $P = KAC = 0.00343 \times 6.49 \times 5588 = 124.4$ ppm. This value is in good agreement with a value of 125 ppm determined by isotope dilution. Good agreement has been obtained for a wide range of specimens.

(Method after Norrish and Chappell, 1967.)

DISCUSSION

The method can be applied to determining most trace elements in a wide variety of rocks, plants, and ashes. It has the advantage of overcoming matrix effects without dilution or calibration since A is determined directly. Calibration is obtained by using one standard rock or mineral or simple synthetic mixtures.

The method was applied successfully by Sweatman et al. (1967) to the analysis of tin in ores and concentrates. A plot of Sn or SnO_2 % of the sample versus $A \times$ counts per second gave a straight calibration curve with the relationship $\%Sn = (3.31 \times 10^{-4}) \times A \times C$ over the range 0 to 3% Sn.

In the absence of major absorption edges in the sample the absorption coefficient for one wavelength may be used in analyzing several elements. Thus in normal rocks and soils the absorption coefficient A for ZnKα radiation could be used in the measurement of Ni, Cu, and Ga, but it could *not* be used for Mn analysis because of the major FeKα absorption edge between ZnKα and MnKα (see Table 10.1).

Reynolds (1963, 1967) introduced a method for determining the mass absorption coefficients by Compton scattering intensity measurement and using it for trace element analysis in the range Ni to Ag.

EXERCISE 10.19

Determination of Rubidium and Strontium in a Rock

Precise determination of rubidium and strontium to a detection limit of 1 ppm in rocks and minerals is frequently done by *X-R-F*. High accuracy is required if the results are to be utilized for radiometric dating techniques.

1.   Rubidium $K\alpha$ at 2θ 26.62 and Sr$K\alpha$ at 25.14° is detected on a spectrometer by an Mo tube at 100 kV and 20 mA and a LiF dispersing crystal ($2d$ 4.028). A tungsten tube can also be used. The primary collimator is 480 μm. A scintillation counter with secondary collimator in front is required, and the pulse height analyser is set at 0.8 to 2.2 V window.

2.   The rock samples are pulverized and made into whole-rock undiluted pressed discs backed and rimmed by boric acid (exericse 10.1); 2.5 g of rock sample is used. The sample area is about 5 cm^2.

3.   Each specimen is counted for 200 seconds on each of the $K\alpha$ lines and background is determined on the same sample at $\pm 0.4°$ 2θ from the peak position for 100 seconds. The background value is subtracted from the peak value.

(Method after Chappell et al., 1969.)

DISCUSSION

The background does not vary linearly with 2θ and the primary radiation from the tube may contain Sr and Rb impurity. Hence the background profile should be determined on, for example, Spec-pure silica glass (Herasil, supplied by Heraeus GmbH, Hanau, West Germany). The glass is counted identically to the rock specimens and any Rb and Sr detected should be subtracted from the rock analysis.

Concentration P of a trace element $= AC/k$, where A is the mass absorption coefficient of the sample for the radiation in use, k is a constant, C is the total X-ray intensity per unit area from the fluorescing radiation, and P is the concentration of the element in the specimen. The values of k for Rb and Sr are determined by analyzing standards.

Standards may be made by grinding acid-washed quartz with Spec-pure rubidium chloride and strontium carbonate converted to strontium chloride with known concentrations of about 2000 ppm of these elements in each standard. Standard rocks and minerals for example, G-2 or W-1, may also be used directly.

Mass absorption coefficients are measured as in exercise 10.16 by pressing 300 mg of each sample in a 0.5 in. diameter hole in a perspex holder and the attenuation of the Rb$K\alpha$ and Sr$K\alpha$ radiation of each sample measured (exercise 10.16). The samples may be mixed with filter paper pulp and the attenuation corrected for the presence of the pulp, as in exercise 10.16. At the 99.7% confidence level the statistical lower level of detection is less than 1 ppm for both Rb and Sr.

Trace Element Analysis by Matrix Calibration Against Al_2O_3

For any particular rock it is found that the relative absorption of the rock, for example dolerite W-1, compared with that of a standard pure Al_2O_3 powder, is relatively constant over a wide range of wavelengths. Figure 10.16A shows a plot of the mass absorption coefficient ratio $\mu/\rho(\text{rock})/\mu/\rho(Al_2O_3)$ versus wavelength.

This constancy of relative absorption of rock/Al_2O_3 over a wavelength range is characteristic of a great variety of silicate rocks. For convenience, the X-ray spectrum is broken up into three regions:

1. Wavelength shorter than the K absorption edge of iron; all wavelengths from nickel and heavier.

2. Wavelength between the K absorption edges of iron and titanium. This region includes the K wavelengths for cobalt, manganese, chromium, and vanadium.

3. Wavelength between the K absorption edges of titanium and calcium. This includes the K wavelength of scandium only.

Region 1 is by far the most important for trace element analysis. The curves in Fig. 10.16 show that the absorption of the analysis sample relative to a standard need be known only at one point in the appropriate region (e.g., 1, 2, or 3) to make a quantitative correction for matrix.

METHOD

1. Prepare the following boric acid rimmed and backed pressed powder discs (method exercise 10.1). Each specimen should have an approximate total weight of 2.5 to 3 g so that the specimen is infinitely thick for X-ray purposes:

    a.   Pure Al_2O_3 (Analar) powder to serve as a background blank.

    b. Al_2O_3 powder to which has been added and thoroughly mixed in a WIG-L-BUG mixing machine a carefully weighed amount of a selected trace element. The trace element should be added in the form of an oxide or simple salt, completely homogenized with the Al_2O_3 and pressed into a disc. As many discs should be made up as there are trace elements to be analyzed, each Al_2O_3 disc containing only one trace element.

As an example, suppose we are measuring a suite of granites for their Ba content and it is already known that it averages around 1500 ppm. We want to make up an Al_2O_3 disc containing approximately 1500 ppm Ba as a standard. We could use $BaCO_3$ (analar) as the additive ($BaCO_3 = 69.6\%$ Ba + 30.4% CO_3). To obtain 1500 ppm Ba the sample must contain 0.1500% Ba, and if we have a total specimen weight of 3 g the amount of $BaCO_3$ to be weighed is $(0.1500 \times 3)/69.6 = 0.00646$ g. To facilitate weighing the mixture could be 0.006 g $BaCO_3$ + 2.994 g Al_2O_3 (= total 3 g), which is equivalent to a Ba content of $(0.006 \times 69.6)/3.00 = 1392$ ppm. This then becomes the concentration in our standard. Suitable chemicals may be obtained from BDH Chemicals Ltd., Poole, Dorset, BH 12 4 NN, England.

2. Whole-rock pressed powder samples are made in the same way (exercise 10.1), rimmed and backed with boric acid. They should also have a total weight of 2.5 to 3 g and contain undiluted whole rock powders.

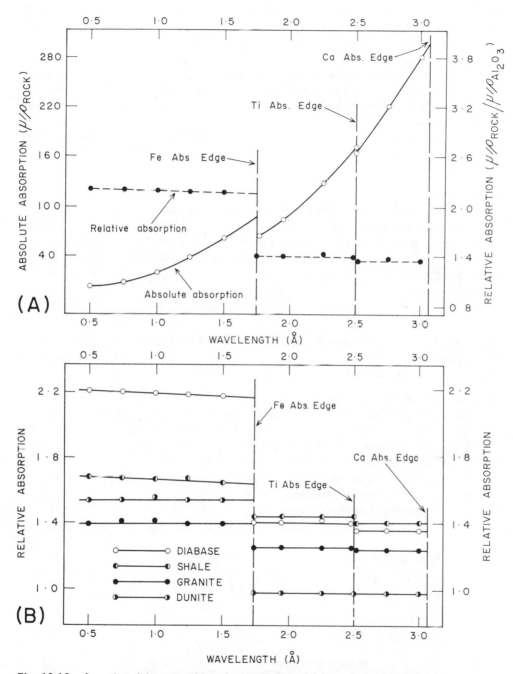

Fig. 10.16 **A** — plot of the mass absorption coefficient of the standard dolerite W-1 (*solid curve*) and the absorption of W-1 relative to Al_2O_3 versus wavelength. **B** — plots of the relative absorption R versus wavelength for granite, diabase, shale, and dunite. (After Hower, 1959.)

3. The spectrometer is set accurately on the 2θ position for the trace element in question; for example, Ba$K\alpha$ at 11.01° for an LiF crystal. The spectrometer is carefully set for HV and pulse height analysis to ensure maximum peak-to-background resolution. Each specimen should be counted for at least 10 seconds and the rates in counts per second recorded and corrected for background. The background for the Al_2O_3 disc is determined at the peak 2θ position on the pure Al_2O_3 disc, and this count rate has to be subtracted from the counts per second on the Al_2O_3 disc which contains the trace element.

The background on the rock specimens must be determined at a suitable $\pm$angular distance on either side of the peak 2θ, averaged, and subtracted from the peak counts per second.

Calculation

4. If $P_{Al_2O_3}$ is the concentration in ppm of the trace element in the Al_2O_3 disc, P_{rock} is the true concentration of the same element in the rock, $C_{Al_2O_3}$ is the rate in counts per second, corrected for background, for that element on the Al_2O_3 disc, C_{rock} is the counts per second rate, corrected for background, on the rock specimen for the same element, and R is the relative absorption $= \mu/\rho_{rock}\big/\mu/\rho_{Al_2O_3}$ (Fig. 10.16), then

$$P_{rock} = \frac{C_{rock}}{C_{Al_2O_3}} \times P_{Al_2O_3} \times R.$$

R is unknown. It must be determined for each of the wavelength regions in Fig. 10.16 One determination of R for a whole region is sufficient because of the constancy of R over any one of the three regions (Fig. 10.16).

Determination of Relative Absorption R

5. If the major element analysis of the rock is known, the value of R can be calculated for each of the regions in Fig. 10.16. Only one calculation is necessary for each of the three wavelength regions. The mass absorption coefficients to be used are given in Table 10.6 (after Hower, 1959). They are average values for oxides for each of the three regions. As an example of the calculation, granite G-1 (analysis taken from Fleischer and Stevens, 1962) gives the following:

$\mu/\rho_{G-1} =$ the sum of the weight fractions
$\qquad \times$ mass absorption coefficients in Table 10.6 (for region 1, suitable for **Ba**
$\qquad\quad$ correction)
$\quad = 78.5 \times 0.0098 + 70.8 \times 0.0087 + 65.0 \times 0.0003 + 38.7 \times 0.0026 + 36.7$
$\qquad \times 0.0139 + 34.9 \times 0.0545 + 11.0 \times 0.009 + 10.1 \times 0.7241 + 9.13 \times 0.1427$
$\qquad + 8.36 \times 0.0041 + 7.56 \times 0.0332 + 3.13 \times 0.0040 + 2.65 \times 0.008$
$\quad = 12.843.$

$$\mu/\rho_{Al_2O_3} = 9.13 \times 1.00 = 9.13,$$

$$R \text{ (region 1)} = \frac{\mu/\rho_{rock}}{\mu/\rho_{Al_2O_3}} = \frac{12.843}{9.13} = 1.40.$$

This value of R allows us to complete the calculation of step 4 above.

6. If the whole rock analysis is unknown, the following method must be followed in order to determine R for each of the three wavelength regions. In each of the three regions an internal trace element standard is added in the same proportions to a pure Al_2O_3 disc and a pure rock disc. The element added should be one that belongs to the correct wavelength region but at the same time is one that is not usually significant

Table 10.6   Mass Absorption Coefficients of the Major
Element Oxides as Average Values for Each of the Three
Wavelength Regions in Fig. 10.16[a]

Oxide	Region 1 (1.000 Å)	Region 2 (1.930 Å)	Region 3 (2.505 Å)
FeO	78.5	60.3	124
Fe_2O_3	70.8	56.3	116
MnO	65.0	54.4	110
TiO_2	38.7	24.8	85.0
CaO	36.7	22.8	462
K_2O	34.9	22.5	450
P_2O_5	11.0	73.6	156
SiO_2	10.1	65.8	135
Al_2O_3	9.13	59.8	129
MgO	8.36	55.1	115
Na_2O	7.56	51.1	106
H_2O	3.13	21.7	44.5
CO_2	2.65	18.2	37.2

[a] After Hower (1959).

in any trace element study. For region 1 arsenic or selenium is suggested, for region 2, cobalt, and for region 3, scandium. The arsenic can be added in the form of As_2O_3, selenium as SeO_2, cobalt as Co_2O_3, and scandium as Sc_2O_3 (BHD Chemicals Ltd.).

The following example for region 1 is identical for regions 2 and 3 except for the choice of trace element added:

a. Prepare a boric acid rimmed and backed disc of Al_2O_3 to which has been added and thoroughly mixed an accurately measured amount of arsenic or selenium in the form of an oxide or simple salt. Compute the As or Se content in ppm. Total weight of powder 2.5 to 3 g.

b. Prepare an identical pure undiluted Al_2O_3 disc of 2.5 to 3 g.

c. Prepare a rock powder disc to which has been added an identical amount of As or Se as in step a.

d. Prepare a pure undiluted rock powder disc to which nothing has been added.

e. If As has been added, set the spectrometer on the arsenic $K\alpha$ line and obtain the counting rate for As in the rock (corrected by subtracting the counts per second obtained at the same 2θ on the specimen prepared in step d which has had no As added). Repeat for Al_2O_3 to obtain the counting rate As$K\alpha$ for the specimen prepared as in step a, minus the counting rate (background) determined on the specimen prepared as in step b.

f. Since both Al_2O_3 and rock have had equal ppm of As (or Se) added, the value of R may be calculated directly; for example, the counting rate on Se$K\alpha$ made on Al_2O_3 containing 1000 ppm (corrected for background) was 1440 cps. The corrected counting rate on granite G-1 to which 1000 ppm Se has been added = 1006 cps. R for G-1 therefore = 1440/1006 = 1.43. This R value, which is used for all trace element determinations in region 1, is preferable to that determined by matrix corrections based on the whole-rock major element analysis in step 5. The value of 1.40 obtained there, however, is in fair agreement with the present value of 1.43; but 1.43, experimentally determined, is to be preferred because of the uncertainty of the published values (Table 10.6) of mass absorption coefficients.

g. The value of R required for region 2 can be determined experimentally in an identical manner by adding Co to Al_2O_3 and the rock disc.

h. The value of R for region 3 (Fig. 10.16) is obtained by adding Sc to the discs in the same concentration.

DISCUSSION

The amount of the element added in each case should never exceed that of a trace element and ideally should be around 1000 to 2000 ppm of the total 2.5 to 3 g disc. Careful mixing of the Al_2O_3 or rock and the added trace element must be done in a mechanical mixer such as a WIG-L-BUG. The powdered chemicals may be weighed directly into the plastic vials of the mixer.

For regions 2 and 3 only semiquantitative matrix corrections can be made for Co, Mn, Cr, V, and Sc. For rocks high in Ti cobalt is a suitable internal standard. For rocks low in Ti scandium is preferable.

Carbonate rocks can be analyzed in the same manner as silicate rocks, but instead of Al_2O_3 discs of $CaCO_3$ can be used to which the trace elements have been added in known amounts. A single internal standard should suffice.

7. *Semiquantitative applications.* The gross mass absorption coefficient of rocks is not greatly variable. It varies usually between $\times 1$ and $\times 2$ of the Al_2O_3. Even when the

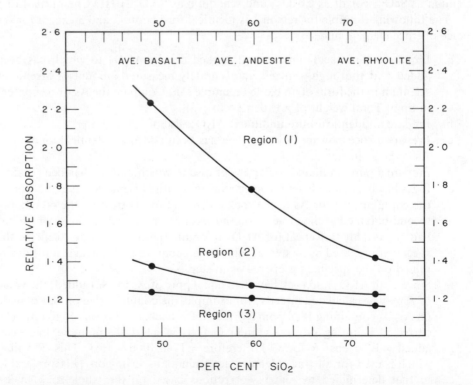

Fig. 10.17 Plots of the relative absorption R versus wavelength for the common igneous rocks. (After Hower, 1959.)

major element analysis is unknown and no internal standard is used, a fairly good analysis can be made by guessing at the absorption of a rock in relation to Al_2O_3. Igneous rocks allow this to be done with a fair degree of accuracy. Figure 10.16B shows a plot of relative absorption R for a few standard rock types for the three regions of wavelength (after Hower, 1959). The variation of relative absorption R among the common igneous rocks is shown in Fig. 10.17. By knowing approximately the rock type we can estimate the relative absorption R from Fig. 10.17 and use it to calculate the trace element concentration in step 4.

(Method after Hower, 1959.)

EXERCISE 10.21

Trace Element Analysis Without Matrix Correction

$$\text{ppm}\, Z_x = \frac{\text{cps}\, Z\alpha_x}{\text{cps}\, Z\alpha_\text{std}} \times \frac{\mu/\rho_x}{\mu/\rho_\text{std}} \times \text{ppm}\, Z_\text{std},$$

where ppm Z is the concentration of a given trace element Z, x is the unknown specimen, std is a standard specimen of mineral or rock of known composition, μ/ρ is the mass absorption coefficient of the sample for the wavelength being used, and cps $Z\alpha$ is the X-ray intensity in counts per second corrected for background.

The equation is valid only if the wavelength is shorter than the wavelength of the absorption edge of the heaviest matrix element. For silicate rocks and minerals this is Fe. Hence the equation applies only to the determination of nickel and heavier elements, that is, to region 1 in Fig. 10.16.

Enhancement will also occur when the radiation of the matrix elements has wavelengths equal to or shorter than the wavelength of the absorption edge of the trace element. This will not occur for trace elements in region 1 of Fig. 10.16 for silicate rocks.

For trace element analysis of certain silicate rocks it should be possible to select several standard rocks whose mass absorption coefficients are closely identical to a suite of unknowns; for example, the trace element (elements heavier than Ni) content of a suite of granites can be compared with several international granite standards (G-2; GR; GA; QMC I-1) (Flanagan, 1970). Because both standards and unknowns are granites and of comparable mass absorption coefficients, the above equation can be assumed to be

$$\text{ppm}\, Z_x = \frac{\text{cps}\, Z\alpha_x}{\text{cps}\, Z\alpha_\text{std}} \times \text{ppm}\, Z_\text{std}.$$

To allow for any fluctuation of mass absorption coefficient from specimen to specimen the ppm Z_x might be computed by comparing it in turn with each of the standard granites. The values so obtained may be averaged. This method will give approximately correct trace element concentrations, but of course it is valid only if the mass absorption coefficient of the unknown is similar to that of the standard. If there is any doubt regarding the similarity of the mass absorption coefficients, one of the methods in this chapter (exercise 10.16 or 10.20) should be used to measure or compute the mass absorption coefficients of the samples. The equation at the beginning of this exercise should then be used.

D. SOME SPECIAL TECHNIQUES

Although space available in a book like this is insufficient to describe all techniques of X-ray fluorescence as applied to petrology and mineralogy, a few sundry applications are mentioned briefly here.

Special Dilution Methods

Many ore minerals do not fuse satisfactorily by the methods in exercise 10.8. Accordingly, for major element analysis of many ore minerals analyses are performed on whole-specimen powdered pressed discs (exercise 10.1). Dilution of the powdered specimen by a suitably chosen chemical before pressing can effect a matrix of nearly constant mass absorption coefficient between standard and unknowns.

Mitchell and Kellam (1968) pointed out that the diluent ideally must have a mass absorption coefficient similar to that of the specimens to be analyzed (i.e., dilution in a neutral matrix) if the dilution is to reduce the matrix effects effectively. Two examples are given to illustrate the methods:

1. Tin ores ranging in Sn content from 55 to 77% may be diluted 20-fold with iron oxide. This is adequate to produce a relatively constant value of μ/ρ that will result in a linear relationship between counts per second and percent Sn. Background count rates are negligible and constant and can be disregarded (Sweatman et al., 1967). The use of Fe_2O_3 to dilute cassiterite has two advantages over other diluents:

    a.   It is important that the diluent have a similar mass absorption coefficient to the specimens being analyzed.

    b.   It is also important that the diluent have a similar specific gravity to facilitate efficient mixing of the specimen and diluent.

Specimens are prepared by weighing directly into the WIG-L-BUG $2 \times \frac{1}{2}$ in. plastic vials 2.5 g Fe_2O_3 and 0.125 g cassiterite ore; adding one spherical plexiglas pestle, closing the lid and mixing for 10 minutes. The mixture is then pressed into a boric acid rimmed and backed disc (exercise 10.1).

2. Chromite ore can be analyzed by dilution in a similar manner. Because of the severe matrix problems between Cr and Fe, dilution in a neutral matrix is advisable (Mitchell and Kellam, 1968). Specimens are prepared similarly by weighing 2.0 g analar manganous carbonate +0.4 g finely powdered chromite and mixing in a WIG-L-BUG for 10 minutes. A boric acid rimmed and backed disc is then made. The chromite specimens are then measured for Cr_2O_3, Al_2O_3, MgO, Fe_2O_3, TiO_2, and SiO_2 by direct comparison against two chromite standards of widely divergent composition (Hutchison, 1972). Although stronger dilution of the specimens would be advisable, the amount of dilution is limited effectively by the desire to analyze the specimens for Mg.

Analysis of Vanadium in the Presence of Titanium

It would be difficult to analyze for vanadium in the presence of titanium because of the near coincidence of $TiK\beta$ ($\lambda = 2.514$ Å) and $VK\alpha$ ($\lambda = 2.505$ Å).

The procedure by Snetsinger et al. (1968) involves determining the ratio of $TiK\alpha$ to $TiK\beta$ (the latter measured at the wavelength of $VK\alpha$) for analar grade

TiO_2. If accurate determinations for vanadium in trace amounts (e.g., in chromite) are required, then $VK\beta$ would be a good line because of its intensity. The ratio of $TiK\alpha/TiK\beta$ for rutile can be used to correct the value of $VK\alpha$ (which includes $TiK\beta$) for any specimen containing both titanium and vanadium.

The Macroprobe Attachment for the X-Ray Spectrometer

This is a relatively inexpensive attachment (for the Philips spectrometer) that allows X-ray fluorescence analysis of crystals in a polished thin section simultaneously with optical viewing. Its performance, however, nowhere approaches that of the electron microprobe or scanning electron microscope. The disadvantage arises from the fact that a curved mica diffraction crystal is used to disperse the various wavelengths. Intensities are low, hence a large irradiated spot must be determined. Applications are limited accordingly to coarse-grained rocks and to elements heavier than potassium. The standard 20 μ thick slide is too thin and it is considered that the critical thickness of the sections to ensure maximum intensity is at least 70 μ. This hinders identification of the minerals in thin section.

The minimum size of the irradiated spot may be 100 μ in diameter for iron and heavier elements, but for elements such as Ca and K it must be as large as 460 μ (Hermes and Raglund, 1967).

Sensitivities are low: for example, for potassium only 16 cps/1 % for a 500-μ aperature; for calcium 33 cps for a 500-μ aperature; for Ti 48 cps for a 500-μ aperature. All three have 50 kV, 45 mA W radiation detected by a flow counter in vacuum. Mn gives 60 cps at 50 kV, 45 mA W radiation in vacuum with a scintillation counter, and Fe gives 200 cps per 1 % with 50 kV, 35 mA W radiation detected in air by a scintillation counter.

EXERCISE 10.22

Determination of Chlorine in Silicate Rocks and Minerals

1. 0.5 g of accurately weighed finely pulverized rock or mineral sample is ground for 10 minutes in a mixer mill (WIG-L-BUG).

2. The ground powder is poured into a mortar and 0.5 g of accurately weighed chromatographic cellulose (Whatman CF 11) is added. The cellulose and the sample are hand ground and well mixed.

3. The mixture is returned to a WIG-L-BUG plastic vial and reground and mixed for 5 minutes. The finely ground and mixed powder (which should now be approximately 400 mesh) is then pressed into a pellet at 30,000 psi with the pellet die in Fig. 10.1. Alternatively, the pellet die of Fabbi (1970) may be used.

4. A PET dispersing crystal, 2θ 65.42°, flow proportional counter at 1400 V, 5 mil collimator, vacuum path, 50 kV, 50 mA, and Cr tube were used on the spectrometer. Interference by second-order $K\alpha$Cr radiation from the target is effectively discriminated against by setting the lower level and window of the pulse height analyzer at 0.70 and 1.00 V, respectively.

5. The pellets are irradiated for 200 seconds and their counts recorded. U.S. Geological Survey standards W-1, G-2, GSP-1, AGV-1, PCC-1, DTS-1 and BCR-1 are analyzed in the same way to establish a calibration curve, cps versus Cl concentration as reported in their analyses, using the least squares regression to establish the line.

(After Fabbi, 1972.)

DISCUSSION

The Cl analysis of rocks and minerals yields results that compare well with the published results. This X-ray method is preferable to any chemical method. The gravimetric procedure of precipitating chlorine as AgCl is not reliable for trace amounts of Cl ($<0.05\%$) (see exercise 12.7). A titration method detects traces of Cl but cannot be used up to 0.20% Cl. Chemical methods are also lengthy and tedious.

By this X-ray method the detection limit is about 10 ppm. The accuracy is good.

Care must be taken to prevent Cl contamination of the pellets. The surface of the pelletizing die must be thoroughly cleaned before use. If the die of Fabbi (1970) is used, the glass lens against which the powdered sample is pressed is a source of Cl contamination and must be cleaned with dilute HNO_3 solution (5%) before use. The pellets after being analyzed in the vacuum of the spectrometer are susceptible to absorption of Cl from the atmosphere, thus preventing their repeated use. Fresh standard rock pellets have to be made up each time when a new lot of specimens has been prepared for analysis. Pellets are stored in a dessicator before use.

EXERCISE 10.23

Determination of Phosphorus in Silicate Rocks and Minerals

1. Use a Cr tube, germanium crystal ($2d$ 6.54 Å), gas flow proportional counter at 1500 V, 5-mil collimator, vacuum path, 55 kV, 50 mA, pulse height analyzer at lower level 2.20 V, and window at 2.00 V.

2. Prepare pellets of 0.5 g of the rocks and of standard rocks (G-1, W-1, AGV-1, etc.) ground to −400 mesh and mixed with 0.5 g of chromatographic cellulose (as in exercise 10.22).

3. Each of the samples and standards is counted for 50 seconds.

4. A calibration curve is established by the least squares regression method by plotting the counts of all the standards per 50 seconds against their reported contents of P_2O_5. The concentrations of the unknown samples are then read off the calibration curve.

(After Fabbi, 1971.)

DISCUSSION

If a PET analyzing crystal is used, the phosphorus $K\alpha_1$ line is overlapped and interfered with by the calcium $K\beta$ second-order lines. Although the pulse height analyser removes much of this interference, it is not 100% efficient.

However, the use of the germanium dispersing crystal, combined with the pulse height analyser, accurately set for the $K\alpha_1$ line, gives almost no second order calcium lines. Hence its use allows $PK\alpha$ to be accurately measured for intensity without interference from calcium.

ATOMIC ABSORPTION SPECTROPHOTOMETRY 11

Atomic absorption spectrophotometry might well be considered one of the better and most popular methods now available for major and trace element analysis of rocks, minerals, ores, and water samples. The popularity of this method results from the large number of elements covered, the simplicity of the operation, its sensitivity and reliability, and the modest monetary outlay.

It is not the purpose of this chapter to discuss the theory of the method. Interested readers are referred to Angino and Billings (1967), Ramírez-Muñoz (1968), and Rubéska and Moldan (1969) for more complete details and wider applications. A condensed summary is given by McLaughlin (1967a). Most manufacturers of the necessary equipment provide manuals which, in addition to giving operating instructions, describe standard analytical conditions for all elements and specific determination methods for a wide range of materials. These companies publish supplements from time to time to update their manuals. An example is "Analytical Methods for Flame Spectroscopy," which has 276 pages and is available from Varian Techtron Pty. Ltd., 679–687 Springvale Road, North Springvale, Victoria, Australia 3171 (or 611 Hansen Way, Palo Alto, California 94303).

Specimen Preparation

A. SILICATE ROCKS AND MINERALS

EXERCISE 11.1

The HF—H_2SO_4 Method

1. Place an accurately weighed portion (0.5 g is recommended) of crushed (150-μ size) rock or mineral sample in a 30-ml platinum crucible.

2. Add about 5 drops each of distilled water and concentrated H_2SO_4 and follow with 5 ml of 50% HF.

3. Place the crucible in a porcelain dish, set on a hot plate, and heat slowly in a fume cupboard until all the H_2SO_4 has been fumed.

4. When the sample is nearly dry, repeat from step 2. Repeat once more from step 2 to ensure complete decomposition of the sample. Heat to complete dryness on the third evaporation.

5. Transfer the metal sulfates to a 250-ml beaker with hot 1 : 9 HCl as needed.

6. Transfer again to a 200-ml volumetric flask, cool, and bring to the mark with 1 : 9 HCl. The solution is now ready to be analyzed and can be aspirated directly into the atomic absorption air–acetylene flame.

(After Angino and Billings, 1967.)

DISCUSSION

If a small residue results, other procedures, like the following, may be used instead to cause complete decomposition of the sample.

EXERCISE 11.2

The Na_2CO_3 Fusion Method

1. Add an accurately weighed 0.5 g of crushed rock or mineral to a 30-ml platinum crucible.

2. Weigh about 4 g pure anhydrous Na_2CO_3, add about 3 g of it to the crucible, and stir with a glass rod until well mixed. Clean the glass rod with the remaining Na_2CO_3 and transfer the latter to the crucible.

3. Place the covered crucible in a muffle furnace and heat until the crucible is dull red, hold the temperature for about 10 minutes, and raise the temperature to 1100°C for 60 minutes.

4. Remove the crucible, place it on asbestos boards, and cool it to room temperature with its cover on.

5. The fused wafer can be removed whole by carefully squeezing the soft platinum crucible. Place the wafer in a 250-ml beaker and cover it with approximately 100 ml of distilled water.

6. Slowly and carefully add 25 ml of concentrated HCl through a funnel set so that its stem is under the water. Using the acid, wash the crucible cover so that the wash solution is added to the beaker.

7. Heat the beaker gently until the wafer dissolves; then filter the contents into a 200 ml volumetric flask. Wash the beaker and filter paper repeatedly by adding the wash to the flask. Bring to the mark. The liquid is now ready for analysis by aspirating the solution directly into the atomic absorption flame.

(After Angino and Billings, 1967.)

EXERCISE 11.3

Fast and Complete Decomposition for Rocks, Refractory Silicates, and Minerals

The procedures in exercises 11.1 and 11.2 often fail to decompose andalusite, kyanite, sillimanite, corundum, tourmaline, rutile, garnet, staurolite, zircon, spinel, graphite, gold, and pyrite in the rocks.

The present method provides complete and fast decomposition of all geological materials. Use Analar grade chemicals and distilled water throughout.

1. Into a platinum crucible of 25-to-30-ml capacity mix 2 g boric acid with 3 g lithium fluoride. Make a small hollow in the center of the mixture and place in it 0.5 g accurately weighed, finely ground rock or mineral powder which has been sieved to pass through the 100 mesh. Cover the rock or mineral powder by gently tapping the sides of the crucible. There is no need to mix the ingredients. Put on the platinum lid.

2.   Heat gently for 2 to 3 minutes over the small gas flame of a Meker burner. Gradually increase the flame to bring the contents to melting and heat for 10 minutes at full flame (about 800 to 850°C). If the rock contains graphite or sulfides, the crucible should be opened periodically to allow oxidation.

3.   Cool the crucible quickly by half immersing it in a stream of tap water flowing along a slightly inclined surface. Make sure that no water enters the crucible.

4.   Add 10 ml of concentrated H_2SO_4 and gently heat on an asbestos plate over a Bunsen or Meker burner until bubbles of gas appear. Regulate the heating so that the evolution of gas from the covered crucible is gentle and not violent and none of the sample is lost by splashing over the lip of the crucible. This step removes excess boric acid as fluoride.

5.   When gas evolution has stopped, increase the temperature until copious fumes of SO_3 rise and maintain this temperature for 2 to 3 minutes (heating with H_2SO_4 should take a total time of 30 minutes).

6.   Cool the crucible in about 150 ml of distilled water in a 250-ml beaker.

7.   Add 5 ml of concentrated HCl or HNO_3 acid and boil the crucible and its lid until all contents are dissolved and the liquid is clear (in the presence of 40 to 45% aluminum and calcium it usually takes 10 to 15 minutes).

8.   Remove the crucible and its lid from the solution in the beaker with a platinum tipped tongs, but make sure that they have been thoroughly washed down with a wash bottle so that all the solution is kept in the beaker.

9.   Transfer the solution from the beaker to a 250- or 500-ml volumetric flask and make up to full volume. We now have a known volume of liquid containing 0.5 g of the rock or mineral to be analysed.

(After Biskupsky, 1965.)

DISCUSSION

The final solution can be used to determine sodium, potassium, calcium, magnesium, iron, aluminum, titanium, and many minor elements. Silicon cannot be determined because it has been removed as volatile fluoride. Silica can be determined by an X-ray fluorescence method (Chapter 10). Although the lithium introduced into the solution colors the atomic absorption flame, this has not been found to have any serious effect on enhancement or depression of the quantities estimated.

By this method most rocks and minerals decompose satisfactorily. No interference was observed in the analysis of sodium, potassium, calcium, magnesium, iron, aluminum, and titanium (Biskupsky, 1965).

EXERCISE 11.4

Decomposition with a Teflon-Lined Bomb

Acid digestion bombs provide a safe and rapid method for dissolving samples of silicate rocks, silicate minerals, and refractory materials. Figure 11.1 illustrates two designs of commercially available bombs. The bomb illustrated in A and B of Fig. 11.1 is

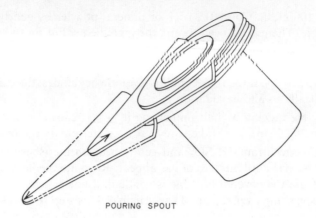

POURING SPOUT

(A)

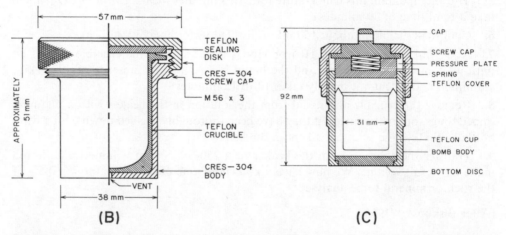

(B)   **(C)**

Fig. 11.1   A and B = details of the Uni-Seal decomposition vessel. C = details of the Parr Instrument Company decomposition vessel. Spare Teflon cups and lids may be purchased separately.

obtainable from Uni-Seal Decomposition Vessels, P.O. Box 9463, Haifa, Israel. The other, illustrated in C of Fig. 11.1 is available from Parr Instrument Co., 211 Fifty-third Street, Moline, Illinois, 61265. A successful Teflon bomb design has recently been reported by Krogh (1971).

The decomposition vessels can be relied on to prevent losses by volatilization of components, reduction and alloying, and contamination. Some commercial bombs (e.g., the Parr design) do leak at appropriate temperatures. The Krogh (1971) design is reported to be leakproof. I have also had excellent results with the Uni-Seal model.

The Teflon (Du Pont TFE fluorocarbon resin)-Teflon contact area which is dimensionally optimized between the crucible rim and the Teflon sealing disc provides a simple and effective means of sealing when the metal cap with the inserted Teflon disc is manually screwed onto the crucible metal body.

Decomposition is achieved in 30 to 40 minutes at even 110°C. This suggests that pressure is built up in the crucible and accounts for the speed of decomposition of the silicates. The vessel undergoes no distortion.

The Teflon pouring spout (Uni-Seal model), which is snapped to the peripheral sides of the crucible rim, prevents any contact between the highly corrosive solution and the metallic parts and facilitates transfer of the decomposed sample. The Parr variety has no pouring spout and the Teflon cup has to be removed from the metal body when the liquid is transferred.

METHOD A

1. Transfer 50 mg (300 mg if vanadium is to be determined) of a representative 150-mesh sample into the Teflon vessel.

2. Add 0.5 ml Aqua Regia as the wetting agent and make sure that the sample is thoroughly wet.

3. Add 3.0 ml 48% hydrofluoric acid and close the vessel by hand-tightening the screw cap containing the Teflon sealing disc.

4. Place the crucible, without tilting, in a drying oven for 30 to 40 minutes at 110°C Then let it cool to room temperature.

5. Unscrew the lid, insert the pouring spout (Uni-Seal design), or remove the Teflon vessel from the metal bomb (Parr design), and transfer the decomposed sample solution with the help of 4 to 6 ml of distilled water from a wash bottle into a 50-ml polystyrene Spex vial. Take care to transfer any precipitated metal fluorides that may have formed. The total volume should not exceed 10 ml.

6. Add 2.2 g of boric acid and stir with a Teflon rod to dissolve it and hasten the reaction. Add 5 to 10 ml of distilled water. Any fluorides will dissolve at this stage and, on dilution to about 40 ml, a clear homogeneous sample solution will result.

7. Transfer to a 100-ml volumetric flask, adjust to volume, and store in a polyethylene container. It must not remain in a glass vessel for more than 2 hours. The solution is then ready for analysis.

(After Bernas, 1968.)

METHOD B

A slight variation of method A is given by Buckley and Cranston (1971). A 100-mg sample of the rock or mineral is used; 1 ml of Aqua Regia is the wetting agent, and 6 ml of concentrated HF are added. The bomb is heated at 100°C for 60 minutes. After cooling, the sample solution is washed into a 125-ml polypropylene bottle containing 5.6 g H_3BO_3 and 20 ml of distilled water. Metal fluorides are dissolved in the boric acid system and shaking the bottle helps dissolution. The sample solution is made up to 100 ml in a volumetric flask and then stored in a polypropylene bottle. Experience has shown that the effect of the solution on the glass volumetric flask is nil for a period of about 1 hour, but the made up solution must *not* be stored in glass.

DISCUSSION

Solutions made in this way have no loss of volatiles and therefore may be analyzed for Si, Al, V, Ti, and all major and minor elements.

If the rocks or clays contain unoxidized carbon, the above preparation technique will leave the carbon as a residue. This introduces little error, for it represents less than 1% of the total sample in most cases (Buckley and Cranston, 1971).

The decomposition method is perfectly satisfactory for granites and gabbroic rocks and presumably for most silicates. Higher temperatures (up to 170°C) will decompose rocks containing refractory minerals. Even pure cassiterite will decompose in HCl in Teflon bombs if the heating temperature is raised to 210°C. Such a high temperature does not distort the Teflon crucibles. A complete ultrapure dissolution procedure for zircon has been given by Krogh (1971).

EXERCISE 11.5

The NaOH Fusion Method for Silica and Alumina Determination

1. 0.5 g of −200-mesh powdered sample is weighed to the nearest 0.1 mg into a 100-ml nickel crucible.

2. About 1 g NaOH pellets (analytical grade) is added and the crucible heated *uncovered* on a *low* Bunsen flame for 2 to 3 minutes until the NaOH melts. This melting must be gentle to avoid spattering. Now increase the heat and occasionally swirl the crucible until a homogeneous clear melt is obtained. Time required is 5 to 10 minutes.

3. Cool the crucible. Add 50 ml distilled water and stir with a nickel spatula.

4. The water is poured off into a polythene 500-ml beaker and a second 50 ml of distilled water added to the crucible.

5. The crucible is heated carefully on a hot plate, without boiling, until (about 10 minutes) the entire fused mass becomes loose.

6. The fused mass is removed from the crucible into the polythene beaker which contains the poured-out solution. The crucible is washed several times into the beaker.

7. The beaker is placed on a magnetic stirrer until all the solid fusion lumps dissolve.

8. Without interrupting the stirring, about 50 ml conc. analar HCl are slowly added and the solution tested with litmus to be distinctly acid.

9. The solution is transferred to a 500-ml volumetric flask, cooled to room temperature, diluted to volume, shaken, and immediately transferred to a 500-ml well-stoppered polythene bottle.

The complete procedure takes about 30 minutes.

(After Katz, 1968).

DISCUSSION

Twenty-five milliliters of this solution, diluted in a 100-ml volumetric flask and immediately transferred to a polythene bottle, serve for silica determination; 50 ml of the solution diluted in a 100-ml volumetric flask, to which 3 ml of 4% KCl solution (in water) are added to repress ionization effects before being made up to the volume, serve for alumina determination.

EXERCISE 11.6

Dissolution of Limestone

If only the elemental content of the carbonate fraction is of interest, the following method applies:

1. Samples of 1 to 2 g are adequate. Place the sample in a 200-ml flask and add 50 ml of 0.1 N HCl or acetic acid to dissolve the sample. Acetic acid is preferable because it lessens the damge to any clays included, if they are to be subsequently analyzed by another method (e.g., X-ray diffraction).

2. If additional acid is required, it is added until effervescence ceases. Bring the solution up to 200 ml with distilled water and it is ready for analysis.

If a *total analysis* of the limestone is desired, the following method applies:

1. A 0.5-g limestone sample is heated in a platinum crucible in a muffle furnace at 1100°C for 1 hour.

2. After cooling, place the sample in a Teflon beaker and treat it with 5 drops of conc. H_2SO_4 and 5 ml of HF. Warm until fumes of SO_3 are evolved. Repeat the acid additions and warm again.

3. The residue is taken up in 1 : 9 HCl and diluted to 100 ml with 1 : 9 HCl. The solution is now ready for analysis.

(After Angino and Billings, 1967)

EXERCISE 11.7

The Basic Instrument and Adjustments

The atomic absorption spectrophotometer may be operated in four modes of spectroscopic analysis: absorbance, transmission, concentration, and emission.

Basically the instrument consists of the following components, in order: the light from a modulated hollow-cathode lamp, which emits a sharp line spectrum of the element to be determined, passes through the flame into which the atomic vapor of the sample to be analyzed is nebulized. The light then passes through a monochromator (wavelength selector) which isolates the required resonance line and into a photomultiplier tube detector to an amplifier and readout facility. A light chopper assembly, synchronized by the electronics system, is automatically activated when the emission mode is selected.

Because of the popularity of atomic absorption spectrophotometry a large range of complete instruments is available commercially. All are basically similar but with design variations. I have chosen to illustrate this chapter with the Varion Techtron model 1200 (Varion Techtron Pty. Ltd., 679 Springvale Road, N. Springvale, Victoria 3171, Australia, or 611 Hansen Way, Palo Alto, California, 94303). The instrument is illustrated in Fig. 11.2, to which the following description refers:

1 = *monochromator*, range 185 to 900 nm, accurate to within 0.2 nm.

2 = *slit width control*; monochromator slit width selector at 0.2, 0.5, 1.0, or 2.0 nm.

3 = *burner system*, a pre-mix system; the burner head is titanium.

4 = *push buttons* for selection of operating mode and other operation parameters.

5 = *readout* (*model 1200*) four-digit numerical readout of absorbance, transmission, or concentration; decimal place can be automatically positioned.

6 = *readout* (*model 1100*); the digital readout is replaced by a linear readout scale.

7 = *window* which permits clear identification of the lamp in the operating position.

8 = *lamp turret* holds four lamps with fingertip adjustment.

9 = *gas control* for fuel and oxidant with flow meters; easy selection of air and nitrous oxide; push-button ignition.

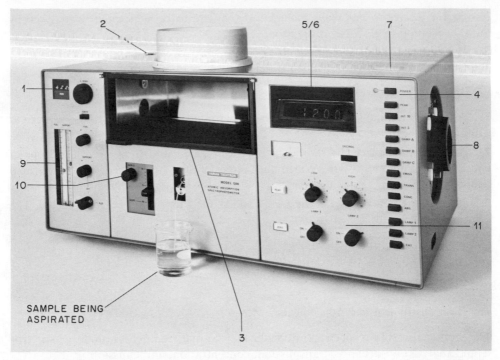

SAMPLE BEING
ASPIRATED

Fig. 11.2 The Varion Techtron Atomic Absorption Spectrophotometer, Model 1200. (By courtesy of Varion Techtron Pty., Ltd., Australia.) See text of exercise 11.7 for description.

10 = *burner positioner* adjustment for vertical and horizontal movement of the burner; the burner head can be manually rotated for operation at any angle.

11 = *amplifier* for linear readout of absorbance provides two recorder outlets of 10 and 100 mV.

Preliminary Adjustments. *Burner assembly.* The position of the bead relative to the mouth of the nebulizer (Fig. 11.3) will affect the sensitivity of the instrument and the detection limits obtainable. The bead should be adjusted with a special screw driver to give maximum absorbance. Anticlockwise rotation of the adjusting screw (Fig. 11.3) increases the gap between the nebulizer and the bead.

If any leakage of air occurs at the junction of the capillary and the polythene tube to the beaker, unsteady and nonreproducible absorption readings will result. Any roughness on the internal surface of the tube will cause clogging of the capillary by foreign particles. In this case the capillary should be cleaned by passing a fine wire through it, being careful not to scratch its internal surface. The polythene tube should be discarded and replaced with a new piece. Clean the nebulizer and spray chamber frequently by spraying them with distilled water.

The constant-level liquid trap (Fig. 11.3) must be kept filled with the appropriate solvent and the overflow taken to a sink by a length of 6-mm diameter polythene tubing.

WARNING.   Never attempt to light the burner without a liquid seal on the drain tube; otherwise an explosion will occur.

If during operation the spray chamber is not draining properly, rinse the spray chamber with an aqueous solution of detergent.

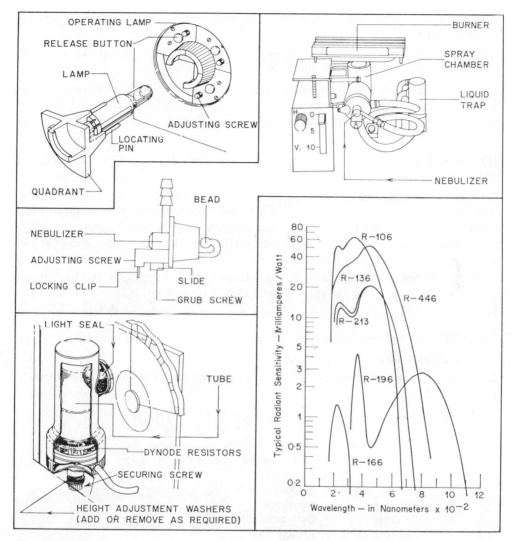

Fig. 11.3 Details of several parts of the atomic absorption spectrophotometer referred to in exercise 11.7.

Burners should be cleaned at regular intervals by washing in hot water and passing a piece of metal along the slot to ensure adequate cleaning.

Burners are interchangeable. Type 02-100035-00 (with a $2\frac{3}{8} \times 0.018$ in. slot) is for nitrous oxide-acetylene, air-acetylene, or nitrous oxide-propane gas mixtures.

Type 02-100036-00 (4″ × 0.020″ slot) is for air-acetylene, air-hydrogen, or nitrous oxide-propane gas mixtures.

Type 02-100037-00 (4″ × 0.050″ slot) is for air-propane, nitrogen/entrained air-hydrogen, or argon/entrained air-hydrogen gas mixtures.

Photomultiplier assembly. The photomultiplier tube base incorporates 10 dynode resistors and a signal preamplifier. A detachable cover encloses the assembly and incorporates a molded rubber light seal which attaches to a protrusion on the monochromator (Fig. 11.3). The assembly is attached to its mounting plate by a device incorporating a height-adjusting screw. The photomultipliers are readily removed and replaced.

CAUTION.   Avoid touching the dynode resistors when changing the tube, as moisture creates electrical leakage paths.

The atomic absorption spectrophotometer comes fitted with HTV tube type R 446 which is sensitive through ultraviolet and is recommended for a range of 193 to 770 nm. Other tubes may be fitted as follows (their spectral ranges are illustrated in Fig. 11.3):

HTV type R 106 is recommended for use below 200 nm, especially for arsenic at 193.7 nm and selenium at 196 nm.

HTV type R 406 (R 196) is recommended for cesium and rubidium. Its relatively low response in the near ultraviolet enables it to overcome second-order spectral interference, when analyzing for potassium, although higher gain must be used, for R 406 has a lower output in the region of 770 nm than R 446. The lower noise level of R 406 allows its use with a higher gain.

HTV type R 166 is sensitive only in the ultraviolet. Its response is 160 to 320 nm which in this range is considerably lower than that of type R 106.

HTV type R 456 has a higher output in the ultraviolet and red than R 446.

Flame types and Lighting Instructions

1.   Use an exhaust hood over the burner flue.

2.   Check that the gas cylinders contain the required gas. The following color code may be used:

Label	Color
Air	Dark admiralty gray with a black top
Acetylene	Maroon
Argon	Peacock blue
Hydrogen	Signal red
Nitrogen	Dark admiralty gray
Nitrous oxide	French blue
Propane	Aluminum

3.   Before connecting the gas supplies to the instrument make sure that the gas cylinders register an adequate pressure, that they are neither leaking nor damaged, and that their valves are operating properly.

4.   Make sure that the connections from the gas cylinders to the instrument are properly made.

5.   Make sure that the constant level liquid trap is filled with water.

6.   Make sure that all internal hose connections are gastight.

WARNING.   Never use OXYGEN as the support gas.

For air-acetylene gas mixture. Burner types 02-100035-00 or 02-100036-00 are suitable. For most elements the flame should be nonluminous, with a fairly hazy blue inner cone of unburnt gas, but for some, particularly tin, molybdenum, and chromium, a fairly luminous flame is required. The best flame can be found for each element by adjusting the gas flow and burner height until maximum absorption is achieved while spraying a test solution that is known to give a reading between 0.1 and 0.2 absorbance units.

LIGHTING THE GAS

a. Turn the AIR-OFF-N$_2$O selector switch to OFF.
b. Make sure that a suitable burner is fitted.
c. Set the line pressures to acetylene 10 psig., air to 60 psig.

CAUTION. Do not use an acetylene cylinder with pressure lower than 60 psig. Always use the acetylene cylinder standing upright and never lying on its side.

d. Turn the support gas selector switch to air.
e. Adjust the support control to give a reading of six units on the flow meter.
f. Adjust the fuel control to give a reading of three to four units.
g. Ignite the burner by means of the push-button igniter and adjust fuel control to obtain the required type of flame.

EXTINGUISHING THE FLAME

a. Turn off the acetylene cylinder control valve.
b. Wait until the flame goes out; then turn off the air cylinder control valve.

For nitrous oxide-acetylene gas mixture. Burner type 02-100035-00 is suitable. This gas mixture allows the determination of many refractory elements not possible with the cooler air-acetylene flame. Provided that the flame conditions and burner positions are optimal, it allows greater sensitivity and freedom from interference than the air-acetylene flame, particularly for magnesium, calcium, strontium, barium, tin, chromium, and molybdenum.

Adjustments are critical. A small departure from optimal settings causes a drastic loss in sensitivity. The best flame conditions are not necessarily the same for all elements. In general, however, a slightly fuel-rich flame is required and the optical path should pass through the flame at about 5 to 10 mm above the burner. The rate at which the solution is drawn through the nebulizer is critical; the optimum is $3\frac{1}{2}$ to $4\frac{1}{2}$ ml per minute.

A number of metals are ionized to an appreciable extent in a nitrous oxide-acetylene flame. This interference may be removed by adding an excess (5000 to 10,000 ppm) of potassium or cesium salt to the solution.

LIGHTING THE FLAME

a. Turn the support gas selector to OFF.
b. Make sure that the correct burner is fitted and the liquid seal is full.
c. Set the line pressures to acetylene—10 psig; nitrous oxide—60 psig; air—60 psig.
d. Turn the support gas selector to N$_2$O
e. Adjust support gas flow to six units.
f. Turn support gas selector to air
g. Adjust fuel control to show three units on the flow meter
h. Wait 5 seconds and light the burner by using the push-button igniter
j. Increase fuel flow to 10 units to obtain a brightly luminous, sooty flame.
k. Turn the support gas selector to N$_2$O.
l. Adjust fuel flow to produce a flame with a red interconal zone approximately 13 mm high.

EXTINGUISHING THE FLAME

    a. Increase the fuel flow to produce a luminous flame.
    b. Quickly rotate the support gas selector to air.
    c. Allow the air-acetylene flame to burn for at least 5 seconds.
    d. Close the acetylene cylinder control valve and WAIT until the flame goes out.
    e. Close the air and nitrous oxide cylinder control valves.

For nitrous oxide-propane flame. Burner type 02-100035-00 or 02-100036-00 is suitable for magnesium and calcium determination in the presence of interfering ions. Suppression of ionization may be effected by adding 5000 to 10,000 ppm of potassium, sodium, or cesium salts when determining elements that ionize easily, although the degree of ionization is little more than in the air-acetylene flame.

TO LIGHT AND EXTINGUISH THE FLAME

The procedure for the nitrous oxide-acetylene flame is used. The flame should be adjusted to give a luminous white interconal zone.

Air-propane and air-coal gas flame. Burner type 02-100037-00 is suitable. The procedure for this mixture is the same as that for air-acetylene.

Air-hydrogen flame. Burner type 02-100036-00 is suitable. This mixture is sometimes used for determining arsenic, selenium, tellurium, and tin. The increased transparency of the flame leads to improved signal/noise ratios and consequently more precise absorption measurements when working in the region below 200 nm. For optimum sensitivity the air-hydrogen flame should be run fuel-rich with a flow meter reading of approximately seven units.

LIGHTING THE GAS

    a. Turn the support gas selector to OFF.
    b. Make sure that the correct burner is fitted and the liquid seal is full.
    c. Connect the hydrogen supply to the fuel gas inlet.
    d. Set the line pressure to hydrogen—10 to 12 psig, air—to 60 psig.
    e. Turn the support gas selector to air.
    f. Adjust the support control to give a flow meter reading of six units.
    g. Adjust the fuel control to give a reading of two to three units.
    h. Light the burner
    j. Adjust the fuel flow to obtain the required sensitivity.

NOTE. The air-hydrogen flame is colorless

EXTINGUISHING THE FLAME

    a. Close the hydrogen cylinder valve and allow 5 to 10 seconds to ensure that the system is adequately purged.
    b. Close the air cylinder valve.

Nitrogen/entrained air-hydrogen flame. Burner type 02-100037-00 is suitable. In determining arsenic the use of this flame increases sensitivity by about three times and improves the signal/noise ratio compared with the air/hydrogen flame. This flame has been used for arsenic, selenium, cadmium, mercury, tin, tellurium, lead, cesium, and zinc. Because of its low temperature, however, chemical interference may be encountered and care should be taken. The flame is used for arsenic and selenium in the application of the vapor-generation technique, in which gas flows are reduced and burner type 02-100036-00 is also suitable.

LIGHTING THE FLAME

a. Turn the support gas selector switch to OFF.
b. Make sure that the correct burner is fitted and the liquid seal, full.
c. Connect the hydrogen supply to the FUEL gas inlet and the nitrogen supply to the N_2O inlet.
d. Set line pressures to hydrogen -10 to 12 psig, nitrogen, -40 psig.
e. Purge the system of air for 5 to 10 seconds by turning the support gas selector control to N_2O.
f. Adjust support control to show a reading of six units.
g. Adjust fuel control to show a reading of three to four units.
h. Light the flame.
i. Adjust the fuel to give the required sensitivity.

EXTINGUISHING THE FLAME

a. Close the hydrogen cylinder valve and wait for the flame to go out.
b. Close the nitrogen cylinder valve.

Argon/entrained air-hydrogen flame. Burner type 02-100037-00 is suitable. The procedure for lighting and extinguishing is the same as for nitrogen/entrained air-hydrogen. This flame is similar to the nitrogen/entrained air-hydrogen flame but more transparent in the far ultraviolet region and less prone to interference problems.

EXERCISE 11.8

Standard Operating Procedures

The lamp current determines not only the intensity of the spectral lines emitted but also their width. To obtain the highest sensitivity in absorption and straight absorbance-concentration curves it is essential that the line to be measured not be broadened by self-absorption. For this reason alkali metal dischange lamps and hollow cathode lamps employing readily volatilized metals such as arsenic, zinc, and cadmium, must generally be run on the minimum current, consistent with adequate signal and stability. The following points are important:

1. Most lamps take a few minutes to warm up.

2. To conserve the life of the lamps it is recommended that only those required for analysis in hand be switched ON.

3. Current to empty lamp quadrants should be switched OFF.

4. In general lamps other than those mentioned above should be operated at maximum current, consistent with maintaining sensitivity. However the current must not exceed the maximum value shown on the lamp.

CAUTION.   Ensure that current controls are turned OFF before fitting lamps into quadrants (Fig. 11.3). This avoids the danger of over-loading lamps whose maximum current rating is low.

5. The strength of the signal, hence the stability of the reading, depends also on the slit width. It should be as wide as possible, consistent with adequate isolation of the required resonance line. Decreased curvature of the calibration graph as the slit width is decreased will indicate insufficient isolation of the resonance line.

Fitting Lamps (Fig. 11.3)

To remove a quadrant grasp the appropriate segment and withdraw the quadrant, taking care to avoid striking the lamp on the case. To remove a lamp depress the release button and withdraw the lamp from its socket. To fit a lamp depress the release button and insert the lamp. Release the button when the lamp locks into position. The lamp quadrants form a knurled drum that can be rotated. The built-in lamp supply permits lamps 1 and 2 to be powered simultaneously.

Adjusting the Photomultiplier

When the photomultiplier is replaced, it may be necessary to adjust its position. Slacken the attachment screw slightly (Fig. 11.3). With the power switch ON and TRANS mode selected, rotate the assembly with the knurled ring at the base until maximum readout is shown on the meter. Adjust the height as required and tighten the attachment screw.

Preliminary settings

Analysis can be performed in the atomic absorption mode, for which three alternative data presentations are available, ABS, CONC, and TRANS, or in the flame emission mode (EMISS).

Atomic Absorption mode

1. Make sure that the gas supplies are correctly connected and that flame and burner type are properly chosen.

2. Make sure that all gas valves are fully closed.

3. Make sure that controls lamp 1 and lamp 2 are OFF; if the model 40 auxiliary lamp supply is used, be certain that lamp 3 and lamp 4 are also OFF.

4. Fit the appropriate lamps in the quadrants.

5. Press the power switch and note that the indicator light is illuminated.

6. Switch on the lamps to be operated and set the currents by using the monitor switches lamp 1 and lamp 2 and their respective controls. Set the currents to lamps 3 and 4 if the auxiliary supply has been fitted.

7. Rotate the turret to bring the required lamp to the operating position (Fig. 11.3).

8. Set the required slit width (see Table 11.1 and the manufacturer's manual).

9. Set the required monochromator wavelength by means of the λ scan control (see Table 11.1).

10. Select Trans mode and A Damp readout. Set the wavelength exactly by rotating the λ scan control to give the maximum signal. If the signal is off scale at either end, press zero until the meter needle returns to any position on the scale. Use the display meter on model 1100 and the small peaking meter on model 1200.

11. Adjust the position of the lamp by using the knurled knob on the quadrant until the cathode is centered in the light path (i.e., until the meter signal is a maximum).

12. Press zero and hold until the light is extinguished and the display reads 100.

13. Recheck the wavelength adjustment as in step 10 above.

The following steps are for atomic absorption measurements and should follow the above preliminary steps:

1. Turn on the gases in the correct order and ignite the flame in accordance with instructions in exercise 11.7.

2. Aspirate a blank solution (distilled water or a reagent blank) into the nebulizer.

3. Select CONC mode and press zero. Hold until the display is set at zero (zero button light goes out).

4. Aspirate a solution containing the analyte element and adjust the flame and burner until the optimum signal is obtained. (B Damp or even C Damp may be required). If necessary, adjust the position of the glass bead for maximum absorbance, using the special screwdriver (Fig. 11.3).

5. Aspirate a standard solution at the low end of the required concentration range and adjust LOW until the displayed result corresponds to the known concentration of the standard.

6. Aspirate a standard solution at the high end of the required concentration range. If necessary, adjust the high control until the displayed result corresponds with the known concentration.

7. Recheck zero, low, and high settings by using the solutions in steps 2, 5, and 6. Make fine adjustments if necessary.

8. Select the required readout mode (usually INT 10) and aspirate the sample. The integrator operates as follows:

Introduce Press sample → READ →	← READ button illuminated ────────→			
	3 sec→	3 or 10 sec →	5 sec →	────→
	signal rise time	selected integration period	display on recorder	recorder returns to zero
	←Instrument and recorder both display zero→		instrument display ←persists until canceled→ by subsequent READ instruction	

9. To switch off the instrument extinguish the flame as detailed in exercise 11.7. Release the power button and check that the indicator light is extinguished.

Flame Emission mode

1. Make sure that the gas supplies are correctly connected.

2. Make sure that all gas valves are fully closed.

3. Press the power switch and check that the indicator light is illuminated.

4. Select EMISS mode and A DAMP readout.

5. Turn on the gas and light the flame in accordance with the instructions in exercise 11.7.

6. Select the required spectral bandpass.

7. Set the approximate wavelength required with λ scan control.

8. Nebulize a solution containing the analyte element and press zero until a reading is obtained.

9. Adjust the monochromator wavelength to give a maximum signal. If the meter goes off scale, press zero until it returns on scale and continue the adjustment. In model 1200 use the peaking meter.

10. Adjust the flame and burner to give optimum signal.

11. Aspirate a blank (distilled water or reagent blank) and adjust the high control until the display reads zero. This control compensates for emission from the flame gases.

12. Aspirate a standard solution at the top of the required concentration range and press zero. In this mode the zero control sets and signals to 100%.

13. Recheck the blank and the standard readings and if necessary readjust.

14. Select the required readout mode (usually INT 10) and aspirate the samples.

EXERCISE 11.9

Making Standard Solutions

Make up separate standard solutions. Use distilled and deionized water and analar chemicals throughout. For each element a 1 liter stock solution is made containing 1000 $\mu g/ml$ (lg/liter) of the element.

Major Elements

ALUMINUM metal (99.999% pure from Halewood Chemicals Ltd., Horton Road, Stanwell Moor, Middlesex, England). It is important that the metal be finely divided, otherwise difficulty will be experienced in dissolving it completely. Dissolve 1.000g in HCl and dilute to 1 liter.

CALCIUM $CaCO_3$. Dissolve 2.497 g in HCl and dilute to 1 liter.

IRON metal (99.5% pure, from British Chemical Standards, Bureau of Analysed Samples, Newton Hill, Middlesborough, Yorkshire, England). Dissolve 1.000 g in HCl and dilute to 1 liter.

POTASSIUM. Dissolve 1.907 g KCl in water and dilute to 1 liter. *Alternatively* dissolve 1.767 g of dried analar or high purity grade-I K_2CO_3 (from Johnson-Matthey and Co. Ltd., Hatton Garden, London W.C.1) in water and dilute to 1 liter.

MAGNESIUM High purity grade-I metal crystals (Johnson-Matthey) or metallic Mg (99.95% pure, from Magnesium Elektron Ltd., Clifton Junction, Near Manchester, England). Dissolve 1.000 g in HCl and dilute to 1 liter.

MANGANESE metal. Dissolve 1.000 g in HCl and dilute to 1 liter.

SODIUM. Dissolve 2.542 g dried NaCl in water and dilute to 1 liter. *Alternatively* dissolve 2.305 g dried analar or high purity grade-I Na_2CO_3 (from Johnson-Matthey) in water and dilute to 1 liter.

SILICON SiO_2 (pure quartz). Dissolve 2.139 g in HF in a Teflon bomb (according to exercise 11.4), followed by H_3BO_3, in the same way that a rock or mineral sample is digested. An alternative to quartz is tetraethyl orthosilicate (TEOS) from Monsanto Chemicals, which is 99.7% SiO_2.

TITANIUM TiO_2. Dissolve 1.668 g in HF in the Teflon bomb (according to exercise 11.4). Alternatively, dissolve in $KHSO_4$ and H_2SO_4.

Minor Elements

COBALT metal. Dissolve 1.000 g in HNO_3 and dilute to 1 liter.

CHROMIUM K_2CrO_4. Dissolve 2.484 g in water and dilute to 1 liter.

COPPER metal. Dissolve 1.000 g in HNO_3 and dilute to 1 liter.

LITHIUM LiCl. Dissolve 6.109 g in water and dilute to 1 liter.

NICKEL metal. Dissolve 1.000 g in HNO_3 and dilute to 1 liter.

LEAD metal. Dissolve 1.000 g in HNO_3 and dilute to 1 liter.

STRONTIUM $Sr(NO_3)_2$. Dissolve 2.415 g in water and dilute to 1 liter.

VANADIUM V_2O_5. Dissolve 1.785 g in HCl and dilute to 1 liter.

ZINC metal. Dissolve 1.000 g in HCl and dilute to 1 liter.

(After Buckley and Cranston, 1971; Bernas, 1968; Hamilton and Henderson, 1968; and Luth and Ingamells, 1965.)

1. A 1 liter "standard A" containing the following element concentrations in $\mu g/ml$—Al 100, Ca 100, Fe 100, K 50, Mg 50, Mn 2, Na 40, Si 350, Ti 20, Co 0.2, Cr 0.5, Cu 0.2, Li 0.1, Ni 0.3, Pb 0.2, Sr 1.0, V 0.5, Zn 0.2—is prepared by adding appropriate volumes of the above individual 1000 $\mu g/ml$ stock solutions to a 1 liter volumetric flask and making it up to the mark with distilled water. This solution is immediately transferred to a polypropylene bottle labeled "Standard A."

2. The volumes of each 1000 $\mu g/ml$ stock solution required in step 1 are easily determined, for example Al requires 100 cc (dilution of 1/10), Si requires 350 cc and so on. Because the amount of minor elements required is too low to be accurately measured (e.g. Li = 0.1 cc), small portions of each of their standard 1000 $\mu g/ml$ solutions may be diluted 1/10 so that 10 times the volume of the diluted solution is then added to make standard A.

3. Three additional standards are made by diluting standard A by 1 : 2, 1 : 10, and 1 : 20 in water. All four standards are prepared in such a way that all contain 6% volume/volume HF, 5.6% weight/volume H_2BO_3, and 1% volume/volume Aqua Regia to match the standards with the sample solution (as made according to Exercise 11.4). The above standard solutions are ideal for method B of Ex. 11.4.

(After Buckley and Cranston, 1971).

DISCUSSION

The advantage of the single stock solution A for all elements is preferable to separate standards for each element. In this way interelement interferences and ionization effects are fully compensated because the combined standard has a matrix similar to that of the sample.

Alternatively separate standards containing the following concentrations in $\mu g/ml$ (=ppm) may be used: Si 1000, Fe 1000, Al 1000, Ti 1800, V 140, Ca 500, Mg 1000, Na 1000, K 700 (Bernas, 1968). Silica standard solutions can be diluted to give 100, 150, and 250 ppm silica; Alumina standard solutions can be diluted to give standard solutions of 20, 50, and 80 ppm alumina (Katz, 1968).

EXERCISE 11.10

Analytical Conditions

The following analytical conditions are recommended for rock or mineral samples that have been digested in a Teflon bomb (method 11.4). The dissolved 100-ml rock sample solution does not need to be diluted further. It can be aspirated directly into the burner. Operating conditions are tabulated in Table 11.1. A ZERO or baseline

Table 11.1 Analytical Conditions for Major and Minor Elements Using a Combined Standard A (Exercise 11.9) and a Teflon Bomb Digested Sample (Exercise 11.4)[a]

Element	Wave-length (nm)	Burner Type	Burner Position	Recorder Expansion	Limits of Oxide Concentration Rock (%)	Lamp Type	Sensitivity (g/ml/1% abs)	Calibration Graph Conc. Limits (ppm)	Spectral Band Slit Width (nm)
Al	309.3	Nitrous oxide acetylene	Parallel to light path	×1	15–28	Shielded cathode	1.2	20–50	0.2
Ca	422.7	Nitrous Oxide acetylene	Perpendicular to light path	×1; ×3	10–14	Multielement shielded cathode	1.3	1–5[b]	0.2
Fe	372.0	Short path air acetylene	Perpendicular	×3	10–14	Shielded cathode	4.0	5–20	0.2
K	766.5	Short path air acetylene	Perpendicular	×3; ×10	5–6	Arc discharge lamp	1.5	0.3–1.0[c]	0.5
Mg	285.2	Nitrous oxide acetylene	Perpendicular	×1	5–8	Multielement shielded cathode	0.6	0.5–2.0[d]	1.0
Mn	279.5	Boling (three-slot)	Parallel	×3	0.3–0.4	Hollow cathode	0.06	1–4	0.2
Na	589.0	Short path air acetylene	Perpendicular	×1; ×3	4–5	Arc discharge lamp	0.5	0.4–1.2[e]	1.0
Si	251.6	Nitrous oxide acetylene	Parallel	×1	35–75	High brightness tube	4.0	100–200	0.2

Element	Wavelength (nm)	Flame	Burner	Sample dilution	Lower limit conc. ppm	Lamp			
Ti	364.3	Nitrous oxide acetylene	Parallel	×10; ×3	3–5	Shielded cathode	2.0	1–6	0.2
Co	240.7	Boling (three slot)	Parallel	×30	20	Hollow cathode	0.15	3–12	0.2
Cr	357.9	Boling (three slot)	Parallel	×30	20	Hollow cathode	0.20	2–8	0.5
Cu	324.7	Boling (three slot)	Parallel	×30	10	Hollow cathode	0.07	2–8	0.2
Li	670.8	Boling (three slot)	Parallel	×30	10	Hollow cathode	0.05	1–4	1.0
Ni	232.0	Boling (three slot)	Parallel	×30	20	Hollow cathode	0.15	3–12	0.2
Pb	217.0	Boling (three slot)	Parallel	×30	50	Hollow cathode	0.50	5–20	1.0
Sr	460.7	Nitrous oxide acetylene	Parallel	×10	100	Hollow cathode	0.12	2–10	0.5
V[f]	318.4	Nitrous oxide acetylene	Parallel	×30; ×10	100	Shielded cathode	1.0	0.5–3.0	0.2
Zn	213.8	Boling (three slot)	Parallel	×10	20	Hollow cathode	0.020	0.4–1.6	0.5

[a] After Buckley and Cranston (1971); Bernas (1968).
[b] Sample dilutions 1 : 4 or 1 : 10.
[c] Sample dilutions 1 : 10 or 1 : 100.
[d] Sample dilutions 1 : 1 or 1 : 4.
[e] Sample dilutions 1 : 10.
[f] Sample weight used = 300 mg.

reading for each element is determined continuously by aspirating an aqueous solution containing HF, H_3BO_3, and Aqua Regia in the same concentrations as in the sample and standard (exercises 11.4 and 11.9).

The operating conditions may be those in Table 11.1, or a nitrous oxide-acetylene flame may be used for Si, Al, Ti, V, Ca, and Mg and an air-acetylene flame for Fe, K, and Na (Bernas, 1968). The concentration may be read directly from the atomic absorption spectrometer or from a calibration graph, established by using the standard solution A and its dilutions.

DISCUSSION

The use of a whole-rock standard solution, made up according to exercise 11.9, is preferable because the standard sample is then comparable in composition to the sample being analyzed (made according to Exercise 11.4). An alternative is to use a comparable rock or mineral standard (exercise 10.7) (Flanagan, 1970) or obtainable from Dr. F. J. Flanagan, U.S. Geological Survey, Washington, D.C. 20242. The rock or mineral standard may be digested, like the sample, in a Teflon bomb (following exercise 11.4). The published analysis of the standard may be used to calibrate the unknown sample.

For operating conditions for other elements the reader is referred to the operating manual supplied with the atomic absorption spectrometer or to Angino and Billings (1967), Ramírez-Muñoz (1968) or to Rubéska and Moldan (1969).

EXERCISE 11.11

Removing Layer Silicates from Quartz and Feldspar

The removal of layer silicates from quartz and feldspar is frequently desirable before examining the minerals optically or analyzing them chemically. The method is as follows:

1. A 200-mg sample of sand or silt-sized sample (dried at 110°C) is weighed into a 50-ml vitreous silica crucible.

2. $Na_2S_2O_7$ powder (12 to 15g) is added and mixed with the sample with a glass rod.

3. The $Na_2S_2O_7$ is fused under a fume hood, with a low flame until vigorous bubbling of the salt stops. Thereafter the full flame of a Meker burner is applied. The fusion is complete when Na_2SO_4 crystals appear on the melt surface while at full heat. The crucible is swirled while cooling to spread the melt on the crucible sides.

4. The solidified melt is transferred as a cake to a 150-ml beaker with 60 ml 3 N HCl and a rubber-tipped glass rod. The cake is slaked by gentle boiling and the resulting suspension transferred to a 70-ml pointed centrifuge tube.

5. The insoluble material is separated by centrifugation and the supernatant liquid discarded. Two further washings with 3 N HCl are employed to complete the transfer and to wash the residue; discard the supernatant liquid each time.

6. The residue from the tube is transferred to a 500-ml nickel or stainless steel beaker with 0.5 N NaOH and the total volume of NaOH in the beaker is made up to 100 to 150 ml. The suspension is boiled rapidly for $2\frac{1}{2}$ minutes and cooled rapidly in a cold bath.

7. The suspension is transferred to 70-ml pointed centrifuge tubes, the insoluble residue from the beaker being transferred quantitatively. The residue is separated by centrifugation and washed three times with 3 N HCl, transferred to a tared platinum (or Teflon) crucible (or beaker), dried at 110°C, and weighed. The residue is now free of micas. Potassium feldspar and quartz are remarkably resistant to attack with this method. The remaining feldspar is now ready for analysis for potassium, being free of mica.

(After Kiely and Jackson, 1964.)

DETERMINATION OF CHEMICAL COMPONENTS NOT ATTAINABLE BY THE FOREGOING METHODS

12

Several important chemical components cannot be determined by the instrumental methods in Chapters 6, 7, 10, and 11 and must be determined by the wet chemical methods described in this chapter.

A. FeO AND Fe₂O₃ DETERMINATION

To the petrologist and mineralogist the ferric/ferrous ratio in a rock or mineral is an important parameter but its correct determination is still a challenge to the analyst. This ratio will, of necessity, change when the sample is fused or put into solution. Even grinding the specimen will oxidize some of the iron. Many analysts work on the +100 mesh material for this very reason, but others determine ferrous/ferric ratios on material ground to −100 mesh.

Total iron is first determined by an X-ray fluorescence method (Chapter 10) or by atomic absorption spectrophotometry (Chapter 11). Ferric iron is not determined separately but by the difference of total iron - ferrous iron.

$$\% \text{ total Fe (as } Fe_2O_3) - (\% \text{ FeO} \times 1.1113) = \% Fe_2O_3.$$

Gravimetric factors: $Fe = Fe_2O_3 \times 0.69944$,
$$Fe = FeO \times 0.77731,$$
$$FeO = Fe_2O_3 \times 0.89981,$$
$$Fe_2O_3 = FeO \times 1.1113.$$

Solution of the rock or mineral must be done in a way that will inhibit the oxidation of the iron.

EXERCISE 12.1

Determination of FeO in a Rock or Mineral

METHOD A

A special Teflon digestion apparatus which will also serve as a titration cell under constant nitrogen flow (Fig. 12.1) has to be constructed. The digestion vessel is connected to a pure nitrogen supply (commercial grade " dry nitrogen " scrubbed with a 15% v/v pyrogallol solution in 50% potassium hydroxide in an acrylic vessel of cylin-

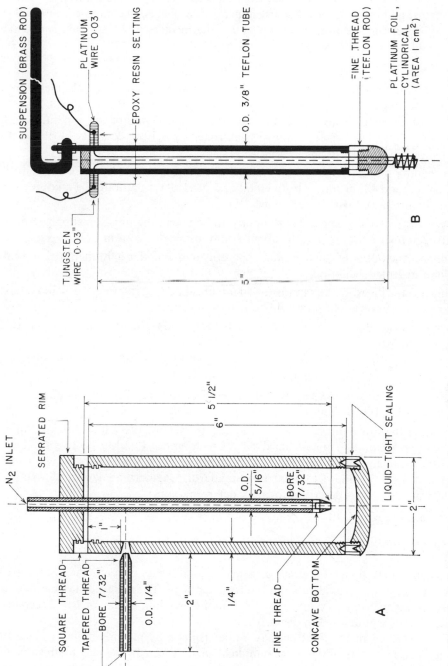

Fig. 12.1 The Teflon digestion apparatus, A, and the platinum-tungsten bimetallic electrode in Teflon housing for titration monitoring, B. (Design after Kiss, 1967.)

SUSPENSION (BRASS ROD)

PLATINUM WIRE 0·03"

EPOXY RESIN SETTING

O.D. 3/8" TEFLON TUBE

FINE THREAD (TEFLON ROD)

PLATINUM FOIL, CYLINDRICAL (AREA 1 cm²)

TUNGSTEN WIRE 0·03"

5"

B

N₂ INLET

SERRATED RIM

SQUARE THREAD

TAPERED THREAD

BORE 7/32"

O.D. 1/4"

FINE THREAD

CONCAVE BOTTOM

O.D. 5/16"

BORE 7/32"

LIQUID-TIGHT SEALING

5 1/2"

6"

2"

2"

1"

1/4"

A

355

drical shape). The manifold of the nitrogen supply is provided with a condensate trap and an outlet for the reagents. The digestion apparatus could be converted to a pressure vessel operated at a maximum of 200°C by constructing two needle valves of Teflon components; all but the most refractory minerals can then be decomposed. For higher temperature heating, heating tape coupled to a thermostat can be wound around the teflon vessel.

PROCEDURE

1. Weigh 0.05 to 0.50 g of finely ground rock or mineral material (240 mesh) into the digestion apparatus (Fig. 12.1). Teflon vessels should never be rubbed because the electrostatic charge may scatter the sample. Wash the wall of the apparatus with about 5 ml of distilled water saturated in nitrogen. Swirl the contents to obtain an even suspension.

2. Depending on the weight of the sample, add 10 to 30 ml of hydrofluoric-sulfuric acid mixture (add 40 ml conc. HF to 60 ml of 5 N H_2SO_4. Saturate the mixture with nitrogen before use). Swirl the contents.

3. Close the top of the apparatus promptly and connect the nitrogen supply. Adjust the nitrogen inlet teflon tube to lie about 7 mm above the bottom of the vessel.

4. Connect the outlet of the digestion apparatus to a cylinder half filled with a weak potassium hydroxide solution.

5. Immerse the digestion apparatus for 60 minutes in a boiling water bath and adjust the nitrogen flow to two to four bubbles per second.

6. Withdraw the digestion apparatus, open the top, and rinse down all the condensed droplets with distilled water saturated with nitrogen.

7. Pull up the threaded lid as far as it can go and, maintaining constant nitrogen flow, transfer to a titration stand.

8. Adjust the nitrogen flow to give moderately fast stirring. The final volume of the solution may be 80 to 100 ml.

9. Immerse the bimetallic electrode (Fig. 12.1B). Record the initial millivolt value and titrate with standard potassium dichromate solution. [(0.0139 N) (1 ml = 1 mg FeO). Dry the finely ground potassium dichromate at 120°C for 2 hours, weigh 0.6826 g, and dissolve and dilute to 1000 ml. Standardize against analar iron^{+2} sulfate heptahydrate exactly as described for rock determination.]

(Method after Kiss, 1967).

DISCUSSION

The response of the bimetallic electrode generally diminishes in routine use because of the formation of a thin oxide film on the tungsten wire. The electrode system can be restored to full response by dipping the tip in molten sodium nitrite for 1 to 2 seconds, after which it is equilibrated in distilled water.

The end point of the titration may be determined by plotting millivolts against the volume in milliliters of the potassium dichromate titrated. The maximum slope of the curve gives the end point. More precise determination of the end point may be determined by constructing a plot of $\Delta E/\Delta V$ (mV/ml) against volume (V) in milliliters; the maximum on the graph gives the end point. For discussion see Vogel (1961, p. 931). The wire outputs from the electrode are attached to a galvanometer to obtain the millivolt readings.

METHOD B

Reagents.

Boric acid solution. Dissolve, with stirring, 100 g boric acid in 1 liter hot water. Cool, and dilute to 2 liters.

Potassium dichromate solution (0.1 N). Weigh 9.810 g finely ground analar grade $K_2Cr_2O_7$ into a weighing bottle, and dry at 110°C in an oven for 2+ hours. Weigh bottle and contents and run the contents through a funnel into a 2-liter volumetric flask. Reweigh the bottle to obtain the true weight of $K_2Cr_2O_7$ used. Wash the funnel and neck of the flask. Dissolve the reagent, and make the flask up to 2 liters with distilled water. The equivalent weight of potassium dichromate is $(K_2Cr_2O_7)/6 = 49.035$.

$$N = \frac{\text{wt } K_2Cr_2O_7 \text{ (g)}}{2 \times 49.035 \text{ (g)}}.$$

A check on the normality may be done by using an aliquot of the standard Fe solution; 10 ml of Fe solution requires approximately 10 ml of 0.1 N $K_2Cr_2O_7$.

Standard iron solution. Weigh accurately about 1.3 g of pure iron, such as wire or chips (washed first in ether and dried in air) and transfer to a 150-ml beaker. Add 50 ml HCl (1 : 1), cover and warm on a steam bath to complete solution. Cool. Transfer to 250-ml flask and make up to volume. Remove aliquot with a pipette.

PROCEDURE

1. Weigh 0.500 g of powdered sample (100 mesh) in a 45-ml platinum crucible with a tight fitting lid and a 2 mm central hole.

2. Add 1 ml of water. Swirl the crucible to spread the sample evenly over the bottom.

3. Add two or three drops of H_2SO_4 (1 : 1) to decompose any carbonates present. Put the cover on and let stand until reaction is complete.

4. To 10 ml of water in a 50-ml platinum or Teflon dish add 5 ml H_2SO_4 (18 M) and 5 ml HF. Place the covered crucible on a silica triangle over a low-flame bunsen. Slide the cover to one side. Quickly and *carefully* add the hot mixture from the platinum or Teflon dish. Replace the cover and immediately begin brushing the sides and cover of the crucible with the flame of a second burner until the contents boil and steam emits from the hole. Adjust the bottom Bunsen to a $\frac{1}{2}$-in. flame so that boiling goes on gently. Continue heating for 10 minutes. Do not prolong heating. Do not evaporate the water to the extent that the H_2SO_4 begins to oxidize the ferrous iron. Use a timer to prevent overheating.

5. To 200 ml of water in a 600-ml beaker add 50 ml of the boric acid solution, 5 ml H_2SO_4 (18 M), and 5 ml 85% H_3PO_4; mix well.

6. At the conclusion of the heating period at once grasp the crucible firmly with the platinum-tipped tongs, press the cover on firmly with a glass rod, and swiftly submerge the crucible beneath the acid solution in the 600-ml beaker. Do *not* allow more than the platinum shoes of the tongs to enter the acid solution.

7. Dislodge the crucible cover with the glass rod and stir all contents. Do not stir too vigorously or air will be taken into the liquid, but make sure that all the crucible contents have been washed out and all the soluble material has been dissolved.

8. Titrate the solution immediately with N/10 $KMnO_4$ (dissolve 3.2 g $KMnO_4$ in 1 liter of water and allow to stand three days at room temperature) to the first appearance of a permanent pink tinge, or with N/10 $K_2Cr_2O_7$ (add 6 drops of 0.2% barium diphenylamine sulfonate to serve as indicator and titrate to the first appearance of a permanent blue-violet color). When the end point is reached, swirl the crucible in the beaker to make sure that the titration is indeed complete

$$\% \, FeO = \frac{\text{milliliters of titrant} \times \text{normality}}{1000} \times \frac{71.85}{1} \times \frac{100}{\text{sample wt (g)}}.$$

(After Maxwell, 1968).

DISCUSSION

Examine the residue, if any, in the beaker with a hand lens. White or grayish grains of quartz which resist decomposition can be ignored.

Brassy yellow grains of pyrite can be corrected for if a sulfur analysis is available. If there are more than a few grains of red to black undecomposed material in the residue, a second decomposition and titration is necessary. All residue must be decanted and rinsed and ground in an agate mortar. Then the complete procedure from step 1 is repeated. Add the volume of the titrant used to that for the first titration. A remaining black residue will most probably be chromite and will resist decomposition.

The titration end point is slow and intervals of decolorization take place before a permanent color change is established. An automatic titrator is far superior and the photocel is more sensitive than the eye to color change at the end point.

METHOD C

Reagents

Sulfuric acid, 9 N: 500 ml of solution contains 125 ml concentrated H_2SO_4, ACS reagent.

Phosphoric acid, 22 N: 500 ml of solution contains 250 ml concentrated H_3PO_4, ACS reagent.

Concentrated HF, ACS reagent.

Sodium diphenylaminesulfonate—0.01%; 100 ml of solution contains 10 mg sodium diphenylaminesulfonate.

Potassium dichromate—0.02000 N: 1 liter contains 980.7 mg U.S. National Bureau of Standards. Standard sample No. 136 $K_2Cr_2O_7$.

Ferrous ammonium sulfate—0.01 N (In H_2SO_4): 1 liter contains 3.922 g $Fe(NH_4)_2$ $(SO_4)_2$ $6H_2O$, ACS reagent, and 28 ml concentrated H_2SO_4, ACS reagent.

Equipment

Stirring bar, magnetic, plastic coated, $\frac{7}{8} \times \frac{3}{8}$ in.

Stirring bar, magnetic, plastic coated, $\frac{3}{8} \times \frac{1}{8}$ in.

Beaker, Berzelius, tall-form, borosilicate glass 100 ml.

Burette, precision bore, class A, 125 ml, graduated in 0.01 ml intervals.

Burettes (2), precision bore, class A, 10 ml, graduated in 0.05 ml or 0.02 ml intervals.

Magnetic stirrers (at least two) approximately 4 in. wide.

Hot plate, semimicro, approximately 70 W, approximately 3 in. in diameter, overall height 2 in. (The overall height of the hot plate should be such that the magnet of the stirrer can drive the stirring bar well.)

Autotransformer, variable, to regulate hot plate temperature.

Crucible and cover, platinum, 10 ml (the bottom should be flat to ensure efficient heating on the hot plate). The cover handle should be bent slightly up.

Platinum stirring rod, heavy gage wire, 0.064 in diameter, with one chisel-pointed end.

Tweezers, with long platinum tips whose ends have been bent at a right angle.

Tongs, crucible, Blair type, with platinum tips shaped to fit the 10-ml platinum crucible.

PROCEDURE (*use distilled water throughout*)

Standardization of 0.01 N Ferrous Ammonium Sulphate

1. Transfer 6.25 ml 9 N H_2SO_4 to a 100-ml Berzelius beaker containing a $\frac{7}{8}$ in $\times$ $\frac{3}{8}$ in magnetic stirring bar.

2. Add 3.75 ml 22 N H_3PO_4 and 25 ml water. Add from a 10-ml burette 9.1 to 9.2 ml 0.02000 N $K_2Cr_2O_7$. Rinse the tip of the burette with water. Add 1 ml concentrated HF, using a plastic pipette. Rinse down the side of the beaker with water and dilute to 75 to 85 ml. Place on a magnetic stirrer and stir well.

3. Titrate with 0.01 N ferrous ammonium sulfate, using the 25-ml burette until the yellow color begins to disappear. Use a piece of white paper under and behind the beaker as a background.

4. Add 0.5 ml of 0.01 % sodium diphenylaminesulfonate. Add the ferrous ammonium sulfate solution drop by drop until the violet or purple color just disappears. Rinse the burette tip with water.

5. Add 0.02000 N $K_2Cr_2O_7$ drop by drop with constant stirring until the purple or violet color reappears. Rinse the burette tip with water.

6. Add the ferrous ammonium sulfate solution, "cracking" the drops added until the purple or violet color just disappears. The tip of the burette is rinsed with water after the addition of each "cracked" droplet.

7. Calculation:

$$N \; Fe^{2+} = \frac{\text{milliliters of } K_2Cr_2O_7 \times \text{normality of } K_2Cr_2O_7}{\text{milliliters of ferrous ammonium sulfate solution}}.$$

The ferrous solution is standardized in duplicate each day it is used.

FeO Determination

1. Weigh accurately 25 to 100 mg of powdered silicate mineral or rock (containing approximately 7 mg FeO) in a 10-ml platinum crucible. Use a semimicrobalance (± 0.01 mg) for 25 to 40 mg samples and a macrobalance (± 0.1 mg) for 40 to 100 mg samples.

2. Place a plastic coated $\frac{3}{8} \times \frac{1}{8}$ in. stirring bar in the crucible. Use a fast-flowing 2-ml pipette to add 2 ml 9 N H_2SO_4 to the crucible. Rinse the inside surfaces of the crucible with water. Rotate the crucible while the acid is being added. Add 1 ml concentrated HF, drop by drop, with a plastic pipette. Put the cover on.

3. Place the crucible on a hot plate whose surface temperature is $350 \pm 10°F$ ($177°C \pm 6$). The hot plate is calibrated with a bimetallic spiral surface thermometer. The hot plate sits on a magnetic stirrer. Heat for 3 minutes. Start the magnetic stirrer and heat for 7 minutes more.

4. While the sample is being treated add to the 100-ml Berzelius beaker a plastic-coated $\frac{7}{8} \times \frac{3}{8}$ in. magnetic stirring bar, 4.25 ml 9 N H_2SO_4, and 3.75 ml 22 N H_3PO_4. Rinse down the inside beaker walls with water.

5. Add 9.1 to 9.2 ml 0.02000 N $K_2Cr_2O_7$ from a 10-ml burette. Rinse the burette tip with water. Dilute the solution with water to a height that will submerge the 10-ml crucible lying on its side in the beaker. Mix the solution well with the magnetic stirrer.

6. Using the heavy platinum wire and the Blair platinum crucible tongs, quickly transfer the crucible (and lid) to the beaker containing the $K_2Cr_2O_7$, plunging it beneath the solution to cover it completely. The crucible is twirled a few times with the platinum wire, always making sure that it is beneath the liquid surface. The crucible is set upright and the solution in it is stirred with the platinum wire; it is again laid on its side in the beaker and twirled a few times with the wire.

7. Using a plastic wash bottle with a fine nozzle, rinse the platinum tips of the tongs with water into the beaker. The tongs are held so that when the water is added drop by drop at the upper portion of the platinum tips it will roll down to the ends of the tips. Each tip is rinsed at least five times.

8. The crucible cover is removed from the solution and rinsed at least three times on both sides with water: the platinum wire is used to lift the handle of the cover onto its edge, when it is grasped by the platinum-tipped tweezers.

9. Set the crucible upright in the beaker with the platinum wire. Rinse the wire into the beaker with water.

10. The platinum tweezers with bent tips are used to lift the crucible out of the solution and the crucible contents are carefully poured into the beaker. Rinse the inner and outer surfaces at least three times with water from a plastic wash bottle and fine nozzle:

a. The platinum tips of the tweezers are rinsed down with drops of water.

b. The inner crucible surface is rinsed from the top down until the crucible is about one-half full.

c. It is then inclined slightly and the outside surface is rinsed from the top down.

d. Empty the crucible with its rim touching the inner surface of the beaker so that the crucible contents are drained as completely as possible.

e. That portion of the inside of the beaker down which the crucible contents have flowed is rinsed drop by drop with a small amount of water.

The above cycle (a to e) is repeated at least twice. Minimum amounts of water are used so that a maximum number of washings can be performed.

11. Rinse down the inside surface of the beaker. There should be approximately 75 to 85 ml of solution in it.

12. Titrate with 0.01 N ferrous ammonium sulfate, using the 25-ml burette if more than 10 ml of 0.01 N Fe^{2+} is needed for back titration. If less than 10 ml of 0.01 N Fe^{2+} is required, use the 10-ml burette. Complete the titration as it has been described for the standardization of 0.01 N ferrous ammonium sulfate.

13. Calculation:

% FeO = [(ml of $K_2Cr_2O_7$ used × normality of $K_2Cr_2O_7$)

$\qquad\qquad\qquad$ − (ml Fe^{2+} used × normality of Fe^{2+})] × 71.84 × 100,

all divided by weight of the sample in milligrams.

(After Meyrowitz, 1963).

METHOD D   *Ferrous: ferric ratios by an electron microprobe*

The relative intensities of Fe$L\alpha$ and $L\beta$ X-ray emission peaks differ significantly with valence state and bond associations, even though the wavelength shift is very small. $L\alpha/L\beta$ intensity increases systematically with increase in

$$\frac{Fe^{2+}}{Fe^{3+}} \quad \text{or} \quad \frac{Mn^{3+}}{Mn^{2+}}$$

in their respective oxides (Albee and Chodos, 1970), in the hematite-ilmenite series, in the magnetite-ulvospinel series, and in anhydrous silicates. $L\alpha/L\beta$ ratios clearly distinguish ferric-rich alkali amphiboles from ferrous-rich calciferous amphiboles. No significant variation has been found in chloritoid, staurolite, and cordierite but variation does occur in chlorite. Ferric-rich muscovite is not distinguishable from ferrous-rich biotite. The study by Albee and Chodos (1970) was done at 15 kV on an electron microprobe with a KAP (or RAP) crystal and pulse height selection. Presumably similar results on the intensity ratios of $L\alpha/L\beta$ may also be obtained by X-ray fluorescence spectrometry.

B.   TOTAL WATER, CARBON DIOXIDE, AND H₂O− DETERMINATION

Loss on ignition at 1000°C

The loss on ignition at 1000°C of a sample is occasionally reported as a substitute for the determination of total water but its use is fraught with danger. Its value will include other volatiles such as carbon dioxide, sulfur, chlorine, fluorine, and, to some extent, the alkalis. During the process ferrous iron is also oxidized, thereby changing the sample weight. Some analysts completely reject the loss on ignition value as meaningless, but if it is to be done 1000°C is the best temperature. High values are caused by loss of volatiles other than water; low values are due to the gain in weight by oxidation of the ferrous to ferric iron, and this oxidation is usually incomplete. Therefore the exact meaning of "loss on ignition" cannot be defined specifically.

EXERCISE 12.2

Simultaneous Determination of Water and Carbon Dioxide

With this method about 12 determinations can be performed per day, for it is not necessary to let the furnace cool between analyses. Water is completely removed from even the most resistant minerals, such as staurolite, within 30 minutes at 1200°C. The method gives a standard deviation of ±0.04% with rocks containing about 5% of water when a 0.5-g sample is used.

METHOD

Apparatus Required. The apparatus required is illustrated diagrammatically in Fig. 12.2.

High Temperature Furnace. This device consists of an alumina tube 22 cm long with an external diameter of 4.2 cm and 84 turns of 0.91 mm Kanthal Al resistance wire on a 15-cm portion of the tube. The turns are more closely spaced at the tube ends than in the center. The resistance wire is bedded in aluminous cement and the space between it and the 10-cm diameter silica outer jacket is maintained by end plates made of sindanyo.

Copper Furnace. This low-temperature furnace consists of a translucent silica tube 9 cm long with an external diameter of 2.3 cm and 16 turns of 1.58 mm × 0.25 mm nichrome tape. The whole is contained in a 6.3-cm diameter brass tube and the intervening space is packed with kieselguhr.

In operation the two furnaces are run in series. The high-temperature furnace is regulated to the desired temperature with a rheostat or Variac autotransformer. The total energy required at 1100°C is about 950 W. Variation of the temperature of the copper furnace over the range 700 to 750°C is of no consequence.

Gas Purification Train. Nitrogen gas (not of "oxygen free" quality) from a cylinder is passed via a multistage regulator and needle valve through a bubble counter containing concentrated H_2SO_4. It is purified by passage through the tubes containing soda lime, fused calcium chloride, and, finally, anhydrous magnesium perchlorate.

Insertion Device. To avoid the necessity for cooling the combustion tube between determinations samples are pushed into the hot zone of the closed combustion tube with an insertion device A (Fig. 12.2) which consists of a brass stuffing box made airtight with a polytetrafluoroethylene (or Teflon) washer, through which slides a stainless steel pusher bar 25 cm long and 6 mm in diameter, with a 12 mm diameter head. The stuffing box is fitted with a brass tube for the introduction of nitrogen into the combustion tube and is connected to the latter by a rubber bung.

Combustion Tube. B in Fig. 12.2 consists of a translucent silica tube 45 cm long with a 1.8-cm internal diameter, fused at one end to a 3-cm length of silica tubing with an external diameter of 5 mm. The combustion tube is supported in the high-temperature furnace so that the 5-mm silica tube is about 10 cm outside it. Although the silica tube is being heated for long periods above 1000°C, no deterioration, except for slight sagging and a small amount of surface devitrification, is experienced even after several months of use. Silica tubing is preferable to impervious alumina tubing because the latter gives variable blank results.

Copper Tube. C in Fig. 12.2 is designed to remove sulphur compounds and oxides of nitrogen. It consists of a translucent silica tube 10 cm long with a 1-cm internal diameter, packed with alternate layers of copper wire and silver pumice (prepared by evaporating 14 mesh pumice with strong silver nitrate solution and igniting strongly) held in place by asbestos plugs. Both ends of the tube are fused to 3-cm lengths of silica tubing with an external diameter of 5 mm. At least once a week a current of coal gas should be passed through the heated tube for 15 minutes to reduce any copper oxide in the metal. The life of the tube is about 3 months with rocks of low sulfur content.

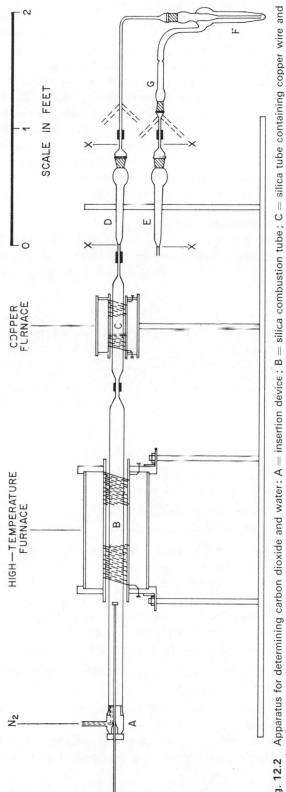
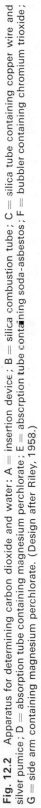

Fig. 12.2 Apparatus for determining carbon dioxide and water: A = insertion device; B = silica combustion tube; C = silica tube containing copper wire and silver pumice; D = absorption tube containing magnesium perchlorate; E = absorption tube containing soda-asbestos; F = bubbler containing chromium trioxide; G = side arm containing magnesium perchlorate. (Design after Riley, 1958.)

Absorption Tubes. These tubes are 14 cm long and have a 1-cm bore; a tube of 5-mm external diameter is sealed at one end and a B 10 socket at the other. They are closed by B 10 cones fused to the 5 mm tubing close to the joint. The narrow tubing at each end is constricted at X (Fig. 12.2) to a capillary with a bore of about 1 mm. If samples containing much carbonate are to be analyzed, an absorption tube of larger diameter (1.5 cm) should be used for the collection of carbon dioxide.

The water absorption tube D (Fig. 12.2) is filled with anhydrous magnesium perchlorate. The carbon dioxide absorption tube E (Fig. 12.2) is packed with an 8-cm layer of soda asbestos (ascarite) and a 2-cm layer of anhydrous magnesium perchlorate. The stoppers of both absorption tubes are mounted in position with hard black wax. Connection of the absorption tubes is made with aged rubber tubing.

Chromium Trioxide Bubbler. With samples containing more than 0.5% sulfur the bubbler F (Fig. 12.2), filled with a saturated solution of chromium trioxide in 85% phorphoric acid, should be interposed between the water absorption tube D and the carbon dioxide tube E. Its side arm G is packed with magnesium perchlorate.

PROCEDURE

1. For the majority of rock and mineral samples the main furnace should be run at 1100°C. If minerals that are difficult to decompose, such as staurolite, are present, the temperature may be raised to 1200°C.

2. The flow of nitrogen should be regulated to about 3 liters an hour. Each day before use allow nitrogen to pass through the apparatus and absorption tubes for about 20 minutes. Remove the absorption tubes, wipe them carefully, and weigh them after 5 minutes.

3. Weigh 0.5 to 1.5 g of the sample (sieved −20 mesh) into a previously ignited 2-ml alumina boat lined with a piece of nickel foil. If much fluoride or sulfur is present, cover the sample with a layer of freshly ignited magnesium oxide.

4. Insert the boat into the end of the combustion tube, replace the insertion device, and allow nitrogen to sweep air out of the apparatus for 5 minutes.

5. Connect the weighed absorption tubes and then push the sample into the furnace with the stainless steel rod. When large amounts of readily decomposable carbonates, such as siderite and magnesite, are present, the boat should not be pushed immediately into the hot part of the furnace, for this leads to such rapid evolution of carbon dioxide that it is not completely absorbed. Such samples should be allowed to decompose in the cooler region just outside the furnace, and only after decomposition is nearly complete should the sample be subjected to the full temperature.

6. After heating for 30 to 40 minutes remove the absorption tubes; wipe them and weigh them after 5 minutes.

7. Carry out a blank determination in the same manner without the sample before the first determination and at the end of the work. The normal blank values for water and carbon dioxide are 0.1 and 0.2 mg/h, respectively. When higher blank values are found for the carbon dioxide, the copper tube is exhausted and should be treated as described above.

(Method after Riley, 1958.)

Determination of Total Water

METHOD

Apparatus. The apparatus is shown in Fig. 12.3. Plywood is used to make the support. The heat shield is made of two pieces of $\frac{1}{4}$-in. asbestos board screwed to the wooden support. A single asbestos board $\frac{1}{2}$-inch. thick will suffice. The overflow tray is plastic to minimize "sweating" of the dish caused by the overflow of ice-cold water. The cooling tray which holds the mixture of ice and water is aluminum but any rust-resistant metal would be suitable. The rubber supports are standard stoppers attached to the two trays with rubber cement.

The height of these supports should be such that, when the overflow tray is in position, the notches in the cooling tray are level with the lower edges of the two slots in the vertical supports of the tray holder. A slight downward inclination toward the open end of the tube is recommended to prevent any condensed water from flowing back into the fusion area of the tube.

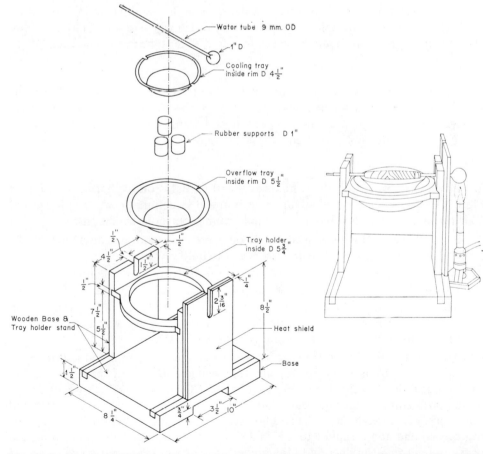

Fig. 12.3 The Penfield apparatus for total water determination. (Design after Courville, 1962.) See exercise 12.3 for details.

During operation the center of the tube is wrapped in a strip of cloth and the tube is supported in the cooling tray and slowly rotated during heating.

(Apparatus design by Courville, 1962)

Equipment

Penfield Tube. The Penfield tube is made of borosilicate tubing of 3 mm outside diameter, walls 1 mm thick, overall length 250 mm, and a bulb of approximately 2.5-cm diameter at one end. The walls of the bulb must be uniform in thickness and the open end must be free of " lip. " A large supply is required.

Funnel. A long-stemmed funnel which, when supported, will reach nearly to the bottom of the sample tube and can be easily inserted and withdrawn.

Capillary Plug. Heat and draw a length of capillary tubing to a short taper which, when cut off and covered with a piece of thin rubber tubing, will fit tightly into the open end of the water tube.

Tube Cap. Insert a short piece of glass rod into a $\frac{1}{2}$-in. length of firm rubber tubing with an internal diameter that will enable it to fit easily but firmly over the open end of the water tube.

Displacement Rod. This displacement device is a glass rod 20-cm long of 6 to 7 mm outside diameter, fine polished at both ends, which will slide easily into the water tube. When not in use, it should be stored in a used water tube to keep it clean.

Reagents

Lead Oxide Flux. Place a quantity of litharge (PbO) in a 90-ml nickel crucible and heat with a moderate flame Bunsen for 30 minutes, stirring frequently with a glass rod to crush any lumps. Cool in a dessicator and transfer to a tightly sealed bottle kept in the dessicator until required.

PROCEDURE

1. Dry a clean water tube for 2 hours at $+110°C$. Put it on the horizontal support (Fig. 12.3). Insert a narrow glass tube into it and, by suction, draw a gentle current of air through it until its interior is in equilibrium with room conditions.

2. Support the dry long-stemmed funnel on a vertical stand and insert it into the water tube so that the end of its stem is near the bulb center.

3. Weigh and transfer 1.000 g of the sample (or less if the water content is known to be greater than 1 %) to a small porcelain crucible and add 2 g of the dry lead oxide flux. Mix with a spatula and transfer via the funnel to the bulb of the water tube. Brush any rock or mineral powder into the funnel and from the walls of the funnel, helping it on its way by tapping the funnel stem with a spatula.

4. Add 0.5 g of the flux to the crucible, rinse the crucible with it and add it to the funnel. If the sample is known to contain appreciable fluorine, it is necessary to add CaO to it as follows: weigh 0.5 g reagent grade $CaCO_3$ into an ignited, weighed, and covered porcelain crucible and heat at $900°C$ for 30 minutes, cool in a dessicator and weigh rapidly (the weight of CaO should be near 0.28 g). Store, covered, in a dessicator until needed and add it rapidly to the final step in the loading of the water tube. Do not try to remove dust adhering to the funnel. If the rock or mineral contains only a small fluorine concentration, the CaO need not be added.

5. Lift the funnel so that the lower end of the stem is at the entrance to the bulb and tap the funnel again to dislodge any powder. Withdraw the funnel. Close the end of the water tube with a capillary plug.

6. Carefully raise the tube to a horizontal position so that no powder escapes from the bulb. Wrap the central 3 to 4 in. of the tube spirally with a strip of wet cloth (about $\frac{1}{2}$-in. wide). Place the tube in the cooling tray, which should contain crushed ice and water, so that the cloth is saturated in ice cold water. The bulb must project about $\frac{3}{4}$ in. beyond the heat shield.

7. Put a small Bunsen flame beneath the bulb. Rotate the bulb slowly to prevent sample caking. Gradually increase the flame to full heat to envelop the bulb (this must be done slowly over a 5-minute period to avoid explosive expulsion of volatiles). Start heating the bulb with the brushlike flame of an oxygen-propane handtorch, and while rotating the tube gradually increase the heat until the walls of the bulb begin to collapse. By this stage the sample and flux should be a molten mixture. Rotate the tube rapidly enough to prevent the bulb from sagging and heat it as strongly as possible. Collapse the walls of the bulb around the fused sample, starting at the bottom and working toward the orifice. This reduces the possibility of blocking the orifice with the fused material and causing the bulb to rupture.

8. Finally, with a sharp flame, fuse the tube at the junction with the bulb, and with the tongs draw off the fused end and discard it. Heat the sealed end of the tube briefly to round any sharp projections.

9. With the tube still horizontal, wipe the hot end cautiously with a wet cloth to cool it before pushing the wet wrapping over it. Place it in the ice bath, pat the other end dry with a lint-free cloth, remove the capillary plug, and quickly insert a clean and dry displacement rod. If there is a space remaining at the upper end of the tube, displace the air in it once or twice with the end of another displacement rod. Close the tube with a tube cap and carefully dry the outside of the tube by patting it with a lint-free cloth. *Do not* rub it, for friction produces static electricity and can cause serious errors in the measurements.

10. Place the tube in a horizontal rack and allow it to stand for 20 minutes near the balance before removing the tube cap and weighing the tube. The weighing is best done by supporting the tube on a wire saddle of known weight.

11. Carefully remove the displacement rod and dry it by heating it cautiously in a low flame. Dry the water tube in the same way. They may be left overnight in an oven at temperature $+110°C$ if this is more convenient. Cool the rod and tube while drawing a gentle current of air through the tube. Return the rod to the tube, wipe the tube with a damp cloth, and pat dry. Place the tube on a horizontal rack, allow to stand for 20 minutes near the balance, and weigh as before. The difference in the two weights equals the total water in the sample.

12. A blank of PbO with no sample and CaO, if added, should be run at regular intervals and always when a new batch of flux is used.

(Method after Maxwell, 1968.)

EXERCISE 12.4

Determination of H_2O-

1. Weigh 1 g of sample (or less if supply is limited) which has been ground to approximately 150 mesh and thoroughly mixed.

2. Transfer it to a clean, weighed, platinum, porcelain, or nickel 30-ml crucible.

3. Weigh the covered crucible and sample. The difference in weight of the sample between steps 1 and 3 should not exceed 0.2 mg.

4. Place the uncovered crucible in an oven, cover it with a 7-cm diameter filter paper, and heat it at 105 to 110°C for 1 hour.

5. Transfer the crucible to a dessicator, cover it, and allow it to cool for 30 minutes. Then weigh.

(After Maxwell, 1968.)

DISCUSSION

If the loss in weight exceeds 1 mg, the heating, cooling, and weighing should be repeated until a constant weight is obtained. If the loss in weight exceeds 5 mg, the crucible should be heated at a higher temperature, for example, 125°C, to see if further loss, which would indicate the presence of a significant amount of hydrous material, occurs.

When a significantly hygroscopic sample is to be weighed, it is essential that it be in equilibrium with the atmosphere. Spread the sample on a clean sheet of glazed paper and allow it to stand overnight (a paper canopy should be erected to protect the sample from dust). Then weigh all portions to be used in the analysis at the same time.

EXERCISE 12.5

Water Determination by Infra-Red Spectroscopy

Only small quantities of sample, less than 0.1 g, are required. Including the sample preparation, it is possible to complete 5 or 6 analyses in 1 hour. The unknowns are compared directly with a standard of known water content.

Powders are ground to pass 250 mesh and dried at 110 to 120°C for 24 hours. The same particle size is used as that in X-ray fluorescence analysis. Ideal particle size should be below 3 μ (3 microns)

$$\frac{\%H_2O}{\%SiO_2} = K \frac{\log I_1}{\log I_2},$$

where I_1 = intensity of the OH absorption band, I_2 = intensity of the 800/cm or 1100/cm Si-O band, and K is a constant that is determined by running a standard. From one instrument scan (e.g., an UNICAM SP 100 spectrometer) it is possible to obtain two determinations by measuring the absorption of the OH vibrational mode at 3450/cm and relating it to both Si-O vibrational modes, using percentage silica as an internal standard.

The powdered samples are mixed with nujol, the mull mounted between KBr windows, and the sample scanned between 650 and 4000/cm. The Si-O vibrational modes are resolved from nearby absorption peaks due to nujol.

Absorption due to free water (3450/cm) and combined water (3680/cm) can also be compared.

(After Aucott and Marshall, 1969.)

DISCUSSION

Breger and Chandler (1969), however, have shown that 1 hour is rarely sufficient for H_2O-determination by the infrared method and that overnight heating is preferable.

C. FLUORINE

Fluorine Determination in Rocks and Minerals

Fluorine is a difficult element to determine in silicate rocks and minerals. The following methods are fairly rapid and reasonably simple.

METHOD A

Apparatus

The fluoride distillation apparatus shown in Fig. 12.4 is assembled. A colorimeter (e.g., Bausch and Lomb, Spectronic 20) is set at 527.5 mμ. Absorption cell of 1 cm light path. Platinum crucibles of 30-ml size. Glass beads. Meker burner.

Reagents

Standard sodium fluoride solution: 0.221 g reagent grade NaF dissolved in 1 liter of distilled water (1 ml = 0.1 mg F). This solution is diluted 100-fold to provide a working solution. A sodium carbonate washing solution consists of 2% Na_2CO_3 aqueous solution. Sodium carbonate (reagent grade) and zinc oxide (reagent grade) are used for the fusion.

Perchloric Acid. Reagent grade $HClO_4$ (60%) is added to three or four volumes distilled water and boiled down to the original volume. The process is repeated and, after cooling, the acid is stored in a Pyrex bottle.

Reagent A. 1.800 g erichrome cyanine R dissolved in distilled water and diluted to 1 liter.

Reagent B. 0.265 g zirconyl chloride octahydrate dissolved in 50 ml distilled water and 700 ml concentrated HCl (reagent grade 1.19 specific gravity) are added to the zirconyl solution. The mixture is diluted to 1 liter with distilled water and cooled to room temperature before use.

Reference Solution. 10 ml of reagent A are added to 100 ml of distilled water. To this solution is added 10 ml of a hydrochloric acid solution prepared by dilution of 7 ml of the concentrated acid (sp. gr. 1.19) to 10 ml with distilled water. The reference solution must be used because of the high absorbance from unreacted eriochrome cyanine R dye in the sample solution.

PROCEDURE

1.   A 50-mg sample of rock or mineral is added to 2.0 g Na_2CO_3 and 0.5 g ZnO in a platinum crucible. The sample is thoroughly mixed with the Na_2CO_3 and ZnO by rotation of the crucible with the fingers.

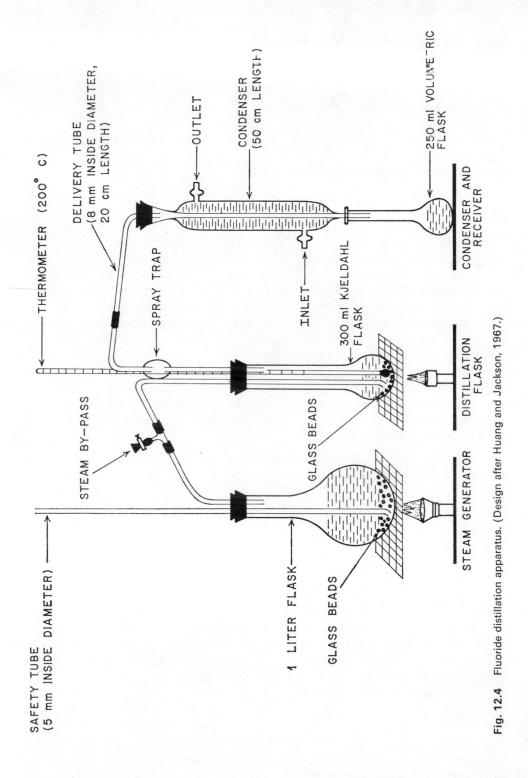

SAFETY TUBE
(5 mm INSIDE DIAMETER)

THERMOMETER (200° C)

DELIVERY TUBE
(8 mm INSIDE DIAMETER,
20 cm LENGTH)

OUTLET

CONDENSER
(50 cm LENGTH)

250 ml VOLUMETRIC
FLASK

STEAM BY-PASS

SPRAY TRAP

INLET

300 ml KJELDAHL
FLASK

CONDENSER AND
RECEIVER

1 LITER FLASK

GLASS BEADS

GLASS BEADS

DISTILLATION
FLASK

STEAM GENERATOR

Fig. 12.4 Fluoride distillation apparatus. (Design after Huang and Jackson, 1967.)

2. Approximately 0.5 g Na_2CO_3 are added on top of the mixture. The crucible is placed in a slanting position on a silica-covered triangle with the lid covering about 0.7 of the top and a low Meker burner flame is applied. The heat is gradually increased until the contents are liquified. The fusion is completed in 30 minutes.

3. The crucible and contents are cooled and placed in a 100-ml beaker and 25 ml of distilled water are added. The beaker is covered and heated on a steamplate until the cake is softened.

4. The crucible and lid are washed and removed after washing down with a wash bottle. The volume of the solution should now be about 50 ml. Any lumps are broken up with a stirring rod flattened at one end and the solution is brought to boiling for several minutes over a burner; stirring is continuous to prevent lumping.

5. The digested cake suspension is allowed to cool and then transferred quantitatively to the distillation flask (Fig. 12.4). (As an alternative for layer silicates, the digested cake suspension is allowed to settle and as much as possible of the supernatant solution is decanted into the distillation flask. The residue is washed with three portions of hot 2% Na_2CO_3 solution, transferred to a filter paper, and again washed with three portions of the same solution. The solution collected from the washings is poured into the flask and the residue is discarded. Recovery of F in silicates is complete by this procedure but that from apatite is not). Distilled water in quantities of 15 to 20 ml are used to make the transfer. The total volume at this point should be about 70 ml.

6. Twenty milliters of 60% perchloric acid is added slowly to the flask, which is kept cool by immersion in cold water; 8 to 10 glass beads are added. The distillation flask is connected to the condenser (Fig. 12.4) and the steam generator and the contents of the flask are mixed by gentle swirling.

7. Heat is applied to the steam generator and the flask with the bypass open. When the temperature of the liquid in the flask reaches 135°C, steam is introduced by closing the bypass. The temperature is increased to 140°C and 150 ml of distillate is collected at the rate of about 5 ml per minute.

CAUTION. The flask temperature must be maintained between 130 and 140°C by adjusting the flame underneath the flask; the upper limit in particular must not be exceeded. The flame underneath the distillation flask should touch no part of the flask that is not in contact with the liquid inside. Local superheating causes the decomposition of the distilling acid and results in the contamination of the distillate. The thermometer bulb and the steam inlet tube should remain immersed in the boiling liquid all through the distillation.

8. On completion of the distillation the flames are removed and the steam bypass is opened. The distillate is diluted to 250 ml with distilled water and mixed thoroughly.

9. After each run the flask and the attachments are thoroughly rinsed with distilled water, treated with 10% NaOH, and finally rinsed again thoroughly with distilled water. The apparatus should be free of contamination and all connections must be leakproof. This can be checked periodically by determinations on the standard fluoride solution.

10. An aliquot of distillate, containing no more than 50 μg of F, is diluted to 50 ml. Then 5 ml of reagent A, followed by 5 ml of reagent B, are added to the sample and mixed well.

11. The colorimeter is set to zero absorbance at 527.5 mμ and the reference solution and absorbance of the sample solution are recorded. The fluorine value of the sample aliquot is determined from a curve prepared by subjecting standard solutions of fluoride to the above procedure and linearly plotting the absorbance values against F standards in the range of 0 to 50 μg of fluorine in 50 ml. The resultant plot is a straight line of negative slope, since zirconyl ions are withdrawn from the colored complex in proportion to F present, thus forming $ZrOF_2$. Because the color reaction reaches equilibrium rapidly and is stable at ordinary temperature, the colorimeter reading can be taken immediately after mixing the reagents. A new calibration curve is prepared at the same time that each set of samples is analyzed. A blank is carried with all reagents used in the procedure.

(Method after Huang and Jackson, 1967)

DISCUSSION

Recovery of F from minerals such as hornblende and apatite is within ± 1 to 3% of the total F present or added as standard NaF. This kind of accuracy is satisfactory for semimicrochemical analysis.

METHOD B

This method is suitable for the analysis of silicate rocks and minerals containing moderate to large amounts ($> 0.1\%$) of fluorine.

Reagents

   Alizarin Fluorine Blue Reagent. Dissolve 25 g of alizarine fluorine blue lanthanum complex preparation (Hopkin and William Ltd., Chadwell Heath, Essex, England) in a mixture of 150 ml of propan-2-ol and 350 ml of distilled water. Filter before use.

   Standard Fluoride Solution. Dissolve 0.0663 g of sodium fluoride in 1 liter of distilled water to give a solution containing 30 μg of fluorine per milliliter.

PROCEDURE

1.   Weigh out exactly 1 g of powdered rock or mineral into a platinum crucible, add six times the sample weight of anhydrous sodium carbonate (AR grade), and mix. Cover the crucible.

2.   Fuse over a Meker burner for 20 minutes. Digest the fusion cake in 100 ml of hot distilled water, filter through Whatman 41 filter paper, and wash the residue with hot water.

3.   Collect the filtrate in a 250-ml beaker. Add approximately 2 g of powdered ammonium carbonate (AR grade), digest on a water bath for 30 minutes, allow to cool, and add another 1 g of ammonium carbonate. Allow to stand for 12 hours.

4.   Filter the mixture through Whatman 41 filter paper into a 1-liter conical flask and wash with dilute ammonium carbonate solution. Add a few drops of methyl orange to the filtrate, and then make the solution just acid by the careful addition of $1 + 1$ hydrochloric acid, stirring vigorously after the effervescence has ceased.

5.   Make the solution up to 500 ml in a graduated flask and store in a polythene bottle until ready for the colorimetric determinations.

6.   Transfer 1 ml of solution to a 50-ml volumetric flask. Add 20 ml of alizarin fluorine blue reagent solution and dilute to volume.

7.   Pipette 1-ml aliquots of standard fluoride solution (containing 30 μg of fluoride) into 50, 100, and 200 ml graduated flasks and treat with corresponding amounts of reagent solution.

8.   Prepare a reagent blank consisting of 20 ml of reagent solution diluted to 50 ml. Allow to stand for 1 hour and then measure the optical densities of the solutions at 630 nm in 1-cm cells. Use distilled water in the reference cell.

9.   Plot a calibration curve from the measurements of the blank and standard solutions. The reagent blank has a high optical density and the similar appearance of unknown, standard, and reagent blank solutions need not cause concern. The reagent is a reddish-purple color, which becomes bluish-purple in the presence of large amounts of fluorine.

(Method after Hall and Walsh, 1969.)

DISCUSSION

The method has been found to give complete recovery of F from basic igneous rocks, granodiorite, aplite, biotite, and muscovite, and no interference is found by other elements. The method is also successful for amphibole F analysis.

Total Chlorine in Silicates

The recommended procedure for chlorine determination by an X-ray fluorescence method was given in exercise 10.22.

The gravimetric determination is tedious and not good for trace amounts. For details of the method refer to Maxwell (1968), p. 447.

A. GRAPHICAL

Triangular Variation Diagrams: General Description

Triangular diagrams are used widely in geology to display the composition of a mineral or rock in terms of three selected components. Figure 13.1 is an example of a diagram in which the three components are named A, B, and C. In an actual example A may represent the wt $\%$ Al_2O_3, B, the wt $\%$ CaO, and C, the weight $\%$ MgO in a rock or mineral. In another case A may represent the modal amount of quartz; B, that of K-feldspar; and C, that of plagioclase in a granite. The triangle need not be equilateral but it usually is for ease in plotting the data.

The particular point plotted on this figure has the value $A_{20}B_{30}C_{50}$; that is, in terms of only these components the mineral or rock represented by this point contains the three components in the ratio $A:B:C$ 20:30:50. It is not said whether the ratio is in terms of weight or volume, but it must be clearly stated in an actual example. If A represented Al_2O_3 by weight, the position of this point would NOT mean that the rock or mineral contains 20% Al_2O_3 but that the weight ratio of Al_2O_3 to the components B and C is 20% $(A + B + C = 100\%)$.

The base line BC represents all points with 0% A and point A represents 100% A. Thus point $A_{20}B_{30}C_{50}$ is two lines up from the base BC when each line is spaced at a 10% interval. It is customary on triangular diagrams to have dark lines at 10% intervals and finer lines in between at either 1 or 2% intervals. Hence between the base line BC and the 20% A line would be faint lines representing all intermediate values. Similarly the 50% C line is parallel to the AB base line and the 30% B line parallel to AC.

Plotting. In the absence of ruled composition lines, the point may be plotted by taking the ratios of any two components in turn.

The A/B ratio is given as 20/30. Hence a point may be plotted on the AB base line to represent this ratio; it is plotted at a distance $20/(20 + 30) \times 100 = 40$ divisions from B (and $100 - 40 = 60$ from A). This point may be joined up by a straight line to apex C. All points that lie anywhere on this line have a ratio $A/B = \frac{20}{30} = \frac{2}{3}$.

The B/C ratio is given as 30/50. The point is plotted on the BC base at a distance $30/(30 + 50) \times 100$ from C $(=37.5$ divisions from C). The line drawn from this point to apex A represents all points with a ratio $B/C = \frac{3}{5}$.

The A/C ratio = 20/50. A point is constructed at a distance $20/(20 + 50) \times 100$ divisions = 28.57 from C and is joined to apex B to represent all points with an A/C

EXAMPLE   A 20%
          B 30%
          C 50%

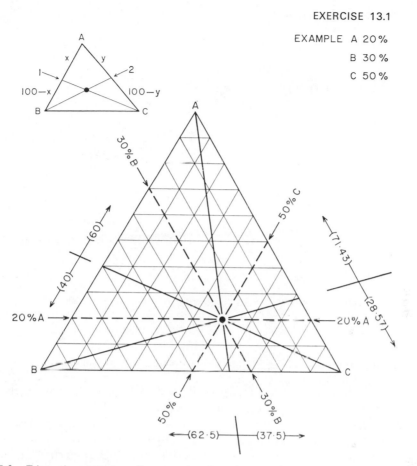

Fig. 13.1 Triangular variation diagram showing the position of a point of composition $A_{20}B_{30}C_{50}$. See exercise 13.1 for description.

ratio of $\frac{2}{5}$. All these three lines intersect at the required point. Of course, in an actual case only two lines need be drawn.

Reading. Occasionally, as in the garnet diagram after Winchell (1958) shown in Figs. 7.13, 7.14, and 7.15, the composition of the point must be read from its position on the triangular diagram. Note that these triangles are not equilateral, but this fact has no effect on the method given here.

The small upper diagram in Fig. 13.1 outlines the method:

1. Draw a straight line from apex C through the point and let it cut the base line AB in point 1. Express point 1 in terms of a percentage distance from $A = x\%$ and from $B\ 100 - x\%$.

2. Draw a line through the point to cut the base line AC in point 2 and express it in terms of a percentage distance from $A = y\%$ and $C\ 100 - y\%$.

The following relationship can be easily shown to hold, where $A + B + C = 100$ total.

$$A\% = \frac{100}{1 + x/(100 - x) + y/(100 - y)}.$$

In our example, $x = 60$, $y = 71.43$. Therefore

$$A\% = \frac{100}{1 + 60/40 + 71.43/28.57} = \frac{100}{1 + 1.5 + 2.5} = \frac{100}{5} = 20\%.$$

Similarly,

$$B\% = \frac{100}{1 + 40/60 + 62.5/37.5} = \frac{100}{3.33} = 30\%.$$

and

$$C\% = \frac{100}{1 + 37.5/62.5 + 28.57/71.43} = \frac{100}{2} = 50\%,$$

giving the point as $A_{20}B_{30}C_{50}$.

EXERCISE 13.2

Graphical Representation of the Metamorphic Paragenesis of Carbonate Rocks

The composition of siliceous carbonate rocks and their constituent minerals can be expressed in terms of the three components CaO, MgO, and SiO_2.

The whole rock and the mineral constituent compositions of any metamorphic assemblage may be graphically displayed on a triangular diagram for which the corners represent the molecular proportions of SiO_2, CaO, and MgO, respectively (Fig. 13.2).

The positions of constituent minerals are plotted directly on the diagram according to their molecular proportions of SiO_2:CaO:MgO recalculated to a percentage. Figure 13.2A shows the positions of the majority of the common minerals which occur in metamorphic calcareous rocks. Their positions are calculated as follows:

Mineral	Formula	Molecular Percentage Ratio		
		SiO_2	CaO	MgO
Quartz	SiO_2	100	0	0
Calcite	$CaCO_3 = CaO.CO_2$	0	100	0
Magnesite	$MgCO_3 = MgO.CO_2$	0	0	100
Dolomite	$CaMg(CO_3)_2 = CaO.MgO.(CO_2)_2$	0	50	50
Diopside	$CaMg(SiO_3)_2 = CaO.MgO.2SiO_2$	50	25	25
Wollastonite	$CaSiO_3$	50	50	0
Forsterite	$Mg_2SiO_4 = 2MgO.SiO_2$	33	0	67
Grossularite	$Ca_3Al_2Si_3O_{12} = 3CaO.3SiO_2.Al_2O_3$	50	50	0
Andradite	$Ca_3Fe_2Si_3O_{12}$	50	50	0
Talc	$Mg_6(Si_8O_{20})(OH)_4$	57	0	43
Vesuvianite	$Ca_{10}(Mg,Fe)_2Al_4(SiO_7)_2$ $(SiO_4)_5(OH,F)_4$	43	47.5	9.5
Tremolite	$Ca_2(Mg,Fe)_5(Si_8O_{22})(OH,F)_2$	53.4	13.3	33.3

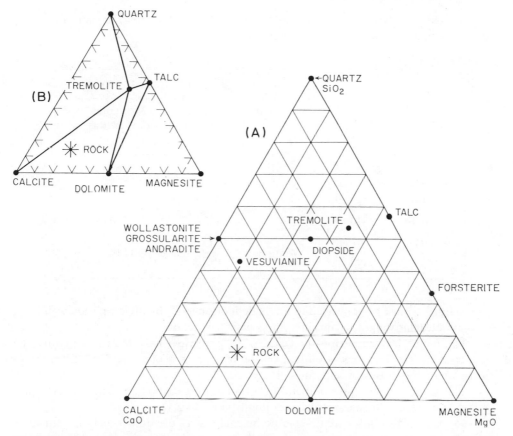

Fig. 13.2 SiO_2-CaO-MgO diagram for graphical representation of metamorphic mineral paragenesis in siliceous carbonate rocks. A = positions of the common metamorphic minerals; B — mineral paragenesis in the albite-epidote hornfels or greenschist facies.

By way of example the formula for tremolite may be rewritten as

	Molecular Proportion ($\times 6.67$)
2 CaO	13.3
5 (Mg, Fe)O[a]	33.3
8 SiO_2	53.4
15, multiplied by $\dfrac{100}{15}$ or 6.67	

[a] Assuming that the amount of FeO is negligible; all has been ascribed to MgO.

These minerals have been plotted on Fig. 13.2 according to the values of SiO_2: CaO:MgO as they have been calculated.

A rock of known chemical analysis may be plotted on the same diagram (Fig. 13.2) As an example, a limestone gave the following analysis:

	1	2	3(=1/2)	4(=3 × 90.114)
	Weight %	Molecular weight	Molecular Proportions	Molecular Proportions %
CaO	39.22	56.08	0.6994	63
SiO_2	10.20	60.09	0.1697	15
MgO	9.70	40.32	0.2406	22
Al_2O_3	10.10		1.1097	100
CO_2	30.79	Multiply by $\dfrac{100}{1.1097}$; that is, 90.114		
	100.00			

The weight % must be converted to an oxide molecular % by dividing each weight % by the molecular weight to obtain a molecular proportion.

The molecular proportions of CaO, SiO_2, and MgO have then to be converted to a total of 100. The rock is plotted on the triangular diagram as CaO : SiO_2 : MgO (molecular proportions) which equal 63 : 15 : 22.

The original sedimentary mineralogy of an impure carbonate rock may be represented by the triangle quartz : calcite : dolomite (Fig. 13.2A).

Metamorphism in the albite-epidote hornfels facies of contact metamorphism or in the greenschist facies of dynamo-thermal metamorphism gives rise to the common mineral assemblages calcite + tremolite + quartz or calcite + tremolite + dolomite. These common assemblages are shown on a facies diagram (Fig. 13.2B) and are indicated by solid tie lines between important mineral combinations. An assemblage of three outlines a triangle.

Also possible in this facies is the assemblage talc + tremolite + quartz and talc + tremolite + dolomite, but because they exist only in rocks of unusual chemistry they will be rather restricted in natural occurrence.

The rock whose composition is plotted in Fig. 13.2 would indicate that metamorphism to the facies of Fig. 13.2B would result in the assemblage calcite + dolomite + tremolite, for its plot lies in that triangle.

With increasing grade of metamorphism, for example, to the hornblende hornfels facies, tremolite becomes unstable and diopside and forsterite become stable phases. The tie lines are now between calcite, diopside, and forsterite, and the rock lies in the paragenetic triangle between these three minerals.

EXERCISE 13.3

The ACF diagram for representing metamorphic paragenesis

The *ACF* diagram is the most generally used to describe mineralogical and chemical variation in metamorphic rocks.

General Description. The ACF diagram is constructed for rocks with excess SiO_2, which crystallize with free quartz. The great majority of metamorphic rocks contain free quartz; hence the ACF diagram for excess SiO_2 is applicable, in which case SiO_2 is not shown as a component and it can be assumed that quartz is always a possible phase. All the mineral phases shown on the ACF facies diagrams are based on the availability of excess SiO_2. If the rocks in question are deficient in SiO_2, a modified ACF diagram showing alternative mineral phases will have to be constructed; for example, in silica-deficient rocks olivine (forsterite) may be a phase but it is never shown in the standard ACF (excess SiO_2) diagrams. Likewise, there is no place in the ACF diagram for sodium and potassium; hence alkali silicates are excluded. This presents a problem for plotting muscovite, paragonite, and biotite micas and for feldspars, which Turner (1968) and Winkler (1967) have unsuccessfully tried to solve.

The A corner represents the molecular $Al_2O_3 + Fe_2O_3$ (Fe_2O_3 is added to Al_2O_3 because normally Fe^{3+} can substitute for Al^{3+}) which is not combined with Na and K; $Al_2O_3 + Fe_2O_3$ which are combined with Na and K, as in the feldspars and the micas, are excluded from the A value; that is,

$$A = Al_2O_3 + Fe_2O_3 - (Na_2O + K_2O).$$

The C corner is the molecular % CaO, corrected for apatite, and for calcite in non-carbonate rocks

$$C = CaO - 3.3 (P_2O_5).$$

The F apex is molecular % $FeO + MgO + MnO$; $A + C + F = 100$.

The accessory minerals are excluded; hence before calculating the ACF values the amounts of A, C, and F components in the accessories are subtracted from the rock chemical analysis.

Strictly speaking, the ACF diagram is not suitable for inclusion of the micas (with the exception of margarite), but many metamorphic rocks contain biotite and muscovite (and perhaps paragonite). Hence some authors (e.g., Turner, 1968) have adopted the convention of plotting muscovite + paragonite at the A apex and biotite at or near the F apex. If this procedure is followed, the calculation of the ACF values differs from that when the micas are excluded from the ACF diagram (e.g., Winkler, 1967).

METHOD

1. From thin sections of the rock estimate or measure the modal content (volume %) of the following accessory minerals: ilmenite, sphene, magnetite, hematite, apatite, pyrite.

2. From thin sections measure the modal % of the following major minerals: hornblende (total amphibole), biotite, white mica (total of muscovite, margarite, and paragonite). No attempt should be made to distinguish the various white micas because there is no clear distinction based on optical properties.

3. Convert the modal volume % of each of the minerals determined above to a weight %. Weight % = modal % × specific gravity. The following specific gravities may be used: magnetite, 5.2; biotite, 3.1; hornblende, 3.3; apatite, 3.1; sphene, 3.5; pyrite, 5.0; ilmenite, 4.7; hematite, 5.3; and white micas, 2.9. Some of these are average values, but we have no alternative than to use only an approximately correct value, for greater precision is both impossible and unwarranted.

4. From the chemical analysis of the rock subtract the following as corrections for the accessory minerals:

From FeO weight %,   50% of the ilmenite weight %,
 30% of the magnetite weight %,
 60% of the pyrite weight %.

From Fe_2O_3 weight %,   70% of the magnetite weight %
 100% of the hematite weight %.

From CaO weight %,   30% of the sphene weight %.

5. In a subsequent correction it is assumed that all the alkalis in the rock are present in the K-feldspar and the albite molecule of the plagioclase. For rocks containing hornblende this is not strictly correct, for hornblende contains a variable amount of Na + K. In steps 2 and 3 the weight percentage of the rock which is composed of hornblende (amphibole) has been determined $(=x\%)$. To ascertain how much $Na_2O + K_2O$ is present in the hornblende an analysis of separated hornblende for Na_2O % $(=n)$ and for $K_2O\%$ $(=k)$ has to be determined. The $Na_2O\%$ of the rock has to be reduced by subtracting $xn/100$ from it, and the K_2O rock percentage has to be reduced by subtracting $xk/100$.

The following steps apply if the micas biotite and muscovite are to be included.

6. It is now necessary to make corrections in the amounts of Al_2O_3 and Fe_2O_3 which occur in mineral phases that are not represented on the *ACF* diagram. These phases are the alkali minerals—albite and K-feldspar. These corrections can be made by subtracting an amount of Al_2O_3 and Fe_2O_3 which combines with K_2O and Na_2O in the feldspars. The micas, however, also contain important concentrations of alkalis so that when the micas are plotted on the *ACF* diagram we must first determine how much alkali is present in the micas and subtract this amount from the total alkalis to find out how much is assumed to belong to the feldspars. This is not an easy task, as we shall see.

7. The $K_2O:(Al_2O_3 + Fe_2O_3)$ ratio in biotite is not fixed and generally can vary from $1:1$ to $1:2$ molecular. We cannot therefore make any assumption regarding the biotite in the rock, and the only way to proceed is to extract some biotite and analyze it for K_2O weight % $(=y)$. In step 3 the weight % of biotite in the rock was determined; let it be x by weight. The weight % of K_2O in the rock which is occurring in the biotite is therefore $xy/100$. This amount is subtracted from the whole rock weight percentage K_2O.

8. A similar correction has to be made for the presence of muscovite and paragonite in the rock. The white mica must be extracted from the rock and analyzed separately for Na_2O weight %$(=n)$ and for K_2O weight % $(=k)$. In step 3 the weight % of white mica in the rock was determined (let it equal x % by weight). Since we do not know the relative amounts of muscovite and paragonite, we will have to assume that they occur in the weight ratio muscovite/paragonite $= k/n$. As a correction for muscovite in the rock, subtract from the total rock weight % K_2O the amount $xk/100$. As a correction for the presence of paragonite, subtract from the total rock weight % Na_2O the amount $xn/100$.

9. The weight percentages of the various oxides in the rock, after correction for accessory minerals, hornblende, and the micas in the preceding steps, are now calculated as molecular proportions by dividing each by its oxide molecular weight. The molecular proportions so obtained are not yet converted to a percentage. This will be done later. SiO_2 and H_2O may be disregarded because they are not used in the calculations.

10. The molecular proportion of CaO is corrected for calcite and apatite by subtracting from it a molecular amount equal to $3.3 (P_2O_5) + CO_2$. However, for calcareous rocks containing modal calcite and/or dolomite as principal phases it is customary to omit the CO_2 correction.

11. A correction is now made for the absence of albite and K-feldspar from the *ACF* diagram. Both have $Na_2O:Al_2O_3$ and $K_2O:Al_2O_3$ molecular ratios of 1:1. The total amount of Na_2O and K_2O molecular proportions is assumed to be present in the feldspars. Therefore the molecular proportions $Na_2O + K_2O$ are added and an amount equal to their sum is subtracted form the molecular proportion of $Al_2O_3 + Fe_2O_3$.

12. The final molecular proportions are grouped as follows:

$$Al_2O_3 + Fe_2O_3 \qquad = A$$
$$CaO \qquad\qquad\qquad = C$$
$$MgO + MnO + FeO = F$$

and are recalculated so that $A + C + F = 100$. These are the values used to plot the rock on an *ACF* diagram on which muscovite and paragonite plot at the *A* apex and biotite at or near the *F* apex.

(After Turner, 1968, with corrections.)

The following steps apply if the micas muscovite and biotite are to be excluded.

Follow steps 1 through 5 as above.

6. Since the micas are to be excluded, the $Al_2O_3 + Fe_2O_3$ which occurs in the micas must be subtracted from the rock analysis. The $K_2O:(Al_2O_3 + Fe_2O_3)$ ratio in biotite is not fixed and generally can range from 1:1 to 1:2. Because this ratio is unknown, we cannot correct for biotite unless a chemical analysis is performed on biotite separated from the rock. Let us assume that the biotite is separated and analyzed for

$$Al_2O_3 \text{ weight } \% = a$$
$$Fe_2O_3 \text{ weight } \% = f$$

In step 3 the weight % biotite in the rock was determined: let it $= x$. The rock Al_2O_3 weight % is now reduced by subtracting $xa/100$. The rock weight % of Fe_2O_3 is reduced by subtracting $xf/100$.

7. The rock Al_2O_3 content must now be corrected for the presence of muscovite and paragonite. The thin section study will have given (steps 2 and 3) a weight % of the total white mica (muscovite + margarite + paragonite) in the rock $= x\%$ by weight. Although muscovite has a molecular ratio of $K_2O:Al_2O_3$ 1:3 and paragonite has a molecular ratio of $Na_2O:Al_2O_3$ 1:2, we cannot make the assumption that the ratio of alkalis to Al_2O_3 is fixed because of the possibility of considerable substitution of Fe_2O_3 for Al_2O_3. Hence, ideally, the white mica in the rock has to be separately analyzed. This, of course, is necessary to determine the amount of muscovite and paragonite. If margarite is present, it is assumed to have all the calcium, with a molecular ratio $CaO:Al_2O_3 = 1:2$. Paragonite is assumed to have all the Na and muscovite all the K. Let the determined weight % of the analyzed white mica be $CaO\% = c$;

$Na_2O\% = n$, $K_2O\% = k$, $Al_2O_3\% = a$, $Fe_2O_3\% = f$. By assuming the perfect separation of K, Na, and Ca to the three white micas (not strictly correct, for muscovite may contain some Na and Ca), the weight ratio of muscovite–paragonite–margarite could be calculated by noting the relative amounts of $k:n:c$ in the analysis. Of the three white micas only margarite will plot on the ACF diagram. An amount of Al_2O_3 has therefore to be left in A for the formation of margarite. The molecular ratio in margarite $Al_2O_3:CaO = 2:1$, which is equivalent to a weight ratio of $3.6:1$. Hence the Al_2O_3 weight $\%$ of the white micas as determined here has to be reduced by an amount $= 3.6 \times c$, and the final $a = a' - 3.6c$. To correct for the muscovite and paragonite subtract from the rock Al_2O_3 weight $\%$ an amount equal to $xa/100$ and subtract from the rock weight $\%$ Fe_2O_3 an amount equal to $xf/100$.

8. The calculation is continued by following steps 9, 10, 11 and 12, exactly in the above scheme.

(After Winkler, 1967, with corrections.)

DISCUSSION

It will be seen from the foregoing discussions that for rocks containing hornblende and micas the calculation of the ACF values necessitates a separate chemical analysis of each mineral phase. Both Turner (1968) and Winkler (1967) have wrongly implied that ACF values can be calculated without this knowledge.

Because white mica is usually called muscovite, the chemical analysis in steps 7 and 8 for CaO, K_2O, and Na_2O will give some indication that margarite and paragonite are present in addition to muscovite. They can be confirmed subsequently by X-ray diffraction analysis of the separated white mica.

ACF rock values are not easy to calculate. The majority of metamorphic rocks contain at least one of hornblende, biotite, or white mica, and separate chemical analyses are required of each of these groups.

In summary, however, it is extremely doubtful if all this effort is really warranted, for the accurate plotting of a rock on an ACF diagram has uncertain petrographic usefulness.

Plotting of minerals on the ACF diagram

The minerals that occur in metamorphic rocks are rather easily plotted on ACF diagrams by calculation of their molecular proportions of A, C, and F: they are listed and their ACF values given on page 384. These minerals are plotted in the ACF diagram in Fig. 13.3 in their appropriate positions.

Use of the ACF Diagram

A certain assemblage of these minerals characterizes a particular metamorphic facies, resulting from a particular range of temperature and pressure conditions. Figure 13.3 gives the example of the staurolite-almandine subfacies of the almandine amphibolite facies of dynamothermal metamorphism. Only a selection of the minerals listed is stable in this range of temperature and pressure, and only these minerals are shown on the ACF facies diagram. Tie lines on the diagram are coexisting mineral pairs. The tie

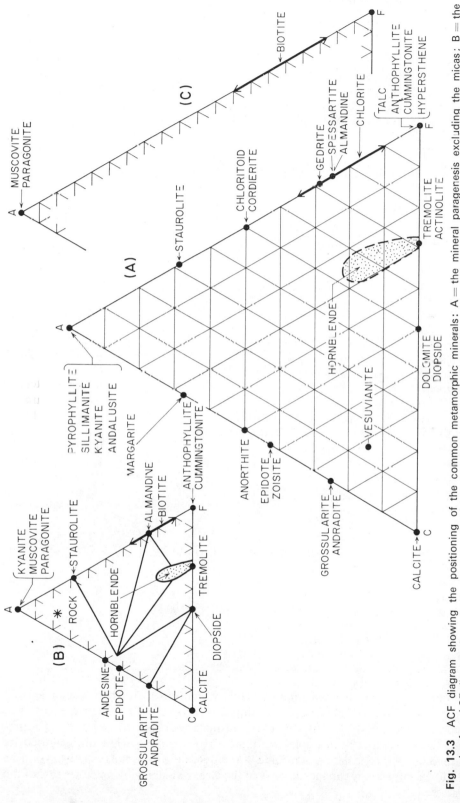

Fig. 13.3 ACF diagram showing the positioning of the common metamorphic minerals: A = the mineral paragenesis excluding the micas; B = the example of an ACF diagram for the staurolite-almandine subfacies of the almandine amphibolite facies; C = additional mineral phases on the AF side of the diagram when the micas are included.

383

	A	C	F
Talc $Mg_6(Si_8O_{20})(OH)_4$	0	0	100
Anthophyllite $(Mg, Fe)_7(Si_8O_{22})(OH, F)_2$	0	0	100
Margarite $Ca_2Al_4(Si_4Al_4O_{20})(OH)_4$	67	33	0
Cummingtonite $(Mg, Fe)_7(Si_8O_{22})(OH)_2$	0	0	100
Gedrite $(Mg, Fe)_5Al_2(Si_6Al_2O_{22})(OH, F)_2$ $= 5(MgO + FeO); 2(Al_2O_3)$	29	0	71
Hypersthene $(Mg, Fe)(SiO_3)$	0	0	100
Tremolite $\Big\}$ $Ca_2(Mg, Fe)_5(Si_8O_{22})(OH, F)_2$ Actinolite	0	29	71
Dolomite $CaMg(CO_3)_2$	0	50	50
Diopside $CaMg(SiO_3)_2$	0	50	50
Calcite $CaCO_3$	0	100	0
Grossularite $Ca_3Al_2Si_3O_{12} = 3CaO; 1Al_2O_3$	25	75	0
Andradite $Ca_3(Fe, Ti)_2Si_3O_{12} = 3 CaO;$ $1(Al_2O_3 + Fe_2O_3)$	25	75	0
Vesuvianite $Ca_{10}(Mg, Fe)_2Al_4(Si_2O_7)_2(SiO_4)_5$ $(OH, F)_4 = 10 CaO; 2(FeO + MgO); 2Al_2O_3$	14	72	14
Epidote $Ca_2(Al_2Fe)_3Si_3(O, OH, F)_{13}$	43	57	0
Zoisite $Ca_2Al_3Si_3(O, OH, F)_{13}$	43	57	0
Anorthite $CaAl_2Si_2O_8$	50	50	0
Pyrophyllite $Al_4(Si_8O_{20})(OH)_4$	100	0	0
Sillimanite, kyanite, andalusite Al_2SiO_5	100	0	0
Staurolite $(Fe, Mg)_2(Al, Fe)_9O_6(SiO_4)_4(OH, O)_2$ $= 2(FeO + MgO); 4\frac{1}{2}(Al_2O_3 + Fe_2O_3)$	69	0	31
Chloritoid $(Fe, Mg, Mn)_2(Al, Fe)Al_3O_2(Si O_4)_2$ $(OH)_4 = 2(FeO + MgO + MnO); 2(Al_2O_3 + Fe_2O_3)$	50	0	50
Cordierite $Al_3(Mg, Fe)_2(Si_5AlO_{18})$	50	0	50
Spessartite $Mn_3Al_2Si_3O_{12}$	25	0	75
Almandine $Fe_3Al_2Si_3O_{12}$	25	0	75
Chlorite $(Mg, Al, Fe)_{12}[(Si, Al)_8O_{20}](OH)_{16},$ variable A and F ratio between	(10 to 35)	0	(90 to 65)
Hornblende $(Na, K)_{0-1}Ca_2(Mg, Fe, Fe, Al)_5$ $(Si_{6-7}Al_{2-1}O_{22})(OH, F)_2;$ the composition is variable, roughly between and therefore hornblende occupies an elongate field	(12 to 20)	(25 to 27)	(53 to 63)
Muscovite $K_2Al_4(Si_6Al_2O_{20})(OH, F)_4$	100	0	0
Paragonite $Na_2Al_4(Si_6Al_2O_{20})(OH, F)_4$	100	0	0
Biotite $K_2(Mg, Fe)_{6-4}(Fe, Al, Ti)_{0-2}$ $(Si_{6-5}Al_{2-3}O_{20})(OH, F)_4$ gives variable ratios of $(MgO + FeO)$ and $(Al_2O_3 + Fe_2O_3)$	(14 to 39)	0	(86 to 61)

lines outline triangles between three minerals which make common assemblages in this particular facies or subfacies; for example, if the rock ACF value were to plot in the upper triangle (Fig. 13.3B), this rock, on metamorphism to this particular subfacies, would be expected to display the mineralogy kyanite + white mica (muscovite + paragonite) + plagioclase + staurolite + (epidote). If a rock plots exactly on or near a tie line, its mineralogy can include that of the two triangles lying on either side of the tie line.

Occasionally in a facies diagram not all tie lines outline triangles, but they may cross one another; for example, in Fig. 13.3 we may have a tie line between andradite and hornblende crossing one from andesine to diopside. The crossing of tie lines may have several alternative explanations:

a.   The phase assemblages, which are linked up by crossing tie lines, may represent disequilibrium.

b.   They may represent univariant equilibrium; for example, a range of pressure and temperature on a conventional reaction curve.

c.   They may represent divariant equilibrium, in which case the crossing of the ties do not express any precise chemical equation. This could be true if the ratio MgO/FeO in the F component was markedly different in the two coexisting phases. The ACF diagram does not show this kind of information, which is better illustrated by the AFM diagram.

EXERCISE 13.4

The A'KF diagram for representing metamorphic paragenesis

Next to the ACF diagram, the $A'KF$ diagram is commonly used to depict mineralogical and chemical variation in metamorphic rocks. It is advantageous to display it in conjunction with the ACF diagram to which it is considered complementary.

General Description. The $A'KF$ diagram is constructed for rocks with an excess of Al_2O_3 and SiO_2, all the rocks portrayed should therefore have the possibility of free feldspars and quartz as constituents. The potassium-bearing minerals (K-feldspar, muscovite, biotite, and stilpnomelane) are represented together with minerals containing (MgO + FeO) and (Al_2O_3 + Fe_2O_3), whereas CaO-bearing minerals cannot be represented. The diagram is based on molecular proportions. The general scheme is as follows:

$A' = Al_2O_3 + Fe_2O_3$ corrected for minerals which combine Na_2O, K_2O, and CaO with Al_2O_3 and Fe_2O_3.

$K = K_2O$. (K-feldspar is given the value 100 K.)

$F = FeO + MgO + MnO$ corrected for minerals containing the F component which are not represented.

$$A' + K + F = 100.$$

METHOD

1.   By thin-section analysis estimate modally the volume % of each of the following accessory minerals: magnetite, ilmenite, hematite, and pyrite.

2.   Estimate the modal volume % of the following major minerals by thin section analysis: Garnet, vesuvianite, epidote (including zoisite), amphibole (tremolite + hornblende), white mica (muscovite + margarite + paragonite), diopside, dolomite, and plagioclase. Staining of the thin section will be necessary for a modal analysis of dolomite and plagioclase (see Chapter 2).

3.   Convert the modal analyses volume % to a weight % of each mineral. Weight % = volume % × specific gravity. The following specific gravities may be used: magnetite, 5.2; ilmenite, 4.7; hematite, 5.3; pyrite, 5.0; garnet, 3.8; vesuvianite, 3.4; epidote, 3.4; amphibole, 3.3; white mica, 2.9; plagioclase, 2.65; diopside, 3.3; dolomite, 2.9.

4. Because CaO is not represented on the $A'KF$ diagram, those minerals combining CaO and Al_2O_3 need to have their Al_2O_3 ($+Fe_2O_3$) contents, which are combined with CaO, deducted from the Al_2O_3 ($+Fe_2O_3$) rock value. The minerals in question are (Fig. 13.3A) the following:

	Molecular C	Proportion A	Necessary Deduction from A as a Fraction of CaO	
			Molecular	Weight
Anorthite (plagioclase)	50	50	1.0	1.8
Epidote (+ zoisite)	57	43	.75	1.4
Grossularite + andradite	75	25	.33	0.6
Vesuvianite	72	14	.20	0.4
Hornblende (variable)	26	15	0.6	1.1
Margarite	33	67	2.0	3.6

Among the minerals listed only epidote, with a weight % CaO of 24%, and vesuvianite, with a weight % CaO 39%, have reasonably fixed compositions.

It is customary to assume that all the CaO is present in the plagioclase; therefore then deduct an equal molecular amount of Al_2O_3 from A' if Turner's scheme (1968) is being followed. This method clearly overdeducts if significant amounts of these minerals (other than plagioclase) are present in the rock.

Suppose it were possible to have a chemical analysis of CaO for plagioclase, garnet, amphibole, and white mica. With this knowledge we proceed to deduct from the rock Al_2O_3 the following weight % amounts:

$$\text{For plagioclase: } 1.8 \times \frac{\text{its modal weight \%}}{100} \times \text{its weight \% CaO}$$

$$\text{For epidote and zoisite: } 1.4 \times \frac{\text{its modal weight \%}}{100} \times 24.$$

$$\text{For garnet: } 0.6 \times \frac{\text{its modal weight \%}}{100} \times \text{its CaO weight \%.}$$

$$\text{For vesuvianite: } 0.4 \times \frac{\text{its modal weight \%}}{100} \times 39.$$

$$\text{For amphibole: } 1.1 \times \frac{\text{its modal weight \%}}{100} \times \text{its CaO weight \%.}$$

$$\text{For white mica: } 3.6 \times \frac{\text{its modal weight \%}}{100} \times \text{its CaO weight \%.}$$

It is extremely unlikely, however, that a good CaO weight % analysis can be performed on each of the above mineral phases, in which case it will be clearly impossible to make a correct deduction from the rock weight % Al_2O_3. The present step therefore can be eliminated if the separate analyses are impossible, in which case all hope of calculating the $A'KF$ values of a rock accurately will be given up.

5.   A correction has to be made to the F value (FeO + MgO + MnO) for minerals that contain F but are not present in the $A'KF$ diagram. These minerals (see Fig. 13.3) are vesuvianite, whose formula is reasonably well fixed with a weight % MgO + FeO of 7.8%, dolomite, with a weight % MgO of 21.8, diopside with 18.6% MgO, and amphibole of variable composition (including hornblende and tremolite-actinolite). Each of these should have been modally analyzed and converted to a modal weight % in step 3. The amphibole has to be separately analyzed for weight % FeO + MgO + MnO; the others are assumed to have constant and known composition. The rock MgO weight % should be reduced by the following amounts:

$$\text{For vesuvianite: } 7.8 \times \frac{\text{its modal weight \%}}{100}.$$

$$\text{For dolomite: } 21.8 \times \frac{\text{its modal weight \%}}{100}.$$

$$\text{For diopside: } 18.6 \times \frac{\text{its modal weight \%}}{100}.$$

$$\text{For amphibole: } \frac{\text{its modal weight \%}}{100} \times \text{its weight \% MgO}.$$

The rock FeO weight % should be reduced by:

$$\text{For amphibole: } \frac{\text{its modal weight \%}}{100} \times \text{its weight \% FeO}.$$

Thus only the amphibole has to be analyzed separately for weight % MgO and FeO. The other minerals can be assumed to have the given weight % composition.

6.   At a subsequent stage we assume that all the Na_2O present in the rock is in the albite molecule of plagioclase. If the white mica includes paragonite, however, we shall have to make a correction. Paragonite, which is not represented on the diagram, has a molecular ratio of $Na_2O : Al_2O_3$ 1:3, which is a weight ratio of 1:4.9.

We assume that any Na_2O in the analysis of the white mica is due exclusively to paragonite (this is not strictly correct, but it is a reasonable assumption).

Subtract from the rock Al_2O_3 weight % an amount = modal weight % white mica/100 × 4.9 × Na_2O weight % in the white mica.

The rock Na_2O weight % must be reduced by the amount of Na_2O in the paragonite. From the rock Na_2O% subtract an amount = modal weight % white mica/100 × Na_2O weight % in white mica analysis. The Na_2O weight % left in the rock can be considered to be entirely in the albite molecule of the plagioclase. This is corrected in a later stage.

7.   The weight percentages of the rock analyses, after all the above subtractions, are converted to molecular proportions by dividing each oxide weight % by its molecular weight.

8.   The molecular proportions are grouped as follows:

$$A' = Al_2O_3 + Fe_2O_3 \text{ molecular.}$$

Subtract the molecular total Na_2O from A' to correct for all the remaining Na_2O present in albite, which is not represented on the $A'KF$ diagram.

9.   If in step 4 the corrections for CaO had not been done, one would now have to make the oversimplified assumption that all CaO is present as anorthite, in which case subtract from A' an amount equal to CaO molecular. This will result in overreduction of A' if other calcium minerals are present, and it is recommended that step 4 be performed and the present step 9 eliminated.

10.   Now subtract from A' the total molecular K_2O. This is done on the basis that all the K_2O is present in K-feldspar (which is plotted at the K apex) and in biotite and muscovite. Because we are subtracting the total K_2O molecular, we are making a correction for $K_2O : Al_2O_3 = 1 : 1$, which is correct for K-feldspar but incorrect for the micas. The micas, however, will be corrected in a subsequent step. So we have found that A' has become equal to $Al_2O_3 + Fe_2O_3$ less an amount equal to

$$(K_2O + Na_2O).$$

11. $F = FeO + MgO + MnO$ remaining in the rock after corrections have been made.

12. $K = K_2O$.

13.   The values of A', K, and F are recalculated so that $A' + K + F = 100$. The position of any rock can now be plotted directly on the diagram.

Plotting minerals on the A'KF diagram

Minerals lying along the A-F line of the ACF diagram (Fig. 13.3) occupy the same positions on the $A'KF$ diagram, with the exception that the micas (Fig. 13.3C) are not plotted on the A'-F line. K-feldspar is plotted at the K apex. Other minerals present are muscovite, biotite, and stilpnomelane.

Muscovite should be located with a $K : A'$ ratio of $1 : 3$. In step 10, however, all the A' values have been reduced by an amount $1 : 1$ of $K_2O : Al_2O_3$, assuming that all the K_2O was present in K-feldspar. Hence, because of the already reduced A' value, muscovite must now be plotted not at $K : A'$ $1 : 3$ but at $1 : 2$ instead, so that its plotted position is $A' = 67$, $K = 33$. In actual fact muscovite may contain significant amounts of Mg + Fe, so that it does not actually lie on the K-A' line. For simplicity, however, this is not done, and it is conventionally plotted at A' 67, K 33 on $A'KF$ diagrams.

Biotite can be regarded as generally having a ratio $K_2O : (Fe, Mg)O : Al_2O_3$ $1 : 6 : 1$, but in step 9 we have already deducted Al_2O_3 of equal amount to K_2O, and the biotite should now be regarded as essentially having Al_2O_3 zero, making it plot at $K : F = 1 : 6$ or K 14, F 86. Some (Fe, Mg) of biotite may be replaced by Al, so that biotite is represented by a small field extending parallel to the $A'F$ side of the diagram from a point at $14\% K$ $86\% F$ to about $14\% K$ $15\% A'$ $71\% F$ (molecular).

Stilpnomelane $(K, Na, Ca)_{0-1.4}$ $(Fe^{3+}Fe^{2+}MgAlMn)_{5.9-8.2}$ $(Si_8O_{20})(OH)_4$ $(O, OH, H_2O)_{3.6-8.5}$ is also represented. Because $K_2O : Al_2O_3$ $1 : 1$ has already been deducted in step 9, we can assume that in the plotting of this mineral K_2O can be taken as zero. Hence this mineral will be plotted in a small field in the vicinity of the F corner.

The positions of the minerals in an $A'KF$ diagram and an actual example of a facies diagram are shown Fig. 13.4.

(modified after Winkler, 1967.)

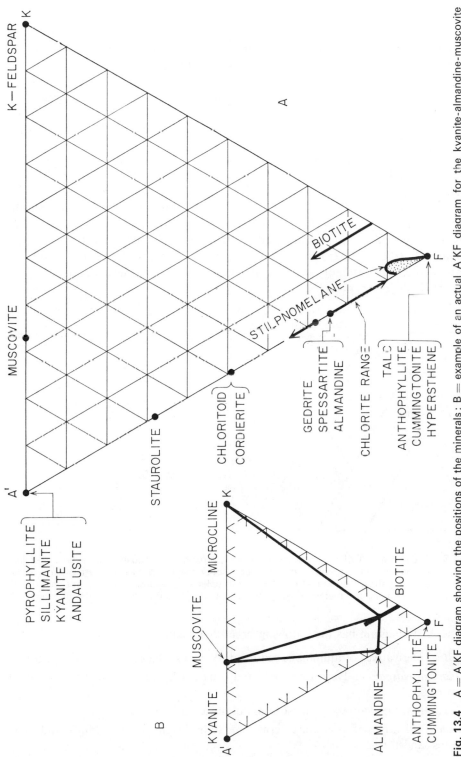

Fig. 13.4 A = A'KF diagram showing the positions of the minerals; B = example of an actual A'KF diagram for the kyanite-almandine-muscovite subfacies of the almandine amphibolite facies. Biotite may coexist with kyanite, but it is not obvious from this diagram.

DISCUSSION

Paragonite is not represented. If it is present, an additional diagram with apices A, F, Na may be produced to portray the sodium minerals.

EXERCISE 13.5

The Use of ACF and A′KF Diagrams

ACF and *A′KF* diagrams have been constructed for most metamorphic facies and subfacies on the basis of petrographic observations (Winkler, 1967; Turner, 1968).

The chemical compositions of igneous and sedimentary rocks are of interest in the study of metamorphic assemblages because they are the parent rocks of metamorphism. Chemical analyses of these rocks may be converted to *A,C*, and *F* and *A′*, *K* and *F* values and the fields of the various sedimentary and igneous rocks displayed diagrammatically (Fig. 13.5). It is useful to know these general fields so that from them we can predict the metamorphic mineral paragenesis under the conditions of any metamorphic facies or subfacies.

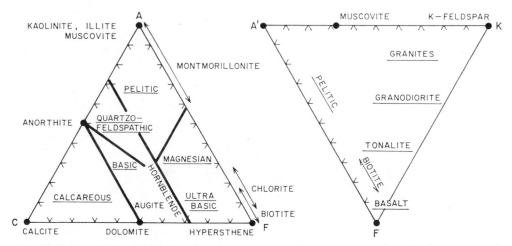

Fig. 13.5   The ACF and A′KF diagrams showing the approximate compositional fields of chemical classes of sedimentary and igneous rocks. Divisions between these classes are arbitrary. Some primary minerals are shown on the ACF diagram. All rocks are presumed to have possible quartz and feldspars as primary and metamorphic constitutens; they are not shown in the diagrams.

The ACF diagram has the following broad field subdivisions:

1. *Pelitic.* Derivatives of aluminous sedimentary rocks; shales muds and mudstones.

2. *Quartzo-feldspathic.* Derivatives of sandstones, arkoses, greywackes, and acid igneous rocks.

3. *Calcareous.* Derivatives of limestones, dolomites, and marls.

4. *Basic.* Derivatives of basic and intermediate to basic igneous rocks, tuffs, and some tuffaceous sandstones.

5. *Magnesian.* Derivatives of ultrabasic igneous rocks and some magnesian sedimentary rocks of unusual chemical composition.

The $A'KF$ diagram can display the range of igneous rocks along and near the K to F edge, but most of the field is applicable to the pelitic rocks.

The two diagrams are complementary. In order to decide what minerals composed of the A and F components may be found in any one facies, attention must be paid to the content of the K component present in the rock. Erroneous deductions may be made by considering only the ACF diagram; for example, the plot of a rock on an ACF diagram in the pelitic field may indicate a mineralogy of andalusite + cordierite + plagioclase + quartz for one particular facies. The rock will have this mineralogy only if it contains no K_2O. The presence of K_2O, however, will also show that on an $A'KF$ diagram muscovite will be an important constituent.

EXERCISE 13.6

The AFM Diagram for Representing Metamorphic Paragenesis

FeO and MgO are two independent components that should not be grouped together but rather considered separately. The AFM projection of Thompson (1957) has this advantage over the ACF and $A'KF$ diagrams: $A = Al_2O_3$, $F = FeO$; and $M = MgO$ (molecular ratios). The AFM diagram is actually a projection onto the AFM face of all points of the tetrahedron, of which the fourth corner is K_2O. Because of the absence of CaO in this tetrahedron, the AFM diagram is suitable only in the representation of mineral assemblages of metamorphic rocks derived from shales and shaly sands, that is, for pelitic rocks in general. These are by far the most common sedimentary rocks and certainly the most useful in describing the critical metamorphic reactions that define the metamorphic facies.

With few exceptions, pelitic metamorphic rocks include quartz and muscovite in their mineral assemblage. The AFM diagram is for rocks with SiO_2 saturation, so that quartz is a phase assumed to be present and not represented. H_2O is also assumed to be present.

Because metamorphic pelitic rocks usually contain essential muscovite, Thompson (1957) chose the muscovite point on the $K_2O - FeO - MgO - Al_2O_3$ tetrahedron as the projection point. All points within the tetrahedron are therefore joined up to the muscovite point and projected onto the AFM plane. An AFM diagram is shown in Fig. 13.6 and the positions of the common minerals are illustrated as they appear in pelitic rocks.

The construction of the AFM diagram is described in Fig 13.6B. Since most pelitic rocks contain muscovite, all points within the tetrahedron are projected onto the AFM plane through the muscovite point.

Biotite has a range of composition within the tetrahedron and plots on the AFM diagram below the FM line. K-feldspar (point K in Fig. 13.6B) can project at point A and Thompson (1957) employs it in this way. Many authors, however, assume that the join between P (muscovite) and K (K-feldspar) can be considered approximately parallel to the AFM base (actually it is not), and it is also assumed that it will intersect the AFM plane downward at infinity. Hence Turner (1968) and Winkler (1967) plot K-feldspar at infinity only along the base of the AFM plane. Although not strictly correct (it plots at point A), it is acceptable for the purposes of metamorphic facies representation.

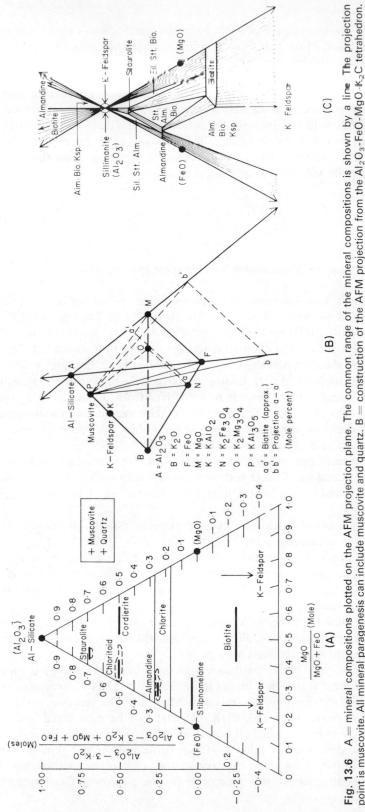

Fig. 13.6 A = mineral compositions plotted on the AFM projection plane. The common range of the mineral compositions is shown by a line. The projection point is muscovite. All mineral paragenesis can include muscovite and quartz. B = construction of the AFM projection from the Al_2O_3-FeO-MgO-K_2C tetrahedron. Muscovite is the projection point. All ratios are molecular. C = example of an AFM diagram for mineral paragenesis in the muscovite-sillimanite zone. (After Thompson, 1957.)

The plotting of the mineral phases is carried out as follows:

1. The mineral analyses in weight $\%$ are converted to a molecular $\%$ of the oxides.

2. The molecular proportions A and M are calculated thus:

$$A = \frac{Al_2O_3 - 3K_2O}{Al_2O_3 - 3K_2O + MgO + FeO},$$

$$M = \frac{MgO}{MgO + FeO}.$$

The $(Al_2O_3 - 3K_2O)$ value in the A calculation is because muscovite is taken as the projection point and has three times as much Al_2O_3 as K_2O molecular.

3. Before calculating the A and M values of garnet the amount of Al_2O_3 required to form the end members spessartite $[Mn_3Al_2(SiO_4)_3]$ and grossularite $[CaAl_2(SiO_4)_3]$, which are not represented on the AFM diagram, has to be subtracted from Al_2O_3 in the garnet analysis.

4. The composition of a metamorphic rock can be shown on an AFM diagram, but the chemical composition of the rock has to be corrected first for chemical constituents in minerals not represented. Corrections have to be made for Ca-bearing minerals, for albite, for plagioclase, and for paragonite. In practice the AFM diagram is not used for plotting whole rocks because of the restriction to the pelitic rocks; it shows the mineral paragenesis only.

5. Other AFM diagrams may be constructed for rocks that do not contain muscovite; for example, Baker (1961) constructed an AFM diagram for pelitic rocks devoid of muscovite but containing K-feldspar. In such a case the projection point may be more conveniently taken as K (Fig. 13.6B). When K-feldspar is used as the projection point, all assemblages are assumed to include K-feldspar and the scale of A changes to

$$A = \frac{Al_2O_3 - K_2O}{Al_2O_3 - K_2O + FeO + MgO}$$

because the ratio of Al_2O_2 to K_2O in feldspar is $1:1$. In this diagram muscovite would plot at the A apex.

EXERCISE 13.7

Classification of Basaltic Rocks

Based on the C.I.P.W. norm values, basalt may be subdivided as follows:

1. **THOLEIITE** characterized by normative hypersthene (Green and Ringwood 1967) = tholeiite (saturated hypersthene basalt) (Yoder and Tilley, 1962).

2. **THOLEIITE** (oversaturated) = quartz tholeiite; basalt characterized by normative quartz + normative hypersthene.

3. **OLIVINE THOLEIITE** (undersaturated); basalts with normative hypersthene + normative olivine.

4. **OLIVINE BASALT** basalt containing normative olivine (Yoder and Tilley 1962); used for brevity for OLIVINE THOLEIITE, but in which normative hypersthene is 0 to 3%. (Green and Ringwood 1967).

5.   ALKALI BASALT contains normative olivine + nepheline (Yoder and Tilley 1962) – ALKALI OLIVINE BASALT; basalt with normative olivine + nepheline. The normative nepheline is less than 5% (Green and Ringwood, 1967).

6.   BASANITE includes basalts characterized by normative olivine + nepheline, with nepheline greater than 5% (Green and Ringwood, 1967).

7.   OLIVINE NEPHELINITE basaltlike rock without normative albite and typically with olivine, diopside, and nepheline as major normative minerals (Green and Ringwood, 1967).

There is, however, a continuum between all basalt types. *Modally*, tholeiite, in the strict sense, has essential augite or subcalcic augite, plagioclase (near An_{50}), and iron oxides. Olivine is subordinate or absent. Characteristically, there is an interstitial vitreous acid residuum, commonly pigmented (Yoder and Tilley, 1962).

EXERCISE 13.8

Alkali Versus SiO_2 Diagram for Volcanic Series

A plot of $Na_2O + K_2O$ weight % against SiO_2 weight % in the rock separates rocks into three basaltic series:

1.   Tholeiitic series (pigeonitic rock series).
2.   High alumina basalt series.
3.   Alkali basalt series (Fig. 13.7).

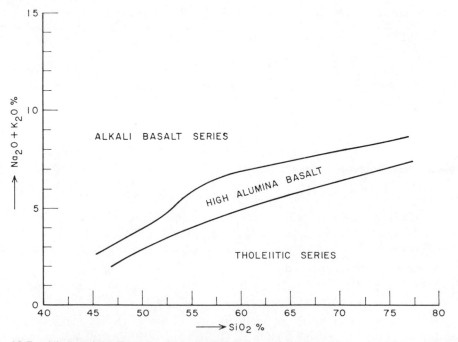

Fig. 13.7   Alkali weight % versus SiO_2 % diagram for separating the basaltic rock series into three divisions. (After Kuno, 1966.)

Associated with tholeiite and high alumina basalt are andesite, dacite, and rhyolite, derived from tholeiite or high alumina basalt by fractionation.

The alkali olivine basalt is associated with hawaiite, mugearite, trachyandesite, trachyte, and alkali rhyolite.

The tholeiite series always occurs toward the oceanic side that is, toward the subduction zone or trench of an island arc. The alkali series always occurs on the continental side.

The magma type is directly related to depth of origin on the Benioff Zone, as it dips down under the continent away from the Oceanic Trench. Depth of magma derivation from the mantle is given as

Tholeiite basalt   100 to 180 km
High alumina basalt   180 to 280 km
Alkali olivine basalt   280 to 380 km (Kuno, 1966).

EXERCISE 13.9

The Solidification Index for Volcanic Series

The parameter

$$\frac{100\,MgO}{MgO + FeO + Fe_2O_3 + Na_2O + K_2O}$$

(all oxides in weight %) is called the solidification index (SI). It is a good index for the separation of rock types in a series. SI for basalt = 30 to 40; for basaltic andesite, 20 to 29; for andesite, 10 to 19; for andesitic dacite, 0 to 9.

The solidification index SI is a better index to use in variation diagrams than SiO_2 wt % (e.g., Fig. 13.7) because it is known that the SiO_2 value does not change steadily throughout fractionation.

On the other hand, when the ternary variation diagram $MgO:FeO + Fe_2O_3:Na_2O + K_2O$ (Fig. 13.10) is used, the fractionation trend of any rock suite is from the MgO-rich side toward the MgO-poor side; in other words, the value of 100 $MgO/(MgO + FeO + Fe_2O_3 + Na_2O + K_2O)$ invariably decreases as the fractionation increases. The SI value represents a measure of the amount of residual liquid relative to crystallized solid phase as the magma fractionates; therefore in variation diagrams the use of SI as the abscissa in place of SiO_2 is preferable.

If for any rock suite we now plot weight % CaO as the ordinate against SI values and on the same diagram weight % $Na_2O + K_2O$ against SI values as the abscissa (see Fig. 13.8) for the same rock suite, the SI value at which the two curves (one for CaO %, the other for $Na_2O + K_2O$) intersect is called the alkali-lime index (after Kuno, 1959).

This alkali-lime index differs from that originally given by Peacock. A plot of the CaO wt % of a rock against the alkali-lime index (Fig. 13.8) gives a good separation of rocks into tholeiitic series, alkali rock series, and calc-alkali rock series (Kuno, 1959).

Total iron (as % $FeO + Fe_2O_3$)or total iron expressed as wt % FeO can also be usefully plotted against SI to illustrate fractionation trends in a magma series.

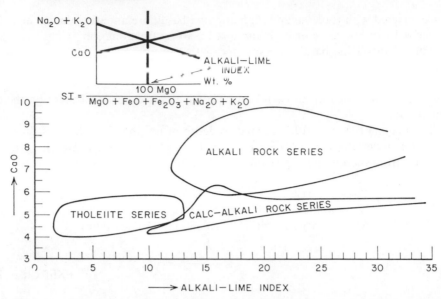

Fig. 13.8 The CaO weight % plotted against the corresponding alkali-lime index, as defined, gives a good separation into tholeiite, alkali, and calc-alkali rock series. (After Kuno, 1959.)

EXERCISE 13.10

The Crystallization Index

The crystallization index (CI) measures the progression of partial magmas or igneous rocks from the primitive system anorthite-diopside-forsterite.

$$CI = \sum (An + Di' + Fo' + Sp')$$

in which An = normative anorthite,

  Di' = magnesian diopside $CaMgSi_2O_6$ calculated from normative diopside,
  Fo' = normative forsterite plus normative enstatite
    converted to forsterite,
  Sp' = magnesian spinel, $MgAl_2O_4$, calculated from normative corundum in
    ultramafic rocks.

The CI is computed preferably after the norm is recalculated to 100% anhydrous. The following conversions may be used:

  Di' = 2.157003 En_{Di} (En of normative diopside)
  Fo' = Fo + 0.700837 En_{Hy} (En of normative hypersthene)
  Sp' = is used *only* for ultramafic rocks. Normative corundum
    in more silicic rocks is not recalculated to spinel.

For ultramafic rocks, in the norm calculation after making anorthite, any excess Al_2O_3 is used to make spinel, $(Mg, Fe)Al_2O_4$ with MgO and FeO in the same ratios normally determined to make diopside. The magnesian spinel $(MgAl_2O_4 = Sp')$ is then included in the CI.

Rocks made exclusively of anorthite, magnesian diopside, forsterite, or their mixtures in the norm have CI = 100%.

Rocks exclusively of quartz, alkali feldspars, feldspathoids, or their mixtures have CI = 0%.

The calculation of the crystallization index is illustrated by an example taken from Poldervaart and Parker (1964):

	Analysis	Molecular Weight		Molecular Weight	Norm	Recalculated Norm
SiO_2	49.12	60.06	Qu	60.06	...	...
TiO_2	0.98	79.90	Or	556.49	3.24	3.25
Al_2O_3	11.82	101.94	Ab	524.29	13.95	13.98
Fe_2O_3	2.19	159.70	An	278.14	23.23	23.27 (1)
FeO	9.62	71.85	Di {Wo	116.14	8.89	8.91
MnO	0.14	70.93	Di {En	100.38	5.76	5.77 (2)
MgO	13.44	40.32	Di {Fs	131.91	2.52	2.52
CaO	9.14	56.08	Hy {En	100.38	18.03	18.06 (3)
Na_2O	1.65	61.99	Hy {Fs	131.91	7.90	7.92
K_2O	0.55	94.19	Ol {Fo	140.70	6.77	6.78 (4)
P_2O_5	0.14	141.96	Ol {Fa	203.76	3.15	3.16
H_2O+	0.99		Il	151.75	1.86	1.86
H_2O-	0.05		Mt	231.55	3.17	3.18
			Ap	310.20	0.30	0.30
	99.83		H_2O		1.04	1.04
					99.81	100.00

(1) $An' = 23.27.$
(2) $2.157003 \times 5.77^a.$ $Di' = 12.45$
(3) 0.700837×18.06^b
(4) 6.78 $Fo' = 19.44$

$CI = 55.16$

$^a 5.77 \times \dfrac{\text{mol wt } CaMgSi_2O_6}{\text{mol wt } MgSiO_3} = 5.77 \times \dfrac{216.52}{100.38}$

$= 5.77 \times 2.157003;$

$^b 18.06 \times \dfrac{\text{mol wt } Mg_2SiO_4}{2 \times \text{mol wt } MgSiO_3} = \dfrac{18.06 \times 140.70}{2 \times 100.38}$

$= 18.06 \times 0.700837.$

Differentiation trends are shown by plotting SiO_2, Al_2O_3, Fe_2O_3, FeO, MgO, CaO, Na_2O, K_2O, and TiO_2 weight % against CI or normative Qu($+$or $-ve$) (the SiO_2 in the norm calculation required to saturate the silica deficient minerals) against CI (Fig. 13.9).

The CI index should have preference over SiO_2 or SI (solidification index) because it is based on theoretical petrological principles.

(After Poldervaart and Parker, 1964.)

DISCUSSION

The differentiation index (see exercise 13.11) is more widely used than the crystallization index and has the advantage of being simpler to compute. The differentiation index is the sum of the salic constituents; the crystallization index is essentially the sum of the mafic constituents minus the iron-bearing normative minerals. Thus there is a definite correlation between the two (Thornton and Tuttle, 1965), and only in a

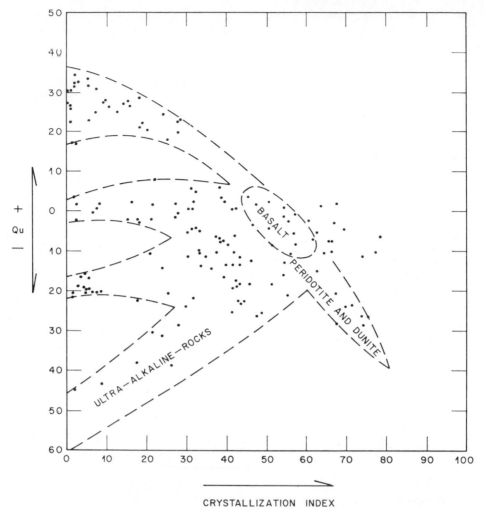

Fig. 13.9 A plot of normative quartz against crystallization index showing how igneous rocks are distributed and illustrating the various differentiating trends. (After Poldervaart and Parker, 1964.)

rock series showing enrichment in iron to an unusual degree (e.g., the Skaergaard) will there be drastic differences. If the sum of these iron-bearing normative minerals is small, the relation between the two indices will be simple and reciprocal.

EXERCISE 13.11

The Differentiation Index

1. The differentiation index is computed simply as the sum of the weight % of normative quartz + orthoclase + albite + nepheline + leucite + kalsilite; in other words, the total salic minerals less anorthite.

2. No more than three of these normative minerals will appear in any given norm; thus the differentiation index is easily computed as the sum of the percentages of three normative minerals.

3. Average rock types have the following differentiation indices (DI): alkali granite = 93; alkali rhyolite = 91; granite = 80; rhyolite = 88; granodiorite = 67; quartz latite = 68; diorite = 48; andesite = 56; gabbro = 30; basalt = 35; olivine gabbro = 27; olivine diabase = 30; peridotite = 6 and picrite = 12.

4. The differentiation index is a natural quantity to use for variation diagrams of rock series because it is a measure of a rock's basicity. Useful variation diagrams plot oxide weight % directly against DI. Commonly plotted oxides are SiO_2, Al_2O_3, Fe_2O_3, FeO, MgO, CaO, Na_2O, and K_2O.

DISCUSSION

The differentiation index is far more widely used than the crystallization index (exercise 13.10). However Poldervaart and Parker (1964) consider the DI less effective than the CI for igneous series that show moderate to extreme iron enrichment. The CI is as good as the DI for igneous series that do not show iron enrichment, but it is more generally applicable because it properly *also* represents magmatic variation when there is iron enrichment.

EXERCISE 13.12

Other Types of Variation Diagram

In addition to the diagrams in Figs. 13.7, 13.8, and 13.9 which contain important chemical information regarding igneous rocks, several other variation diagrams in common use are illustrated in Fig. 13.10

A favorite way of illustrating certain significant aspects of the fractionation of basic magmas is by plotting the percentage of Mg, Fe^{+2} + Mn, and Na + K as atoms or oxides at the corners of a triangular diagram (Wager and Brown, 1967). The trend in composition of both average rocks and liquids is from a point near the Mg corner toward the Fe^{+2} + Mn corner, then toward the Na + K corner (Fig. 13.10). An alternative is to plot FeO + MnO versus Na_2O + K_2O versus MgO as weight percentages.

Another useful way of illustrating fractionation (Wager and Brown, 1967) is to plot the albite and iron ratios (Fig. 13.10). The albite ratio is defined as the molecular percentage of albite in the normative plagioclase. If nepheline is present in the norm, it is converted to albite, and in deriving the Ab ratio the nepheline converted to albite is included with the albite. The iron ratio is defined as

$$\frac{(Fe^{+2} + Mn) \times 100}{Fe^{+2} + Mn + Mg}.$$

A plot of the albite ratio versus the iron ratio should give the order in which the rocks were formed when a single fractionation series is being considered.

A commonly used diagram for illustrating chemical variation is shown in Fig. 13.7 to describe the differences in alkali/silica ratios in basalt series. Another is the *AFM* plot of oxide whole rock percentages by weight (Fig. 13.10) in which

$$A = Na_2O + K_2O,$$
$$F = FeO \ (Fe^{+2} \text{ and } Fe^{+3} \text{ calculated as FeO})$$
$$M = MgO.$$

This diagram shows the differentiation of a basaltic magma toward rhyolite or trachyte, for example, by MacDonald and Katsura (1964). Some other types of variation diagram are illustrated—plotting $K_2O : Na_2O : CaO$ and $FeO : Na_2O + K_2O : MgO$ may also be used to show differentiation from basalt to trachyte (Fig. 13.10). A plot of

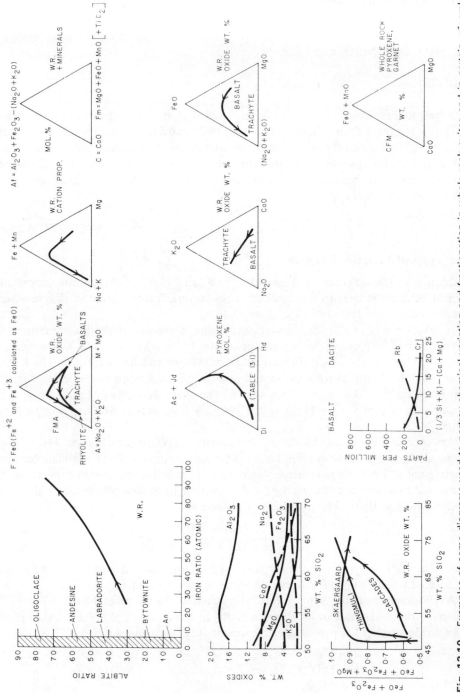

Fig. 13.10 Examples of some diagrams used to portray chemical variation and fractionation in whole rock suites and in certain mineral groups. W.R. means whole rock.

(FeO + Fe$_2$O$_3$)/(FeO + Fe$_2$O$_3$ + MgO) versus weight % SiO$_2$ has been used by Car-michael (1964) to compare the Thingmuli Tertiary volcano of Iceland with other basaltic provinces (Fig. 13.10).

Simple plots of weight % of various oxides against SiO$_2$ (Fig. 13.10) have been used by numerous authors to show chemical variation in a rock series. The contents of trace elements in parts per million versus ($\frac{1}{3}$Si + K) − (Ca + Mg) can be used to illustrate differentiation in a rock series from basalt to dacite. These are not the only possibilities for constructing variation diagrams. Many others will be found in the literature, but Fig. 13.10 illustrates a selection of the most common.

EXERCISE 13.13

Calculation of Pyroxene Analyses in Terms of End Members

METHOD A

The basic formula for the pyroxenes is $X_{1-p}Y_{1+p}Z_2O_6$, where $X =$ Ca, Na, K, $Y =$ Mg, Fe^{2+}, Mn, Li, Ni, Al, Fe^{3+}, Cr, Ti, and $Z =$ Si, Al. In the orthopyroxene series p is (approx.) 1, for the monoclinic pyroxenes p may vary from 0 to 1.

The following rules for computation are given by Yoder and Tilley (1962).

The end members are Di [diopside CaMg(Si$_2$O$_6$)], Hd [hedenbergite Ca(F^{+2}, Mn)(Si$_2$O$_6$)], Ac [acmite Na Fe^{+3}(Si$_2$O$_6$)], Jd [jadeite NaAl(Si$_2$O$_6$)], and Tsch [Tschermak's molecule (Ca,Mg)Al(AlSiO$_6$)].

1. Convert the chemical weight % analysis of the pyroxene to a structural formula (a worked example is given in Table 13.1).

Column 1 lists the reported weight % chemical analysis.

Column 2 is a listing of molecular oxide weights.

Column 3 shows the molecular proportions of the oxides and is computed by dividing column 1 by 2.

Column 4 gives the atomic proportions of oxygen. It is obtained by multiplying column 3 by the number of oxygens in each oxide.

Column 5 lists the number of anions on the basis of six oxygens (the basic pyroxene formula unit). This is done by multiplying column 4 by 6/(total of column 4).

Column 6 gives the number of ions in the formula. It is calculated by multiplying column 5 by the ratio of cation to oxygen; for example, in the case of Si multiply column 5 by $\frac{1}{2}$ to obtain column 6; multiply by $\frac{2}{3}$ to obtain Al and Fe^{+3}. In the case of Na multiply by 2.

2. Al (and Fe^{+3}, then Ti only if needed) is assigned with Si to complete the Z position.

3. Na, K, and Ca are assigned with the requisite Mg (and Fe^{+2} if needed) to complete the X position.

4. The remaining Fe^{+3}, Fe^{+2}, Mg, Cr, Al, Mn, and Ti are assigned to the Y position. At this stage the complete structural formula may be written.

5. The numbers of ions in the formula (column 6) may now be grouped as follows: Ca; Mg; $\sum$ Fe = Fe^{+2} + Fe^{+3} + Mn, totaled, then recalculated as a percentage; Ca + Mg + $\sum$ Fe = 100. The values so computed are used to plot the pyroxene on the Wo (wollastonite)CaSiO$_3$–En (enstatite) Mg SiO$_3$–Fs (ferrosilite)FeSiO$_3$ triangular diagram (Fig. 13.11). It will plot within the Di(diopside)–Hd (hedenbergite)–En-Fs part of the triangle. The apices of the complete triangle are Wo = 100% Ca,

Table 13.1 Calculation of the Structural Formula of a Pyroxene from the Given Chemical Analysis[a]

Oxide	(1) Weight %	(2) Molecular Weight	(3) Molecular Proportions of Oxides	(4) Atomic Proportions of Oxygen	(5) Number of Anions on basis of six Oxygens	(6) Number of Ions in Formula
SiO_2	48.61	60.09	0.8090	1.6180	3.635	Si = 1.817 $\left.\right\}$ Z = 2
Al_2O_3	4.80	101.94	0.0471	0.1413	0.317	Al = 0.211 = $\left\{\begin{array}{l}0.183\\0.028\end{array}\right.$
TiO_2	1.91	79.90	0.0239	0.0478	0.107	Ti = 0.054
Cr_2O_3	0.09	152.02	0.0006	0.0018	0.004	Cr = 0.003
Fe_2O_3	2.75	159.70	0.0172	0.0516	0.116	Fe^{+3} = 0.077
FeO	7.14	71.85	0.0994	0.0994	0.223	Fe^{+2} = 0.223
MnO	0.20	70.94	0.0028	0.0028	0.006	Mn = 0.006
NiO	...	74.71	...	...	...	
MgO	13.42	40.32	0.3328	0.3328	0.748	Mg = 0.748 = $\left\{\begin{array}{l}0.616\\0.132\end{array}\right.$
CaO	20.38	56.08	0.3634	0.3634	0.816	Ca = 0.816
Na_2O	0.63	61.98	0.0102	0.0102	0.023	Na = 0.046
K_2O	0.11	94.20	0.0012	0.0012	0.003	K = 0.006
H_2O+	n.d.	18.02	...			
H_2O-	0.03	18.02	0.0017		0.004	Ca = 0.816 43.6%
						Mg = 0.748 40.0%
Total	100.07	Total oxygens		2.6703		

Y = 1.007

X = 1

$$\frac{6}{2.6703} = 2.2469$$

$Fe(Fe^{+2} + Fe^{+3} + Mn) = 0.306$ 16.4

1.870

$$\frac{100}{1.870} = 53.476$$

100%

[a] Example from Yoder and Tilley (1962, No. 20, 346). See exercise 13.13 for details. n.d. = not determined.

402

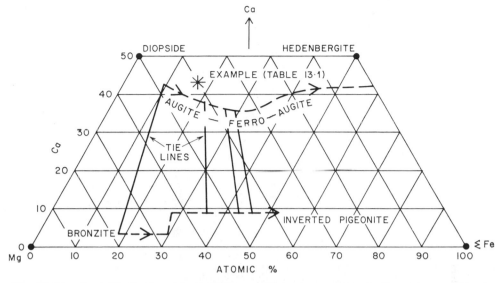

Fig. 13.11 Crystallization trend for the Skaergaard intrusion as shown by the pyroxenes. Atomic % Ca is plotted versus Mg and Fe (Fe^{+2} + Fe^{+3} + Mn) (from Wager and Brown, 1967, p. 39) The clinopyroxene from Table 13.1 is plotted as an example.

En = 100% Mg, and Fs − 100% Fe (molecular proportions). The example calculated in Table 13.1 is plotted on the diagram of Fig. 13.11.

6. The Na + K is taken as Jd + Ac. An amount of Al from the Y position is assigned to an equal amount of K and twice that amount of Si from Z to form Jd. An amount of Fe^{+3} from the Y position is assigned to an equal amount of Na and twice the amount of Si from Z to form Ac. In other words, the amount of Si assigned to Ac + Jd = 2(Na + K) See worked example to follow.

7. The total Al (and Fe^{+3}) in the Z position is taken as Tsch. An amount equal to Al (+ Fe^{+3}) in Z of Mg (+ Fe^{+2} + Ca, if needed) from the X position is assigned to Tsch. An amount equal to Al (+ Fe^{+3}) in Z of Al (and Fe^{+3}, if needed, then only + Ti, then only + Mg, and then only Fe^{+2}) from the Y position is assigned to Tsch. An amount equal to Al (+ Fe^{+3}) in Z of Si is assigned to Tsch. See worked example to follow.

8. The Fe^{+2} + Mn in the Y position is taken as Hd. An equal amount of Ca from the X position and an amount twice this of Si from the Z position are also assigned to Hd.

9. The Mg + Cr in the Y position is assigned to Di. An equal amount of Ca (or whatever is left) from the X position is assigned to Di. An amount equal to twice this amount (or whatever is left) of Si in the Z position is also assigned to Di.

DISCUSSION

Pyroxene analyses rarely reduce to a complete complement of atoms for each position. Obviously some personal judgment must be exercised in the ultimate assignment of the atoms to the various end members. As an example, the pyroxene quoted in Table 13.1 (from Yoder and Tilley, 1962, p. 364, 367) may be reduced to the end members as follows:

	$X\,(=100)$	$Y\,(=100.7)$	$Z\,(=200)$	Total	%
	$(Na_{4.6}K_{0.6}Ca_{81.6}Mg_{13.2})$	$(Mg_{61.6}\,Mn_{0.6}\,Fe^{+2}_{22.3}\,Fe^{+3}_{7.7}\,Cr_{0.3}\,Ti_{5.4}\,Al_{2.8})$	$(Al_{18.3}\,Si_{181.7})$		
Ac + Jd	$Na_{4.6}K_{0.6}$	$Fe^{+3}_{4.6}$ $Al_{0.6}$	$Si_{10.4}$	20.8	5.2
Tsch	$Ca_{5.1}Mg_{13.2}$	$Mg_{7.6}\quad Fe^{+3}_{3.1}\quad Ti_{5.4}\,Al_{2.2}$	$Al_{18.3}\,Si_{18.3}$	73.2	18.3
Hd	$Ca_{22.9}$	$Mn_{0.6}\,Fe^{+2}_{22.3}$	$Si_{45.8}$	91.6	22.8
Di	$Ca_{53.6}\quad Mg_{54.0}$	$Cr_{0.3}$	$Si_{107.2}$	215.1	53.7
				400.7	

Formula: $Tsch_{18.3}Ac + Jd_{5.2}Hd_{22.8}Di_{53.7}$. The last three total 81.7, therefore $\times \dfrac{100}{81.7}$ their % values can be plotted as $Ac + Jd_5Hd_{28}Di_{66}$.

It can be seen from this calculation that a perfect balance of $Y(54.3)$, $X(53.6)$, and $Si(2 \times 53.6)$ was not possible (for Di) but the error is small and can be neglected. The position of this pyroxene on the Ac + Jd: Di: Hd diagram is shown in Fig. 13.10 and also on the Di-Hd-En-Fs diagram in Fig. 13.11. Pyroxene plots on both diagrams are used to show differentiation trends in basalts and basic rocks generally.

(Method after Yoder and Tilley, 1962.)

It is also commonly done to calculate the norms of pyroxenes separated from basic igneous rocks. The appearance of Ol (olivine) in the norm of an analyzed hypersthene and other pyroxenes is due to solid solutions involving the Tschermak molecule. Normative Ne (nepheline) appears in the pyroxenes from the critically undersaturated alkali basalts.

If separate analyses of orthopyroxene (hypersthene) or pigeonite and clinopyroxene (augite) can be made available from the same rock, it is customary to plot both coexisting pyroxenes on a Di-Hd-En(Mg)-Fs(Fe) diagram. The two plotted pyroxenes from the same rock are then joined by a tie line on the diagram (Fig. 13.11). The clinopyroxene will belong to the augite-ferroaugite series; the orthopyroxene will belong to the bronzite-hypersthene and inverted pigeonite series.

If pyroxene analyses from a range of basalts in a differentiated series are available, this diagram (Fig. 13.11), in addition to the Ac + Jd: Di: Hd diagram (Fig. 13.10), can be used to show the fractionation trend of the magma.

METHOD B

1. The chemical composition of the clinopyroxene is computed first as cationic proportions from the chemical analysis.

2. K is added to Na, Mn to Fe^{2+}, and Cr to Al.

3. The cations are allocated to the various end members in the following order:

a. $NaFe^{3+}Si_2O_6$ Acmite
b. $NaAlSi_2O_6$ Jadeite
c. $CaTiAl_2O_6$
d. $CaFe^{3+}AlSiO_6$
e. $CaAl_2SiO_6$ Ca-Tschermak's component
f. $CaFe_2^{3+}SiO_6$
g. $CaSiO_2$ Wollastonite
h. $MgSiO_2$ Enstatite
i. $Fe^{2+}SiO_3$ Ferrosilite

(Method after Kushiro, 1962)

DISCUSSION

The order of calculation after forming $NaAlSi_2O_6$ follows the generally accepted assumption that the deficiency in the Si tetrahedral position of clinopyroxene is first filled with Al and then with Fe^{3+}, if it is still deficient. It is a problem whether NaFeSi$_2$O$_6$ or $NaAlSi_2O_6$ should be calculated first. If $NaAlSi_2O_6$ is done first, then all the clinopyroxenes will contain it. Igneous clinopyroxenes formed under low pressure would not contain $NaAlSi_2O_6$ and therefore it would be better to calculate $NaFeSi_2O_6$ first.

$NaFeSi_2O_6$ is calculated for most clinopyroxenes; $NaAlSi_2O_6$ is not calculated for most igneous clinopyroxenes; $NaAlSi_2O_6$ is incompatable with Fe^{3+}-containing components with the exception of $NaFeSi_2O_6$; varying amounts of $CaTiAl_2O_6$ are calculated for most clinopyroxenes; $CaFeAlSiO_6$ is calculated for igneous clinopyroxenes, $CaAl_2SiO_6$ is calculated for most clinopyroxenes; and $CaFe_2SiO_6$ is rarely calculated for igneous clinopyroxenes.

EXERCISE 13.14

Relationship of Pyroxene Composition with Metamorphic Grade

When clinopyroxene analyses are recast into molecular % of the various end members Ac, Jd, Tsch, Hd, and Di, in that order, a clear distinction can be made between a clinopyroxene from an eclogite and one from a basic granulite.

Undoubted eclogite-facies clinopyroxenes have a molecular ratio $Jd/Tsch > \frac{4}{5}$, whereas those from basic rocks of lower pressure facies (such as basic granulites) have a Jd/Tsch ratio $< \frac{1}{2}$; hence the nature of the clinopyroxene is diagnostic of the grade.

The number of cations in each pyroxene is calculated on the basis of 6 or 12 anion equivalents, as in exercise 13.13, according to the general formula XYZ_2O_6 and assigned to the hypothetical end members Ac, Jd, Tsch [Tschermak's molecule (Ca, Mg)$AlAlSiO_6$, no distinction being made between Ca and Mg], Hd, and Di, in that order, following the convention of Yoder and Tilley (1962 pp. 366–367), as given in exercise 13.13, except that all Fe^{+3} in the Y position is first assigned to acmite ($NaFe^{+3}Si_2O_6$) and the remaining Na + K in the X position is allotted to an equal number of Al atoms to form jadeite ($NaAlSi_2O_6$).

The example worked in exercise 13.13 has the ratio

$$\frac{Jd}{Tsch} = \frac{4.6}{18.3} = \frac{1}{4},$$

which would assign it to the basic granulite facies and definitely exclude it from the eclogite.

(Method after White, 1964.)

EXERCISE 13.15

Classification and Display of Amphibole Structural Formulas

The calciferous amphiboles are defined as having Ca > 1.50 in the half-unit cell which is always referred to 24(O, OH, F, and Cl). Subcalciferous amphiboles have Ca > 1.00 and < 1.50. The naming of amphiboles is confused, but the following scheme, as described by Leake (1968), may be generally adopted. The classification is based on the well known magnesian end members.

Tr	TREMOLITE	$Ca_2Mg_5Si_8O_{22}(OH)_2$
Ri	RICHTERITE	$Na_2CaMg_5Si_8O_{22}(OH)_2$
Ed	EDENITE	$NaCa_2Mg_5Si_7AlO_{22}(OH)_2$
Pg	PARGASITE	$NaCa_2Mg_4Al_6SiAl_2O_{22}(OH)_2$
Ts	TSCHERMAKITE	$Ca_2Mg_3Al_2Al_2Si_6O_{22}(OH)_2$

Three important variables define the principal name given to an amphibole: the amounts of Si, Ca + Na + K, and

$$\frac{Mg}{Fe^{+2} + Fe^{+3} + Mn + Mg}$$

(= Niggli's mg) in the half-unit cell.

Figure 13.12 shows the classification scheme adopted here (Leake, 1968).

In the general formula of amphibole $X_{2-3} Y_5 Z_8 O_{22}(OH,F,Cl,O)_2$, where X = Ca, Na, K, Mn, Y = Mg, Fe^{+2}, Fe^{+3}, Al, Ti, Mn, Cr, Li, Zn, and Z = Si, Al, the quality of a chemical analysis may be considered superior if the following criteria are met:

1. The chemical analysis total should be between 99.40 and 100.60

2. The Si in the half-unit cell must not exceed 8.08.

3. Si + Al in the half-unit cell must not be less than 7.92.

4. The sum of Ca + Na + K in the half-unit cell must lie between 1.75 and 3.05.

5. The sum of $Al^{+6} + Fe^{+3} + Fe^{+2} + Mn + Mg + Cr + Ni$ (the Y group) must lie between 4.75 and 5.25.

6. The sum of OH + F + Cl should lie between 1.00 and 2.99.

7. As a rough guide Al^{+6} in the half-unit cell should not exceed $0.6 \times Al^{+4} + 0.25$. This rule does not apply to glaucophane, in which Ca is below 1.00 in the half-unit cell

The following rules may be helpful in the rapid detection of gross errors in amphibole analyses (Leake, 1968):

1. A check will show that the H_2O weight % value should be approximately equal to the calculated OH in the half-unit cell. Usually the OH is a little (0.02 to 0.06) below the H_2O weight % value, but in iron-rich amphiboles it may be a little above.

2. Titania, divided by 10 nearly always gives Ti in the half-unit cell within ± 0.04, unless it exceeds 3 to 4%, when a range of ± 0.08 is common.

3. Ferric oxide, divided by 10, usually gives the approximate Fe^{+3} in the half-unit cell within ± 0.10.

4. The lime content usually shows the following relationships:

$$6.0 \ \% \ CaO \simeq 1.0 \ Ca$$
$$10.0 \ \% \ CaO \simeq 1.5 \ Ca$$
$$13.0 \ \% \ CaO \simeq 2.0 \ Ca$$

Values of CaO in excess of 13% require special scrutiny.

5. Silica contents in excess of 47 to 49% usually give Si 7 or more in the half-unit cell, whereas values of 38 to 40% give Si less than 6 as a rough guide.

6. If an error is suspected, the positive charges should be computed by summing the products of the cations and the valencies of the elements concerned. This sum should be equal to the sum of the anion charges within approximately ± 0.05. If the difference is greater than ± 0.10, there is certainly an error. Unless the sum of $H_2O + \frac{1}{2}F$ is appreciably less than one or greater than three, the sum of the charges will lie within the range 45.0 to 47.0.

(After Leake, 1968.)

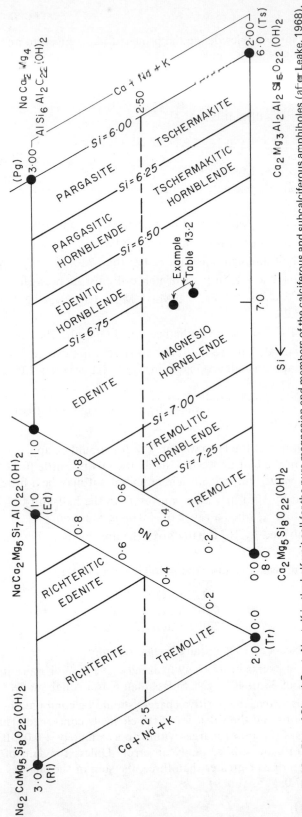

Fig. 13.12 Plot of Si and Ca + Na + K in the half unit cell for the pure magnesian end members of the calciferous and subcalciferous amphiboles (after Leake, 1968).

Computation of the structural half unit cell formula. The conversion of a weight %
chemical analysis to a structural formula is illustrated in Table 13.2. It closely follows
the example of the pyroxene of Table 13.1. Column 1 is the given chemical analysis.
Column 2 gives standard molecular weights of the oxides. Column 3 gives the molecular proportions of the oxides (=column 1 divided by column 2). Column 4 gives the
atomic proportions of oxygen (=column 3 multiplied by the number of oxygen in
each oxide). This column is totaled and 24 is divided by its total. Column 5 is the number
of anions on the basis of 24(O + OH) (=column 4 multiplied by 24/the total of column
4). Column 6 is the number of ions in the formula (=column 5 multiplied by the
number of cations/number of oxygens in the oxide; e.g., for SiO_2, multiplied by $\frac{1}{2}$, for
Al_2O_3, multiplied by $\frac{2}{3}$, for Na_2O multiplied by 2).

The amphibole computed in Table 13.2 may be positioned on the diagram in
Fig. 13.12. From Table 13.2 the Si value is 6.853 and the Ca + Na + K = 2.304.
These two values allow the amphibole to be positioned as the lower point in the field
of magnesio hornblende.

Alternatively, we may plot the Na value (the amount of Na required to make
Na + Ca up to 2) against the Si value of 6.853. This is 0.368 in Table 13.2, plotted in
Fig. 13.12 as the upper point. The two methods of plotting are not self-consistent and
in any one exercise the Ca + Na + K should be adopted in preference to the Na
value.

Formula Based on End Members. Yoder and Tilley (1962) proposed the following
arbitrary scheme for assigning an amphibole half-unit cell formula to end members.
They assume tremolite $Ca_2Mg_5Si_8O_{22}(OH)_2$ as the basis and then make type sub-
stitutions in the following order:

Glaucophane: $NaAl^6$ for CaMg.
Kupfferite Mg for Ca.
Edenite $NaAl^4$ for Si.
Tschermakite Al^6Al^4 for MgSi.
Soda tremolite NaNa for Ca.
Oxyhornblende Fe^{+3} for $Fe^{+2}H$.
Ferrihornblende $2Fe^{+3}$ for $3Fe^{+2}$.
Titanhornblende Ti for 2Mg.

The rules are as follows:

1. Na + K are assigned to the Ca position to make a total of 2 (Table 13.2). In the
example Na added = 0.368.

2. Half this amount × 100 = glaucophane mol %. In the example glaucophane
Gp = 18.4.

3. If there is inadequate Na + K in step 1, Mg is assigned as kupfferite.

4. If Na + K is in excess, it is assigned to the vacant site and the full amount × 100 =
Ed mol %. In our example (Table 13.2) the excess Na + K = 0.304 and Ed = 30.4.

5. An amount of Al equivalent to Na + K in the vacant site is then assigned to the Si
position. An additional amount of Al is added to Si to make the total = 8.000. Half the
latter amount of Al × 100 is the tschermakite content in molecular %. In our
example the additional amount of Al added = 1.138 − 0.304 = 0.824 and Ts =
0.824/2 × 100 = 41.2.

Table 13.2 Calculation of the Structural Half Unit Cell Formula of an Amphibole[a]

	(1) Weight %	(2) Molecular Weight Oxides	(3) Molecular Proportions of Oxides	(4) Atomic Proportions Oxygen	(5) Number of Anions on Basis 24 (O + OH)	(6) Number of Ions in Formula
SiO_2	48.04	60.09	0.7995	1.5990	13.706	$Si = 6.853$
Al_2O_3	13.02	101.94	0.1277	0.3831	3.284	$P = 0.019$
Fe_2O_3	1.28	159.70	0.0080	0.0240	0.206	$Al = 2.189$
FeO	9.49	71.85	0.1321	0.1321	1.132	$Cr = 0.010$
MnO	0.16	70.94	0.0023	0.0023	0.020	$Fe^{+3} = 0.137$
MgO	10.17	40.32	0.2522	0.2522	2.162	$Ti = 0.240$
CaO	10.68	56.08	0.1904	0.1904	1.632	$Mg = 2.162$
Na_2O	2.10	61.98	0.0339	0.0339	0.290	$Fe^{2+} = 1.132$
K_2O	0.51	94.20	0.0054	0.0054	0.046	$Mn = 0.020$
P_2O_5	0.16	141.95	0.0011	0.0055	0.047	$Ca = 1.632$
TiO_2	2.24	79.90	0.0280	0.0560	0.480	$Na = 0.580$
Cr_2O_3	0.09	152.02	0.0006	0.0018	0.015	$K = 0.092$
H_2O+	2.06	18.02	0.1143	0.1143	0.980	$OH = 1.960$
Total	100.00		Total 2.8000	2.8000		

Column (6) groupings:

$$Si = 6.853 \left. \begin{array}{r} 6.853 \\ 0.019 \end{array} \right\} 8.000$$

$$Al = 2.189 \left. \begin{array}{r} 1.128 \\ 1.061 \end{array} \right\}$$

$$\left. \begin{array}{r} Cr = 0.010 \\ Fe^{+3} = 0.137 \end{array} \right. \quad \left. \begin{array}{r} 0.010 \\ 0.137 \end{array} \right\} 1.278$$

$$Ti = 0.240 \quad 0.240 = 4.762$$

$$\left. \begin{array}{r} Mg = 2.162 \\ Fe^{2+} = 1.132 \end{array} \right. \quad \left. \begin{array}{r} 3.314 \end{array} \right\} 3.314$$

$$\left. \begin{array}{r} Ca = 1.632 \\ Na = 0.580 \end{array} \right. \quad \left. \begin{array}{r} 1.632 \\ 0.368 \end{array} \right\} 2.000$$

$$K = 0.092 \left. \begin{array}{r} 0.212 \\ 0.092 \end{array} \right\} 0.734$$

$$\frac{24}{2.8000} = 8.5714$$

[a] Example after Yoder and Tilley (1962), p.456.

6. Half the Ca not consumed by the edenite and tschermakite molecules, $\times 100$, is computed as tremolite. In our example the Ca consumed by edenite $= 2 \times (Na + K)$ in edenite $= 2 \times 0.304 = 0.608$. The Ca consumed by tschermakite $=$ the additional Al added, $= 0.824$. Hence the total Ca used to form edenite $+$ tschermakite $= 0.608 + 0.824 = 1.432$. Hence Ca left over $= 1.632 - 1.432 = 0.200$. Half this $\times 100 = Tr = 10.0$. Hence the amphibole in Table 13.2 may be written as $Gp_{18.4}Ed_{30.4}Tsch_{41.2}Tr_{10.0}$.

DISCUSSION

The example in Table 13.2 sums to 100 % without recourse to additional molecules. Riebeckite and Hastingsite, however, can be formulated from these molecules if desired. Account is taken of soda tremolite and oxyhornblende, where the presence of these molecules may be evident in the summation of the Mg position. A deficiency of R^{+3} may indicate soda tremolite; an excess may indicate oxyhornblende.

The plotting of amphiboles in a figure such as 13.12 and the recasting into end member formulas by the above methods, though imperfect and arbitrary, may help to illustrate petrographic trends.

In addition, the chemical analyses of complete basaltic rocks have been computed in an identical way to illustrate what happens when basalts are converted to amphibolites. The chemical analysis of a basalt is analogous to that of a hornblende; all that is required is the addition of a suitable amount of water to the analysis and a recalculation to 100% total. The treatment of a basalt analysis as an amphibole has been used by Yoder and Tilley (1962) to make deductions about the conversion of basaltic rocks to amphibolites and hornblendites.

Robinson and Jaffe (1969) have used a projection like Thompson's AFM diagram (exercise 13.6) to project amphibole analyses.

The tetrahedron $(Al_2O_3 - Na_2O) : CaO : MgO : (FeO + MnO)$ (molecular proportions) is used as the basis. Since many amphibole-bearing rocks contain plagioclase, anorthite is a suitable projection point. The projection assumes that the mineralogical assemblage will

   (i)   contain quartz,
   (ii) contain plagioclase,
   (iii) Show that the composition of the plagioclase has no effect on that of the amphibole,
   (iv) show that other components are negligible.

These requirements are probably never fulfilled but nevertheless, the projection is useful for comparing assemblages. The projection and its construction are shown in Fig. 13.13.

Tie lines can be drawn to join minerals that coexist in any rock in the same way as the *AFM* diagram is used to portray metamorphic facies.

EXERCISE 13.16

Relation Between Garnet Chemistry and Metamorphic Grade

It has been found that the composition of garnets varies systematically with metamorphic grade. $(CaO + MnO)$ wt $\% = 39.5 - 1.012 (FeO + MgO)$ wt $\%$ as a linear relationship (Fig. 13.14).

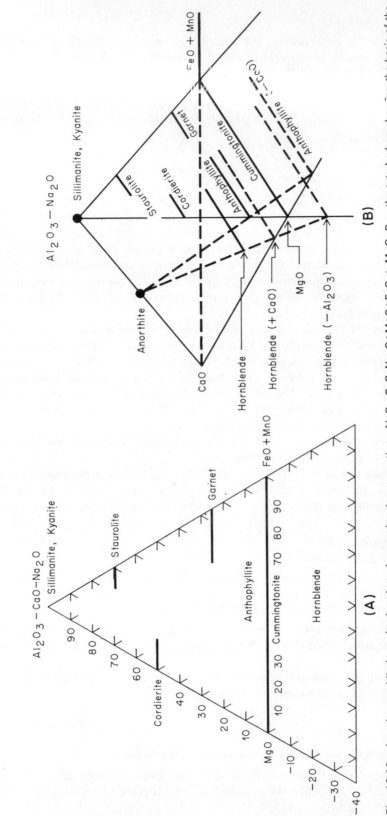

Fig. 13.13 A = the amphibole projection, based on molecular proportions Al_2O_3-CaO-Na_2O : MgO : $FeO + MnO$. B = the tetrahedron showing the basis of the projection from anorthite. (After Robinson and Jaffe, 1969.)

412

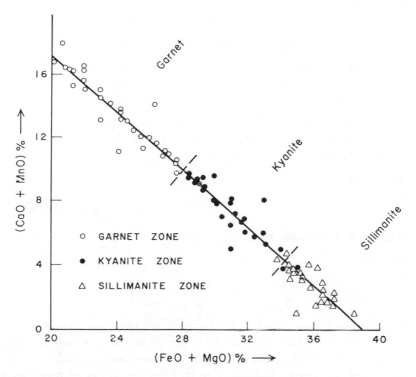

Fig. 13.14 The relationship between garnet composition weight % CaO + MnO versus FeO + MgO and metamorphic grade. (After Nandi, 1967.)

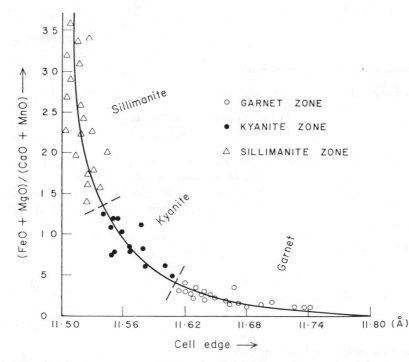

Fig. 13.15 Relationship between metamorphic grade and a plot of weight % (FeO + MgO) / (CaO + MnO) versus unit cell edge in Å. (After Nandi, 1967.)

The FeO + MgO % content increases with metamorphic grade:

$(FeO + MgO)\% < 28$ garnet zone,
$(FeO + MgO)\% > 28 < 34.5$ kyanite zone,
$(FeO + MgO)\% > 34.5$ sillimanite zone.

It has also been shown that (FeO + MgO wt %)/(CaO + MnO wt %) varies systematically with the garnet unit cell edge in Å (Fig. 13.15). The cell edge reduces with increasing metamorphic grade.

(After Nandi, 1967.)

EXERCISE 13.17

Weathering Index of a Silicate Rock

The expression

$$\left[\frac{(Na)a}{0.35} + \frac{(Mg)a}{0.9} + \frac{(K)a}{0.25} + \frac{(Ca)a}{0.7} \right] \times 100$$

is proposed as the weathering index of a silicate rock. $(X)a$ indicates the atomic percentage of element X divided by its atomic weight. The denominators in the function are the bond strength between the element and oxygen in each case.

For any particular rock the weathering index progressively decreases from fresh rock to weathered rock; for example, fresh granite may have a value of about 80, completely weathered granite, 30. Basalt may go from 90 to 20 on weathering. The index therefore provides a chemical index for mathematically describing weathering.

(After Parker, 1970.)

EXERCISE 13.18

The Alkalinity Ratio for Igneous Rocks

The alkalinity ratio

$$\frac{Al_2O_3 + CaO + \text{total alkalis}}{Al_2O_3 + CaO - \text{total alkalis}}$$

(all in weight %) is proposed. When SiO_2 exceeds 50% and $K_2O:Na_2O > 1 < 2.5$, then $2(Na_2O)$ is used in place of the total alkalis in the expression.

The alkalinity ratio is plotted against SiO_2 and found to give a useful separation between strongly alkaline, alkaline, and calc-alkaline rocks over the full range of silica contents.

(After Wright, 1969.)

B.   NORM CALCULATIONS

EXERCISE 13.19

Rules for Calculating the C.I.P.W. Norm

The following normative minerals are used:

Mineral	Abbreviation	Formula	Molecular Weight
		Salic Group	
Quartz	(Q)	SiO_2	60.08
Corundum	(C)	Al_2O_3	101.96
Zircon	(Z)	$ZrO_2 . SiO_2$	183.30
Orthoclase	(Or)	$K_2O. Al_2O_3.6SiO_2$	556.64
Albite	(Ab)	$Na_2O.Al_2O_3.6SiO_2$	524.42
Anorthite	(An)	$CaO.Al_2O_3.2SiO_2$	278.20
Leucite	(Lc)	$K_2O.Al_2O_3.4SiO_2$	436.48
Nepheline	(Ne)	$Na_2O.Al_2O_3.2SiO_2$	284.10
Kaliophilite	(Kp)	$K_2O.Al_2O_3.2SiO_2$	316.32
Halite	(Hl)	$NaCl$	58.44
Thernardite	(Th)	$Na_2O.SO_3$	142.04
Sodium carbonate	(Nc)	$Na_2O.CO_2$	105.99
		Femic Group	
Acmite	(Ac)	$Na_2O.Fe_2O_3.4SiO_2$	461.99
Sodium metasilicate	(Ns)	$Na_2O.SiO_2$	122.06
Potassium metasilicate	(Ks)	$K_2O.SiO_2$	154.28
Diopside	(Di)	$CaO.MgO.2SiO_2$	216.55
		$CaO.FeO.2SiO_2$	248.09
Wollastonite	(Wo)	$CaO.SiO_2$	116.16
Hypersthene	(Hy)	$MgO.SiO_2$	100.39
		$FeO.SiO_2$	131.93
Olivine	(Ol)	$2MgO.SiO_2$	140.70
		$2FeO.SiO_2$	203.78
Dicalcium silicate	(Cs)	$2CaO.SiO_2$	172.24
Magnetite	(Mt)	$FeO.Fe_2O_3$	231.54
Chromite	(Cm)	$FeO.Cr_2O_3$	223.84
Ilmenite	(Il)	$FeO.TiO_2$	151.75
Hematite	(Hm)	Fe_2O_3	159.69
Sphene	(Tn)	$CaO.TiO_2.SiO_2$	196.06
Perovskite	(Pf)	$CaO.TiO_2$	135.98
Rutile	(Ru)	TiO_2	79.90
Apatite	(Ap)	$3CaO.P_2O_5.1/3CaF_2$	336.21
Fluorite	(Fr)	CaF_2	78.08
Pyrite	(Pr)	FeS_2	119.98
Calcite	(Cc)	$CaO.CO_2$	100.09

The following are the step-by-step rules for this calculation. *Amount* means molecular proportion = weight %/formula weight.

1. Calculate the amounts (molecular proportions) of the oxides and elements present by dividing the given weight percentages by the appropriate formula weights. The following formula weights may be used:

SiO_2	60.08	TiO_2	79.90
Al_2O_3	101.96	ZrO_2	123.22
Fe_2O_3	159.69	MnO	70.94
FeO	71.85	NiO	74.71
MgO	40.31	BaO	153.34

CaO	56.08	SrO	103.62
Na_2O	61.98	Cr_2O_3	151.98
K_2O	94.20	Cl	35.45
SO_3	80.06	P_2O_5	141.94
F	19.00	CO_2	44.01
S	32.06		

2. Add the (MnO + NiO) amount to the FeO amount.

3. Add the (BaO + SrO) amount to the CaO amount.

4. Set $Z = ZrO_2$; set $Y^a = Z$.

5. Set $Ap = P_2O_5$ and subtract an amount equal to 10/3 Ap from CaO. If fluorine is present and if $F > \frac{2}{3}$ Ap, subtract an amount equal to $\frac{2}{3}$ Ap from F. If $F < \frac{2}{3}$ Ap, all the fluorine is allocated to Ap.

6. Set $Fr = F/2$ and subtract an amount equal to Fr from CaO.

7. Set $Hl = Cl$ and subtract an amount equal to H1/2 from Na_2O.

8. If SO_3 is to be calculated as SO_3 and not as S, set $Th = SO_3$; subtract an amount equal to Th from Na_2O.

9. If S is present or if SO_3 is to be calculated as S, put $Pr = S/2$ or $SO_3/2$ and subtract an amount equal to Pr from FeO.

10. a. If the rock contains cancrinite, put $Nc = CO_2$, and subtract an amount equal to Nc from Na_2O.

b. If the rock contains calcite, set $Cc = CO_2$; subtract an amount equal to Cc from CaO. If the modal calcite is secondary or from associated limestone, it is *not* included in the norm.

11. Set $Cm = Cr_2O_3$ and subtract an amount equal to Cm from FeO.

12. If $FeO > TiO_2$, set $Il = TiO_2$ and subtract an amount equal to Il from FeO. There is no further TiO_2 available to form other normative minerals. If $FeO < TiO_2$, set $Il = FeO$ and subtract an amount equal to Il from TiO_2. There is no further FeO available to form other normative minerals.

13. If $Al_2O_3 > K_2O$, set $Or'* = K_2O$ and subtract an amount equal to Or' from Al_2O_3; allocate an amount equal to 6 Or' to Y. If $Al_2O_3 < K_2O$, set $Or' = Al_2O_3$ and subtract an amount equal to Or' from K_2O. There is no further Al_2O_3 available to form other normative minerals. Set $K_2O = Ks$ and add an amount equal to (6 Or' + Ks) to Y.

14. If $Al_2O_3 > Na_2O$, set $Ab' = Na_2O$ and subtract an amount equal to Ab' from Al_2O_3. There is no further Na_2O available to form other normative minerals. Add an amount equal to 6Ab' to Y (the same Y as in step 13). If $Al_2O_3 < Na_2O$, set $Ab' = Al_2O_3$ and subtract an amount equal to Ab' from Na_2O. There is no further Al_2O_3 available to form other normative minerals. Add an amount equal to 6Ab' to Y.

15. If $Na_2O > Fe_2O_3$, set $Ac = Fe_2O_3$ and subtract an amount equal to Ac from Na_2O. There is no further Fe_2O_3 available to form further normative minerals. Set $Ns = Na_2O$ and add an amount equal to (4Ac + Ns) to Y. If $Na_2O < Fe_2O_3$, set $Ac = Na_2O$ and subtract an amount equal to Ac from Fe_2O_3; add an amount equal to 4Ac to Y.

---

[a] Throughout the calculation, amounts will be allocated to Y. The final total Y is required at rule 21.
* Or' means orthoclase (Or) formed only provisionally at this stage; similarly Ab' means albite formed provisionally.

16. If $Al_2O_3 > CaO$, set $An = CaO$ and subtract an amount equal to An from Al_2O_3. There is no further CaO available to form other normative minerals. Add an amount equal to 2An to Y; set $C = Al_2O_3$. If $Al_2O_3 < CaO$, set $An = Al_2O_3$, subtract an amount equal to An from CaO, and add an amount equal to 2 An to Y.

17. If $CaO > TiO_2$, set $Tn' = TiO_2$, subtract an amount equal to Tn' from CaO, and add an amount equal to Tn' to Y. If $CaO < TiO_2$, set $Tn' = CaO$ and subtract an amount equal to Tn' from TiO_2. There is no further CaO available to form other normative minerals. Set $Ru = TiO_2$ and add an amount equal to Tn' to Y.

18. If $Fe_2O_3 > FeO$, set $Mt = FeO$ and subtract an amount equal to Mt from Fe_2O_3; There is no further FeO available to form other normative minerals. Set $Hm = Fe_2O_3$. If $Fe_2O_3 < FeO$, set $Mt = Fe_2O_3$ and subtract an amount equal to Mt from FeO.

19. Add MgO to FeO to form (Mg, Fe)O. Calculate the ratios $MgO/(MgO + FeO)$ and $FeO/(MgO + FeO)$ and use these relative proportions to calculate the weight percentages of diopside, hypersthene, and olivine.

20. If $CaO > (Mg, Fe)O$, set $Di' = (Mg, Fe)O$, subtract an amount equal to Di' from CaO, set $Wo' = CaO$, and add an amount equal to $(2Di' + Wo')$ to Y. If $CaO < (Mg,Fe)O$, set $Di' = CaO$, subtract an amount equal to Di' from (Mg, Fe)O, set $Hy' = (Mg, Fe)O$, and add an amount equal to $(2Di' + Hy')$ to Y.

21. Y now gives the amount of silica required for all the normative minerals formed so far. If $SiO_2 > Y$, set $Q = SiO_2 - Y$. The calculation of the norm is completed by conversion of the molecular proportions to weight percentages of normative minerals by multiplying the amount of each normative mineral molecule by its molecular weight (given at the beginning of this exercise). If $SiO_2 < Y$, however, the norm calculation must proceed. The deficiency $D = Y - SiO_2$ and the rules from 22 have to be applied. The rules continue to be applied until the deficiency D has been reduced to zero, at which stage the calculation is completed by conversion of the molecular proportions to weight percentages of the normative minerals.

22. If $D < Hy'/2$, set $Ol = D$ and $Hy = Hy' - 2D$. The silica deficiency is now zero and the norm is complete. If $D > Hy'/2$ set $Ol = Hy'/2$ and $Hy = 0$; put $D_1 = D - Hy'/2$ and proceed.

23. If $D_1 < Tn'$, set $Tn = Tn' - D_1$ and $Pf = D_1$. The silica deficiency is now zero. If $D_1 > Tn'$, set $Pf = Tn'$ and $Tn = 0$, put $D_2 = D_1 - Tn'$.

24. If $D_2 < 4Ab'$, set $Ne = D_2/4$ and $Ab = Ab' - D_2/4$. The silica deficiency is now zero. If $D_2 > 4Ab'$, set $Ne = Ab'$ and $Ab = 0$ and put $D_3 = D_2 - 4Ab'$.

25. If $D_3 < 2Or'$, set $Lc = D_3/2$ and $Or = Or' - D_3/2$. The silica deficiency is now zero. If $D_3 > 2Or'$, set $Lc' = Or'$ and $Or' = 0$; put $D_4 = D_3 - 2Or'$,

26. If $D_4 < Wo'/2$, set $Cs = D_4$ and $Wo = Wo' - 2D_4$. The silica deficiency is now zero. If $D_4 > Wo'/2$, set $Cs = Wo'/2$ and $Wo = 0$, put $D_5 = D_4 - Wo'/2$.

27. If $D_5 < Di'$ add an amount equal to $D_5/2$ to the amounts of Cs and Ol already in the norm; set $Di = Di' - D_5$. The silica deficiency is now zero. If $D_5 > Di'$, add an amount equal to $Di'/2$ to the amounts of Cs and Ol already in the norm; put $Di = 0$ and $D_6 = D_5 - Di'$.

28. Set $Kp = D_6/2$ and $Lc = Lc' - D_6/2$.

(After Johannsen, 1931; modified by Kelsey, 1965.)

As an example, a Bostonite gives the following calculation:

	Weight %	Amount (with progressive deductions)
SiO_2	61.15	1.019
Al_2O_3	22.07	0.217, 0.143, 0.048, −
Fe_2O_3	1.05	0.006, −
FeO	1.02	0.014, 0.011, 0.005 ⎫ 0.015, −
MgO	0.40	0.010 ⎭
CaO	0.75	0.014, −
Na_2O	5.86	0.095, −
K_2O	7.01	0.074,
H_2O+	0.71	−
TiO_2	0.21	0.003, −

Normative Minerals

		Y
Il	0.003	
Or′	0.074	0.444
Ab′	0.095	0.570
An	0.014	0.028
C	0.034	
Mt′	0.006	
Hy′	0.015	0.015
		$\overline{1.057}$

$$\frac{MgO}{(Mg, Fe)O} = \tfrac{2}{3},$$

$$\frac{FeO}{(Mg, Fe)O} = \tfrac{1}{3}.$$

$D = 0.038$, Hy′/2 = 0.008; therefore Ol = 0.008, Hy = 0.

$$D_1 = 0.030.$$

4Ab′ = 0.380; therefore Ab = 0.087; Ne = 0.008.

Norm

C	3.47	Ol	1.21
Or	41.14	Mt	1.39
Ab	45.59	Il	0.46
An	3.89		
Ne	2.27		

Rules for Calculating the Niggli Molecular Norm (Catanorm)

The following normative minerals are used:

Mineral	Abbreviation	Formula
	Salic Group	
Quartz	(Q)	SiO_2
Corundum	(C)	$AlO_{1\frac{1}{2}}$
Zircon	(Z)	$ZrO_2 \cdot SiO_2$
Orthoclase	(Or)	$KO_{\frac{1}{2}} \cdot AlO_{1\frac{1}{2}} \cdot 3SiO_2$
Albite	(Ab)	$NaO_{\frac{1}{2}} \cdot AlO_{1\frac{1}{2}} \cdot 3SiO_2$
Anorthite	(An)	$CaO \cdot 2AlO_{1\frac{1}{2}} \cdot 2SiO_2$
Leucite	(Lc)	$KO_{\frac{1}{2}} \cdot AlO_{1\frac{1}{2}} \cdot 2SiO_2$
Nepheline	(Ne)	$NaO_{\frac{1}{2}} \cdot AlO_{1\frac{1}{2}} \cdot SiO_2$
Kaliophilite	(Kp)	$KO_{\frac{1}{2}} \cdot AlO_{1\frac{1}{2}} \cdot SiO_2$
Halite	(Hl)	$NaCl$
Plagioclase	(Plag)	$Ab + An = Ab_x An_{100-x}$
	Femic Group	
Acmite	(Ac)	$NaO_{\frac{1}{2}} \cdot FeO_{1\frac{1}{2}} \cdot 2SiO_2$
Sodium metasilicate	(Ns)	$2NaO_{\frac{1}{2}} \cdot SiO_2$
Potassium metasilicate	(Ks)	$2KO_{\frac{1}{2}} \cdot SiO_2$
Wollastonite	(Wo)	$CaO \cdot SiO_2$
Enstatite	(En)	$MgO \cdot SiO_2$
Ferrosilite	(Fs)	$FeO \cdot SiO_2$
Forsterite	(Fo)	$2MgO \cdot SiO_2$
Fayalite	(Fa)	$2FeO \cdot SiO_2$
Calcium orthosilicate	(Cs)	$2CaO \cdot SiO_2$
Magnetite	(Mt)	$FeO \cdot 2FeO_{1\frac{1}{2}}$
Chromite	(Cm)	$FeO \cdot 2CrO_{1\frac{1}{2}}$
Hematite	(Hm)	$FeO_{1\frac{1}{2}}$
Ilmenite	(Il)	$FeO \cdot TiO_2$
Titanite	(Tn)	$CaO \cdot TiO_2 \cdot SiO_2$
Perovskite	(Pf)	$CaO \cdot TiO_2$
Rutile	(Ru)	TiO_2
Apatite	(Ap)	$9CaO \cdot 6PO_{2\frac{1}{2}} \cdot CaF_2$

(if there is no fluorine in the analysis, $Ap = 5CaO \cdot 3PO_{2\frac{1}{2}}$)

Fluorite	(Fr)	CaF_2
Pyrite	(Pr)	FeS_2
Calcite	(Cc)	$CaO \cdot CO_2$
Cassiterite	(Ct)	SnO_2
Diopside	(Di)	$Wo_{50}En_xFs_{50-x}$
Hypersthene	(Hy)	En_xFs_{100-x}
Olivine	(Ol)	Fo_xFa_{100-x}

The following are the rules for calculation.

1. Determine the cation proportions of each chemical component by dividing the percentage weight of each oxide by its equivalent molecular weight and multiply each result by 1000 for convenience of computation. The following equivalent molecular weights may be used:

SiO_2	60.08;	$AlO_{1\frac{1}{2}}$	50.98;	TiO_2	79.90;	$FeO_{1\frac{1}{2}}$	79.85;
FeO	71.85;	MnO	70.94;	NiO	74.71;	MgO	40.31;
CaO	56.08;	BaO	153.34;	SrO	103.62;	$NaO_{\frac{1}{2}}$	30.99
$KO_{\frac{1}{2}}$	47.10;	$PO_{2\frac{1}{2}}$	70.97;	CO_2	40.01;	ZrO_2	123.22;
F	19.00;	S	32.06;	$CrO_{1\frac{1}{2}}$	75.99;	Cl	35.45;
SnO_2	150.69.						

2. Convert the cation proportions to a percentage by multiplying each by 100/the total of the cation proportions.

3. Make FeO $=$ FeO $+$ MnO $+$ NiO.

4. Make CaO $=$ CaO $+$ BaO $+$ SrO.

5. Make calcite $= 2 \times (CO_2)$. Subtract the CO_2 amount from CaO.

6. Make cassiterite $= SnO_2$.

7. a. If $PO_{2\frac{1}{2}} < 3 \times F$, make apatite $= 3 \times PO_{2\frac{1}{2}}$. Reduce CaO by an amount $=$ 1.667 of $PO_{2\frac{1}{2}}$. Reduce F by 0.33 of $PO_{2\frac{1}{2}}$.

    b. If $PO_{2\frac{1}{2}} \gtrsim 3 \times F$, make apatite $= (2.667 \times PO_{2\frac{1}{2}}) + F$. Reduce CaO by 1.667 of $PO_{2\frac{1}{2}}$; F becomes zero.

8. Make fluorite $= 1.5 \times F$. Reduce CaO by $\frac{1}{2}F$; F becomes zero.

9. Make halite $=$ twice Cl. Reduce Na by an amount $=$ Cl; Cl becomes zero.

10. Make pyrite $= 1.5 \times S$. Reduce FeO by an amount $= \frac{1}{2}S$.

11. Make chromite $= 1.5 \times CrO_{1\frac{1}{2}}$. Reduce FeO by an amount $= \frac{1}{2} CrO_{1\frac{1}{2}}$.

12. Make zircon $=$ twice ZrO_2. Reduce SiO_2 by an amount $= ZrO_2$.

13. a. If $TiO_2 \leq FeO$, make ilmenite $=$ twice TiO_2. Reduce FeO by an amount equal to TiO_2

    b. If $TiO_2 > FeO$, make ilmenite $=$ twice FeO. Reduce TiO_2 by an amount $=$ FeO; FeO becomes zero.

14. a. If $KO_{\frac{1}{2}} \leq AlO_{1\frac{1}{2}}$, make orthoclase $= 5 \times KO_{\frac{1}{2}}$. Reduce $AlO_{1\frac{1}{2}}$ by an amount $= KO_{\frac{1}{2}}$. Reduce SiO_2 by an amount $= 3 \times KO_{\frac{1}{2}}$; $KO_{\frac{1}{2}}$ becomes zero.

    b. If $KO_{\frac{1}{2}} > AlO_{1\frac{1}{2}}$, make orthoclase $= 5 \times AlO_{1\frac{1}{2}}$. Reduce $KO_{\frac{1}{2}}$ by an amount $= AlO_{1\frac{1}{2}}$. Reduce SiO_2 by an amount $= 3 \times AlO_{1\frac{1}{2}}$; $AlO_{1\frac{1}{2}}$ becomes zero.

15. The $KO_{\frac{1}{2}}$ remaining is used to make potassium metasilicate of amount $= 1.5 \times KO_{\frac{1}{2}}$. SiO_2 is reduced by an amount $= \frac{1}{2}$ of $KO_{\frac{1}{2}}$; $KO_{\frac{1}{2}}$ becomes zero.

16. a. If $NaO_{\frac{1}{2}} \lesssim AlO_{1\frac{1}{2}}$, make albite $= 5 \times NaO_{\frac{1}{2}}$. Reduce $AlO_{1\frac{1}{2}}$ by an amount $= NaO_{\frac{1}{2}}$. Reduce SiO_2 by an amount $= 3 \times NaO_{\frac{1}{2}}$; $NaO_{\frac{1}{2}}$ becomes zero.

    b. If $NaO_{\frac{1}{2}} > AlO_{1\frac{1}{2}}$, make albite $= 5 \times AlO_{1\frac{1}{2}}$. Reduce $NaO_{\frac{1}{2}}$ by an amount $= AlO_{1\frac{1}{2}}$. Reduce SiO_2 by an amount $= 3 \times AlO_{1\frac{1}{2}}$; $AlO_{1\frac{1}{2}}$ becomes zero.

17. a. If $NaO_{\frac{1}{2}} \leq FeO_{1\frac{1}{2}}$, make acmite $= 4 \times NaO_{\frac{1}{2}}$. Reduce $FeO_{1\frac{1}{2}}$ by an amount $= NaO_{\frac{1}{2}}$. Reduce SiO_2 by an amount $=$ twice $NaO_{\frac{1}{2}}$.

    b. If $NaO_{\frac{1}{2}} > FeO_{1\frac{1}{2}}$, make acmite $= 4 \times FeO_{1\frac{1}{2}}$. Reduce SiO_2 by an amount $=$ twice $FeO_{1\frac{1}{2}}$; $FeO_{1\frac{1}{2}}$ becomes zero.

18. Make sodium metasilicate from the sodium left over. $NS = 1.5 \times NaO_{\frac{1}{2}}$. Reduce SiO_2 by an amount $= \frac{1}{2}$ the $NaO_{\frac{1}{2}}$.

19. a. If $CaO \lesssim \frac{1}{2}AlO_{1\frac{1}{2}}$, make anorthite $= 5 \times CaO$. Reduce $AlO_{1\frac{1}{2}}$ by an amount $=$ twice CaO. Reduce SiO_2 by an amount $=$ twice CaO; CaO becomes zero.

    b. If $CaO > \frac{1}{2}AlO_{1\frac{1}{2}}$, make anorthite $= 2\frac{1}{2} \times AlO_{1\frac{1}{2}}$. Reduce CaO by an amount $= \frac{1}{2}AlO_{1\frac{1}{2}}$. Reduce SiO_2 by an amount $= AlO_{1\frac{1}{2}}$; $AlO_{1\frac{1}{2}}$ now becomes zero.

20. a. If $TiO_2 \lesssim CaO$, make titanite (sphene) $= 3 \times TiO_2$. Reduce CaO by an amount $= TiO_2$. Reduce SiO_2 by an amount $= TiO_2$; TiO_2 becomes zero.

    b. If $TiO_2 > CaO$, make titanite $= 3 \times CaO$. Reduce TiO_2 by an amount $= CaO$. Reduce SiO_2 by an amount $= CaO$; CaO becomes zero.

21. Make rutile $=$ the TiO_2 remaining.

22. Make corundum $=$ the $AlO_{1\frac{1}{2}}$ remaining.

23. a. If $FeO_{1\frac{1}{2}} \lesssim$ twice FeO, then make magnetite $= 1\frac{1}{2}$ times $FeO_{1\frac{1}{2}}$. Reduce FeO by an amount $= \frac{1}{2}FeO_{1\frac{1}{2}}$; $FeO_{1\frac{1}{2}}$ becomes zero.

    b. If $FeO_{1\frac{1}{2}} >$ twice FeO, make magnetite $= 3 \times FeO$. Reduce $FeO_{1\frac{1}{2}}$ by an amount $-$ twice FeO; FeO becomes zero.

24. Make hematite $= FeO_{1\frac{1}{2}}$.

25. Make wollastonite $=$ twice CaO. Reduce SiO_2 by an amount $= CaO$; CaO becomes zero.

26. Make enstatite $=$ twice the MgO remaining. Reduce SiO_2 by an amount $=$ the MgO; MgO becomes zero.

27. Make ferrosilite $=$ twice the FeO remaining. Reduce SiO_2 by an amount $= FeO$.

28. Make hypersthene $=$ enstatite $+$ ferrosilite.

29. Calculate the ratio $En_x Fs_{100-x}$, where $x = (En \times 100)/(En + Fs)$.

30. a. If hypersthene $<$ wollastonite, make diopside $=$ twice the hypersthene amount. Reduce wollastonite by an amount $=$ the hypersthene used; Hy becomes zero.

    b. If hypersthene $\geq$ wollastonite, make diopside $=$ twice the wollastonite. Reduce hypersthene by an amount $=$ the wollastonite; Wo becomes zero.

31. a. If the total SiO_2 remaining is zero, quartz $=$ zero and the norm calculation is complete.

    b. If the total SiO_2 remaining is more than zero, make quartz $= SiO_2$ remaining. The norm is now complete.

    c. If the total SiO_2 remaining is less than zero (having a negative value), make quartz $= -SiO_2$ (i.e., an amount of negative quartz) and proceed to the next steps until the negative quartz is destroyed by reclaiming SiO_2 by substituting minerals that contain less SiO_2 than those already formed.

32. a. If hypersthene $\gtrsim 4 \times$ the negative quartz, make olivine $= 3 \times$ the quartz value (Q). Reduce hypersthene by an amount $= 4 \times Q$; Q becomes zero and the norm is complete.

    b. If, however, hypersthene $< 4 \times$ the negative quartz, make olivine $= \frac{3}{4}$ the amount of hypersthene. Reduce the negative quartz by an amount $= \frac{1}{4}$ the hypersthene; Hy becomes zero.

33. a.   If titanite $\geq 3 \times$ the remaining negative quartz, make perovskite = twice the remaining negative quartz. Reduce titanite by an amount $= 3 \times Q$; Q becomes zero and the norm is complete.

b.   However, if titanite $< 3 \times$ the negative quartz, make perovskite $= 0.667 \times$ Tn. The negative quartz is then reduced by an amount $= 0.333$ times titanite; Tn becomes zero.

34. a.   If albite is greater or equal to $2\frac{1}{2}$ times the negative quartz, make nepheline = $1\frac{1}{2} \times$ the negative quartz. Reduce albite by an amount $= 2\frac{1}{2} \times$ the negative quartz; Quartz becomes zero and the norm is complete.

b.   If albite is less than $2\frac{1}{2} \times$ the negative quartz, make Ne $= 0.6 \times$ Ab. Reduce the negative quartz by an amount $= 0.4 \times$ Ab; Ab becomes zero.

35. a.   If orthoclase $\gtrless 5 \times$ the negative quartz, make leucite $= 4 \times$ the negative quartz. Reduce Or by an amount $= 5 \times$ the negative quartz; Q becomes zero and the norm is complete.

b.   If orthoclase $< 5 \times$ the negative quartz, make leucite $= 0.8 \times$ the Or amount. Reduce the negative quartz by an amount $= 0.2 \times$ the Or; Or becomes zero.

36. a.   If leucite $\gtrless 4 \times$ the amount of negative quartz, make Kp $= 3 \times$ the negative quartz. Reduce Lc by an amount $= 4 \times$ the negative quartz; Q becomes zero and the norm is complete.

b.   If Lc $< 4 \times$ the negative quartz, make Kp $= 0.75$ times Lc. Reduce the negative quartz by an amount $= 0.25 \times$ Lc; Lc becomes zero.

37. a.   If wollastonite $\gtrless 4 \times$ the negative quartz, make Cs $= 3 \times$ the negative quartz. Reduce Wo by an amount $= 4 \times$ the negative quartz; Q becomes zero and the norm is complete.

b.   If Wo $< 4 \times$ the negative quartz, make Cs $= 0.75$ the Wo amount. Reduce the negative quartz by an amount 0.25 times Wo; Wo becomes zero.

38. a.   If diopside $\gtrless 4 \times$ the negative quartz, make Cs $= (1\frac{1}{2} \times$ the negative quartz) + the Cs already formed. Make olivine $= (1.5 \times$ the negative quartz) + the Ol already formed. Reduce the amount of diopside by an amount $= 4 \times$ the negative quartz; Q becomes zero and the norm is complete.

b.   If diopside $< 4 \times$ the negative quartz, make Cs $= (0.375 \times$ Di) + the Cs already formed. Make Ol $= (0.375 \times$ Di) + the Ol already formed. Reduce the amount of negative quartz by 0.25 of Di; Di becomes zero.

By this stage the negative quartz should have been reduced to zero and the norm calculation completed.

(Method after Barth, 1962a; modified by Hutchison and Jeacoke 1971).

DISCUSSION

From step 31 on the following equations are used in converting to less SiO_2 saturated minerals: $4 \; Hy = 3 \; Ol + Q$; $5Ab = 3Ne + 2Q$; $5Or = 4Lc + Q$; $4Lc = 3Kp + Q$; $4Wo = 3Cs + Q$; $4Di = 3Cs + 3Ol + 2Q$. These rules have been written in the form of a complete FORTRAN IV computer program specifically for an I.B.M. 1130 computer (Hutchison, and Jeacocke, 1971). A complete listing of the program may be obtained from me on request. On rare occasions the program will give a negative value to An. This is a result of insufficient CaO in the analysis to combine with both P_2O_5 and CO_2. It is recommended that the norm be recalculated to omit either or

```
OXIDE     WEIGHT      CATION      GROUP CATION
        PER CENT    PER CENT      PER CENT

SI       46.53       44.57        44.57
AL       14.00       16.79        16.79
TI        2.18        1.57         1.57
FE3       2.93        2.11         2.11

FE2      10.10        8.09)
MN        0.11        0.08)        8.17
NI        0.00        0.00)

MG        6.97        9.95         9.95

CA        8.82        9.05)
BA        0.02        0.00)        9.05
SR        0.00        0.00)

NA        2.62        4.86         4.86
K         1.09        1.33         1.33
P         0.43        0.34         0.34
C         0.72        0.94         0.94
ZR        0.00        0.00         0.00
F         0.00        0.00         0.00
S         0.05        0.08         0.08
CR        0.02        0.01         0.01
CL        0.10        0.16         0.16
SN        0.00        0.00         0.00
LI        0.00        0.00
CU        0.00        0.00
```

SALIC

```
QTZ     COR     ZIR     OR      AB      AN      LC      NE      KP      HL          TOTAL

0.000   0.000   0.000   6.659  23.517  26.908   0.000   0.000   0.000   0.324       57.409
```

FEMIC

```
AC      NS      KS      WO      EN      FS      FO      FA      CS      MT      CM      HM      IL      TN

0.000   0.000   0.000   0.000   0.000   0.000   0.000   0.000   0.000   3.167   0.022   0.000   3.140   0.000


PF      RU      AP      FR      PR      CC      CT      DI      HY      OL          TOTAL

0.000   0.000   0.929   0.000   0.134   1.883   0.000   8.619  18.982   5.709      42.590
```

DI =   8.619 WO(50)EN(32) FS(18)

HY =   18.982EN(64)FS(36)

OL =   5.709 FO(64)FA(36)

PLAG   50.425 AB(47)AN(53)

SALIC(57)FEMIC(43)

QTZ + OR + PLAG =   57.085   QTZ(0)OR(12)PLAG(88)

  AN + AB + OR =   57.085   AN(47)AB(41)OR(12)

QTZ + AB + OR =   30.176   QTZ(0)AB(78)OR(22)

Fig. 13.16 Actual example of the norm and cation-percentage computer output following the rules of exercise 13.20 and the program of Hutchison and Jeacocke (1971).

both CO_2 and P_2O_5 to allow An to remain positive. The program is designed to supply additional information, and Fig. 13.16 gives an example of the actual printout of the norm following the rules of this exercise.

EXERCISE 13.21

The Mesonorm for Metamorphic and Granitic Rocks

The Mesonorm is a modification of the Niggli molecular or catanorm in exercise 13.20. It includes the following additional normative minerals.

Mineral	Abbre-viation	Formula	Number of cations
Biotite	(Bi)	$K(Mg + Fe)_3AlSi_3O_{10}(OH)_2$	8
Actinolite	(Act)	$Ca_2(Mg + Fe)_5Si_8O_{22}(OH)_2$	15
Edenite	(Ed)	$NaCa_2(Mg + Fe)_5AlSi_7O_{22}(OH)_2$	16
Riebeckite	(Ri)	$Na_2Fe_2^{+3}Fe_3^{+2}Si_8O_{22}(OH)_2$	15
Sphene	(Tn)	$CaTiSiO_5$	3
Spinel	(Sp)	$MgAl_2O_4$	3

Hornblende Ho = Act + Ed + Ri. Except for Il (ilmenite) and acmite (Ac), all the other mesonorm minerals are the same as those for the catanorm. The rules are as follows:

1. Perform rules 1 to 12, as given in exercise 13.20.

2. Make sphene $Tn = 3 \times TiO_2$. Reduce CaO by an amount $= TiO_2$. Reduce $Si\,O_2$ by an amount equal to TiO_2.

3. Perform steps 14, 15, and 16 as in exercise 13.20.

4. If in step 16 there is an excess of Na over Al, it is combined with an equal amount of Fe^{+3} to form riebeckite $Na_2Fe_2^{+3}Fe_3^{+2}Si_8O_{22}(OH)_2$. When $NaO_{\frac{1}{2}} \leq FeO_{1\frac{1}{2}}$, make riebeckite $= 7\frac{1}{2}$ times the $NaO_{\frac{1}{2}}$. Reduce the $FeO_{1\frac{1}{2}}$ by an amount $= NaO_{\frac{1}{2}}$. Reduce FeO by an amount $= 1\frac{1}{2}$ times the amount of $NaO_{\frac{1}{2}}$. Reduce SiO_2 by an amount $= 4$ times the $NaO_{\frac{1}{2}}$; $NaO_{\frac{1}{2}}$ becomes zero. When $NaO_{\frac{1}{2}} > FeO_{1\frac{1}{2}}$, make Ri $= 7\frac{1}{2}$ times the $FeO_{1\frac{1}{2}}$. Reduce $NaO_{\frac{1}{2}}$ by an amount $= FeO_{1\frac{1}{2}}$. Reduce FeO by an amount $= 1\frac{1}{2}$ times $FeO_{1\frac{1}{2}}$. Reduce SiO_2 by an amount $= 4$ times $FeO_{1\frac{1}{2}}$; $FeO_{1\frac{1}{2}}$ becomes zero.

5. If sodium is left over, perform step 18 in exercise 13.20. Usually, however, $FeO_{1\frac{1}{2}}$ is left over and is assigned to magnetite. For this assignment apply steps 23 and 24 of exercise 13.20.

6. The remaining FeO is combined with MgO and called (Mg + Fe); the other remaining cations of $AlO_{1\frac{1}{2}}$ and CaO are called Al' and Ca', respectively; Or becomes the fourth constituent.

7. Make anorthite (An) from the Al' by following step 19 of exercise 13.20 exactly.

8. If there is an excess of Al' over Ca', it is calculated as corundum, C = Al', as in step 22 of exercise 13.20.

9. The available (Mg + Fe) is used to convert the orthoclase formed in step 3 to biotite:

$$5 \text{ Or} + 3(Mg + Fe) = 8 \text{ Bi.}$$

If (Mg + Fe) $\gtrless 0.6 \times$ the amount of Or, then Bi, $= 2.667$ times the (Mg + Fe); Or is reduced by an amount $= 1.667$ (Mg + Fe) and (Mg + Fe) becomes zero. If (Mg + Fe) $> 0.6 \times$ the amount of Or, then Bi $= 1.6$ times the amount of Or. The (Mg + Fe) is reduced by an amount $= 0.6$ times the Or content; Or becomes zero.

10. *Alternative 1* for rocks in which amphibole is appropriate (e.g., amphibolites and intermediate igneous rocks).

Actinolite $Ca_2(Mg + Fe)_5Si_8O_{22}(OH)_2$ may now be made from the remaining (Mg + Fe) and CaO. If (Mg + Fe) $\gtrless 2.5 \times$ the CaO remaining, form Act $= 3 \times$ (Mg + Fe). Reduce CaO by an amount $= 0.4 \times$ (Mg + Fe). Reduce SiO_2 by an amount $= 1.6 \times$ (Mg + Fe); (Mg + Fe) becomes zero. If (Mg + Fe) $> 2.5 \times$ CaO, form Act $= 7.5$ times CaO. Reduce (Mg + Fe) by an amount $= 2\frac{1}{2} \times$ CaO. Reduce SiO_2 by an amount $= 7\frac{1}{2}$ times CaO; CaO becomes zero. If (Mg + Fe) remains, it is calculated as Hy. Hy $= 2$ (Mg + Fe). Reduce SiO_2 by an amount $=$ (Mg + Fe); (Mg + Fe) becomes zero. If CaO remains, it is calculated as wollastonite. Wo $= 2CaO$. Reduce SiO_2 by an amount $=$ CaO; CaO becomes zero.

11. *Alternative 2* for rocks in which pyroxene is more appropriate (basic igneous rocks).

Instead of actinolite, pyroxene may be made. Follow steps 25, 26, 27, 28, 29 and 30 of exercise 13.20 to form wollastonite + hypersthene + diopside.

12.   If the total SiO_2 remaining $\gtrless 0$, the norm calculation is complete. Quartz is made to equal the SiO_2 remaining.

13.   If SiO_2 has a negative value, too much silica has been used and desilification has to be effected. Convert Act into edenite

$$5Ab + 15Act = 16\ Ed + 4Q$$

If Act $\gtrless 3.75$ times the negative SiO_2, then make Ed $= 4$ times the negative SiO_2. Reduce Ab by an amount $= 1.25$ the negative SiO_2. Reduce Act by an amount $= 3.75$ the negative SiO_2. Negative SiO_2 becomes 0. The norm is complete. If Act < 3.75 times the negative SiO_2, make Ed $= 1.067$ times Act. Reduce Ab by an amount $= 0.333$ Act. Reduce the negative SiO_2 by an amount $= 0.267$ Act. Act becomes zero. Negative SiO_2 still exists.

14.   Proceed with desilification according to the relationship $4Hy = 3\ Ol + Q$ by employing step 32 in exercise 13.20.

15.   If negative SiO_2 still remains, employ the relationship

$$3Ol + 4C = 6\ Sp + Q$$

16.   If desilification is still required, proceed with $5\ Ab = 3Ne + 2Q$ according to step 34 in exercise 13.20.

17.   Finally hornblende Ho can be made $=$ Act $+$ Ed $+$ Ri.

(after Barth, 1962.)

DISCUSSION

The use of biotite in the norm greatly enhances its value in portraying granitic rocks, for without it all the potassium is allocated to orthoclase and Or is overestimated. Also, pyroxene in a granite norm is of little value, and biotite is a better mafic component. Parslow (1969) recommends that if biotite is the main mafic mineral in a granite Bi with an average formula $K_2(Ti, Mg, Fe^{+2})_6 (Si_6Al_2O_{20})(OH, F)_4$ be taken as Bi to replace Il and Hy in the norm. The biotite step is introduced after the magnetite step. He also proposes the introduction of muscovite in the granite norm by using an average formula $K_2Al_4(Si_2Al_2O_{20})(OH, F)_4$. The amount of normative muscovite formed will depend on a modal analysis of primary muscovite in rock thin sections.

Computer programs for calculating norms and related petrographic and mineralogical values are obtainable from The Department of Geology, University of Auckland, Auckland 1, New Zealand (Rodgers et al., 1970); from the Geology Department, Indiana University, Bloomington, Indiana, (Vitaliano et al., 1965); Geology Department, The University, Bristol, England (Howarth, 1966); and from the Geology Department, University of Malaya, Kuala Lumpur, Malaysia (Hutchison and Jeacocke, 1971).

C.   STATISTICAL ANALYSIS OF DATA

A comprehensive description of statistical methods is impossible here because of the limitations of space. Readers are referred to more complete books, for example, Parl (1967) and Koch and Link (1970).

Frequency Distributions

It is common in geological research to accumulate a large number of data of a particular type; for example, one may have more than 30 chemical analyses of granites or basalts from a petrographic province or the modal percentage of quartz in more than 30 granite specimens collected from a batholith. Large numbers of measurements need to be displayed as frequency distributions. The method is best illustrated by an example.

   An ore body has been sampled. Analyses have been performed on 224 specimens collected systematically from the body. If we listed all 224 analyses, it would be hard to see any pattern. Hence it is common to rearrange this kind of data into a frequency distribution (the example is taken from Koch and Link, 1970):

Oxide Weight % Interval	Interval midpoint W	Frequency f	Relative Frequency rf %	Cumulative Frequency cf	Relative Cumulative Frequency rcf %
14–16	15	1	0.45	1	0.45
16–18	17	1	0.45	2	0.90
18–20	19	8	3.57	10	4.47
20–22	21	21	9.37	31	13.84
22–24	23	44	19.64	75	33.48
24–26	25	54	24.12	129	57.60
26–28	27	56	25.00	185	82.60
28–30	29	30	13.39	215	95.99
30–32	31	7	3.12	222	99.11
32–34	33	2	0.89	224	100.00

   In preparing this table from the original analyses, we had a choice of the interval size. The interval of 2% chosen resulted in 10 groups. Practical experience and theoretical considerations show that all the essential information is preserved if the interval is chosen so that the number of groups lies between 10 and 50. If, however, a set of data contains extreme values, it is sometimes convenient to widen the width of the intervals with small or large midpoints. Otherwise it is essential within the middle range of a distribution that the intervals be of equal size.

   Column 1 of the table lists the weight % determinations tabulated in intervals of 2% oxide to include the complete range of analyses from lowest to highest. The intervals chosen must be equal. Column 2 gives the midpoint of each interval. Column 3 is the number of analyses in each interval = *frequency f*. Column 4 is the frequency converted to a percentage of the 224 total analyses = 100f/224. Column 5 is the cumulative frequency and is obtained by summing successive lines in column 3 up to a total of 224. Column 6 gives the values of column 5 converted to a percentage.

Histograms. This kind of data may be usefully displayed as a histogram. The base length corresponds to the interval width and the height to the frequency or number of observations in each interval. Figure 13.17 shows the histogram for the above example.

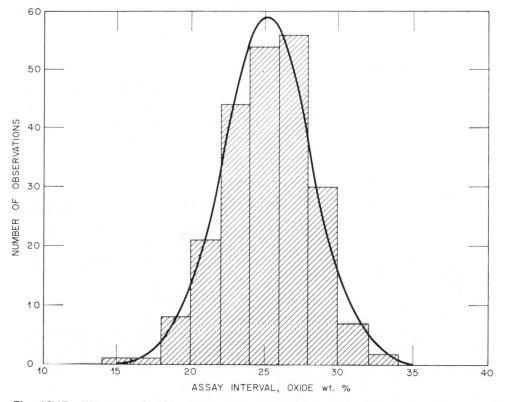

Fig. 13.17 Histogram showing the frequency distribution of the 224 oxide analyses and a smooth curve fitted to the histogram.

Alternatively, we could plot a histogram of the relative frequency % on the vertical scale or we could plot relative cumulative frequency against the oxide interval.

If a large number of analyses were available and the intervals were taken as very small, the histogram could be replaced by a smooth curve, called a theoretical frequency distribution. Such a curve is shown fitted to the histogram of Fig. 13.17. It has been found from experience that most natural phenomena can be fitted by a *normal* distribution or *Gaussian* curve. The equation for the normal distribution is

$$f(w) = \frac{1}{\sigma\sqrt{2\pi}} \exp\left[-\frac{1}{2}\left(\frac{w-\mu}{\sigma}\right)^2\right],$$

where w is an observation, μ and σ are variable constants which must be evaluated for any single example, and μ and σ are called *parameters*; 2, π and exp are fixed constants.

Populations and Samples. Generally a set of data represents only a small fraction of the total number of observations that might be made; for example an infinitely large number of specimens would characterize a pluton, yet we have, of necessity, to collect a limited number of hand specimens which we then consider as characteristic of the whole pluton. The total set of potential observations is called a *population*, and the set of actual observations is called a *sample* from this population. On the other hand, we may redefine the population to mean all the hand specimens that we have collected from the pluton and the sample to be a smaller number of collected hand specimens which we have analyzed modally. Sample in the above context does not agree with the usage of geological sample, which means a specimen collected. To distinguish them sample should be called *statistical* sample, and the single collected hand specimen should be called *geological sample*. The two should not be confused.

Mean, Variance, Standard Deviation, and Coefficient of Variation. For samples that follow the normal distribution, such as the analyses shown in the histogram in Figure 13.17, the important summary statistical expressions are *mean* (a measure of the central tendency), the *variance* (a measure of the spread of a distribution), the *standard deviation* (the square root of the variance), and the *coefficient of variation* (the ratio of standard deviation to the mean).

As is nearly always the case, the set of observations is a sample rather than a population and the *mean* is computed from the formula:

$$\bar{w} = \frac{\sum w}{n}.$$

Likewise, when the observations constitute a sample, the variance is computed from the expression

$$S^2 = \frac{\sum (w - \bar{w})^2}{n - 1},$$

where S^2 is the sample variance and S, its square root, is the sample *standard deviation*.

The *coefficient of variation* is the ratio of standard deviation to mean. For a sample the coefficient of variation is computed from

$$C = \frac{S}{\bar{w}}.$$

By way of illustration, these summary expressions will now be derived from a fictitious sample of five observations:

Observations (w)	Observations Squared (w^2)	Deviation from Mean $(w - \bar{w})$	Squared Deviation from Mean $(w - \bar{w})^2$
2	4	-3	9
4	16	-1	1
6	36	1	1
6	36	1	1
7	49	2	4
sum $(\sum)$ 25	141	0	16

Calculations

$$\bar{w} = \frac{25}{5} = 5$$

$$S^2 = \frac{16}{4} = 4. \quad S = \sqrt{4} = 2$$

$$C = \frac{2}{5} = 0.400$$

This method of calculating the sample variance is inconvenient in practice because the sample mean has to be calculated before the sum of the squares can be calculated. The following short cut procedure is generally used:

$$SS \text{ (sum of squares)} = \sum (w - \bar{w})^2$$

$$= \sum w^2 - \frac{(\sum w)^2}{n}$$

$$= 141 - \frac{25^2}{5}$$

$$= 141 - 125$$

$$= 16,$$

$$S^2 = \frac{16}{4} = \left(\frac{SS}{n-1}\right)$$

$$= 4.$$

The same calculations can also be performed on classed data such as the oxide analyses displayed in Fig. 13.17:

Oxide Weight %	Interval Midpoint (w)	Frequency (f)	fw	fw²
14–16	15	1	15	225
16–18	17	1	17	289
18–20	19	8	152	2,888
20–22	21	21	441	9,261
22–24	23	44	1,012	23,276
24–26	25	54	1,350	33,750
26–28	27	56	1,512	40,824
28–30	29	30	870	25,230
30–32	31	7	217	6,727
32–34	33	2	66	2,178
Total		224	5652	144,648

Calculation

$$\text{mean } \bar{w} = \frac{\sum fw}{\sum f} = \frac{5652}{224} = 25.23,$$

$$SS = \sum fw^2 - \frac{(\sum fw)^2}{\sum f} = 144,648 - \frac{(5652)^2}{224} = 2036,$$

$$\text{variance, } S^2 = \frac{SS}{\sum f - 1} = \frac{2036}{223} = 9.13,$$

$$\text{standard deviation } S = \sqrt{9.13} = 3.02,$$

$$\text{coefficient of variation } C = \frac{S}{\bar{w}} = \frac{3.02}{25.23} = 0.119.$$

Interval Estimation. There are two disadvantages of a point estimate. The first is that it is almost certain to be wrong; for example, we may say that a granite contains 65.1 % by weight SiO₂. The very first chemical analysis we perform on this granite will be certain not to give 65.1 % but some other value. The second disadvantage of the point estimate is that it conveys no information about how wrong it is likely to be. Because of these disadvantages, the *confidence interval* estimate was devised to incorporate three basic attributes of a sample: the mean, the variability as measured by the standard deviation, and the sample size. In interval estimates the parameter being estimated lies between two values called *confidence limits* for a specified percentage of the intervals calculated; for example, instead of saying that the granite contains 65.1 % SiO_2, which is certain to be wrong, it is best to say that the granite contains with 90 % confidence not less than 60.1 nor more than 70.1 % SiO_2. Confidence intervals are of two kinds: two sided, in which the interval is bounded on both sides by a calculated value, or one-sided; we might say that with a 90 % confidence the granite contains not less than 62.8 % SiO_2.

If $\bar{w}$ is the mean of a random sample from a normal distribution with a mean of μ and a variance of σ^2, the probability is 90 % (9 chances out of 10) that $\bar{w}$ is not more than 1.645 standard deviations larger or smaller than μ. In other words, in 9 random

95% Confidence

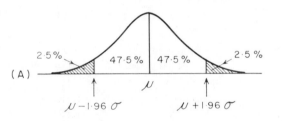

90% Confidence

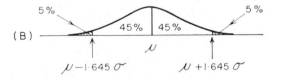

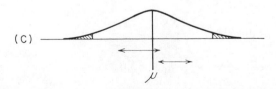

Fig. 13.18 A = 95% confidence interval for a mean of μ and a standard deviation of σ; B = 90% confidence interval; C = two confidence intervals, one containing μ and the other incorrect.

samples out of 10 the sample mean will lie in the correct calculated interval. This relationship may be written

$$\bar{w} - 1.645 \frac{\sigma}{\sqrt{n}} < \mu < \bar{w} + 1.645 \frac{\sigma}{\sqrt{n}}$$

for 90% confidence. The expression then becomes

$$\bar{w} - 1.96 \frac{\sigma}{\sqrt{n}} < \mu < \bar{w} + 1.96 \frac{\sigma}{\sqrt{n}}$$

for 95% confidence.

Figure 13.18 presents these intervals graphically for both 90 and 95% confidence coefficients. If σ is known, the interval may be evaluated to give a 90 or 95% confidence interval for μ. If such intervals are calculated repeatedly for random samples from a normal distribution, 9 out of 10 (for 90% confidence) will be correct (i.e., they will contain μ), whereas 1 in 10 will be incorrect (they will not contain μ). Figure 13.18 shows two intervals, one containing and one not containing μ. At 95% confidence 95 random samples in 100 will have the correct intervals.

In any particular situation σ is usually unknown, hence it must be replaced in the calculation by the sample standard deviation S using the expression

$$t = \frac{\bar{w} - \mu}{S/\sqrt{n}},$$

t is called the Student's t or the t-distribution. A selection of t values is given in Table 13.3. It will be seen that when the degrees of freedom (number of observations in a

Table 13.3   The t-Table for 80, 90, and 95% Double-Sided Confidence Intervals

Degrees of Freedom $(n-1)$	Level of Significance tn			Degrees of Freedom $(n-1)$	Level of Significance tn		
	10%	5%	2.5%		10%	5%	2.5%
1	3.078	6.314	12.706	18	1.330	1.734	2.101
2	1.886	2.920	4.303	19	1.328	1.729	2.093
3	1.638	2.353	3.182	20	1.325	1.725	2.086
4	1.533	2.132	2.776	21	1.323	1.721	2.080
5	1.476	2.015	2.571	22	1.321	1.717	2.074
6	1.440	1.943	2.447	23	1.319	1.714	2.069
7	1.415	1.895	2.365	24	1.318	1.711	2.064
8	1.397	1.860	2.306	25	1.316	1.708	2.060
9	1.383	1.833	2.262	26	1.315	1.706	2.056
10	1.372	1.812	2.228	27	1.314	1.703	2.052
11	1.363	1.796	2.201	28	1.313	1.701	2.048
12	1.356	1.782	2.179	29	1.311	1.699	2.045
13	1.350	1.771	2.160	30	1.310	1.697	2.042
14	1.345	1.761	2.145	40	1.303	1.684	2.021
15	1.341	1.753	2.131	60	1.296	1.671	2.000
16	1.337	1.746	2.120	120	1.289	1.658	1.980
17	1.333	1.740	2.110	∞	1.282	1.645	1.960

sample) approaches infinity the variance of S^2/σ^2 approaches zero. Therefore for infinity degrees of freedom S becomes equal to σ. Thus for an infinite number of degrees of freedom the t-distribution is in fact the standardized normal distribution As a rule of thumb, if the sample size exceeds 30 items, the difference between the normal and the t-distribution figures is considered insignificant for most practical purposes.

Let us now calculate the two-sided confidence interval for a population mean for which σ is unknown. The expression is

$$\bar{w} - t_{5\%} \frac{S}{\sqrt{n}} < \mu < \bar{w} + t_{5\%} \frac{S}{\sqrt{n}}$$

for a 90% confidence interval.

For our fictitious sample of five observations illustrated above

$$\bar{w} = 5,$$
$$S = 2,$$
$$n = 5,$$
$$\sqrt{n} = 2.236,$$
$$\frac{S}{\sqrt{n}} = 0.8945,$$

degrees of freedom $(= n - 1) = 4,$
$$t_{5\%} = 2.132, \text{ (from Table 13.3)},$$

$$t_{5\%} \frac{S}{\sqrt{n}} = \frac{2.132 \times 2}{2.236} = 1.9,$$

$$\mu_L = \bar{w} - t_{5\%} \frac{S}{\sqrt{n}} = 5 - 1.9 = 3.1,$$

$$\mu_U = \bar{w} + t_{5\%} \frac{S}{\sqrt{n}} = 5 + 1.9 = 6.9.$$

Thus at the 90% confidence level our confidence limits are lower 3.1 and upper 6.9. The bigger sample in Figure 13.17 is calculated in the same way:

$$\bar{w} = 25.23,$$
$$S = 3.02,$$
$$n = 224,$$
$$\sqrt{n} = 14.967.$$
$$\frac{S}{\sqrt{n}} = 0.2018,$$

degrees of freedom $= 223,$

$$t_{5\%} = 1.651,*$$

* $t_{5\%}$, when $n - 1 = 223$, can be obtained by extrapolation from Table 13.3: $\frac{1}{120} = 0.008333$, $t_{5\%} = 1.658$, $\frac{1}{223} = 0.004484$, $t_{5\%} = 1.651$ (by interpolation), $\frac{1}{\infty} = 0.00000$, $t_{5\%} = 1.645$.

$$t_{5\%} \frac{S}{\sqrt{n}} = 0.33,$$

$$\mu_L = 24.90,$$

$$\mu_U = 25.56.$$

At the 90% confidence level, our confidence limits are lower 24.90 and upper 25.56.

In calculating the confidence intervals, a confidence coefficient of 90% was used. If another confidence coefficient is adopted, the value of t is different and the confidence intervals have a different width for the same data. By selecting a lower confidence coefficient we narrow the interval, but narrowing the interval provides less chance of catching the population mean. The confidence coefficient is chosen by the investigator. A good choice for geological data is 90% because the variability is larger than for most data derived from laboratory-controlled experiments. However, a confidence coefficient of 95% may be used if the data are considered to be less variable.

One-sided confidence intervals may also be calculated by using the t-distribution; for example, the confidence statement may have the form "with a specified confidence of being correct, the population mean is not lower than a specified value." The words "one-sided" refer to the fact that only one side is calculated. Actually the other side is present but known; for example, one side may go as low as 0%, and we are interested only in the upper limit. A one-sided 90% lower confidence interval is calculated from the relation

$$\bar{w} - t_{10\%} \frac{S}{\sqrt{n}} < \mu < \infty$$

and a one-sided 90% upper confidence interval is calculated from the relation

$$-\infty < \mu < \bar{w} + t_{10\%} \frac{S}{\sqrt{n}}.$$

In mining geology, for example, it is usually important that the ore be above cut-off grade rather than an upper limit set to the grade. As examples, one-sided 90% lower confidence intervals are calculated for the same two examples as performed above.

For our fictitious sample of five observations

$$\bar{w} = 5,$$

$$\frac{S}{\sqrt{n}} = 0.8945,$$

$$(n - 1) = 4,$$

$$t_{10\%} = 1.533 \text{ (from Table 13.3)},$$

$$t \frac{S}{\sqrt{n}} = 1.4,$$

$$\mu_L = 3.6,$$

and for the example in Figure 13.17

$$\bar{w} = 25.23,$$

$$\frac{S}{\sqrt{n}} = 0.2018,$$

$$(n - 1) = 223,$$

$$t_{10\%} = 1.286,$$

$$t\frac{S}{\sqrt{n}} = 0.26,$$

$$\mu_L = 24.97.$$

It will be seen that the one-sided confidence limits are substantially higher than the two-sided limits. This difference is critical in practice, for example, in deciding the cutoff limit of ore in mining.

The Confidence Interval for the Difference Between Two Population Means

Frequently we are required to compare observations from two related geological phenomena. As an example, the following iron analyses were obtained from six marine shales: 6.9, 6.5, 3.8, 3.5, 4.9, and 3.7. From these observations

$$\sum w = 29.3; \quad n = 6, \quad \bar{w} = 4.883, \quad (\sum w)^2 = 858.49;$$

$$\frac{(\sum w)^2}{n} = 143.08; \quad \sum w^2 = 154.25; \quad SS = 11.17,$$

$$\text{degrees of freedom (df)} = 5, \frac{1}{n} = 0.1667$$

The following iron analyses were obtained on seven nonmarine shales: 4.8; 3.9; 5.9; 5.0; 4.7; 6.0 and 3.7. From these observations we obtain

$$\sum w = 34.0,$$

$$n = 7,$$

$$\bar{w} = 4.857,$$

$$(\sum w)^2 = 1156.00,$$

$$\frac{(\sum w)^2}{n} = 165.14,$$

$$\sum w^2 = 169.84,$$

$$SS = 4.70,$$

$$df = 6,$$

$$\frac{1}{n} = 0.1429.$$

The marine shales give a mean of 4.883, and the nonmarine shales, 4.857. To enable us to make some comment on the difference of these two means we must now combine the analyses:

$$\bar{w}_1 - \bar{w}_2 = 0.026,$$

$$\text{pooled } SS = 15.87,$$

$$\text{pooled df} = 11, \text{ pooled } S^2 = 1.4427,$$

(pooled S^2 is calculated by adding the two SS values and dividing by the sum of the numbers of degrees of freedom),

$$\frac{1}{n_1} + \frac{1}{n_2} = 0.3096,$$

$$S_p^2\left(\frac{1}{n_1} + \frac{1}{n_2}\right) = 0.4467,$$

$$\left[S_p^2\left(\frac{1}{n_1} + \frac{1}{n_2}\right)\right]^{1/2} = 0.6683;$$

$t_{5\%}$ with 11 df $= 1.796$,

$$t_{5\%}\left[S_p^2\left(\frac{1}{n_1} + \frac{1}{n_2}\right)\right]^{1/2} = 1.200,$$

$$\mu_L = -1.174,*$$

$$\mu_U = 1.226.*$$

As the last step the upper and lower confidence limits for the difference $(\mu_1 - \mu_2)$ are calculated. The general conclusion may be drawn from this exercise that there are no important differences between the means of the two samples because the limits -1.174 and 1.226 include 0 and moreover the upper and lower limits are not far removed from zero. If both interval limits are close to 0, there can be no large differences between the means.

EXERCISE 13.23

Simple Linear Regression

Frequently in mineralogical and petrological research we accumulate data from two parameters which are presumed to have a linear relationship. By plotting one measurement against another we can fit a line to the points by visual estimation. It is best, however, to calculate the equation to the line that gives the best fit by means of the least squares method.

$$* S_p^2 = \frac{SS_1 + SS_2}{(n_1 - 1) + (n_2 - 1)}, \quad \mu_L = (\bar{w}_1 - \bar{w}_2) - t_{5\%}\left[S_p^2\left(\frac{1}{n_1} + \frac{1}{n_2}\right)\right]^{1/2},$$

$$\mu_U = (\bar{w}_1 - \bar{w}_2) + t_{5\%}\left[S_p^2\left(\frac{1}{n_1} + \frac{1}{n_2}\right)\right]^{1/2}.$$

The calculation of a regression equation may be illustrated by an example. Hutchison (1972) gave the relationship between chemical analysis and reflectivity of chromite. His data are as follows:

X Reflectivity % at 590 nm	Y Number of Al Cations per Unit Cell	XY	X^2	Y^2
13.7	3.0	41.10	187.69	9.00
13.4	4.1	54.94	179.56	16.81
12.5	4.6	57.50	156.25	21.16
12.6	5.3	66.78	158.76	28.09
11.9	5.5	65.45	141.61	30.25
12.4	5.7	70.68	153.76	32.49
11.4	8.1	92.34	129.96	65.61
10.7	8.1	86.67	114.49	65.61
10.2	9.1	92.82	104.04	82.81
108.8 (total)	53.5	628.28	1326.12	351.83
12.09 (mean)	5.94		$n = 9$	

For a straight-line relationship the equation is $Y = a + bX$. The most convenient calculation (Parl, 1967) is based on the relationships

$$b = \frac{\sum XY - \overline{X} \sum Y}{\sum X^2 - \overline{X} \sum X}$$

and

$$a = \overline{Y} - b\overline{X}.$$

For our example

$$b = \frac{628.28 - 12.09(53.50)}{1326.12 - 12.09(108.80)}$$

$$= \frac{-18.54}{10.73} = -1.728,$$

$$a = 5.94 + 1.728(12.09) = 26.832.$$

The equation becomes

$$Y = 26.832 - 1.728\ X.$$

Figure 13.19 shows the data of the above measurements plotted one against the other. The straight line is not drawn in by visual estimation, but is the line computed as above. It may be constructed by determining any two points on it, for example, when X = 10.0, Y = 26.832 − 17.28 = 9.55. When X = 13, Y = 26.832 − 22.464 = 4.368.

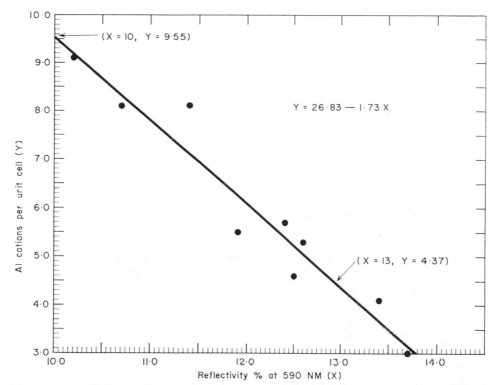

Fig. 13.19 Fitting a straight line $Y = 26.83 - 1.73\,X$ to a plot of aluminum cations per unit cell (Y) versus reflectivity % (X) for chromite. (After Hutchison, 1972.)

This calculation is performed by the quick method. Alternatively the values of a and b may be determined by solving the two simultaneous equations:

$$\sum Y = na + b \sum X$$

and

$$\sum XY = a \sum X + b \sum X^2.$$

The regression line represents only a measure of the average relationship between the dependent and the independent variables. Some *estimating errors* will be involved unless there is perfect correlation. The greater the dispersion, or scatter, of the points around the regression line, the greater the estimating error. The magnitude of this estimating error which is measured by the standard deviation about the estimating line is termed the *standard error of estimate* and is designated $Sy \cdot X$

$$Sy \cdot X = \left(\frac{\sum Y^2 - a \sum Y - b \sum XY}{n} \right)^{1/2} \qquad \text{(Parl, 1967).}$$

For our example

$$Sy \cdot X = \left(\frac{351.83 - 26.832(53.5) + 1.728(628.28)}{9} \right)^{1/2}$$

$$= \left(\frac{1.99}{9} \right)^{1/2}$$

$$= \sqrt{0.22} = 0.47.$$

THERMAL ANALYSIS TECHNIQUES $\quad$ 14

A. DIFFERENTIAL THERMAL ANALYSIS

On heating, reactions involving crystallization, oxidation, and some other chemical reactions are EXOTHERMIC. An exothermic peak means that the temperature of the specimen being analyzed has become higher than that of the inert reference material (usually alumina) because the reaction produces heat.

Phase changes, dehydration, decomposition, and crystalline inversions are generally ENDOTHERMIC. An endothermic peak indicates that the specimen has absorbed more heat than the reference material during the reaction.

The difference in temperature between the specimen and the inert reference material will cause a current to flow in the differential thermocouple. The direction of current flow determines whether the deflection is upward (exothermic) or downward (endothermic) on the chart. The more energetic the reaction, the greater the current flow, hence the greater the peak.

The DTA curve (differential temperature ΔT versus furnace temperature t) will be fairly reproducible for any mineral on a particular machine under identical operating conditions. The mass of the reacting material is approximately proportional to the area under the curve. The recognition of departure from and return to base line is most important when quantitative estimations are required.

Figure 14.1 shows a typical DTA curve for anglesite, $PbSO_4$, which contains a small amount ($< 5\%$) of pyrite impurity. Peak A is exothermic and is due to the oxidation of the pyrite. Peak B is endothermic and is due to the crystalline inversion of orthorhombic to monoclinic anglesite. Like many inversions the change is rapid, hence the peak is sharp. Peak C represents decomposition and, like A, is initially slow but gains impetus rapidly. The base line (when no reactions are taking place) is easily seen and extrapolated under the peaks.

The DTA method, unlike X-ray powder diffractometry, ordinarily cannot be used on a completely unknown material to identify its mineralogy from the record because the curves of different components overlap and there is also a chance of reaction between components. Any reaction in the sample will give a peak.

DTA, however, can quickly identify the presence of small quantities of, for example, carbonate in a rock or kaolinite or other clays, but it cannot be considered as the same "fingerprinting" method that X-ray diffraction is.

If any peak can with certainty be assigned to a single mineral component, the volume of that component in the specimen is best related to peak area, measured by a planimeter from the extrapolated baseline up or down.

438

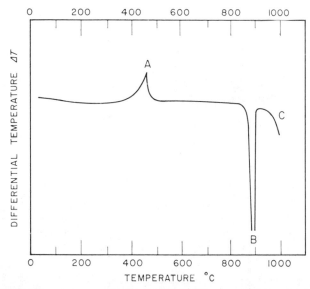

Fig. 14.1 Differential thermal curve for anglesite, $PbSO_4$, containing approximately 5% pyrite contamination: A = oxidation of pyrite; B = crystallographic inversion of the anglesite; C = decomposition of the anglesite. (After McLaughlin, 1967b.)

No attempt to discuss the complete theory and practice of DTA has been made in this chapter. The unfamiliar reader is referred to the monograph edited by Mackenzie (1957) and (1970a) and to a shorter summary by McLaughlin (1967b).

Apparatus

A wide range of instruments is commercially available. Details of some of the more popular models can be found in Mackenzie and Mitchell (1970a). I have chosen to illustrate the method by reference to the well-known Stone DTA system (Fig. 14.2) (available from Columbia Scientific Industries, 3625 Bluestein Boulevard, P.O. Box 6190, Austin, Texas 78762). This particular instrument is the most sensitive thermal analytical research instrument available. It combines a sensitivity of 0.5 $\mu V/in$. with a complete selection of sample holders and dynamic gas choice and allows a pressure operation of 10^{-6} Torr to 3000 psig and temperature operation of -160 to $+1600°C$.

EXERCISE 14.1

Choice of Sample Holders and Thermocouples

Three basic designs of sample holders are available. Each is appropriate for a specific sample type and temperature range.

Type SH-8BE (Fig. 14.3A). This type has reference and sample cavities drilled into a solid block. Differential thermocouple beads extend into the cavities and are in direct contact with the sample and reference material. Gas flow is directed through the sample for excellent atmospheric control. Powder samples that *do not melt or sinter* are to be used in this type of holder, but sulfide samples cannot be because of their reaction with the thermocouple wiring. An appropriate choice of thermocouple is *Platinel II* (P) which operates to a maximum of 1100°C and cannot be used with H_2 or H_2S dynamic gas.

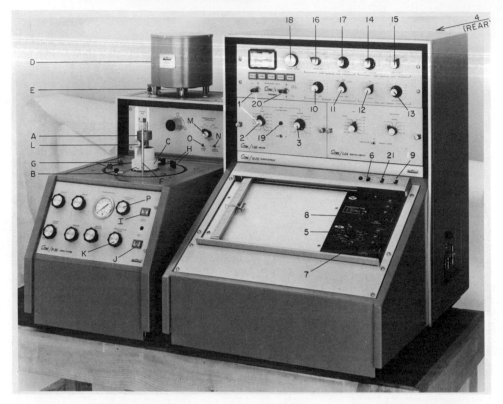

Fig. 14.2   C.S.I.-STONE model 202 DTA system consisting of recorder-controller model LB 202 and furnace platform JP 202. (By courtesy of Columbia Scientific Industries, Austin, Texas.)

Type SH-11 BP (Fig. 14.3B). Differential thermocouple beads are attached to ceramic posts to permit dynamic gas flow around the cups. Liquid samples or samples that melt can be tested with this type of holder. The cups are of platinum, and if the samples melt during the DTA run the melted sample can be removed from the cup just as it can from any platinum crucible.

An appropriate thermocouple for this holder is *platinum-platinum 10% rhodium* (S) which operates to a maximum temperature of 1600°C. It must not be used with H_2 or H_2S dynamic gas.

Type SH-11 BR (Fig. 14.3C). This sample holder has a ring differential thermocouple which is highly sensitive and is used with small samples of 0.1 to 10 mg. The samples are contained in small foil dishes that sit on the holder rings. Dynamic gas flows around the sample. This sample holder is suitable for sulfides because the specimen does not touch the thermocouple. An appropriate thermocouple is *Platinel II*, in which case the sample holder may be used only to a maximum 1100°C.

Sample holder bodies of aluminum (maximum temperature 550°C), nickel (maximum temperature 1100°C), stainless steel (maximum temperature 1000°C), palladium ruthenium (PD) (maximum temperature 1370°C), or platinum 10% rhodium (PT) (maximum 1600°C) can be obtained.

Sample dishes of aluminum, nickel, or platinum for sample holder type SH-11 BR may be bought or made by using a special press. The choice of sample dish material will decide the maximum temperature of the DTA run. Nickel (1100°C) and platinum

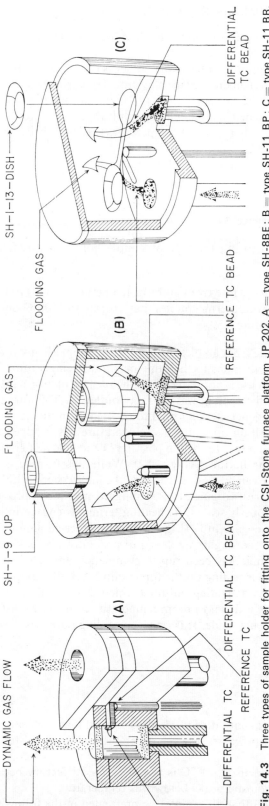

Fig. 14.3 Three types of sample holder for fitting onto the CSI-Stone furnace platform JP 202. A = type SH-8BE; B = type SH-11 BP; C = type SH-11 BR. When the sample holder is fitted to the furnace platform (A in Fig. 14.2), the reference material is put into the left-hand and the sample into the right-hand cavity or dish.

(1600°C) will be the most commonly used. Always make sure that the maximum temperature is not exceeded.

Once chosen, the appropriate sample holder is easily pushed into the corresponding sockets (A, Fig. 14.2) on the furnace platform. The wiring terminal from the sample holder is plugged into the socket (C, Fig. 14.2) and an earthing wire is secured under knob B (Fig. 14.2). When handling the sample holders, never hold them by the ceramic base. Handle only the metal top part (A, Fig. 14.2) because the ceramic part absorbs water readily from the fingers.

EXERCISE 14.2

Choice of Heating Furnace

The furnace is chosen to suit the type of sample holder in use. Only two types are required.

Type F-1D, which can operate from ambient temperature to a 1100°C, is used for all sample holders fitted with thermocouples that operate to 1100°C only. Do not exceed the maximum temperature.

Type F-1 C can operate from ambient to 1600°C. It should be used for sample holder SH-11 BP, which also operates to the same maximum temperature.

The furnace not in use is normally set on the furnace platform (D, Fig. 14.2) where it can be attached to the cooling water supply outlets E. This furnace may have just been run to heat a specimen. It is allowed to cool down in the heating position and is removed from the position only after its temperature has dropped *below 500°C* and not before. It is then removed to the top position D and a flow of cold air can be directed into it by activating button I (Fig. 14.2). While this furnace is cooling, another already cooled furnace can be mounted around the loaded sample holder. It is carefully slid downward, with the two guide rails F plugged into the appropriate holes in the furnace. Once it is properly seated, its heating terminals engage in sockets H and its water-cooling sockets engage in G (Fig. 14.2). The water-cooling hose connections G must be carefully and thoroughly tightened to prevent water leakage.

After a DTA sample has been run to completion and allowed to cool down to 500°C and the furnace is resting on the top cooling rack the sample holder A may be more rapidly cooled by activating button J which directs a flow of cold air over the sample holder to bring it down to room temperature. Button J must not be activated when a DTA run is being made. It is only for cooling purposes.

EXERCISE 14.3

Sample Pretreatment

For comparative and quantitative work it is essential that each sample have the same pretreatment, for fine powders have the capacity to absorb moisture from the atmosphere.

Drying the specimens at 100°C is normally unsatisfactory because it causes the disappearance of diagnostic peaks below this temperature.

It is recommended that all samples be pretreated in the following way.

1. Place a beaker or dish of a saturated solution of $Mg(NO_3)_2 . 6H_2O$ in the lower compartment of a dessicator. This gives a relative humidity of about 55% at 18°C with a low temperature coefficient, so that even at 30°C the humidity is 51.5%. There is therefore little change in water vapor pressure.

2. All powdered specimens are placed in this dessicator on the wire gauze in open containers and kept in this controlled atmosphere for several hours before being analyzed.

(After Mackenzie and Mitchell, 1970b.)

EXERCISE 14.4

Loading Type SH-11 BR Sample Holder

1. A small dish of aluminum (for maximum temperature 550°C) or platinum (for maximum temperature 1600°C) containing the sample is placed on the right-hand ring of the thermocouple (when the sample holder is fitted on the furnace platform). An identical dish containing the reference inert sample (alumina) is placed on the left-hand ring.

2. To obtain zero base line drift an exact balance must be obtained between sample and reference. For a sample of known specific heat calculate the required weight of reference from

$$\text{weight of reference} = \frac{\text{weight of sample} \times \text{specific heat of sample}}{\text{specific heat of reference}}.$$

3. For a sample with unknown specific heat, a trial-and-error method is used. First weigh the sample and estimate its specific heat. Then calculate the amount of reference material necessary from the above formula. Set the amplifier at 40 μV and begin the run. Use less reference if the thermogram drifts in an exothermic direction (up) and more reference material if the drift is endothermic (down). Proper balance of sample to reference weight will give a thermogram with little or no base line drift.

4. Once a thermogram has been obtained with zero base-line drift,

$$\text{sample specific heat} = \frac{\text{reference weight} \times \text{reference specific heat}}{\text{sample weight used to give zero drift}}$$

5. The amount of sample can be as little as 0.1 mg but may be as much as 10 mg for some materials (e.g. plastics). In general the smaller the sample, the sharper the peaks and melting points on the thermogram.

6. Reference material may be an aluminum foil disc for temperatures less than 550°C. For higher temperatures, glass beads, Al_2O_3, or MgO may be used.

7. A sample pan former and encapsulator tool can be used to make the sample dishes from aluminum or platinum foil.

8. Extreme care should be taken in handling the dishes. Use tweezers and never touch the thermocouple wires with the fingers or the tweezers.

9. Place the cap on the sample holder.

EXERCISE 14.5

Loading the SII-11 BP Sample Holder

1. For powder samples and sinterable material the test cup (front) can be filled completely. For liquids or substances that melt the cup must not be filled more than one-third.

2. The reference material put in the other cup is usually alpha alumina.

3. Minimum zero drift is obtained by balancing the relative amounts of sample and reference. Follow steps 3 and 4 of exercise 14.4 to obtain a balance of sample and reference.

4. Place the cap on the sample holder.

EXERCISE 14.6

Loading the Sample Holder SH-8 BE

This holder must be reserved for materials that will neither melt nor sinter during a run.

1. Before filling the cavities check that the thermocouples lie in the cavity centers. Readjust if necessary.

2. Make sure that the disc in the bottom of each cavity is positioned flat and not clogged or broken.

3. Using a weighing spatula, fill the reference cavity (left side when fitted to platform) with powdered inert material (-100 $+200$ mesh). Leave enough space on top of the sample for the top porous disc to fit flush with top of the sample holder block.

4. Do not pack the cavity. Settle the powder by tapping the side of the block.

5. Fill the test material cavity (right side) with powdered (-48 $+200$ mesh) material. Slight packing may be necessary. A 2-in. glass rod of $\frac{7}{32}$-in. diameter and flat ends would be suitable. Be careful not to bend the thermocouple. Use only the weight of the rod to pack the material.

6. Using tweezers, carefully place the porous disc on top of the powder in the cavities. Press the disc down so it is flush with the top of the sample holder.

7. Place the cap on the sample holder.

EXERCISE 14.7

Selection of Heating Rate

A *slow* heating rate has the following effects: little base-line drift, near equilibrium conditions, broad shallow peaks on $\Delta T/t$ curves, and long time per determination.

A *fast* heating rate has the following effects: base-line drift may be appreciable, conditions are far from equilibrium, large narrow peaks on $\Delta T/t$ curves, and short time per determination. (Mackenzie and Mitchell, 1970b.)

For normal work, rates of 8 to 12°C per minute are commonly used, and unless specified otherwise a rate of 10°C per minute will be the standard. This gives peaks of a satisfactory size, overlap of neighboring peaks is not excessive, and the time per determination is reasonable.

It is most essential, of course, that the heating rate be linear, and the heating apparatus must be reliable in this respect.

Temperature Calibration

Since the peak temperature is probably the most widely quoted criterion in DTA, it is wise to ensure that the thermocouple temperature reading is accurate. The following list of materials can be run on the instrument to check the exact temperature reading and calibrate the instrument.

Transition point of $Na_2SO_4 \cdot 10\ H_2O \rightleftharpoons Na_2SO_4$ + solution at 32.38°C.
Boiling of water at 100°C.
Solidification of tin (on cooling) at 231.85°C.
Melting of potassium dichromate at 397.5°C.
The quartz $\alpha \rightleftharpoons \beta$ transition at 573°C is commonly used because many natural samples may contain some quartz.
Melting of potassium chloride at 770.3°C.
Melting of sodium sulphate at 884°C.
Melting of Li_2SiO_3 at 1201°C.
Solidification of iron in a reducing atmosphere (upon cooling) at 1530°C.

(List after Mackenzie and Mitchell, 1970b.)

EXERCISE 14.9

Operating Procedure

This procedure is specifically for the model illustrated in Fig. 14.2, but it may be modified, by reference to the manufacturer's instruction, to suit any other make of apparatus.

1. Place OFF-ON-STANDBY switch to ON position (1, Fig. 14.2).

2. Place the sample (right-hand cavity) and standard reference material (left-hand cavity) in sample holder A (Fig. 14.2). Put on the lid.

3. Install a cold furnace D to surround the sample holder and make sure that the water connections G are tight. Turn on the cooling water by fully opening tap K counterclockwise. Allow this water to run during the complete heating and cooling operation until you are ready to remove the furnace at the end.

4. If pressure or vacuum is used, place the pressure metal bell over the furnace and secure it to the furnace platform with the bolts. This bell is not attached for normal runs at atmospheric pressure.

5. Apply pressure or vacuum as needed (only if the bell top has been fitted).

6. If the DTA curve is to be run with a flow of dynamic gas, open the valve on the gas cylinder (e.g., it may be oxygen-free nitrogen) (pressure 10 psig) and adjust knob L (Fig. 14.2) until the gas flow on the meter above this knob registers a constant at 0.075 (standard cubic feet per hour). Check that the flow is constant. Allow it to stabilize, otherwise erratic flow of gas will cause spurious peaks on the DTA record.

7. Set amplification for differential thermocouples on the dc amplifier (switch 2 for coarse, switch 3 for fine) (Fig. 14.2). The total amplification is found by multiplying the value 2 by the value 3. The correct amplification is determined by trial and error based on previous experience. If the DTA curve is flat and featureless, then on the next run a lower value of knob 2 would be appropriate to give greater amplification of the

curve. As an average setting, knob 2 may be set at 80 and 3 at 1.5. If the peaks give too large a deflection, 2 may be adjusted to 150 or even higher numbers. If the peaks are too small, 2 can be adjusted to 40 or even 20 on the next run.

8. Turn the reference THERMOCOUPLE TYPE switches, located on the rear of the programmer (4, Fig. 14.2), to the positions appropriate to the thermocouple in the sample holder: K = chromel-alumel, P is for platinel II, S for platinum-platinum 10% rhodium, and J = iron-constantan. It is essential that the position of these switches agree with the type of thermocouple in use.

9. Turn the switch M to the appropriate thermocouple in use. C/A = chromel-alumel, PN = platinel II, PT/RH = platinum-10% rhodium, and I/C = iron constantan.

10. Turn switch 1 to STANDBY.

11. Set the recorder x-axis for the desired temperature range (5, Figure 14.2) and the appropriate thermocouple in use. The choice of ranges is the following: for chromel-alumel or platinel II, 0° to 275, 0° to 550°, or 0° to 1100°C; for platinum-10% rodium, 0° to 550°, 0° to 1100°, or 0° to 1600°C; for iron-constantan, −200 to 120° or −200 to 300°C. Appropriate chart paper is used to agree with the temperature range in use.

12. Use switch 6 to turn the recorder ON.

13. Zero the x-axis of the recorder. This is normally done by selecting 0°C on switch N, holding button 0 depressed, and, while it is depressed, adjusting knob 7 until the pen stylus coincides exactly with the 0°C vertical line on the chart paper. Release button 0. The needle of the pen will move to ambient room temperature.

14. Adjust the y-axis pen position with knob 8 until the pen is somewhere central on the y-axis of the chart.

15. Install a sheet of chart paper. It can be held in position by moving switch 9 to HOLD position. Check again that the x-axis zero of the pen (as in step 13) co-incides with the 0°C line on the chart. Adjust as necessary.

16. Set the rate °C per minute control (10, Fig. 14.2). Normally for most DTA curves this will be chosen at 10°C per minute. The heating rate must be specified on a DTA chart. For special purposes other heating rates may be selected, but 10°C per minute is standard.

17. Select the INITIAL MODE switch (11). In nearly all operations this will be set to HEAT.

18. Set the INITIAL RATE STABILIZER. Set switch 12 to 1 or 2; set to position 1 for operation with F-1 C furnaces; set to position 2 for operating with F-1 D furnaces. Set switch 13 as follows: to 28 seconds for type SH-11 sample holders made of all materials except aluminum; to 35 seconds for type SH-8 sample holders of all materials except aluminum; to 35 seconds for type SH-12 sample holders; to 12 seconds for any type of sample holder made of aluminum.

19. Set LIMIT switches as follows:

Upper Limit. Switch 14 can be set to stop the heating program at a certain percentage of the temperature scale. Normally it is set to something just under 100%; for example, if the sample holder and knob 5 are set to give a full scale of 1100°C, then by setting knob 14 to 90% the heating operation will stop at 90% of 1100, or at 990°C. The setting of knob 14 will depend on the highest temperature we are interested in, but normally it should not be set higher than about

98%. Switch 15 instructs the instrument what to do when the highest temperature is reached. If knob 15 is set as follows,

STOP—the instrument will stop operating when the highest temperature is reached;

STANDBY—the instrument will not switch off, but wait for further instructions;

COOL—a cooling program will start as soon as the highest temperature is reached at the same rate as set on knob 10;

HOLD—means the specimen will be held constant at the highest temperature;

CYCLE—means a cooling programme will start, and once the specimen cools to the lowest temperature it will be heated again, then cooled, and so on in a cycle, until told to stop.

Lower Limit Switches 16 and 17 instruct the instrument what to do if, after reaching the upper temperature limit, the instruction had been COOL or CYCLE, in which case once the specimen has cooled to the lower temperature (selected on switch 17 to be a percentage of the reading on the x-axis; e.g., 0% would mean 0°C, 10%, when the scale is 1100°C, would mean 110°C) knob 16 would instruct the instrument what to do, that is, STOP, heat again, and so on. If the instruction STOP is set on 15 or 16, the instrument will switch off by itself and need not be attended.

20. Set MANUAL VOLTAGE control 18 *fully counterclockwise*. This step must not be forgotten.

21. Turn on the dc amplifier, using switch 19.

22. Place STOP-START or CONTINUE switch to START OR CONTINUE position. The heating program is now initiated and the DTA curve will be automatically traced out.

23. Lower the recorder pen so that it starts writing (switch 21, Fig. 14.2). At the end of the program the instrument will switch off according to the instructions. The furnace is removed from around the sample holder only after it has cooled below 500°C. To remove it close knob K, release washers G, and move the furnace vertically upward. It can now be set on top to cool in position D. Secure washers E, switch on cooling water supply P, and switch on the air supply I. The exposed sample holder may now be cooled rapidly to room temperature by operating the air supply J.

EXERCISE 14.10

Reporting the Conditions of an Experiment

On completion of the DTA thermogram it is important to report the conditions of the experiment. Only in this way can an experiment be reproduced in the same laboratory with the same equipment and results obtained in one laboratory compared meaningfully with those obtained in another. The following information should be recorded and reported with every DTA thermogram:

1. Identification of all substances (sample, reference, diluent) by a definite name, empirical formula, or equivalent compositional data.

2. A statement of the source of all substances, details of their histories, pretreatment, and chemical purities, as far as known.

3.   Measurement of the average rate of linear temperature change over the temperature range involving the phenomena of interest.

4.   Identification of the sample atmosphere by pressure, composition, and purity; whether the atmosphere is static, self-generated, or dynamic through or over the sample. When applicable, the ambient atmospheric pressure and humidity should be specified. If the pressure is other than atmospheric, full details of the method of control should be given.

5.   A statement of the dimensions, geometry, and materials of the sample holder and the method of loading the sample when applicable.

6.   Identification of the abscissa scale in terms of time or of temperature at a specified location. Time or temperature should be plotted to increase from left to right.

7.   A statement of the methods used to identify the intermediate and final products.

8.   Faithful reproduction of all original records.

9.   Whenever possible, each thermal effect should be identified and supplementary supporting evidence stated.

10.   Sample weight and dilution of the sample recorded.

11.   Identification of the apparatus, including the geometry and materials of the thermocouples and the location of the differential thermocouple and the temperature-measuring thermocouples.

12.   The ordinate scale should indicate deflection per degree centigrade at a specified temperature. Preferred plotting will indicate upward deflection as a positive temperature differential and downward deflection as a negative temperature differential with respect to the reference. Deviations from this practice should be clearly marked.

(After Mackenzie and Mitchell, 1970b.)

EXERCISE 14.11

Identification of the Mineralogy of a Specimen by the DTA Thermogram

As mentioned above in the introduction, the DTA method cannot usually be used on a completely unknown material to identify its mineralogy from a thermogram because the curves of different components overlap and there is also a chance of reaction between components. Nevertheless the method may be used for confirmation and in some cases for estimation of the concentration of a particular mineral in the specimen.

The shape and intensity of DTA curves of a mineral are strongly influenced by amorphous coatings and disordered structures on the surface of the particles (the Beilby layer) and by differences in particle and crystallite size, degree of crystallinity of the crystalites, and ion substitution in the structure. Especially in natural clay minerals, variations in these phenomena are numerous. Hence an accurate quantitative analysis by DTA of the amount of a clay mineral in a sample is in many cases extremely difficult if not impossible (Marel, 1956).

A listing of the most important peak temperatures for a wide range of minerals is given in Table 14.1 and may be used to identify, or confirm the identification of a mineral that is suspected of being present in the sample. It should be noted that although many minerals have characteristic peaks at fixed temperatures others have their reaction peaks within a broad range of temperature. Also, it must be noted that a peak present on a thermogram run in air may be completely absent or greatly reduced in intensity if a nitrogen dynamic gas is used.

Table 14.1   Listing in Order of Temperature Increase of Characteristic DTA Peak Temperatures for a wide Range of Common Silicate, Carbonate, Oxide and Sulfide Minerals[a]

DTA Reaction Peak Temperature °C	Reaction Type (see footnote for symbols)			Mineral	Other Peaks (°C)
65	N	S	W	Opal	115, 300, but may be single at 140
90	N	S	W	Tridymite	107, 162
100–150	N	S	M	Illite	550, 900
107	N	S	I	Tridymite	90, 162
110–140	N	B	I	Montmorillonite	700, 850–950
100–150	N	B	I	Allophane	1000
115	N	S	W	Opal	65, 300, but may be single at 140
120	N	S	I	Halloysite	560, 950–1050
140–160	N	S	I	Vermiculite	270–280, 800–900
150	N	S	I	Sepiolite	300, 500, 800
162	N	S	I	Tridymite	90, 107
180	N	S	I	Palygorskite	280, 350–600, 800
229–253	N	S	I	Cristobalite	may be double or single
270–280	N	B	W	Vermiculite	150–160, 800–900
280	X	B	W	Epidote	960–980, 1200–1220
280	N	S	W	Palygorskite	180, 350–600, 800
300	N	S	W	Opal	65, 115, but may be single at 140
300	N	B	W	Sepiolite	150, 500, 800
300–330	N	S	I	Gibbsite	only peak
300–350	X	B	W	Biotite	700–800, 1000–1200
340–360	N	S	I	Lepidocrocite	only peak
350	X	B	W	Muscovite	800–900, 1100
350–400	X	S	W	Magnetite	600–1000
350–400	X	B	W	Phlogopite	1050–1200
370	X	S	I	Covellite	520
380 (380–405)	X	S	I	Goethite	only peak
380	N	S	W	Manganite	950–960
350–550	X	B	W	Carbon (natural)	only peak
350–600	N	B	W	Palygorskite	180, 280, 800
400–800	X	B	W	Amphibole	750–850, 1000–1100
450	X	S	W	Gersdorffite	680
460	X	S	I	Marcasite	1000
500	X	S	W	Niccolite	680
500	N	S	W	Aragonite	860–1010
500	N	B	W	Sepiolite	150, 300, 800
500–700	N	B	I	Kaolinite	950–1050
520	X	S	I	Covellite	370
520	X	S	I	Chalcocite	120
530	X	S	I	Chalcopyrite	600
530	X	B	W	Proustite	840
540–585	N	S	I	Diaspore	only peak
538–550	X	S	I	Pyrite	only peak
550–616	N	B	I	Chlorite	745–810

449

Table 14.1—*continued*

DTA Reaction Peak Temperature °C	Reaction Type (see Footnote for symbols)			Mineral	Other Peaks (°C)
550	X	S	I	Carrollite	620
550	X	S	I	Arsenopyrite	only peak
550	N	B	W	Illite	100–150, 900
560	N	S	I	Halloysite	120, 950–1000
560	N	B	I	Rhodochrosite	660, 670
565	N	S	I	Stibnite	
570	X	S	I	Siegenite	750, 840
573	N	S	I	Quartz	only peak
580–600	N	B	W	Leucite	
585	N	S	I	Siderite	610, 900
600	X	S	I	Pyrrhotite	
600	X	S	I	Chalcopyrite	530
600–630	X	B	I	Bornite	
600	X	S	I	Skutterudite	
600–1000	X	B	W	Magnetite	350–400
600–800	N	B	W	Pyrophyllite	750–850
610	X	S	I	Siderite	(air only), 585, 900
620	X	S	I	Carrollite	550
625–725	N	S	I	Pyrolusite	950–1050
630–660	X	B	I	Bornite	
630–820	X	B	M	Graphite	
650	X	S	I	Loellingite	
660	N	B	I	Rhodochrosite	560, 670
660–690	N	B	I	Magnesite	only peak
670	X	B	W	Argentite	
670	X	S	W	Rhodochrosite	560–660
680	X	S	I	Niccolite	500, 800
680	X	S	I	Gersdorffite	450
680	N	S	W	Hematite	only peak
700	N	B	M/W	Montmorillonite	120–950, 850–950
700	N	S	I	Serpentine	830
700–800	X	B	M	Biotite	300–350, 1000–1200
721	N	S	I	Teallite	
730	X	S	I	Millerite	
750	X	S	I	Siegenite	570, 840
750–850	X	S	I	Cobaltite	(composite peak)
745–810	X	S	I	Chlorite	550–616
780–820	N	B	W	Plagioclase	
750–850	N	B	I	Hornblende	
750–850	N	B	M	Pyrophyllite	600–800
770–820	X	B	I	Sphalerite	
790	N	S	I	Dolomite	940
800	X	S	W	Niccolite	500, 680
800	N/X	S	W	Palygorskite	180, 280, 350–600
800–900	N	B	W	Muscovite	350, 800–900
800–900	N/X	S	W	Vermiculite	150–160, 270–280
810–820	N/X	S	I	Sepiolite	150, 300–500

Table 14.1—*continued*

DTA Reaction Peak Temperature °C	Reaction Type (see Footnote for symbols)			Mineral	Other Peaks (°C)
820–900	N	B	W	Albite	
830	X	S	I	Tremolite	1050–1150
840	X	S	M	Millerite	730
840	X	S	I	Siegenite	570, 750
840	X	S	I	Galena	
840	X	B	I	Proustite	530
830–840	X	S	I	Serpentine	700
850–950	N/X	S	W	Montmorillonite	120–140, 700
860	X	B	W	Richterite	1050
870	N	S	I	Teallite	
890–910	N	S	I	Calcite	only peak
860–1010	X	B	(variable)	Zircon	(metamict)
900	N/X	S	W	Illite	100–150, 550
900	N	B	W	Albite	820
940	N	S	I	Dolomite	790
950–960	N/X	B	M	Manganite	380
950–1000	N	S	I	Talc	only peak
950–1050	X	S	M	Halloysite	120, 560
950–1050	N	S	M	Pyrolusite	625–725
950–1050	X	S	I	Kaolinite	500–700
960	N	S	M	Epidote	1200
987	X	S	I	Epidote	280, 938
998	N	S	M	Zoisite	
1000	N	B	M	Tourmaline	
1000	X	S	M	Allophane	100–150
1000–1200	N	B	M	Biotite	300–350, 700–800
1050–1150	N	B	I	Tremolite	830
1050	N	B	W	Richterite	860
1050–1200	N	S	W	Phlogopite	350–400
1100	N	B	W	Muscovite	350, 800–900
1200	N	S	M	Epidote	960
1254	N	B	W	Nepheline	

[a] Data from various sources: X = exothermic peak (convex upward); N = endothermic peak (convex downward); S = sharp (i.e., over a narrow temperature range); B = broad; I = intense; W = weak; M = moderate. Composite peaks are noted as X/N.

The method of using this table is to compare the peak temperatures on your thermogram in increasing order of temperature and tentatively assign any peak to a particular mineral. Then refer to the following brief descriptions of the individual mineral DTA records for confirmation or rejection of the identification. In many cases DTA thermograms of a particular mineral are characterized by a single peak, others by two or more peaks, in which case it will be necessary to find the additional peaks as a confirmation based on one peak only. It is recommended that data from an additional method, such as X-ray diffraction, be employed and the DTA result used only as a confirmation.

DTA THERMOGRAMS OF MINERAL GROUPS

SILICA MINERALS

QUARTZ. The inversion of $\alpha \rightleftarrows \beta$ quartz at 573°C ± 1°C (endothermic) has become a standard "reference temperature" in DTA work. More than 95% of natural quartz specimens invert within a range of 2.5°C. The area of the 573° peak, which is the only feature of a quartz DTA curve, can be used to estimate the amount of natural quartz in a specimen containing more than 1% quartz. When this is attempted however, calibration should be done with quartz from the same locality. Particle size of quartz should be the same for all determinations and experimental conditions reproduced accurately. Langer and Kerr (1967) have warned that DTA curves vary systematically with sample, reference, and instrumental variables. The quartz inversion peak area increases with sample weight per unit volume but the reaction temperature remains constant. There is negligible shift in temperature of inversion with variable grain size. Packing of the sample sharpens the reaction peak. It is important that the reference material be of similar grain size and mass specific heat to the sample so that the baseline slope can be controlled (Langer and Kerr, 1967). Change in heating rate produces no observable shift in the quartz inversion temperature. Quartz thermograms are described by Dawson and Wilburn, 1970; Grimshaw and Roberts, 1957, and McLaughlin, 1967b.

TRIDYMITE. Tridymite has a weak endothermic peak at about 90°C, a strong endothermic peak about 107°, and a fairly strong endothermic peak at about 162°C, but the temperatures vary, usually within the range 86 to 166°C. Curves show a sharp progressive drop from 0° to 90°C. Although tridymite can be detected by DTA, it is clear that the results must be treated with caution and quantitative work is not practicable. (Details in Dawson and Wilburn, 1970; Grimshaw and Roberts, 1957; and McLaughlin, 1967b). There are no further peaks above 164°C.

CRISTOBALITE. Cristobalite undergoes inversion between 200 and 260°C. The actual temperature depends on the thermal history of the specimen, but individual endothermic peaks may be displayed at 229 and 253°C. For general use only heating curves may be used. Some cristobalites show only one peak at about 253°C. (Grimshaw and Roberts, 1957; Dawson and Wilburn, 1970; McLaughlin, 1967b). Curves show a sharp progressive drop from 0 to 210°C.

OPAL. Peaks occur at 65, 115 and 300° but may be replaced by a single peak at 140°C. Thermograms are variable with source of material. (Dawson and Wilburn, 1970).

FELDSPARS

Albite. Albite shows a small endothermic peak at 820 or at 900°C. The peak is likely to be due to hydrous phases present as contaminants.

Intermediate Plagioclases (oligoclase to labradorite). These minerals give an endothermic peak in the range 780 to 820°. The peak is unlikely to be due to inversion from a low to a high temperature state because such an inversion is too sluggish to show on DTA records (Glasser, 1970).

FELDSPATHOIDS

Nepheline. There is no thermal activity below 1200°C. At 1254° nepheline converts to carnegieite, but the reaction is sluggish. On cooling the carnegieite converts at 687°. On reheating the carnegieite inverts at 692°C (Glasser, 1970).

Leucite. An endothermic peak commences at 580° and ranges over 580 to 600° (McLaughlin, 1967b).

AMPHIBOLES

Tremolite. An exothermic peak occurs at about 830°C, which may represent a polymorphic transition. Tremolite recrystallizes at 1000 to 1100° to yield a monoclinic pyroxene, equivalent to a broad endothermic peak at 1050 to 1150°C.

Richterite. A broader exothermic peak appears at about 860° and an endothermic peak at 1050°.

Hornblende. Hornblende shows an endothermic peak in the range 750 to 850° which probably represents dehydroxylation.

Most common amphiboles contain significant quantities of oxidizable constituents, of which Fe^{2+} is the most important. DTA curves obtained in an oxidizing atmosphere almost always give an exothermic effect arising from Fe^{2+} oxidation somewhere in the region 400 to 800°C. Decomposition temperatures of amphiboles, except for tremolite, are rarely higher than 800 to 850°. The decomposition reaction is weakly endothermic and yields a pyroxene (Glasser, 1970).

EPIDOTE GROUP

Zoisite. A simple sharp endothermic peak at 998°C corresponds to a loss of water from the structure (McLaughlin, 1957).

Epidote. The positions of two characteristic endothermic peaks remain nearly constant at 960 and 1200°C for the high aluminum end members and at 985 and 1220°C for the high iron end members. Melting occurs at 1240° for high iron epidote (Glasser, 1970). McLaughlin (1957) shows an epidote DTA curve with a weak broad exothermic peak at 280° and exothermic peaks at 938 and 987° (main sharp peak).

TOURMALINE. Tourmaline gives a conspicuous broad endothermic peak at about 1000 ±10°C.

ZIRCON. Metamict zircon gives a single or more usually double exothermic peak in the range 890 to 910°C, an effect that is not reversible on cooling. The effect is associated with partial recrystallization. The area under the peak can be used as a direct measure of the total amount of radioactive damage stored within the individual crystal. (Glasser, 1970).

ZEOLITES. Zeolites show intense endothermic peaks caused by the loss of zeolitic water.

CARBONATES

Calcite. A single strong and sharp endothermic peak may vary from 860 to 1010°, the variation being due to particle size and technique.

Aragonite. In addition to the strong endothermic peak at 860 to 1010°, aragonite shows a small endothermic peak at about 500°. This peak represents the irreversible transition to calcite.

Magnesite. Magnesite shows an intense endothermic peak at 660 to 690°C. The peak begins at 580 to 630°.

Dolomite. Dolomite's two intense endothermic peaks have temperatures of 790 and 940°C. (Webb and Krüger, 1970.)

Siderite. Siderite is characterized by a strong endothermic peak at about 585°C, both in air and N_2, and a strong sharp exothermic peak at 610° when the DTA is run in air. When run in N_2, this 610° peak is completely absent. A broad exothermic peak occurs at about 900° in air and is absent in N_2 (Bayliss and Warne, 1972). The detection of siderite in mixtures with kaolinite is inhibited by the superposition of their peaks in the 500 to 600° range. Up to 30% well crystallized siderite in a mixture may be overlooked (Bayliss and Warne, 1972).

Rhodochrosite. Typically a double endothermic peak at about 560 to 660°C is followed by an exothermic peak at 670°C (Webb and Krüger, 1970).

ORE MINERALS

Sulfide and arsenide minerals will attack thermocouple junctions. Therefore DTA analysis of these minerals must be done by a method in which the mineral powder is not in contact with the thermocouple but sits in a small containing pan, such as sample holders of type SH-11 BP or SH-11 BR (exercise 14.1).

Pyrite. In an oxidizing atmosphere this mineral shows a sharp and intense exothermic peak at 550°C due to oxidation. The sample is best diluted and the complexity of the DTA depends on the amount of dilution.

Marcasite. Marcasite's initial oxidation peak is at approximately 460° (sharp exothermic), whereas pyrite is at a higher temperature. Marcasite oxidizes at a lower temperature because of its relative instability. The thermal reaction products for both marcasite and pyrite at 1000°C are hematite and sulfur dioxide (Kopp and Kerr, 1958).

Pyrrhotite. A sharp and intense exothermal peak appears at 600°.

Covellite. Covellite has a sharp exothermic peak at 370° and a larger peak (exothermic) at 520°. (Bollin, 1970).

Chalcocite. Chalcocite shows an exothermic peak at 520°.

Carrollite. A double intense exothermal peak occurs at 550 and 620°.

Chalcopyrite. Chalcopyrite has a double intense exothermal peak at 530 and 600°.

Bornite. An exothermal broad peak appears at 600 to 630°.

Millerite. A very sharp and intense exothermic peak shows at 730° and a less intense but equally sharp one at 840°.

Siegenite. Siegenite has exothermic peaks at 570, 750, and 840°.

Sphalerite. A broad but intense exothermic peak occurs at about 850°. Using an internal standard of quartz (573°C), the peak temperature of sphalerite decreases with increasing iron content. The unit cell a (Å) can also be related to iron content:

% Fe in Sphalerite	a, Å	DTA exothermic peak temperature $\pm 3°C$
<0.1	5.4082	821
0.3	5.4087	811
2.5	5.4103	793
6.9	5.4160	770
13.2	5.4196	769

(After Kopp and Kerr, 1958.)

Galena. A sharp and intense exothermic peak appears at 840°C.

Loellingite. Loellingite's exothermic peak at 650° is very sharp and intense.

Arsenopyrite. Arsenopyrite has a very sharp intense exothermic peak at 550°.

Gersdorffite. Two intense exothermic peaks at 450 and 680° are fairly broad.

Niccolite. Niccolite has a small sharp exotherm at 500°, a large sharp exotherm at 680, and a very small broad exotherm at 800°.

Teallite. Endotherm sharp peaks appear at 721 and 870°.

Stibnite. A single sharp endotherm peak shows at 565°.

Cobaltite. Cobaltite has a composite intense exothermic peak at 750 to 850°.

Skutterudite. Skutterudite has an intense exothermic peak at 600°.

Proustite. Proustite has two broad, weak exotherms at 530 and 840°.

Argentite. Argentite has a broad, weak exotherm at 670°.

(Bollin, 1970.)

OXIDES AND HYDROXIDES

Hematite. The DTA curve is generally featureless but with a very small reversible peak (endothermic) at about 680°.

Magnetite. A small exotherm at 350 to 400° is due to surface oxidation.

Goethite. A large and sharp endothermic peak at about 380°C may range from 385 to 405°.

Lepidocrocite. Lepidocrocite exhibits a more variable DTA pattern than goethite. DTA cannot be used to distinguish lepidocrocite and goethite because it is found that curves resembling goethite, lepidocrocite, and mixtures of the two can be produced by goethite alone. The variations depend on the degree of crystallization in the minerals.

For reliable identification of the mineral species of hydrous ferric oxides present in natural aggregates it is necessary to use either X-ray diffraction or immersion media of high refractive index (Kelly, 1956).

Diaspore. Diaspore shows a large endothermic peak at about 540 to 585°.

Boehmite. Boehmite shows a large endothermic peak at about 450 to 580°. Both diaspore and boehmite show a single large peak only.

Gibbsite. A large endothermic peak is shown at 300 to 330°C.

(Mackenzie and Berggren, 1970)

Pyrolusite. A large endothermic peak appears at 625 to 725° and another less large endotherm at 950 to 1050°.

Manganite. Manganite gives a sharp intense endotherm at 375 to 380°.

CLAY MINERALS, MICAS, AND CHLORITES

Pyrophyllite. This mineral has a single broad and weak endothermic peak in the 600 to 800° region or it may have two broad endothermic peaks, one at 600 to 800° and the other at 750 to 850°.

Talc. Talc consistently gives a single endothermic strong and sharp peak at 950 to 1000°.

Montmorillonite. A variety of thermograms is yielded by montmorillonite. Usually a large fairly broad endothermic peak occurs at about 120 to 140°. There is also usually a medium-small endothermic peak at about 700° and a small S-shaped endothermic/exothermic peak at 850 to 950°.

Saponite. An uncomplicated curve shows a large, sharp, low-temperature endothermic peak at 136 to 218° and a single large to moderate endothermic peak at 850 to 900°.

Vermiculite. A large endothermic peak at 150 to 160° is followed by a small endothermic peak at 270 to 280°, then a small S-shaped endothermic/exothermic peak at 800 to 900°C.

Illite. A medium-sized endothermic peak appears at 100 to 150° and an S-shaped endothermic/exothermic peak at about 900°. There is usually also a small broad endothermic peak at 550°.

Muscovite. Muscovite typically yields two relatively small endothermic peaks at 800 to 900° and about 1100° and a very small exothermic peak at 350°.

Phlogopite. Phlogopite usually shows a relatively small endothermic peak at 1050 to 1200° and a very small exothermic peak in the 350 to 400° region (usually more pronounced than in muscovite).

Biotite. Biotite is variable. Generally it has an endothermic peak at 1000 to 1200°. Two other peaks occur, a small exothermic peak at 300 to 350°, and a larger exothermic peak at about 700 to 800°.

Lepidolite. A single endothermic peak which shows at about 900° is large and sharp (Mackenzie, 1970b).

Chlorite. The DTA curve is characterized by a strong but broad endothermic peak at 550 to 616° and a very strong, very sharp exothermic peak at 745 to 810° (Rahden, 1972).

Kaolinite. Kaolinite typically shows a broad intense endothermic peak in the region 500 to 700° and a very sharp exothermic peak between 950 and 1050°, usually around 1000°C. As the particle size of kaolinite decreases, the temperature of the endothermic peak decreases correspondingly. The heating rate produces a marked effect on the area and temperature of the endothermic peak. Peak area $(in^2) = 0.031 \times$ heating rate (°C per minute) + 0.005. Incremental differences between successively determined peak temperatures are greater at heating rates of 10 to 20°C per minute and become less for 40 to 50°C per minute. (Langer and Kerr, 1967).

Halloysite. A strong endothermic peak occurs at about 120°C. Halloysite's main endothermic peak is strong at 560°. It also has an exothermic sharp peak at 950 to 1050°. Halloysite may be distinguished by saturating it with ethylene glycol first. This induces an endothermic peak to appear at 500°C.

Serpentine. Both antigorite and chrysotile give an endothermic strong peak at about 700°. Both also give a strong sharp exothermic peak at about 830°.

Allophane. A large broad endothermic peak occurs at 100 to 150°C and a sharp exothermic peak at about 1000° (Mackenzie, 1970b).

Palygorskite. A large endothermic peak shows at 180°, and usually a smaller endotherm at 280°. From 350 to 600° a broad weak endothermic peak is present, sometimes single, sometimes double. At 800° it is common to have an endothermic effect (weak) followed by a strong sharp exothermic effect.

Sepiolite. Sepiolite has a strong, sharp endothermic peak at about 150°. Often two weak broad endothermic effects occur at about 300 and 500°. All show a strong sharp endothermic peak at 800°, followed immediately by a strong sharp exothermic peak.

(Vivaldi and Hach-Ali, 1970).

CARBON AND GRAPHITE

Graphite. Graphite gives a broad exothermic weak peak in air with peak temperatures ranging from 600 to 850°.

Carbon. Carbon in coal, peat, or as ash gives broad, more intense exothermic peaks in air in a temperature range of about 300 to 550°C. Thus natural carbon may be easily distinguished from natural graphite by the temperature of oxidation of the material (Swaine, 1969).

EXERCISE 14.12

Qualitative and Quantitative Estimation

Assuming that the experimental variables are kept constant, DTA records should be reproducible in a qualitative way, but quantitative data require greater standardization. Excellent reproducibility is usually obtained from sharp crystallographic inversions (e.g., for quartz at 573°C). Changes involving release of volatiles or oxidation are obviously less reproducible.

In describing DTA thermal curves, the term "characteristic temperature" is often used at the point at which a line drawn at 45° to the base line is tangential to the beginning of the peak on the curve (Fig. 14.4). This "characteristic temperature" is of considerable use in qualitative work, for it is claimed to be less subject to variation than peak temperature.

Quantitative estimations are concerned with the area outlined by the peak; hence it is important to decide when the peak leaves and returns to the base line. With minerals of reasonably good crystallinity it is possible to correlate the concentration of the mineral with peak area (e.g., for quartz and carbonates). Of necessity, any peak must be entirely due to a single mineral. With minerals in which structure disorder is frequent (e.g., clays) great care must be exercised in quantitative work.

Other peak parameters are shown in Figure 14.4: peak height, cosecant of the angle formed between two lines tangential to the sides of the curve, and slope ratio in which the peak is divided into two angles by a vertical from the base line to the peak and straight lines from the peak and tangential to the sides of the curve. The slope ratio is $\tan \alpha / \tan \beta = a/b$ (Fig. 14.4). The slope ratio is related to the degree of structure disorder. (After McLaughlin, 1967b.)

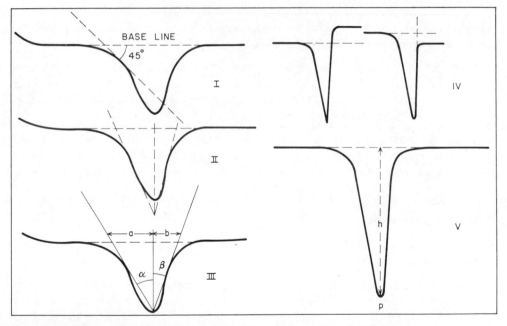

Fig. 14.4 Measurement on DTA curves. I = determination of the "characteristic temperature"; II = determination of peak angle; III = determination of slope ratio of the peak; IV = measurement of areas under the thermal peak when the base line after the peak is above or below the original; V = measurement of the peak area in the normal case. (After McLaughlin, 1967b.)

B.   THERMOGRAVIMETRIC ANALYSIS

This technique consists of studying the loss in weight of a substance as it is being heated. It is abbreviated TG.

DTA alone, although yielding valuable information on a mineral or mixture, is usually not sufficient on its own for most investigative purposes. Considerably more information can be obtained by combining it with an X-ray diffraction analysis

(Chapters 6 and 7). By using TG in conjunction with DTA it becomes possible to distinguish crystalline transitions, second-order transitions, and solid-phase reactions occurring without weight change on the one hand from phenomena such as loss of volatiles and dehydroxylation involving weight loss, or gain on the other. For a description of the theory of the method the reader is referred to Redfern (1970).

A TG record, called a thermogravimetric or TG curve, records the weight of the specimen in an environment heated or cooled at a controlled linear rate as a function of temperature.

Thermogravimetric Apparatus

Several instruments are available commercially. Figure 14.5 illustrates the TGA model 1050-RGP Stone (supplied by Columbia Scientific Industries Inc., Analytical and Industrial Division, 3625 Bluestein Boulevard, P.O. Box 6190 Austin, Texas 78702),

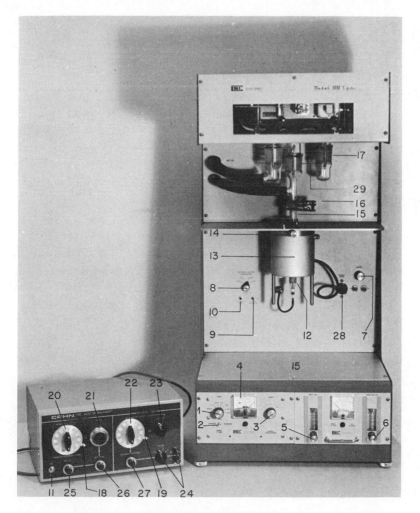

Fig. 14.5   Model 1050 Stone TGA apparatus of 0–1600°C temperature range and 10^{-6} g weight loss sensitivity incorporating a Cahn electrobalance. The instrument output may be plugged into the CSI-Stone recorder-controller model LB 202 (Fig. 14.2). (By courtesy of Columbia Scientific Industries Inc., Austin, Texas.) See exercise 14.13 for details.

which operates over a temperature range of 0 to 1600°C with a weight loss sensitivity of 1×10^{-6}g per recorder chart division under optimum operating conditions. Maximum specimen capacity is 2.5 g, maximum weight change measurable, 1.0 g, sample pan size, 5- or 9-mm diameter, with temperature program rates selected at 1, 2.5, 5, 10, 25 or 50°C per minute, temperature increasing, holding, or decreasing as desired.

The Cahn balance, which is the top part of Fig. 14.5, is a nullpoint instrument and may be employed under controlled atmosphere, pressure, or vacuum. From the schematic diagram (Fig. 14.6) it will be seen that a photocell serves as a detector for a light passing through a slit behind the balance beam and interrupted by a flag fixed to the beam. Any movement of the balance beam causes a change in photocell voltage which is amplified and then applied to the electromagnet coil attached to the balance. The coil is in a magnetic field and the current generated by the amplified voltage exerts a torque on the beam, instantaneously restoring it to the null position.

EXERCISE 14.13

Routine Operation of the TGA Apparatus

   1.   Place MODE knob 1 (Fig. 14.5) to STANDBY.

   2.   Place toggle switch to INCREASE (2, Fig. 14.5).

   3.   Place RESET clockwise to its stop position (3).

   4.   *Limit switch.* Grasp the small black knob (4) on the limit switch meter face and turn clockwise; the black needle on the meter releases and indicates zero temperature. Turn the knob counterclockwise until the red needle on the meter indicates the shutoff temperature desired.

   5.   INERT GAS (5) valve closed (clockwise).

   6.   REACTIVE Gas (6) valve closed (clockwise).

   7.   Coolant (7) valve open (counterclockwise).

   8.   Select the proper scale on the X-Y recorder (Fig. 14.2). Turn knob 8 (Fig. 14.5) to the correct thermocouple used in the TGA furnace. Turn switch 9 to position 0°C, depress button 10, and, holding it depressed, zero the recorder pen by adjusting the recorder zero potentiometer (Fig. 14.2).

   9.   Turn the Cahn balance control cabinet power switch 11 (Fig. 14.5) to ON.

  10.   Remove the furnace as follows:

       a.   Unplug the reference (environmental) thermocouple plug (12).

       b.   Hold the furnace (13) with the left hand and pull the furnace locking pin (14) out to its stop position.

       c.   Slowly lower the furnace until it clears the guide rods and the reference (environmental) thermocouple plug (12).

       d.   Set the furnace upright on the shelf (15, Fig. 14.5). It is not necessary to disconnect the furnace cables or coolant hoses.

  11.   Remove the hangdown tubes as follows:

       a.   Hold the lower hangdown tube with the left hand (15) and unclamp the swivel joint clamp (16) with the right hand and set the swivel joint clamp aside.

b. Carefully bring the lower hangdown tube (15) down until it is clear of the hangdown wire and stirrips. Set the lower hangdown tube aside.

c. Remove the retaining spring from the Cahn 1387 cap (17) on the " C " loop and lower the cap away from the stirrip. Set the cap and spring aside.

12. Adjust and calibrate the Cahn balance system (Fig. 14.6) and the recorder according to sections 1.4, 1.5, and 1.6 of the Cahn instruction manual for the sample size to be investigated.

The following are abbreviated instructions for calibration of the Cahn *balance*. Calibration of loop A and loop B (Fig. 14.6) is the same, with the exception that weights used on loop A are only one-fifth the value of those used on loop B. Control settings must be changed accordingly. The following procedure is for calibrating loop B:

a. Adjust RANGE and MULTIPLIER controls of the S-601 amplifier of the LB 202 controller (Fig. 14.2) for 2000 μV.

b. Position the recorder pen exactly to midscale, using the base-line positioning potentiometer (Fig. 14.2) on the y-axis.

c. Apply power to Model LB 202 recorder controller (Fig. 14.2) and the S 601 amplifier or the P 1000 programmer by turning Mode to RESET (1, Fig. 14.5).

d. Turn MASS DIAL RANGE (18) and RECORDER RANGE (19, Fig. 14.5) tabs to B position.

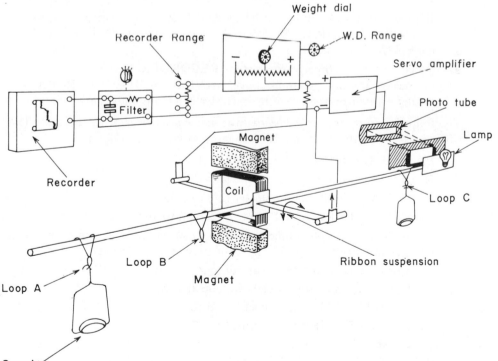

Fig. 14.6 Schematic diagram of a null-point TGA balance. The Cahn balance is incorporated in the model 1050 Stone TGA system. (After Redfern, 1970.) See Exercise 14.13 for details.

e. Turn MASS DIAL RANGE control (20) to 100.

f. Rotate MASS DIAL potentiometer (21) to 0.5000.

g. Turn RECORDER RANGE control (22) to Z.

h. Turn FACTOR switch (23) to 1.

i. Turn reference junction compensator THERMOCOUPLE TYPE switch (8) to the type of thermocouple being used in the furnace.

j. Turn REFERENCE TEMPERATURE switch (9) to 0°C. When the recorder controller (Fig. 14.2) is used, the filter (24, Fig. 14.5) must be adjusted to achieve desired noise level without sacrificing accurate response.

k. Rotate SET 5 (25) and SET 0/10 potentiometers (26) to their approximate midpoints.

l. Place a 50 mg calibration weight on sample pan of loop B (Fig. 14.6).

m. Apply power to the TGA apparatus and allow 2 hours for equilibration.

n. Turn RECORDER RANGE control (22 and 23) to 1000 (the recorder pen should move upscale).

o. Add weight to loop C pan Fig. 14.6) until the recorder pen is at midscale. Grains of alumina, platinum wire, or other inert minerals may be used as the counterweights in loop C.

p. Place RECORDER RANGE control (22 and 23) to consecutively lower ranges, stopping when the recorder pen approaches midscale.

q. Add or remove weight to or from loop C until recorder pen is at midscale.

r. Adjust SET 5 (25) control to effect very fine adjustment of the recorder pen.

s. Repeat steps p through r until RECORDER RANGE control is on MINIMUM RECORDER RANGE setting indicated by MASS DIAL RANGE control (20, Fig. 14.5).

t. When step s is properly completed, RECORDER RANGE (22, 23) can be placed in any position above MINIMUM RECORDER RANGE setting without movement of the recorder pen from midscale.

The following steps are for *calibrating* the *meter movement* of the balance. This procedure must follow the balance *procedure* given in steps a through t.

a. Place RECORDER RANGE control (22) to Z.

b. Remove the 50-mg weight from the loop "B" pan.

c. Rotate MASS DIAL potentiometer (21) to 0.0000.

d. Turn RECORDER RANGE control (22, 23) to 100.

e. Adjust SET 0/10 potentiometer (26) to position recorder pen to midscale (the potentiometer should be approximately at its midpoint).

f. Turn RECORDER RANGE control to Z.

g. Replace 50 mg calibration weight in loop B sample pan.

h. Rotate MASS DIAL potentiometer to 0.5000.

i. Turn RECORDER RANGE control to 100.

j. Adjust SET 5 (25) potentiometer to position recorder to midscale (the potentiometer should be approximately at its midposition).

k. Repeat steps a through j until no recorder pen movement is noted when switching RECORDER RANGE control from Z to 100.

l.   Repeat steps a through k, switching RECORDER RANGE control from Z to 50, Z to 10, Z to 5, etc., until RECORDER RANGE control is in MINI-MUM RECORDER RANGE setting indicated by MASS DIAL RANGE control.

The following steps are required to *calibrate* the *recorder* This procedure must follow the steps for calibrating the balance and the meter.

a.   Turn RECORDER RANGE control to Z. The 50-mg calibration weight should still be in loop B of the sample pan.

b.   Position recorder pen to zero, using the recorder base-line positioner (Fig. 14.2). Zero is the bottom index of the X-Y recorder.

c.   Rotate MASS DIAL potentiometer to 0.4000.

d.   Turn RECORDER RANGE control to 10.

e.   Recorder pen should indicate full scale. If it does not, adjust CALIBRATE RECORDER potentiometer (27) on the TGA balance control unit panel for full-scale deflection.

f.   Repeat steps a through e until no adjustment is needed to position the re-corder pen.

13.   When the balance system is adjusted and calibrated according to the Cahn balance instructions or the above abbreviated steps, reassemble the hangdown tubes as follows:

a.   Install the Cahn 1387 (17, Fig. 14.5) cap over the C loop hangdown wire and stirrup and secure with a Cahn retaining spring 2170.

b.   Carefully slip the lower hangdown tube up over the A hangdown wire and stirrup and mate the upper end with the swivel connection. Place the swivel joint clamp in position to hold the joints together (16).

14.   The specimen, which should be pretreated exactly as for a DTA analysis (exercise 14.3) and must be less than 2.5 g, is now loaded onto the sample pan B of Fig. 14.6.

15.   Install the furnace as follows:

a.   While holding the furnace in one hand with the coolant fittings up and to the right, feed the reference (environmental) thermocouple plug (12) through the furnace cavity and at the same time start pushing the furnace up on the furnace guide rails. The furnace can be adjusted to centrally enclose the specimen.

b.   Pull the furnace locking pin (14) out to its stop position and carefully slip the furnace up on the lower hangdown tube and seat the furnace against the stand. Push the furnace locking pin in to lock the furnace in place.

16.   Plug the reference (environmental) thermocouple plug into the jack marked REFERENCE T.C. (28).

17.   Install a Cahn 1871 vacuum elbow (29) on the B loop fitting on the Cahn glass vacuum bottle and secure with a Cahn retaining spring 2170.

18.   To start the furnace turn the MODE control (1) to the desired rate. The furnace voltage meter should immediately indicate a low voltage in the range 5 to 15 V. If no voltage is shown, check the COOLANT valve (7) for extreme counterclockwise position and the COOLANT DRAIN hose for water flowing, for the furnace voltage cannot be applied unless the coolant is flowing. Depending on the heating rate selected, one to several minutes may elapse before the furnace temperature begins to rise. As the temperature rises, the pilot light on the gas control panel will increase in intensity and the voltage will increase on the furnace voltmeter.

19.   On completion of the run the following steps should be performed:

a.   Turn the MODE control to RESET.

b.   Turn the RESET control clockwise to stop.

c.   Grasp the small black knob on the limit switch-meter face (4) and turn clockwise until the black needle on the meter releases and indicates the existing furnace temperature. Then turn the small black knob counterclockwise until the red needle indicates the desired shut off temperature for the next run.

20.   When the furnace temperature is lower than 50°C, return to paragraph 10 of this section to remove the furnace in readiness for the next run.

DISCUSSION

For isothermal modes of temperature control set the programmer for a normal run. Increase the furnace temperature at a rate of 10°C per minute until the indicated temperature is 5 % below the desired hold temperature. Then set the toggle switch to HOLD position (2, Fig. 14.5) and make a slight adjustment of the RESET control until the desired temperature is reached.

Once the run has reached maximum temperature the toggle switch may be set at DECREASE and a run will be made under cooling conditions. When the furnace temperature cools to ambient, turn the MODE control to RESET position.

Runs may also be made with reactive gas, inert gas, and vacuum or pressure.

For use of the A loop rather than the B loop in the Cahn balance proceed with the normal installation and operation instructions but use the B loop where A is designated and the A loop where B is designated. The maximum sample weight on the A loop is 0.5g.

EXERCISE 14.14

Results and Interpretation of TG Curves

The results of some TG determinations on a variety of minerals are shown in Fig. 14.7. The TG curves of this figure may be compared with the DTA peak temperatures given in exercise 14.11. A DTA peak may correlate with a sharp loss of weight on the TG curve which indicates that the reaction caused a loss of volatiles from the specimen; for example, the chlorite TG curves show a progressive loss of weight in the range 400 to 600°C; thereafter the curve is flat (Rahden and Rahden 1972). This shows that the endothermic DTA peak at 550 to 616° is connected with a loss of volatile material (dehydroxylation). On the other hand, the sharp exothermic chlorite peak at 745 to 810° does not correlate with any change in weight on a TG curve, so that this reaction is a phase change not associated with any weight loss.

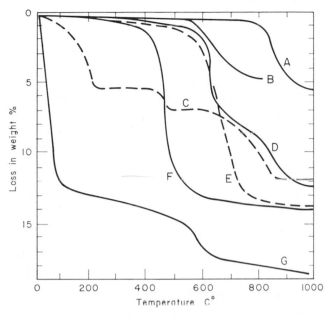

Fig. 14.7 Thermogravimetric curves for a variety of minerals: A = talc; B = pyrophyllite; C = calcium and magnesium carbonate precipitate; D — chlorite; E = serpentine; F = kaolinite; G = montmorillonite. (After McLaughlin, 1967b.)

C. THERMOLUMINESCENCE

Thermoluminescence is the thermal release of electrons which have been trapped at metastable high-energy levels within a crystal structure.

Electrons that are bound to ions in the crystal structure may receive additional energy from any form of ionizing radiation such as UV, X-rays, gamma rays, or beta particles. If the energy so absorbed by a valence electron is of sufficient intensity, that electron may be freed from the ion and, by its higher energy state, is capable of freely wandering throughout the crystal structure. It is said to have been raised to the conduction band of energy. Between the valence and conduction band of energy there is in all semiconductors a forbidden range of energies. Ions in a perfect crystalline lattice occupy the lowest energy levels possible for that particular crystalline substance. Crystalline imperfections are common in minerals and represent points of higher energy. Electrons bound to such imperfections are themselves at a higher energy level than the valence band.

An electron, when raised to the conduction band by absorption of energy, will normally combine with a hole somewhere in a lattice position and thereby lose its extra energy. While wandering throughout the crystal, however, a conduction band electron may combine with a hole in a lattice imperfection and be trapped metastably. There is no way for this electron to return to the valence band until it is again raised in energy to the conduction band. How can an electron be freed from such a trap? Natural thermal motion of the crystal does not impart enough energy to free trapped electrons, but with increasing temperature thermal motion does impart enough energy to free trapped electrons. Traps have differing energy depths below the conduction band and to free electrons so trapped will need a heating of the specimen to higher and higher temperatures as the traps are deeper.

Natural thermoluminescence in minerals is believed to be produced mainly by the alpha radiation from radioactive impurities. Tectonic pressure is also undoubtedly a contributing factor. Even very pure quartz and calcite crystals may contain trace amounts of uranium and thorium ions. Alpha activity from these impurities continually drives valence electrons into the conduction band where many are trapped in crystalline imperfections. Imperfections are of various kinds: Frenkel defects, vacant lattice sites, impurity ions, dislocations, and twin planes.

The thermoluminescence of a specimen can be released and measured by heating the specimen at a linear rate from room temperature to about 450°C. The release of metastably trapped electrons gives off energy in the form of light, which can be measured continuously as temperature increases. The plot of light intensity against temperature is called a glow curve. The theory and applications of thermoluminescence are described fully in many papers to be found in McDougall (1968).

EXERCISE 14.15

Apparatus and Procedure for Glow Curve Plotting

A schematic diagram of the apparatus is illustrated in Fig. 14.8. With the exception of the heating and measuring chamber, all components can be bought from manufacturers of electronic equipment. The chamber is an aluminum container for the photomultiplier. At its base is a camera shutter leading into a fused quartz rod. The photomultiplier assembly is securely attached to the heating assembly. Heating elements are removable for easy exchange of specimens. The thermoluminescence produced by the specimen while it is being heated travels up the quartz rod to the photomultiplier. The quartz rod acts as a heat filter as well as a light collector. The various components have the following specifications:

Power Supply. Capable of giving a range of 1000 to 2000 stabilized dc voltage, negative output.

Photomultiplier. R.C.A. 6810-A or E.M.I. 9558B or equivalent. The base must be specially wired to hold the photocathode at −1000 to −2000 V. The anode is isolated from the chain of dynodes. Its output is fed directly to a micro-micro ammeter dc amplifier. The last dynode before the anode is held at a suitable negative potential with respect to the anode by earthing it through a suitable resistance. The anode goes to earth through the amplifier.

Amplifier. A micro-micro dc ammeter, similar to Leeds and Northrup model 9836A or AVO type 1388B is suitable.

Heating Controller. Should be capable of giving precise control of the heating rate of the hot plate specimen-well at 1°C per second, from room temperature to 500°C. A Leeds and Northrup set point controller 10877 is suitable. The output from the controller is fed to a final control amplifier to supply about 1000 W to the hot plate heating coil.

X-Y Recorder. This is used for recording the glow curve.

Alpha scintillation counter. Suitable for recording low intensity natural radioactivity of the specimens.

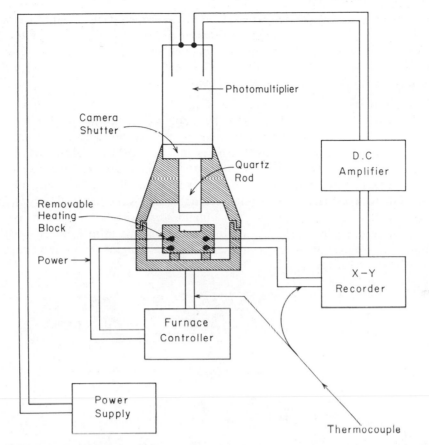

Fig. 14.8   Schematic diagram of the equipment used to measure thermoluminescence glow curves. (After Hutchison, 1965.)

METHOD (*natural glow curve*)

1.   For each glow curve a standard volume of the $-120 +230$ U.S. sieved sample is taken. A small hole 0.17 in. in diameter drilled to a depth of 0.15 in. in the corner of a brass plate serves as a measure. The powdered specimen is shaken into this hole and the brass plate tapped gently to ensure even settling of the powder. Excess powder is scraped off level with a spatula blade.

2.   The measured volume is poured into the shallow well of the heating plate (suitable heating well 0.5 in. in diameter by 0.05 in. deep). The powder is evenly spread over the well floor with a small metal rod of slightly smaller diameter than the well. Ensure that the camera shutter is closed when the heating chamber is opened, otherwise daylight will spoil the photomultiplier.

3.   The natural thermoluminescence glow curve is run at a heating rate of 1°C per second from ambient to 500°C. Beyond 450° blackbody radiation may obscure the thermoluminescence completely. The unwanted blackbody radiation may be filtered by imposing a suitable blue filter in front of the photomultiplier tube.

4.   The curve representing blackbody radiation plus dark current (background) must be determined over the whole temperature range to give a base line for the glow curve.

This may be done by plotting light output as the specimen cools down to room temperature. The background may also be recorded on the same specimen, after it has cooled back to room temperature, on a return from ambient to maximum at the same heating rate. All its thermoluminescence will have gone and the light output will now be entirely due to blackbody plus dark current.

5. The sensitivity of the photomultiplier should be calibrated frequently, at least at the end of each glow curve. This is done by replacing the heating stage with a small "isolite" capsule. The isolite is a small aluminum planchet coated with carbon-14 and a phosphor layer superimposed. The isolite should have approximately 1 μL intensity. It must be kept in darkness for at least 1 hour before use because the phosphor glows when excited by light. An ideal arrangement is to enclose the isolite in a light-tight capsule with a small camera shutter on top, which is then opened only when the capsule is placed in the light-tight thermoluminescence heating chamber. With the isolite in place, the camera shutters on top of the isolite and in front of the photo-multiplier are opened. With the photomultiplier operating at the same voltage as it

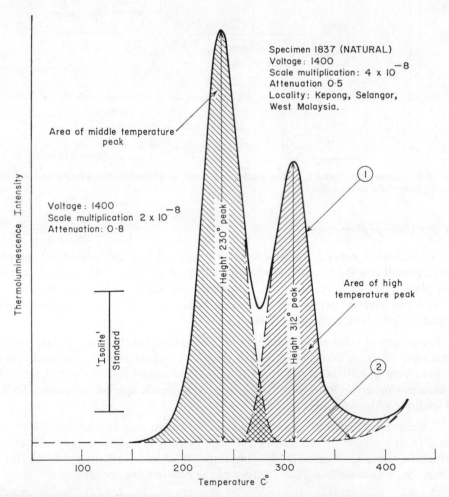

Fig. 14.9 Typical natural thermoluminescence glow curve obtained from calcite. (After Hutchison, 1966.)

was set during the glow curve run, the deflection caused by the isolite is recorded on the same chart paper as the glow curve.

Results. Figure 14.9 illustrates a typical glow curve for calcite. It is characterized by two peaks, one at 230°C, referred to as the middle temperature peak, and a high temperature peak at 312°C. The number of trapped electrons released during heating is obviously in the ratio of the peak area or peak height. In this figure (1) is the actual glow curve and (2) is the background curve due to dark current and blackbody radiation.

1. The glow curve is resolved graphically into its composite peaks (as shown in Fig. 14.9). The maximum height of the 312°C peak is measured in centimeters and converted to standard isolite units by the relation

$$
\begin{array}{l}
\text{height of peak} \\
\text{(in isolite units)}
\end{array}
=
\frac{
\begin{array}{c}
\text{height of peak (cm)} \times \text{peak scale multiplication} \\
\times \text{ isolite attenuation}
\end{array}
}{
\begin{array}{c}
\text{height of isolite (cm)} \times \text{isolite scale multiplication} \\
\times \text{ peak attentuation}
\end{array}
}.
$$

It is essential that both peak and isolite have been measured at the same photomultiplier voltage so that the sensitivity is identical. In Fig. 14.9 the height of the 312°C peak can be calculated as 7.4 isolite units.

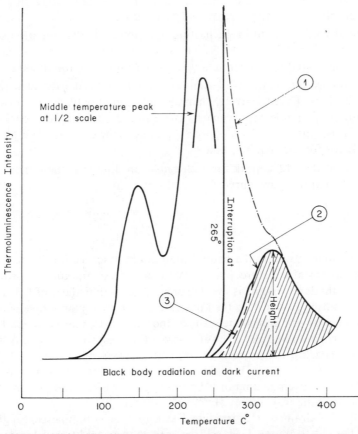

Fig. 14.10   Normal (1) and interrupted glow curve (2) of a calcite limestone. The height of the high-temperature peak can now readily be determined (curve 3). (After Hutchison, 1968a.)

2.   Normally the high temperature peak is quite distinct from the middle temperature peak. If the high-temperature peak is small however, accurate resolution is impossible, as shown in Fig. 14.10. An interrupted glow curve may be run on such specimens. The original curve is shown in Fig. 14.10 as curve (1). A repeat glow curve is run on a fresh identical sample and heating is abruptly terminated in the range 255 to 265°. The sample is allowed to cool slightly by about 10 or 20°C, during which time the thermoluminescence output drops to zero. When heating is restarted at the same rate, an interrupted glow curve [(2) in Fig. 14.10] is obtained, which gives the high-temperature peak modified on its low-temperature side by thermoluminescence belonging to the middle-temperature peak, which has not been completely drained thermally. This part, however, is easily resolved, and the shape of the high-temperature peak can now easily be drawn in [(3) in Fig. 14.10].

EXERCISE 14.16

The Artificial Glow Curve

1.   Approximately 2 g of mineral powder is placed in a crucible which is then maintained at 310°C for 90 minutes. This ensures that all the natural thermoluminescence, including the calcite 312°C peak, is completely drained. A check can be made by running a glow curve on this heated material. If drainage is complete, only the black-body radiation plus dark current curve will be produced.

2.   A measured volume is put in the heating well, under a beta radiation source and subjected to approximately 1×10^5 R. The glow curve is then run. A suitable beta capsule is a plate coated with strontium-90 and yttrium-90 of about 600 mCi.

3.   Glow curves are run on identical samples which have been irradiated to 3×10^5, 9×10^5 and 20×10^5 R, respectively. Gamma radiation may be used in place of beta. If so, an amount of sample approximately twice that used for a glow curve is placed in a small gelatine capsule and sealed with a piece of labeled tape, then placed in the gamma irradiator for the required time.

4.   The height of the 312°C peak for each radiation dosage is then measured in the same way as the natural glow curve.

DISCUSSION

Many workers prefer to drain the natural thermoluminescence not by heating, which may damage the crystal structure, but by UV radiation. An amount of powder slightly more than the standard volume is evenly spread over a thin sheet of fused quartz of high UV transmitting power. It is then placed between two quartz envelope mercury-arc germicidal lamps one immediately above and one below the specimen. Ultraviolet irradiation is carried on in a closed box (to shield the eyes) for 12 hours. A glow curve is then run to measure the amount of residual thermoluminescence, which will be small, but which will have to be subtracted as part of the background. Similarly drained samples are given gamma or beta doses of increasing amounts and their glow curves run in the same way. (After Hutchison, 1965.)

An artificially drained and irradiated calcite thermoluminescence glow curve is shown in Fig. 14.10 (curve 1). It will be seen that an additional low-temperature peak at 110°C is present, which is not found in most natural samples because it is drained naturally at ambient atmospheric temperature. Calcite from frigid climates, for example, Antarctica, may show a low-temperature natural peak because of the low ambient temperature.

Thermoluminescence Calibration Curve

1. A calibration curve can now be drawn on graph paper by plotting height of the 312°C peak against dosage in roentgens (Fig. 14.11). For thermal drainage the curve starts at the origin O. For UV drainage the curve starts at a point D, where OD is the amount of residual thermoluminescence after drainage; OD will be quite small. This curve can be extrapolated with accuracy to the base line to establish point C. For curve 3, therefore, the radiation dosage scale should have its origin at point C. Alternatively, as in Fig. 14.11, all determined points on curve 3 must be moved horizontally to the right by a distance CO, and the final curve 2 drawn through the origin.

2. The value of R_a, the relative equivalent radiation dosage, as defined by Zeller et al. (1957) may be determined from the natural thermoluminescence peak height, as in Fig. 14.11. Points A and B are the intersections of the natural level with the UV and heat-drained curves, respectively. The R_a value converts the natural thermoluminescence of a specimen, which is dependent on lithology and impurities in the specimen, to the basic property of equivalent radiation storage accumulated by the specimen in its lifetime, which should be independent of lithology and impurities.

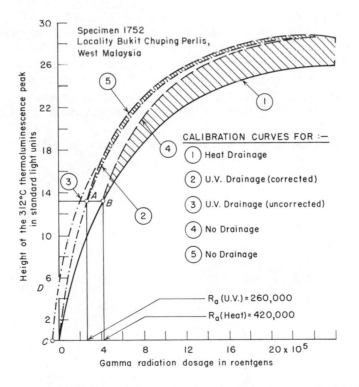

Fig. 14.11 Thermoluminescence calibration curves for calcite. See exercise 14.17 for details. (After Hutchison, 1965, 1968a.)

3.   The R_a value of a specimen is, of course, proportional to the natural radioactivity of the specimen. Accordingly, each powdered specimen must be evenly spread over an aluminum planchette and counted in an alpha scintillation detector for about 24 hours. The specimen's natural alpha activity in counts per hour is designated α (detailed method in Hutchison, 1965).

4.   The property R_a/α is called the thermoluminescence index of the mineral. This index is directly proportional to the time that has elapsed since the mineral last crystallized or since its thermoluminescence was reduced by natural heating during a metamorphic event. This index therefore may be used for age determination of limestones, which have had no tectonic history after deposition, or to date tectonic events (Hutchison, 1968b).

EXERCISE 14.18

Other Applications of Thermoluminescence

Several authors (see McDougall, 1968) have shown that in addition to calcite the following minerals usually exhibit natural thermoluminescence: feldspars, dolomite, nepheline, quartz, apatite, zircon, halite, and fluorite.

The thermoluminescence of carbonates has been used in stratigraphic studies to "fingerprint" limestone horizons of similar geological history and provide a means of stratigraphic correlation.

The thermoluminescence property of feldspars means that the majority of igneous rocks, especially the granitic rocks, give glow curves that can be made use of to show anomalous thermoluminescence halos around mineralized zones and ore bodies resulting from impurity metals migrating into the country rocks. The stress history of rocks has been deduced by studying the thermoluminescence of halite deposits.

One of the most successful application of thermoluminescence has been the dating of archaeological pottery by a method similar to that for calcite given above.

The middle-temperature thermoluminescence peak has been used in palaeoclimatological studies because its height is in equilibrium with the ambient atmosphere temperature.

Many of these applications are described in the papers in McDougall (1968).

REFERENCES

Abramowitz, M., and I. A. Stegun, Eds. (1965). *Handbook of mathematical functions: with formulas, graphs, and mathematical tables.* Dover, New York. 1046 p.

Adams, J. K., and R. E. Arvidson (1969). Thin sections of free grains using two mounting mediums. *Geol. Mag.*, **106**, No. 6, 598–599.

Agterberg, F. P. (1964). Statistical analysis of X-ray data for olivine. *Mineral. Mag.*, **33**, 742–748.

Albee, A. L., and A. A. Chodos (1970). Semiquantitative electron microprobe determination of Fe^{2+}/Fe^{3+} and Mn^{2+}/Mn^{3+} in oxides and silicates and its application to petrologic problems. *Am. Mineralogist*, **55**, 491–501.

Allen, W. C. (1966). An X-ray method for defining composition of a magnesium spinel. *Am. Mineralogist*, **51**, 239–243.

Angino, E. E., and C. K. Billings (1967). *Atomic Absorption Spectrometry in Geology.* Elsevier, Amsterdam. 144 p.

Arnold, R. G., and L. E. Reichen (1962). Measurement of the metal content of naturally occurring, metal-deficient, hexagonal pyrrhotite by an X-ray spacing method. *Am. Mineralogist*, **47**, 105

A.S.T.M. (American Society for testing materials) *Powder Data File*, now known as *Joint Committee for Powder Diffraction Standards*. See J.C.P.D.S.

Atchley, F. W. (1958). Low magnification thin section photography. *Am. Mineralogist*, **43**, 997–1000.

Aucott, J. W., and M. Marshall (1969). Quantitative determination of water in granites by infra-red analysis. *Mineral. Mag.*, **37**, 256–261.

Azároff, L. V., and M. J. Buerger (1958). *The powder method in X-ray crystallography.* McGraw-Hill, New York. 342 p.

Bailey, E. H., and R. E. Stevens (1960). Selective staining of K-feldspar and plagioclase on rock slabs and thin sections. *Am. Mineralogist*, **45**, 1020–1025.

Baker, F. (1961). Phase relations in cordierite-garnet-bearing Kinsman quartz monzonite and the enclosing schist, Lovewell Mountain Quadrangle, New Hampshire. *Am. Mineralogist*, **46**, 1166–1176.

Bambauer, H. U., M. Corlett, E. Eberhard, and K. Viswanathan (1967). Diagrams for the determination of plagioclases using X-ray powder methods. *Schweiz. Mineral. Petrog. Mitt.*, **47**, 333–349.

Barth, T. F. W. (1962a). *Theoretical Petrology*, 2nd ed. Wiley, New York. 416 p.

——— (1962b). A final proposal for calculating the mesonorm of metamorphic rocks. *J. Geol.*, **70**, 497–498.

Bayliss, P., and S. St. J. Warne (1972). Differential thermal analysis of siderite-kaolinite mixtures. *Am. Mineralogist*, **57**, 960–966.

Benjamin, R. E. K. (1971). Recovery of heavy liquids from dilute solutions. *Am. Mineralogist*, **56**, 613–619.

Bensch, J. J., and H. J. Brynard (1972). New approach to density measurements using Archimedes' Principle. *Nature, Physical Science*, **239**, 96.

Berek, M. (1923). Neue Wege zur Universalmethode. *Neues Jahr. Min. Geol. Palaeontol.*, **48**, 34–62.

Berman, M., and S. Ergun (1969). *Angular positions of X-ray emission lines of the elements for common analysing crystals.* U. S. Dept. of the Interior, Bureau of Mines Information Circular 8400, 309 p. Superintendent of Documents. U. S. Government Printing Office, Washington D.C. 20402.

Bernas, B. (1968). A new method for decomposition and comprehensive analysis of silicates by atomic absorption spectrometry. *Anal. Chem.*, **40**, 1682–1686.

Berry, L. G., and B. Mason (1959). *Mineralogy: concepts, descriptions, determinations.* Freeman, San Francisco. 612 p.

Bertin, E. P. (1970). *Principles and practice of X-ray spectrometric analysis.* Plenum, New York. 679 p.

Biscaye, P. E. (1964). Distinction between kaolinite and chlorite in recent sediments by X-ray diffraction. *Am. Mineralogist*, **49**, 1281–1289.

Diskupsky, V. D. (1965). Fast and complete decomposition of rocks, refractory silicates and minerals *Anal. Chim. Acta*, **33**, 333–334.

Bloss, F. D. (1969). Indexing powder patterns for cubic materials. *Am. Mineralogist*, **54**, 924–930.

———, G. Frenzel, and P. D. Robinson (1967). Reducing preferred orientation in diffractometer samples. *Am. Mineralogist*, **52**, 1243–1247.

Bollin, E. M. (1970). Chalcogenides. Chapter 7, p. 193–236. In R. C. Mackenzie, Ed., *Differential Thermal Analysis*, Vol. 1. Academic, London. 775 p.

Boone, G. M., and E. P. Wheeler (1968). Staining for cordierite and feldspars in thin section. *Am. Mineralogist*, **53**, 327–331.

Borg, I. Y., and D. K. Smith (1968). Calculated powder patterns: I Five plagioclases. *Am. Mineralogist*, **53**, 1709–1723.

———, ——— (1969a). Calculated X-Ray powder patterns for silicate minerals. *Geol. Soc. Am. Mem.*, **122**. 896 p.

———, ——— (1969b). Calculated powder patterns, Part II. Six potassium feldspars and barium feldspar. *Am. Mineralogist*, **54**, 163–181.

Borley, G., and M. T. Frost (1963). Some observations on igneous ferrohastingsites. *Mineral. Mag.*, **33**, 646.

Bowie, S. H. U. (1967). Microscopy: reflected light. In *Physical methods in determinative mineralogy*, Chapter 3, p. 103–159, J. Zussman, Ed. Academic, London. 514 p.

———, and M. Font-Altaba (1970). International Tables for the microscopic determination of crystalline substances absorbing in visible light. *Commission on Ore Microscopy of the International Mineralogical Association*. Department of Cristalografia y Mineralogia, University of Barcelona, Spain. 37 p.

Bradshaw, P. M. D. (1967). Measurement of the modal composition of a granitic rock powder by point-counting, infra-red spectroscopy, and X-ray diffraction. *Mineral. Mag.*, **36**, 94–100.

Breger, I. A., and J. C. Chandler (1969). Determination of fixed water in rocks by infrared absorption. *Anal. Chem.*, **41**, 506–510.

Brindley, G. W., and F. H. Gillery (1956). X-ray identification of chlorite species. *Am. Mineralogist*, **41**, 169.

Brown, G., Ed. (1961). *The X-ray identification and crystal structures of clay minerals*. 2nd ed. Mineralogical Society (Clay Minerals Group), London. 544 p.

Brunton, G. (1955). Vapour pressure glycolation of oriented clay minerals. *Am. Mineralogist*, **40**, 124–126.

Buckley, D. E., and R. E. Cranston (1971). Atomic absorption analyses of 18 elements from a single decomposition of aluminosilicate. *Chem. Geol.*, **7**, 273–284.

Burley, B. J., E. B. Freeman, and D. M. Shaw (1961). Studies on scapolite. *Can. Mineralogist*, **6**, 670.

Cabri, L. J. (1969). Density determinations: accuracy and application to sphalerite stochiometry. *Am. Mineralogist*, **54**, 539–548.

Carmichael, I. S. E. (1964). The petrology of Thingmuli, a Tertiary volcano in eastern Iceland. *J. Petrol.*, **5**, 435–460.

Carroll, D. (1970). Clay minerals: a guide to their X-ray identification. *Geol. Soc. Am.*, *Spec. Paper* **126**, 80 p.

Cervelle, B. (1967). Contribution à l'étude de la séries ilménite-geikielite. *Bur. Rech. Géol. Minières Bull.*, **6**, 1–26.

Chappell, B. W., W. Compston, P. A. Arriens, and M. J. Vernon (1969). Rubidium and strontium determinations by X-ray fluorescence spectrometry and isotope dilution below the part per million level. *Geochim. Cosmochim. Acta*, **33**, 1002–1006.

Chayes, F. (1953). In defense of the second decimal. *Am. Mineralogist*, **38**, 784–793.

Chromý, S. (1969). Photoelectric apparatus for refractive index determination by the immersion method. *Am. Mineralogist*, **54**, 549–553.

Clark, G. L. (1967). Properties of X-rays. Chapter 1, p. 1–31 in E. F. Kaelble, Ed., *Handbook of X-Rays: for diffraction, emission, absorption and microscopy*. McGraw-Hill, New York.

Clark, L. A. (1962). X-ray method for rapid determination of sulfur and cobalt in loellingite. *Can. Mineralogist*, **7**, 306–311.

Cochran, M. C., and J. R. Jensen (1965). An improved holder for grinding thin sections. *Am. Mineralogist*, **50**, 2092–2094.

Cole, J. F., and H. Villiger (1969). A FORTRAN IV computer programme for the rapid computation of indexed X-ray d-spacings, Q values, and Bragg angles. *Mineral. Mag.*, **37**, 300–301.

Colucci, S. L. (1970). A new technique in thermoluminescence photography. *Am. Mineralogist*, **55**, 1797–1800.

Compton, R. R. (1962). *Manual of field geology*. Wiley, New York. 378 p.

Copeland, D. A. (1965). A simple apparatus for trimming thin sections. *Am. Mineralogist*, **50**, 1128–1130.

Courville, S. (1962). Apparatus for total water determination by the Penfield method. *Can. Mineralogist*, **7**, 326–329.

Davies, T. T., and P. R. Hooper (1963). The determination of the calcite: aragonite ratio in mollusc shells by X-ray diffraction. *Mineral. Mag.*, **33**, 608.

Dawson, J. B., and F. W. Wilburn (1970). Silica minerals. Chapter 17, p. 477–495. In R. C. Mackenzie, Ed., *Differential Thermal Analysis*, Vol. 1. Academic, London, 775 p.

Dawson, K. R., and W. D. Crawley (1963). An improved technique for staining potash feldspars. *Can. Mineralogist*, **7**, 805–808.

Deer, W. A., R. A. Howie, and J. Zussman (1962a). *Rock-forming minerals, Vol. 1, (Ortho- and ring silicates)*. Longmans, Green, London. 333 p.

——, ——, —— (1962b). *Rock-forming minerals, Vol. 3, (sheet silicates)*. Longmans, Green, London. 270 p.

——, ——, —— (1962c). *Rock-forming minerals, Vol. 5, (nonsilicates)*. Longmans, Green, London. 371 p.

——, ——, —— (1963a). *Rock-forming minerals, Vol. 2, (chain silicates)*. Longmans, Green, London. 379 p.

——, ——, —— (1963b). *Rock-forming minerals, Vol. 4, (framework silicates)*. Longmans, Green, London. 435 p.

——, ——, —— (1966). *An introduction to the rock-forming minerals*. Wiley, New York, Longmans, Green, London. 528 p.

Desborough, G. A., and H. J. Rose, Jr. (1968). X-ray and chemical analysis of orthopyroxenes from the lower part of the Bushveld Complex, South Africa. *U. S. Geol. Surv. Profess. Paper*, **600-B**, B1–B5.

Dickson, J. A. D. (1965). A modified staining technique for carbonates in thin section. *Nature*, **205**, No. 497, 587.

Dodge, T. A. (1934). The determination of optic angle with the Universal stage. *Am. Mineralogist*, **19**, 62.

Domanska, E., and J. Nedoma (1969). X-ray powder data for idocrase. *Mineral. Mag.*, **37**, 343–348.

Embrey, P. G. (1969). Density determination by titration. *Mineral. Mag.*, **37**, 523.

Emerson, D. O. (1958). A stage for macro point counting. *Am. Mineralogist*, **43**, 1000–1003.

Emmons, R. C. (1959). The universal stage (with five axes of rotation). *Geol. Society Am. Mem.* **8** (reprint). 205 p.

Fabbi, B. P. (1970). A die for pelletizing samples for X-ray fluorescence analysis. *U. S. Geol. Surv., Profess. Paper* **700-B**, B187–B189.

—— (1971). Rapid X-ray fluorescence determination of phosphorus in geologic samples. *Appl. Spectr.*, **25**, 41–43.

—— (1972). A refined fusion X-ray fluorescence technique, and determination of major and minor elements in silicate standards. *Am. Mineralogist*, **57**, 237–245.

——, and L. F. Espos (1972). X-ray fluorescence determination of chlorine in standard silicate rocks. *Appl. Spectr.*, **26**, No. 2, 293–295.

Fang, J. H., and F. D. Bloss (1966). *X-ray diffraction tables. Determination of d, sin² θ, or Q for 0.01° intervals of 2θ for the wavelengths Kα, Kα₁, Kα₂, and Kβ of copper, iron, molybdenum, and chromium radiations and for Lα₁ of tungsten.* Southern Illinois University Press, Carbondale and Edwardsville.

Finger, L. W. (1971). Determination of dead-time in X-ray detector systems. *Carnegie Inst. Yearbook*, **70**, 275–277.

Fisher, G. W., and L. G. Medaris, Jr. (1969). Cell dimensions and X-ray determinative curve for synthetic Mg-Fe olivines. *Am. Mineralogist*, **54**, 741–753.

Flanagan, F. J. (1967). U. S. Geological Survey Silicate rock standards. *Geochim. Cosmochim. Acta*, **31**, 289–308.

────── (1969). U.S. Geological Survey standards-II. First compilation of data for the new U.S.G.S. rocks. *Geochim. Cosmochim. Acta*, **33**, 81–120.

────── (1970). Sources of geochemical standards II. *Geochim. Cosmochim. Acta*, **34**, 121–125.

Fleischer, M. (1969) U.S. Geological Survey standards I. Additional data on rocks G-1 and W-1 1965–1967. *Geochim. Cosmochim. Acta*, **33**, 65–79.

────── , and R. E. Stevens (1962). Summary of new data on rock samples G-1 and W-1. *Geochim. Cosmochim. Acta*, **26**, 525–543.

Flinter, B. H. (1959). The magnetic separation of some alluvial minerals in Malaya. *Am. Mineralogist*, **44**, 738–751.

Ford, A. B., and E. L. Boudette (1968). On staining of anorthoclase. *Am. Mineralogist*, **53**, 331–334.

Franzini, M., and L. Schiaffino (1965). On the X-ray determination of the iron-magnesium ratio of biotites. *Z. Krist.*, **122**, 100–107.

Friedman, G. M. (1959). Identification of carbonate minerals by staining methods. *J. Sediment. Petrol.*, **29**, 87–97.

Frost, M. T. (1963). Amphiboles from Younger Granites of Nigeria: Part II X-ray data. *Mineral. Mag.*, **33**, 377.

Geological Society of America Memoir 122. (1969). Calculated X-ray powder patterns for silicate minerals. I. Y. Borg and D. K. Smith. 896 p.

Gibbs, R. J. (1965). Error due to segregation in quantitative clay mineral X-ray diffraction mounting techniques. *Am. Mineralogist*, **50**, 741–751.

Glasser, F. P. (1970). Other silicates. In R. C. Mackenzie, Ed., *Differential Thermal Analysis*, Vol. 1, Chapter 21, p. 575–608. Academic, London.

Gleason, S. (1960). *Ultraviolet Guide to Minerals*. Van Nostrand, Princeton, New Jersey. 244 p.

Goldsmith, J. R., and F. Laves (1954). The microcline-sanidine stability relations. *Geochim. Cosmochim. acta*, **5**, 1–19.

────── , D. L. Graf, and O. I. Joensuu (1955). The occurrence of magnesian calcites in nature. *Geochim. cosmochim. Acta*, **7**, 212–230.

Graham, A. R. (1969). Quantitative determination of hexagonal and monoclinic pyrrhotites by X-ray diffraction. *Can. Mineralogist*, **10**, 4–24.

Green, D. H., and A. E. Ringwood (1967). The genesis of basaltic magmas. *Contrib. Mineral. Petrol.*, **15**, 103–190.

Grimshaw, R. W., and A. L. Roberts (1957). The silica minerals. Chapter XI, p. 275–298, in *The Differential Thermal Investigation of Clays* Ed. by R. C. Mackenzie. Mineralogical. Society (Clay Minerals Group), London. 456 p.

Gulbrandsen, R. A. (1960). A method of X-ray analysis for determining the ratio of calcite to dolomite in mineral mixtures. *U.S. Geol. Surv. Bull.* **1111-D**, 147–152.

Gunatilaka, H. A., and R. Till (1971). A precise and accurate method for the quantitative determination of carbonate minerals by X-ray diffraction using a spiking technique. *Mineral. Mag.*, **38**, 481–487.

Hagni, R. D. (1966). The preparation of thin sections of fragmental materials using epoxy resin. *Am. Mineralogist*, **51**, 1237–1242.

Haines, M. (1968). Two staining tests for brucite in marble. *Mining Mag.*, **36**, 886–888.

Hall, A., and J. N. Walsh (1969). A rapid method for the determination of fluorine in silicate rocks and minerals. *Anal. Chim. Acta*, **45**, 341–342.

Hamilton, D. L., and C. M. B. Henderson (1968). The preparation of silicate compositions by a gelling method. *Mineral. Mag.*, **36**, 832–838.

────── , and A. D. Edgar (1969). The variation of the $\bar{2}01$ reflection in plagioclase. *Mineral. Mag.*, **37**, 16–25.

Harker, A. (1950). *Metamorphism. A study of the transformation of rock-masses*, 3rd ed. Methuen, London. Dutton, New York. 362 p.

Hartshorne, N. H., and A. Stuart (1970). *Crystals and the polarising microscope*. 4th ed. Arnold, London, 614 p.

Harwood, D. S., and R. R. Larson (1969). Variation in the delta-index of cordierite around the Cupsupic pluton, west-central Maine. *Am. Mineralogist*, **54**, 896–908.

Heinrich, E. W. (1965). *Microscopic identification of minerals*. McGraw-Hill, New York. 414 p.

Heinrich, K. F. J. (1966). X-ray absorption uncertainty. In T. D. McKinley, K. F. J. Heinrich, and D. B. Wittry, Eds., *The Electron Microprobe*, Wiley, New York, p. 296–377.

Herber, L. J. (1969). Separation of feldspar from quartz by flotation. *Am. Mineralogist*, **54**, 1212–1217.

Hermes, O. D., and P. C. Ragland (1967). Quantitative chemical analysis of minerals in thin-section with the X-ray macroprobe. *Am. Mineralogist*, **52**, 493–508.

Hey, M. H. (1954). A new review of the chlorites. *Mineral. Mag.*, **30**, 277.

Himmelberg, G. R., and E. D. Jackson (1967). X-ray determinative curve for some orthopyroxenes of composition Mg_{48-85} from the Stillwater Complex, Montana. *U.S. Geol. Surv. Profess. Paper* **575-B**. B101–B102.

Holmes, A. (1921). *Petrographic methods and calculations*. Murby, London. 515 p.

Hooper, P. R., and L. Atkins (1969). The preparation of fused samples in X-ray fluorescence analysis *Mineral. Mag.*, **37**, 409–413.

Hormann, P. K., and G. Morteani (1969). On a systematic error in the X-ray determination of the iron content of chlorites and biotites: a discussion. *Am. Mineral.*, **54**, 1491–1494.

Horn, E. E., and H. Schulz (1968). Bestimmung des Tourmalin-chemismus auf röntgenographischem Wege. *Neues Jahrb. Mineral. Abhand.*, **108**, 20–35.

Hosking, K. F. G. (1958). The identification—largely by staining techniques—of coloured mineral grains in composite samples. *Camborne School of Mines Mag.*, **58**, 5–14.

——— (1964). Rapid identification of mineral grains in composite samples. *Mining Mag.*, Jan.–Feb., 1–12.

Howarth, R. J. (1966). Calculation of mineral unit cell contents: FORTRAN computer programme. *Mineral. Mag.*, **35**, 787.

Hower, J. (1959). Matrix corrections in the X-ray spectrographic trace element analysis of rocks and minerals. *Am. Mineralogist*, **44**, 19–32.

Huang, P. M., and M. L. Jackson (1967). Fluorine determination in minerals and rocks. *Am. Mineralogist*, **52**, 1503–1507.

Hutchison, C. S. (1965). The calibration of thermoluminescence measurements. *New Zealand J. Sci.*, **8**, 431–445.

——— (1966). *Tectonic and petrological relations within three rock associations of orogenic zones in Malaysia*. Unpublished Ph.D. thesis, University of Malaya, Kuala Lumpur. 270 p.

——— (1968a). The dating by thermoluminescence of tectonic and magmatic events in orogenesis. Chapter 6.3, p. 341–358. In D. J. McDougall, Ed., *Thermoluminescence of Geological Materials*. Academic, London. 678 p.

——— (1968b). Dating tectonism in the Indosinian-Thai-Malayan Orogen by thermoluminescence. *Geol. Soc. Am. Bull.*, **79**, 375–386.

——— (1972). Alpine-type chromite in North Borneo, with special reference to Darvel Bay. *Am. Mineralogist*, **57**, 835–856.

———, and J. E. Jcacocke (1971). FORTRAN IV computer programme for calculation of the Niggli Molecular Norm. *Geol. Soc. Malaysia, Bull.* **4**, 91–95.

Ingamells, C. O. (1970). Lithium metaborate flux in silicate analysis. *Anal. Chim. Acta*, **52**, 323–334.

J.C.P.D.S. *Powder Data File* (various dates). Both organic and inorganic sections and various indexes. Published and frequently revised by the Joint Committee on Powder Diffraction Standards, 1601 Park Lane, Swarthmore, Pennsylvania, 19081.

Jahanbagloo, I. C. (1969). X-ray diffraction study of olivine solid solution series. *Am. Mineralogist*, **54**, 246–250.

Jambor, J. L., and C. H. Smith (1964). Olivine composition determination with small diameter X-ray powder cameras. *Mineral. Mag.*, **33**, 730–741.

Jenkins, R. (1972). Recent developments in analysing crystals for X-ray spectrometry. *X-ray Spectr.*, **1**, 23–28.

———, and J. L. De Vries (1970a). *Practical X-ray spectrometry*, 2nd ed. Philips Technical Library, N. V. Philips, Eindhoven, Netherlands; Springer-Verlag, New York. 183 p.

———, ——— (1970b). *Worked examples in X-ray spectrometry*. Philips Technical Library, N. V. Philips, Eindhoven, Netherlands. Springer-Verlag, New York. 129 p.

Johannsen, A. (1918). *Manual of petrographic methods*. McGraw-Hill, New York, 649 p. (Reprinted 1968 by Hafner, New York.)

——— (1931). *A descriptive petrography of the igneous rocks. Vol. 1. Introduction, textures, classifications and glossary*. University of Chicago Press, Chicago. 318 p.

Johnson, W., and K. W. Andrews (1962). Quantitative X-ray examination of aluminosilicates. *Trans. Brit. Ceram. Soc.*, **61**, 724.

Jones, F. T. (1968). Spindle stage with easily changed liquid and improved crystal holder. *Am. Mineralogist*, **53**, 1399–1403.

Jones, J. B., R. W. Nesbitt, and P. G. Slade (1969). The determination of the orthoclase content of homogenized alkali feldspar using the $\bar{2}01$ X-ray method. *Mineral. Mag.*, **37**, 489–496.

Katz, A. (1968). The direct and rapid determination of alumina and silica in silicate rocks and minerals by atomic absorption spectroscopy, *Am. Mineralogist*, **53**, 283–289.

Keil, K. (1965). Mineralogical modal analysis with the electron microprobe X-ray analyser. *Am. Mineralogist*, **50**, 2089–2092.

Kelly, W. C. (1956). Application of differential thermal analysis to identification of the natural hydrous ferric oxides. *Am. Mineralogist*, **41**, 353–355.

Kelsey, C. H. (1965). Calculation of the C.I.P.W. norm. *Mineral. Mag.*, **34** (Tilley Volume), 276–282.

Kerr, P. F. (1959). *Optical Mineralogy.*, 3rd ed. McGraw-Hill, New York, 442 p.

Kiely, P. V., and M. L. Jackson (1964). Selective dissolution of micas from potassium feldspars by sodium pyrosulfate fusion of soils and sediments. *Am. Mineralogist*, **49**, 1648–1659.

Kinter, E. B., and S. Diamond (1956). Preparation and treatment of oriented specimens of clays. In A. Swineford, Ed., *Clays and Clay Minerals*, Publication 456 Nat. Acad. Sci.-Nat. Res. Coun., 21, Washington, D.C. See also *Soil Science* (1956), **81**, 111–120.

Kiss, E. (1967). Chemical determination of some major constituents in rocks and minerals. *Anal. Chim. Acta*, **39**, 223–234.

Klein, C. Jr., and D. R. Waldbaum (1967). X-ray crystallographic properties of the cummingtonite-grunerite series. *J. Geol.*, **75**, 379–392.

Koch, G. S., Jr., and R. F. Link (1970). *Statistical analysis of geological Data*. Wiley, New York. 374 p.

Kopp, O. C., and P. F. Kerr (1958). Differential thermal analysis of sphalerite. *Am. Mineralogist*, **43**, 732–748.

——, —— (1958). Differential thermal analysis of pyrite and marcasite. *Am. Mineralogist*, **43**, 1079–1097.

Krock, T. E. (1971). A low-contamination method for decomposition of zircon and extraction of U and Pb for isotope age determinations. *Carnegie Inst. Yearbook*, **70**, 258–266.

Kuellmer, F. J. (1959). X-ray intensity measurements on perthitic materials I: theoretical consider-ations. *J. Geol.*, **67**, 648–660.

—— (1960). X-ray measurements on perthitic materials II-Data from natural alkali feldspars. *J. Geol.*, **68**, 307–323.

Kuno, H. (1959). Origin of Cenozoic petrographic provinces of Japan and surrounding areas. *Bull. Volcanol.* **Ser. II** 37–76.

—— (1966). Lateral variation of basaltic magma type across continental margins and island arcs. *Bull. Volcanol.*, **XXIX**, 195–222.

Kushiro, I. (1962). Clinopyroxene solid solutions. Part I. The $CaAl_2SiO_6$ component. *Japan. J. Geol. Geography*, **33**, 213–220.

Laduron, D. M. (1971). A staining method for distinguishing paragonite from muscovite in thin section. *Am. Mineralogist*, **56**, 1117–1119.

Lahee, F. H. (1961). *Field Geology*, 6th ed. McGraw-Hill, New York. 926 p.

Langer, A. M., and P. F. Kerr (1967). Evaluation of kaolinite and quartz differential thermal curves with a new high temperature cell. *Am. Mineralogist*, **52**, 509–523.

Langer, K., and W. Schreyer (1969). Infrared and powder X-ray diffraction studies on the poly-morphism of cordierite, $Mg_2(Al_4Si_5O_{18})$. *Am. Mineralogist*, **54**, 1442–1459.

Leake, B. E. (1968). A catalog of analysed calciferous and subcalciferous amphiboles together with their nomenclature and associated minerals. *Geol. Soc. Am. Spec. Paper* **98**. 210 p.

Le Maitre, R. W., and M. T. Haukka (1973). The effect of prolonged X-ray irradiation on lithium tetraborate glass discs as used in X.R.F. analyses. *Geochim. Cosmochim. Acta*, **37**, 708–710.

Levin, S. B. (1950). Genesis of some Adirondack garnet deposits. *Bull. Geol. Soc. Am.*, **61**, 519–565.

Long, J. V. P. (1967). Electron probe microanalysis. Chapter 5, 215–260. In J. Zussman, Ed., *Physical methods in determinative mineralogy*. Academic, London, 514 p.

Luth, W. C., and C. O. Ingamells (1965). Gel preparation of starting materials for hydrothermal ex-perimentation. *Am. Mineralogist*, **50**, 255–258.

Lyons, P. C. (1971). Staining of feldspars on rock-slab surfaces for modal analysis. *Mineral. Mag.*, **38**, 518–519.

McAndrew, J. (1972). Differential dispersion measurement of refraction index. *Am. Mineralogist*, **57**, 231–236.

MacDonald, G. A., and T. Katsura (1964). Chemical composition of Hawaiian lavas. *J. Petrol.*, **5**, 82–133.

McDougall, D. J., Ed. (1968). *Thermoluminescence of Geological Materials.* Academic, London. 678 p.

Macgregor, I. D., and C. H. Smith (1963). The use of chrome spinels in petrographic studies of ultramafic intrusions. *Can. Mineralogist*, **7**, 403–412.

Mackenzie, R. C., Ed. (1957). *The differential thermal investigation of clays.* Mineralogical Society, London. 456 p.

———, Ed. (1970a). *Differential thermal analysis, Vol. 1. Fundamental aspects.* Academic, London. 775 p.

—— (1970b). Simple phyllosilicates based on gibbsite- and brucite-like sheets. Chapter 18, 497–537. In R. C. Mackenzie, Ed., *Differential Thermal Analysis, Vol. 1.* Academic, London. 775 p.

———, and G. Berggren (1970). Oxides and hydroxides of high valency elements. Chapter 9, p. 271–302. In R. C. Mackenzie, Ed., *Differential Thermal Analysis, Vol.* 1. Academic, London. 775 p.

———, and B. D. Mitchell (1970a). Instrumentation. Chapter 3, 63–99. In R. C. Mackenzie, Ed., *Differential Thermal Analysis*, Vol. 1. Academic, London. 775 p.

———, ——— (1970b). Technique, Chapter 4, 101–122. In R. C. Mackenzie, Ed., *Differential Thermal Analysis*, Vol. 1. Academic, London. 775 p.

McKinley, T. D., K. F. J. Heinrich, and D. B. Wittry (1966). *The electron microprobe.* Wiley, New York.

McLaughlin, R. J. W. (1957). Other minerals, Chapter XIV, 364–388. In R. C. Mackenzie, Ed., *The differential thermal investigation of clays.* Mineralogical Society (Clay Minerals Group). London. 456 p.

—— (1967a). Atomic absorption spectrometry, Chapter 13, 475–486. In J. Zussman, Ed., *Physical methods in determinative mineralogy.* Academic, London, 514 p.

—— (1967b). Thermal techniques, Chapter 9, 405–444. In J. Zussman, Ed., *Physical methods in determinative mineralogy.* Academic, London. 514 p.

McLeod, C. R., and J. A. Chamberlain (1968). Reflectivity and Vickers microhardness of ore minerals; chart and tables. *Geol. Surv. Canada, Paper* 68–64.

Marel, H. W. van der (1956). Quantitative differential thermal analyses of clay and other minerals. *Am. Mineralogist*, **41**, 222–240.

Mason, B., and L. G. Berry (1968). *Elements of mineralogy.* Freeman, San Francisco. 550 p.

Maxwell, J. A. (1968). *Rock and mineral analysis.* Wiley, New York. 584 p.

May, I., and J. Marinenko (1966). A micropycnometer for the determination of the specific gravity of minerals. *Am. Mineralogist*, **51**, 931–934.

Mertie, J. B., Jr. (1942). Nomograms of optic angle formulae. *Am. Mineralogist*, **27**, 538–551.

Meyrowitz, R. (1963). A semimicroprocedure for the determination of ferrous iron in nonrefractory silicate minerals. *Am. Mineralogist*, **48**, 340–347.

———, F. Cuttitta, and N. Hickling (1959). A new diluent for bromoform in heavy liquid separation of minerals. *Am. Mineralogist*, **44**, 884–885.

Mitchell, B. J., and J. E. Kellam (1968). Unusual matrix effects in X-ray spectroscopy: a study of the range and reversal of absorption enhancement. *Appl. Spectr.*, **22**, 742–748.

Miyashiro, A. (1957). Cordierite-Indialite relations. *Am. J. Sci.*, **255**, 43–62.

Möllring, F. K. (1968). *Microscopy from the very beginning.* Carl Zeiss, Oberkochen, West Germany. 64 p.

Moore, A. C. (1969). A method for determining mineral compositions by measurement of the mass absorption coefficient. *Am. Mineralogist*, **54**, 1180–1189.

Moorehouse, W. W. (1959). *The study of rocks in thin-section.* Harper & Row, New York. 514 p.

Moreland, G. C. (1968). Preparation of polished thin sections. *Am. Mineralogist*, **53**, 2070–2074.

Morelli, G. L. (1967). Determinazione della composizione delle fasi trigonali nel sistema $MgCO_3$—$FeCO_3$—$CaCO_3$ mediante la diffrazione del raggi. *Rend. Soc. Mineral. Italiana*, **23**, 315–332.

Morimoto, N., and L. A. Clark (1961). Arsenopyrite crystal-chemical relations. *Am. Mineralogist*, **46**, 1448.

Morse, S. A. (1968). Revised dispersion method for low plagioclase. *Am. Mineralogist*, **53**, 105–115.

Muir, I. D. (1967). Microscopy: transmitted light. Chapter 2, 31–102. In *Physical methods in determinative mineralogy*, J. Zussman, Ed. Academic, London. 514 p.

Müller, G. (translated by H-U. Schmincke) (1967). *Methods in sedimentary petrology*. Hafner, New York. 283 p.

Müller, J. D. (1967). Laboratory methods of mineral separation. Chapter 1, 1–30. In J. Zussman, Ed. *Physical methods in determinative mineralogy*. Academic, London. 514 p.

Müller, R. O. (translated by K. Keil) (1972). *Spectrochemical analysis by X-ray fluorescence* Plenum, New York. 326 p.

Nandi, K. (1967). Garnets as indices of progressive regional metamorphism. *Mineral. Mag.*, **36**, 89–93.

Nesbitt, R. W. (1964). Combined rock and thin section modal analysis. *Am. Mineralogist*, **49**, 1131–1136.

Noble, D. C. (1965a). A rapid conoscopic method for measurement of 2V on the spindle stage. *Am. Mineralogist*, **50**, 180–185.

——— (1965b). Determination of the composition and structural state of plagioclase with the five-axis universal stage. *Am. Mineralogist*, **50**, 367–381.

——— (1968). Optic angle determined conoscopically on the spindle stage II: selected rotation method. *Am. Mineralogist*, **53**, 278–282.

Nold, J. L., and K. P. Erickson (1967). Changes in K-feldspar staining methods and adaptations for field use. *Am. Mineralogist*, **52**, 1575–1576.

Norrish, K., and B. W. Chappell (1967). X-ray fluorescence spectrography. Chapter 4, 161–214. In J. Zussman, Ed., *Physical methods in determinative mineralogy*. Academic, London. 514 p.

———, and J. T. Hutton (1969). An accurate X-ray spectrographic method for the analysis of a wide range of geological samples. *Geochim. Cosmochim. Acta*, **33**, 431–453.

———, and R. M. Taylor (1962). Quantitative analysis by X-ray diffraction. *Clay Minerals Bull.*, **5**, 98–109.

Orville, P. M. (1963). Alkali ion exchange between vapor and feldspar phases. *Am. J. Sci.*, **261**, 201–237.

——— (1967). Unit-cell parameters of the microcline-low albite and the sanidine-high albite solid solution series. *Am. Mineralogist*, **52**, 55–86.

Østergaard, T. V. (1968). A continuous density separator for mineral separation. *Mineral. Mag.*, **36**, 890–891.

Owen, L. B. (1971). A rapid sample preparation method for powder diffraction cameras. *Am. Mineralogist*, **56**, 1835–1836.

Parker, A. (1970). An index of weathering for silicate rocks. *Geol. Mag.*, **107**, 501–504.

Parker, R. L. (1956). A stereographic contruction for determining optic axial angles. *Am. mineralogist*, **41**, 935–939.

Parl, B. (1967). *Basic Statistics*. Doubleday, New York. 364 p.

Parrish, W. (1965). Advances in X-ray diffractometry of clay minerals. *X-Ray Analysis Papers*. W. Parrish, Ed. Centrex, Eindhoven, Holland. p. 105–129.

———, and M. Mack (1963a). *Data for X-ray analysis*, 2nd edn. Vol. 1. *Charts for solution of Bragg's equation (d versus θ and 2θ for copper K radiation)*. Philips Technical Library. Philips, Eindhoven, Holland. 125 p.

———, ——— (1963b). *Data for X-ray analysis*, 2nd edn. Vol. II. *Charts for solution of Bragg's equation (d versus θ and 2θ for molybdenum K, Cobalt K, and Tungsten L radiations)*. Philips Technical Library. Philips, Eindhoven, Holland. 141 p.

———, and K. Lowitzsch (1965). Geometry, alignment and angular calibration of X-ray diffractometers. *X-Ray Analysis Papers*. W. Parrish, Ed. Centrex, Eindhoven, Holland. p. 82–104.

Parslow, G. R. (1969). Mesonorms of granitic rock analyses. *Mineral. Mag.*, **37**, 262–269.

Parsons, I. (1968). Homogeneity in alkali feldspars. *Mineral. Mag.*, **36**, 797–804.

Partridge, A. C., and C. W. Smith (1971). Small-sample flotation testing: a new cell. *Institute of Mining and Metallurgy Trans.* Bull. 778, **80**, C199–C200 (September).

Petruk, W. (1964). Determination of the heavy atom content in chlorite by means of the X-ray diffractometer. *Am. Mineralogist*, **49**, 61.

Poldervaart, A., and A. B. Parker (1964). The crystallization index as a parameter of igneous differentiation in binary variation diagrams. *Am. J. Sci.*, **262**, 281–289.

———, ——— (1965). Reply to discussion by Dr. Thornton and Dr. Tuttle. *Am. J. Sci.*, **263**, 279–283.

Powers, M. (1960). *X-ray fluorescent spectrometer conversion tables for topaz, LiF, NaCl, EDDT, and ADP crystals*. Philips Electronic Instruments, Mount Vernon, New York.

Przibram, K., and J. E. Caffyn (1956). *Irradiation colours and luminescence*. Pergamon, London. 332 p.

Rahden, H. V. R. von, and M. J. E. von Rahden (1972). Some aspects of the identification and characterization of 14 Å chlorites. *Minerals Science and Engineering*, 4, No. 3, 43–54.

Ramdohr, P. (1969). *The ore minerals and their intergrowths*. Pergamon, Oxford. 1174 p.

Ramírez-Muñoz, J. (1968). *Atomic Absorption Spectroscopy*. Elsevier, Amsterdam. 493 p.

Redfern, J. P. (1970). Complementary methods. Chapter 5, 123–158. In R. C. Mackenzie, Ed. *Differential Thermal Analysis*, Vol. 1. Academic, London, 775 p.

Reid, W. P. (1969). Mineral staining tests. *Mineral Ind. Bull.*, 12, No. 3, 1–20.

Reynolds, R. C. (1963). Matrix corrections in trace element analysis by X-ray fluorescence: estimation of the mass absorption coefficient by Compton scattering. *Am. Mineralogist*, 48, 1133–1143.

———— (1967). Estimation of mass absorption coefficients by Compton scattering: improvements and extensions of the method. *Am. Mineralogist*, 52, 1493–1502.

Ribbe, P. H., and P. E. Rosenberg (1971). Optical and X-ray determinative methods for fluorine in topaz. *Am. Mineralogist*, 56, 1812–1821.

Riley, J. F. (1968). The cobaltiferous pyrite series. *Am. Mineralogist*, 53, 293–295.

Riley, J. P. (1958). Simultaneous determination of water and carbon dioxide in rocks and minerals. *Analyst*, 83, 42–49.

Robertson, F. (1961). Knoop hardness for 127 opaque minerals. *Bull. Geol. Soc. Am.*, 72, 621–638.

Robinson, P., and H. W. Jaffe (1969). Chemographic exploration of amphibole assemblages from Central Massachusetts and Southwestern New Hampshire. *Pyroxenes and Amphiboles: Crystal Chemistry and Phase Petrology*, J. J. Papike, Ed. *Mineral. Soc. Am. Spec.* Paper 2, p. 251–274.

Rodgers, K. A., R. H. A. Cochrane, and P. C. Le Couteur (1970). FORTRAN II and FORTRAN IV programs for petrochemical calculations. *Mineral. Mag.*, 37, 952–953.

Rosenblum, S. (1958). Magnetic susceptibilities of minerals in the Frantz isodynamic magnetic separator. *Am. Mineralogist*, 43, 170–173.

Roy, N. N. (1965). A modified spindle stage permitting the direct measurement of 2V. *Am. Mineralogist*, 50, 1441–1449.

Royse, C. F., J. S. Wadell, and L. E. Petersen (1971). X-ray determination of calcite-dolomite: an evaluation. *J. Sediment. Petrol.*, 41, 483–488.

Rubéska, I., and B. Moldan (translated by P. T. Woods) (1969). *Atomic Absorption Spectrophotometry*. Iliffe, London. 189 p.

Ruperto, V. L., R. E. Stevens, and M. B. Norman (1964). Staining of plagioclase feldspar and other minerals with F., D., and C. Red No. 2. *U.S. Geol. Surv. Profess. Paper* 501B, B-152–B-153.

Rutstein, M. S., and R. A. Yund (1969). Unit-cell parameters of synthetic diopside-hedenbergite solid solutions. *Am. Mineralogist*, 54, 238–245.

Sahama, Th. G., K. J. Neuvonen, and K. Hytönen (1956). Determination of the composition of kalsilicates by an X-ray method. *Mineral. Mag.*, 31, 200.

Schenk, R., and G. Kistler (translated by F. Bradley) (1962). *Photomicrography*. Chapman & Hall, London, 132 p.

Schoen, R. (1962). Semi quantitative analysis of chlorites by X-ray diffraction. *Am. Mineralogist*, 47, 1384.

————, and D. E. Lee (1964). Successful separation of silt-size minerals in heavy liquids. *U.S. Geol. Surv. Profess. Paper* 501-B, B154–B157.

Schouten, C. (1962). *Determinative tables for ore microscopy*. Elsevier, Amsterdam. 242 p.

Schryver, K. (1968). Precision and components of variance in the modal analysis of a coarse-grained augen gneiss. *Am. Mineralogist*, 53, 2036–2046.

Shand, S. J. (1939). On staining of feldspathoids and on zonal structure of nepheline. *Am. Mineralogist*, 24, 508–513.

Short, M. A., and E. G. Steward (1959). Measurement of disorder in zinc and cadmium sulphides. *Am. Mineralogist*, 44, 189.

Short, M. N. (1940). Microscopic determination of the ore minerals. *U.S. Geol. Surv. Bull.*, 914, 311 p.

Sine, N. M., W. O. Taylor, G. R. Webber, and C. L. Lewis (1969). Third report of analytical data for CAAS sulphide ore and syenite rock standards. *Geochim. Cosmochim. Acta*, 33, 121–131.

Sinkankas, J. (1968). High pressure epoxy impregnation of porous materials for thin-section and microprobe analysis. *Am. Mineralogist*, 53, 339–342.

Skinner, B. J. (1956). Physical properties of end-members of the garnet group. *Am. Mineralogist*, 41, 428–436.

————, P. B. Barton, and G. Kullerud (1959). Effect of FeS on the unit cell edge of sphalerite: A revision. *Econ. Geol.*, 54, 1040.

Skinner, H. C. W. (1968). X-ray diffraction analysis techniques to monitor composition fluctuations within the mineral group, apatite. *Appl. Spectr.*, **22**, No. 5, 412–414.

Slemmons, D. B. (1962). Determination of volcanic and plutonic plagioclases using a three- or four-axis universal stage. Revision of the Turner method. *Geol. Soc. Am., Spec. Paper* **69**. 64 p.

Smith, J. R., and H. S. Yoder (1956). Variations in X-ray powder diffraction patterns of plagioclase feldspars. *Am. Mineralogist*, **41**, 632–647.

Smith, J. V. (1966). X-ray emission microanalysis of rock-forming minerals II Olivines. *J. Geol.*, **74**, 1–16.

———, and Th. G. Sahama (1954). Determination of the composition of natural nephelines by an X-ray method. *Mineral. Mag.*, **30**, 439.

Smith, O. C. (1953). *Identification and qualitative chemical analysis of minerals*. Van Nostrand, Princeton, New Jersey. 385 p.

Smithson, S. B. (1963). A point-counter for modal analysis of stained rock slabs. *Am. Mineralogist*, **48**, 1164–1166.

Snetsinger, K. G., T. E. Bunch, and K. Keil (1968). Electron microprobe analysis of vanadium in the presence of titanium. *Am. Mineralogist*, **53**, 1770–1773.

Spry, A. (1969). *Metamorphic textures*. Pergamon, Oxford. 350 p.

Stevens, R. E. (1944). Composition of some chromites of the Western Hemisphere. *Am. Mineralogist*, **29**, 1–34.

Straumanis, M. E. (1953). Density determination by a modified suspension method; X-ray molecular weight, and soundness of sodium chloride. *Am. Mineralogist*, **38**, 662–670.

——— (1959). Absorption correction in precision determination of lattice parameters. *J. Appl. Phys.*, **30**, 1965–1969.

Sunderman, H. C. (1970). Refractive index determination by orientation variation. 1. Uniaxial crystals. *Am. Mineralogist*, **55**, 1405–1415.

Swaine, D. J. (1969). The identification and estimation of carbonaceous materials by DTA. *Thermal Analysis*, **2**, Academic, London, 1377–1386.

Swanson, H. E., M. C. Morris, and E. H. Evans (1966). Standard X-ray diffraction powder patterns. *Nat. Bur. Stand. (U.S.)*, *Monograph* **25**, Section 4.

Sweatman, T. R., Y. C. Wong, and K. S. Toong (1967). Application of X-ray fluorescence analysis to the determination of tin in ores and concentrates. *Trans. Inst. Mining Met., Section B*, **76**, B-149–B-154.

———, and J. V. P. Long (1969). Quantitative electron-probe microanalysis of rock-forming minerals. *J. Petrol.*, **10**, No. 2, 332–379.

Tennant, C. B., and R. W. Berger (1957). X-ray determination of the dolomite-calcite ratio of a carbonate rock. *Am. Mineralogist*, **42**, 23–29.

Thompson, J. B., Jr. (1957). The graphical analysis of mineral assemblages in pelitic schists. *Am. Mineralogist*, **42**, 842–858.

Thornton, C. P., and O. F. Tuttle (1960). Chemistry of igneous rocks. I. Differentiation index. *Am. J. Sci.*, **258**, 664–684.

———, ——— (1965). The Crystallization Index as a parameter of igneous differentiation in binary variation diagrams; a discussion. *Am. J. Sci.*, **263**, 277–279.

Tobi, A. C. (1956). A chart for measurement of optic axial angles. *Am. Mineralogist*, **41**, 516–519.

Tröger, W. E., H. U. Bambauer, F. Taborszky, and H. D. Trochim (1969). *Optische Bestimmung der gesteinbildenden Minerale*. Vol. 2 *Textband*, 2nd ed. E. Schweizerbart'sche Verlagsbuchhandlung (Nagele u Obermiller), Stuttgart. 822 p.

———, ———, ———, ——— (1971). *Optische Bestimmung der gesteinbildenden Minerale*. Vol. 1, *Bestimmungstabellen*, 4th ed. E. Schweizerbart'sche Verlagsbuchhandlung (Nagele u Obermiller), Stuttgart. 188 p.

Turner, F. J. (1968). *Metamorphic petrology: mineralogical and field aspects*. McGraw-Hill, New York 403 p.

Tuttle, O. F., and N. L. Bowen (1958). Origin of granite in the light of experimental studies in the system $NaAlSi_3O_8$—$KAlSi_3O_8$—SiO_2—H_2O. *Geol. Soc. Am. Mem.*, **74**, 153 p.

Uytenbogaardt, W. (1968). *Tables for microscopic identification of ore minerals*. Hafner, New York. 242 p.

Viswanathan, K. (1966). Unit cell dimensions and ionic substitution in common clinopyroxenes. *Am. Mineralogist*, **51**, 429–442.

—— (1968). A systematic approach to indexing powder patterns of lower symmetry using De Wolff's principles. *Am. Mineralogist*, **53**, 2047–2060.

—— (1971). A new X-ray method to determine the anorthite content and structural state of plagioclases. *Contrib. Mineral. Petrol.*, **30**, 332–335.

——, and S. Ghose (1965). The effect of Mg^{2+}—Fe^{2+} substitution on the cell dimensions of cummingtonite. *Am. Mineralogist*, **50**, 1106–1112.

Vitaliano, C. J., R. D. Harvey, and J. H. Cleveland (1965). Computer program for norm calculation. *Am. Mineralogist*, **50**, 495–498.

Vivaldi, J. L. M., and P. F. Hach-Ali (1970). Palygorskites and Sepiolites (Hormites), Chapter 20, 553–573. In R. C. Mackenzie, Ed., *Differential Thermal Analysis*, Vol. 1, Academic, London. 775 p.

Vogel, A. I. (1961). *A textbook of quantiative inorganic analysis including elementary instrumental analysis*, 3rd ed. The English Language Book Society and Longmans, Green, London. 1216 p.

Wager, L. R., and G. M. Brown (1967). *Layered Igneous rocks*. Freeman, San Francisco. 588 p.

Wahlstrom, E. E. (1969). *Optical crystallography*, 4th ed. Wiley, New York. 489 p.

Waite, J. M. (1963). Measurement of small changes in lattice spacing applied to calcites of a Pennsylvanian age limestone. *Am. Mineralogist*, **48**, 1033–1039.

Warne, S. St. J. (1962). A quick field or laboratory staining scheme for the differentiation of the major carbonate minerals. *J. Sediment. Petrol.*, **32**, 29–38.

Webb, T. L., and J. E. Krüger (1970). Carbonates. Chapter 10, 303–341, in *Differential thermal analysis*, Vol. 1, Ed. by R. C. Mackenzie. Academic, London. 775 p.

Welday, E. E., A. K. Baird, D. B. McIntyre, and K. W. Madlem (1964). Silicate sample preparation for light-element analysis by X-ray spectrography. *Am. Mineralogist*, **49**, 889–903.

Westbrook J. H., and P. J. Jorgensen (1968). Effects of water desorption on indentation microhardness anisotropy in minerals. *Am. Mineralogist*, **53**, 1899–1909.

White, A. J. R. (1964). Clinopyroxenes from eclogites and basic granulites. *Am. Mineralogist*, **49**, 883–888.

White, E. W., and G. G. Johnson, Jr. (1970). *X-ray emission and absorption wavelengths and two theta tables*. 2nd ed. A.S.T.M. Data Series DS 37A. American Society for Testing Materials, 1916 Race Street, Philadelphia, Pa. 19103 U.S.A. 293 p.

Whittaker, E. J. W. (1960). The crystal chemistry of amphiboles. *Acta Cryst.*, **13**, 291.

——, and J. Zussman (1956). The characterization of serpentine minerals by X-ray diffraction. *Mineral. Mag.*, **31**, 107.

Whitten, E. H. T. (1966). *Structural geology of folded rocks*. Rand McNally, Chicago. 663 p.

Wilcox, R. E. (1959). Use of the spindle stage for determination of principal indices of refraction of crystal fragments. *Am. Mineralogist*, **44**, 1272–1293.

——, and G. A. Izett (1968). Optic angle determined conoscopically on the spindle stage: 1 micrometer ocular method. *Am. Mineralogist*, **53**, 269–282.

Wilkinson, J. F. G. (1963). Some natural analcime solid solutions. *Mineral. Mag.*, **33**, 498.

Williams, H., F. J. Turner, and C. M. Gilbert (1954). *Petrography: an introduction to the study of rocks in thin sections*. Freeman, San Francisco. 406 p.

Winchell, A. N. (1937). *Elements of optical mineralogy: an introduction to microscopic petrography. Part I. Principles and methods*. 5th ed. Wiley, New York, 263 p.

——, and H. Winchell (1951). *Elements of optical mineralogy Part II*, 4th ed. Wiley, New York. 551 p.

Winchell, H. (1946). A chart for measurement of interference figures. *Am. Mineralogist*, **31**, 43–50.

—— (1958). The composition and physical properties of garnet. *Am. Mineralogist*, **43**, 595–600.

——, and R. Tilling (1960). Regressions of physical properties on the compositions of clinopyroxenes. *Am. J. Sci.*, **258**, 529.

Winkler, H. G. F. (translated by N. D. Chatterjee and E. Froese) (1967). *Petrogenesis of metamorphic rocks*, rev. 2nd ed. Springer, Berlin. 237 p.

Woo, C. C. (1964). Heavy media column separation: a new technique for petrographic analysis. *Am. Mineralogist*, **49**, 116–122.

Woodbury, J. L., and T. A. Vogel (1970). A rapid, economical method for polishing thin sections for microprobe and petrographic analyses. *Am. Mineralogist*, **55**, 2095–2102.

Wright, J. B. (1969). A simple alkalinity ratio and its application to questions of non-orogenic granite genesis. *Geol. Mag.*, **106**, 370–384.

484   REFERENCES

Wright, T. L. (1968). X-ray and optical study of alkali feldspars: II an X-ray method for determining the composition and structural state from measurement of 2θ values for three reflections. *Am. Mineralogist,* **53**, 88–104.

———, and D. B. Stewart (1968). X-ray and optical study of alkali feldspars: II determination of composition and structural state from refined unit-cell parameters and 2V. *Am. Mineralogist,* **53**, 38–87.

Yoder, H. S., and H. P. Eugster (1955). Synthetic and natural muscovites. *Geochim. Cosmochim. Acta,* **8**, 225.

———, and T. G. Sahama (1957). Olivine X-ray determinative curve. *Am. Mineralogist,* **42**, 475–491.

———, and C. E. Tilley (1962). Origin of basalt magmas: an experimental study of natural and synthetic rock systems. *J. Petrol.,* **3**, 342–532.

Young, B. B., and A. P. Millman (1964). Microhardness and deformation characteristics of ore minerals. *Bull. Inst. Mining Met.,* **73**, 437–466.

Zeck, H. P. (1969). Measurement of the distortion index (delta) of cordierite. *American Mineralogist,* **54**, 1728–1731.

Zeidler, W. (1968). Preparations of thin film sections. *Am. Mineralogist,* **53**, 1773–1774.

Zeller, E. J., J. L. Wray, and F. Daniels (1957). Factors in age determination of carbonate sediments by thermoluminescence. *Bull. Am. Assoc. Petrol. Geologists,* **41**, 121–129.

Zen, E-AN, and A. L. Albee (1964). Coexisting muscovite and paragonite in pelitic schists. *Am. Mineralogist,* **49**, 904–925.

Zussman, J., Ed. (1967a). *Physical methods in determinative mineralogy.* Academic, London. 514 p.

——— (1967b). X-ray diffraction, Chapter 6, 261–334. In J. Zussman, Ed., *Physical methods in determinative mineralogy.* Academic, London. 514 p.

APPENDIX

Selected alphabetical list of companies that supply equipment and materials for the techniques described in this book. The numbers against each company refer to the product type code given at the end of the list.

CODE

1. AGFA-GEVAERT, 509 Leverkusen, West Germany

2. Allied Chemical Corporation, Specialty Chemicals Division, P.O. Box 70, Morristown, New Jersey 07960.

3. Allied Electronics, 2400 W. Washington Boulevard, Chicago, Illinois 60680.

4. Alminrock Indscer Fabriks, Bangalore 3, Mysore, India.

2. Alpha Inorganics Inc., 8 Congress Street, P.O. Box 159 Beverly, Massachusetts 01915.

2. American Cyanamid Co., Cyanamid International, Wayne, New Jersey, 07470.

5. Atomic Energy of Canada Ltd., Commercial Products, P.O. Box 93, Ottawa, Canada

3. AVO Ltd., Avocet House, 92-96 Vauxhall Bridge Road, London S.W.1, England

6. 7. Bausch & Lomb Inc., 820 Linden Avenue, Rochester, New York 14625

6. 7. Beckman Instruments Inc., Scientific Instrument Division, 2500 Harbor Blvd., Fullerton, California 92634

8. B.D.H. Chemicals Ltd., Poole, Dorset BH 12 4 NN, England

9. Bethlehem Instrument Co. Inc., Bethlehem, Pennsylvania

2. British Chemical Standards, Bureau of Analysed Samples, Newton Hill, Middlesborough, Yorkshire, England

10. 11. 13. Buehler Ltd., & Adolph I. Buehler Inc., 2120 Greenwood Street, Evanston, Illinois, 60204

10. 11. 13. Buehler-Met AG., Postfach 4000, Basel 23, Switzerland

12. Cahn Division/Ventron Instrument Corporation, 7500 Jefferson Street, Paramount, California 90723

13. 14. Cargille-R. P. Cargille, Laboratories Inc., Cedar Grove, New Jersey 07009

15. Central Scientific Company, 2600 S. Kostner Avenue, Chicago, Illinois, 60623

2.	Chemical & Engineering Co Inc., 221 Brook Street, Media, Pennsylvania 19063
2.	CIBA Ltd., Basle, Switzerland
4. 10.	Colorado Geological Industries Inc., 1244 East Colfax Avenue, Denver, Colorado 80218
16.	Columbia Scientific Industries, 3625 Bluestein Boulevard P.O. Box 6190, Austin, Texas, 78762
17.	Cook-Chas. W. Cook & Sons Ltd., 97 Walsall Road, Perry Barr, Birmingham 22B, England
18.	Crescent Manufacturing Co., 7750 West 47th Street, Lyons, Illinois 60534
10. 11.	Cutrock Engineering Co Ltd., 35 Ballards Lane, London, England
9. 17.	Eberbach Co., P.O. Box 1024, Ann Arbor, Michigan, 48106
2.	Elektron Ltd., Clifton Junction, Near Manchester, England.
10. 11.	Engis Ltd., Park Road Trading Estate, Maidstone, Kent, England
19.	Englehard Industries Inc., 113 Astor Street, Newark, New Jersey, 07114
19.	Englehard Industries Inc., Baker Platinum Division, St. Nicolas House, St. Nicolas Road, Sutton, Surrey, England
20.	Enraf-Nonius-N. V. Verenigde Instrumenten Fabrieken, Enraf-Nonuis, Rontgenweg 1, P.O. Box 483, Delft, Holland
21.	Endecotts Test Sieves Ltd., Lombard Road, London S.W. 19, England
2.	Farbwerke Hoechst, Frankfurt a.M. Germany
4.	Filer's, P.O. Box 487-U, Yucaipa, California 92399
15.	Fisher Scientific Co., 633 Greenwich Street, New York 1, New York
22.	Frantz, S. G., 339 East Darrah Lane, P.O. Box 1138, Trenton, New Jersey 08606
17.	Fritsch, D 6580, Idar-Oberstein, West Germany
15.	Gallenkamp, P.O. Box 290, Technico House, Christopher Street, London E.C. 2, England
14.	Gemological Industries Ltd., Saint Dunstan's House, Carey Lane, London E.C. 2, England
10. 11.	Geoscience Instruments Corp., 110 Beckman Street, New York 10038
17. 18.	Glen Greston, The Red House, 37 The Broadway, Stanmore, Middlesex, HA 7 4 DL, England
4.	Gregory, Bottley & Co., 30 Old Church Street, Chelsea, London S.W. 3, England
15.	Griffin & George Ltd., Ealing Road, Alperton, Wembley, Middlesex HA 0 1 HJ England.
2.	Halewood Chemicals Ltd., Horton Road, Stanwell Moor, Staines, Middlesex, England
2.	Heraeus GmbH., Hanau, West Germany
24.	Heyden & Son Ltd., 64 Vivian Avenue, London N.W. 4
14.	Hooker Chemicals Co., Niagara Falls, N.Y.
8.	Hopkin and Williams Ltd., Chadwell Heath, Essex, England

2.   I.C.I., I.C.I. House, Nicholson Street, Melbourne, Victoria 3001, Australia

1.   Ilford Ltd., Ilford, Essex, England

10.   Ingram Laboratories, Griffin, Georgia

8.   Johnson-Matthey & Co Ltd., Hatton Garden, London W.C. 1, England

27. 33.   Joint Committee on Powder Diffraction Standards, 1601 Park Lane, Swarthmore, Pennsylvania 19081

4.   Krantz-Dr. F. Krantz, Bonn, Germany

2.   K. and K. Laboratories, 121 Express Street, Engineers Hill, Plainsview, New York 11803

2.   K and K. Laboratories of California Inc., 6922 Hollywood Boulevard, Hollywood, California 90028

1.   Kodak, Kodak Park, Rochester, New York 14650

1.   Kodak Ltd., Box 14, Hemel Hempstead, Hertfordshire, England

4. 10.   Kyoto Scientific Specimens Ltd., 378 Ichinofunairicho, Kawaramachi-Nijo Minami, Nakagyo-ku, Kyoto, Japan

3.   Leeds and Northrup Co., 4901 Stenton Avenue, Philadelphia, Pennsylvania 19144

25. 26. 32. Leitz-Ernst Leitz GmbH., D 6330 Wetzlar, Germany

25. 26. 32. Leitz-E. Leitz Inc., Rockleigh, New Jersey 07647

20.   Lordon Ceramics, 326 South Road, Croydon Park, South Australia 5008.

15.   Matheson Scientific, 1600 Howard Street, Detroit, Michigan 48216

26.   McCrone Research Associates, 493 East 31st Street, Chicago, Illinois 60616

26.   McCrone Research Associates, 2 McCrone Mews, Belsize Lane, London N.W. 3, England

27.   M. E. L. Equipment Co., Manor Royal, Crawley, Sussex, England.

8.   Merck-E. Merck AG., 61 Darmstadt, West Germany

10.   Microtec Development Laboratory, P.O. Box 1441, Grand Junction, Colorado 81501

2.   National Bureau Of Standards, Office of Standard Reference Material, Washington D. C. 20234

28.   Parr Instrument Co., 211 Fifty third Street, Moline, Illinois 61265

18. 21.   Pascall Engineering Co. Ltd., Gatwick Road, Crawley, Sussex, England

6. 7.   Perkin-Elmer Corp., 803 Main Avenue, Norwalk, Connecticut 06852

20. 23.   Philips-N. V. Philips Gloeilampenfabrieken, Eindhoven, The Netherlands

20. 23.   Philips Electronic Instruments, 750 South Fulton Avenue, Mount Vernon, New York 10550

27. 33.   Powder Diffraction Standards, 1601 Park Lane, Swarthmore, Pennsylvania 19081

29.   Rawplug Co. Ltd., Cromwell Road, London S.W. 7, England

25.   Reichert-C. Reichert optische Werke AG., Wein XVII Hernalser Haupstrasse 219, A 1171 Vienna, Austria

17. 18. 21. Retsch KG, 5657 Haan b., Dusseldorf, Newermarkt 25, West Germany

24,   Ringsdorf-Werke GmbH, 5320 Bad Godesberg, P.O. Box 9087, Mehlem, West Germany

2.   Shell Chemical Co., P.O. Box 2392, Church Street Station, New York, New York 1000

21.   Soiltest Inc., 2205 Lee Street, Evanston, Illinois

2.   Southwestern Analytical Chemicals, P.O. Box 485, Austin, Texas 78767

25. 30.   Swift-James Swift & Sons, Joule Road, Houndmills Industrial Estate, Basingstoke, Hampshire, England

6.   Techtron-Varian Techtron Pty. Ltd., 679–687 Springvale Road, North Springvale, Victoria 3171 Australia

6.   Techtron-Varian Techtron Inc., 611 Hansen Way, Palo Alto, California 94303

18.   Tema-N. V. Tema, Riouwstraat 200, The Hague, Holland

24.   Ultra Carbon Co., P.O. Box 747, Bay City, Michigan, 48709

31.   Ultraviolet Products Inc., 5114 Walnut Grove Ave., San Gabriel, California 91778

2.   Union Carbide Corporation, Chemicals Division, P.O. Box 6112, Cleveland, Ohio

28.   Uni-Seal Decomposition Vessels, P.O. Box 9463, Haifa, Israel

5.   United Kingdom Atomic Energy Authority, The Radiochemical Center, Amersham, Buckinghamshire, England

25. 32.   Vickers Instruments, Haxby Road, York YO 3 7 SD, England

25. 32.   Vickers Instruments Inc., 15 Waite Court, Malden, Massachussetts, 02148

4. 10. 11. Wards Natural Science Establishment Inc., P.O. Box 1712, Rochester, New York 14603

4. 10. 11. Ward's of California, P.O. Box 1749, Monterey, California 93940

25. 26.   Zeiss-Carl Zeiss, 7082 Oberkochen, West Germany.

25. 26.   Zeiss-Carl Zeiss Inc., 444 5th Avenue, New York, N.Y. 10018

EXPLANATION OF CODE

1 = film, both for X-ray diffraction and for photomicrography; photographic paper, and processing chemicals.

2 = specialized chemicals and chemical products

3 = general electrical and electronic units, components, and spares

4 = general geology teaching specimens (rocks, fossils, minerals) and teaching models

5 = gamma and beta irradiators

6 = atomic absorption spectrometers

7 = spectrophotometers, both visible and UV

8 = general analytical chemical supply company

9 = specific gravity balance

10 = general rock cutting, grinding, and thin-section equipment

11 = general supplies of abbrasives, laps, and all petrographic preparation materials.

12 = balance for thermogravimetric analysis

13 = refractive index immersion oils

14 = specific gravity heavy liquids and density standards

15 = general scientific laboratory supplies (hardware, not chemicals)

16 = DTA and TGA apparatus

17 = rock-crushing, pulversizing, and milling equipment

18 = laboratory mixing mills

19 = platinum laboratory ware

20 = X-ray diffraction equipment

21 = sieves, shakers, and sample splitters

22 = magnetic isodynamic separator

23 = X-ray fluorescence spectroscopy equipment

24 = graphite crucibles and all graphite preparations for X-ray fluorescence spectro-scopy and general laboratory use.

25 = polarizing microscopes and microscope accessories

26 = microscope rotation stages (spindle and universal)

27 = X-ray fluorescence spectroscopy conversion charts of 2θ to wavelength.

28 = Teflon decomposition bombs

29 = glue

30 = point counters and tabulators for petrographic modal analysis

31 = ultraviolet lamps and products

32 = micro hardness indenters

33 = powder diffraction data cards, books and indexes.

Units of measure: their conversion and abbreviation

Unit	Symbol	Equivalence
Length		
kilometer	km	1000 m (10^3 m)
meter	m	
decimeter	dm	0.1 m (10^{-1} m)
centimeter	cm	0.01 m (10^{-2} m)
millimeter	mm	0.001 m (10^{-3} m)
micrometer (micron)	μm	10^{-6} m
nanometer	nm	10^{-9} m
angstrom	Å	10^{-10} m
1 inch	in.	2.540 cm
1 foot	ft	0.3048 m
1 mile		1.609 km
Area		
1 square inch	in.^{2}	6.4516 cm^2
Weight		
kilogram	kg	10^3 g
gram	g	
milligram	mg	0.001 g (10^{-3} g)
microgram	μg	10^{-6} g
1 pound	lb	0.4536 kg
Volume		
liter	l	1000.028 cm^3
milliliter	ml	1.000028 cm^3
1 gallon (Imperial)		4.546 l
1 gallon (U.S.)		3.785 l
Density		
1 pound/ft^3		16.02 kg/m^3
		0.01602 g/cm^3
Temperature		
1 degree Fahrenheit	1°F	5/9 degrees Celsius (°C)
Water freezing point		0°C, 32°F
Water boiling point		100°C, 212°F
Pressure		
1 lb/in.^{2}		703.07 kg/m^2
psi		lb/in.^{2}
psig		lb/in.^{2} gage
Dilutions		
1 μg/ml		1 ppm
		0.001 g/l
		1 mg/l

Notes: The symbol μ (micron) should give place to μm (micrometer).
The symbol mμ should give place to nm (nanometer).
ppm = part per million aqueous solution.

INDEX

Bold number: means major reference
F after number: means reference to illustration
Authors' names are in capital letters.

a, determination, 178
AB Epo-Mix epoxide, 13
  Microcloth, 11
  petro-thin slide holder, 9
  Texmet lap cloth, 10
Abbe refractometer, 262, 262F
  procedure, 263
ABRAMOWITZ, M., 320, 473
Absorbance mode, 339
Absorption band, 368
Absorption edge, 286, **288, 289,** 321, 324, 325F
  K-spectrum, 276, 277
  L-Spectrum, 279
Absorption mode, 346
Absorption tubes, 364
  weighing, 364
Accessory minerals, correction, 380
Acetic acid, 338
Acetone, 121
  slurry, 144
Acetylene, 341, 342
ACF calculation, 381
ACF diagram, 383F, **378-385,** 390F
  description, 379
  use, **382-385**
ACF and A'KF diagram use, 390, 391
ACF plotting method, **379-382**
ACF and A'KF similarities, 388
Achromats, 66
Acid digestion bomb, **335-337**
Acmite, 405, 406
  allocation, 403
  jadeite, diopside, hedenbergite diagram, 405
  normative, 415, 416, 419, 420
Acmite end member, 401
Actinolite, on ACF diagram, 383F, 384
  normative, 423, 424, 425
ADAMS, J. K., 14, 473
Adhesives, 80
ADP crystal, 302
AFM corrections, 393
AFM diagram, 392F, **391, 393,** 411, 482

AFM plotting, 393
AFM projection, 391
Age determination, 472
Agfa CT, 18, 71
Agfa-Gevaert, 485
AGTERBERG, F. P., 220, 473
Air flame, 341, 342
Air-acetylene flame, 352
Air-acetylene gas mixture, **342**
Air-coal gas flame, 344
Air-hydrogen flame, 344
Air path, 275
Air propane, 344
A'KF description, 385
A'KF diagram, **385-390,** 389F, 390F
A'KF difficulties, 386
A'KF values, 388
A'KF values for rocks, 387
Alabandite reflectivity and hardness, 41F
Alamine, 130
ALBEE, A. L., 232, 361, 473, 484
Albite, 180, 196F
  high, 195
  high-composition graph, 187F
  intermediate, 195
  low, 195
  low-composition graph, 183F, 185F
  low-diffractogram, 181F
  normative, 398, 415, 419, 420
  normative, provisional, 416
  optic axial angle, 111F
  powder diffraction indexing, 194
Albite($\bar{2}$01), 182
Albite correction, 381, 387, 393
Albite diffraction, 231
Albite diffraction data, 232
Albite diffractogram, 198F
Albite D.T.A. characteristics, 452
Albite D.T.A. peak temperature, 451
Albite-epidote hornfels facies, 377F, 378
Albite-iron ratio plot, 399
Albite ratio, 400F

normative, 399
Albite staining, 18
Albite·twin orientation, 104
Alcohol, 121
Alfa Inorganics, 290
Aligning device 150F, 152, 153
Alignment, camera, 134-136
  X-Ray beam, 149
Alignment tools, 148, 148F
Alizarin fluorine blue reagent, 372
Alizarin fluorine blue lanthanum complex, 372
Alizarin red S, 25, 27F, 28
Alkali basalt, 394, 405
Alkali basalt series, 394, 394F
Alkali feldspar, anomalous, 195
  bulk composition, 182-186
  composition, 111F, 184, 186
  composition determination, 195-197
  high-level, 189
  homogenization, 183
  homogenized-diffractogram, 184
  indexing, 191
  mode, 64
  plutonic, 188
  potassium content, 388
  powder patterns, 474
  Puye, 197
  stability, 476
Alkali feldspars, 480
Alkali feldspar (002), 189, 195
Alkali feldspar (060), 189, 195
Alkali feldspar (131), 188, 189
Alkali feldspar (1̄13), 189, 195
Alkali feldspar (2̄01), 181, 182, 189, 195, 197
Alkali feldspar (2̄04), 189, 195
Alkali feldspar AFM plot, 391, 392F
Alkali feldspar A'KF position, 389F
Alkali feldspar composition graph, 183F, 185F, 187F
Alkali Feldspar correction, 381
Alkali feldspar diffraction, 180
Alkali feldspar diffraction peaks, 206
Alkali feldspar diffractogram, 181F, 189
Alkali feldspar diffractogram indexing, 190-191
Alkali feldspar extraction, 181
Alkali feldspar optic axial angle, 111F
Alkali feldspar peak positions, 195
Alkali feldspar projection point, 391, 393
Alkali feldspar series, 195
Alkali feldspar SH 1070, 197
Alkali feldspar Spencer B, 197
Alkali feldspar staining, 16-20, 22, 475, 480
Alkali feldspar structural state, 110-111, 189-197, 196F
Alkali feldspar structural state, in pluton, 188
Alkali feldspar temperature-structural state, 186-197
Alkali feldspar triclinicity, 188
Alkali feldspar X-Ray determination, 478

Alkali feldspar X-Ray properties, 484
Alkali granite differentiation index, 398
Alkali, Iron, Magnesia diagram, 399
Alkali-lime index, 395, 396F
Alkali olivine basalt, 394
Alkali rhyolite, 395
Alkali rhyolite differentiation index, 398
Alkali rock series, 395, 396F
Alkali:silica diagram, 394F, 394, 395
Alkali:silica ratios, 399
Alkaline rocks, 414
Alkalinity ratio, 414, 483
  plot versus SiO$_2$, 414
Allanite magnetic susceptibility, 119F
ALLEN, W. C., 233, 473
Allied Chemical Corporation, 294, 485
Allied Electronics, 485
Allophane, 227
  distinctive properties, 229
  D.T.A. characteristics, 457
  D.T.A. peak temperature, 449, 451
  heated, 228
Almandine, 209, 212
  ACF positions, 383F, 384
  AFM positions, 392F
  A'KF positions, 389F
Almandine physical properties, 208
Almandine amphibolite facies, 382, 383F, 389F
Alminrock Indscer Fabriks, 485
Alpha Inorganics Inc., 485
Alpha radiation, 466
Alpha scintillation counter, 466, 472
Altaite reflectivity and hardness, 41F
Alumina, 3, 8, 11, 12
  determination, 338
  disc, 296
  boat, 364
  mass absorption coefficient, 327
  matrix calibration, 324-329
  pressed disc, 324
  reference material, 444
Aluminum analysis, 335
Aluminum Analysis settings, 350
Aluminum detection, 274, 281
Aluminum dish, 443
Aluminum mass absorption coefficient, 288
Aluminum matrix correction coefficient, 315
Aluminum plunger, 292F
Aluminum polished sheet, 290
Aluminum sample holders, 440
Aluminum silicates, diffraction data, 231
Aluminum silicates, X-ray data, 477
Aluminum specimen holders, 145
Aluminum standard solution, 348
Aluminum X-Ray data, 276
Aluminum X-Ray detection settings, 305
Amaranth stain solution, 19, 20, 21, 22
American Cyanamid Co., 131, 485

American Society for Testing Materials, 473
Ammonium carbonate, 372
Ammonium di hydrogen sulfate data, 282
Ammonium hydroxide, 226
Ammonium sulfide, 30
Amorphous coatings, 448
Amphibole, 361, 481
  analyses, 387, 478
  analyses, errors, 407
  analyses, test, 407
  analyses for fluorine, 373
  charge balance, 407
  classification, 408F
  correction, 387
  crystal chemistry, 483
  diffraction, 231
  D.T.A. characteristics, 453
  D.T.A. peak temperatures, 449
  end members, 409
  half unit-cell, 407, 409
  mode, 385
  Nigerian, 479
  projection, 412F
  specific gravity, 385
  structural formulae, 406
  structural formula calculation, 410, 411
  tetrahedral projection, 411
Amphibolite, 411
Amphibolite, norm, 424
Amplifier, thermoluminescence, 466
Analcite, 483
  diffraction, 231
  staining, 24
Analytical standards, 287
Anatase magnetic susceptibility, 119F
Andalusite, ACF position, 383F, 384
  A'KF position, 389F
Andalusite decomposition, 334
Andalusite diffraction, 231
Andalusite magnetic susceptibility, 119F
Andesine diffractogram, 198F
Andesite, 395
  differentiation index, 399
  relative absorption, 328F
  solidification index, 395
Andesitic dacite solidification index, 395
Andradite, 212
  ACF position, 383F, 384
  carbonate rocks, 376
  correction, 386
  facies diagram position, 377F
  physical properties, 208
ANDREWS, K. W., 231, 477
Anglesite, crystalline inversion, 438
Anglesite, D.T.A. curve, 438, 439F
Anglesite fluorescence, 32
ANGINO, E. E., 333, 334, 339, 352, 473
Angle correction, 252

Angular calibration of goniometer, 154-155
Angular internal standard, 168, 186
Angular precision, 159
Angular standards, 166-168
Anhydrite staining, 27F
Anion numbers, 401
Anions, number in amphibole, 410
Anisotropic minerals, 248
Ankerite staining, 27F, 28, 29, 30
Anorthite, ACF position, 383F, 384
  calcium content, 388
  normative, 396, 415, 417, 419, 421, 424
  normative-negative, 422
Anorthite correction, 386
Anorthite diffraction data, 232
Anorthite diffractogram, 198F
Anorthoclase, 189, 195, 196F
  diffraction indexine, 192, 193
  optic axial angle, 111F
  staining, 20, 476
Anthophyllite, 412F
  ACF position, 383F, 384
  A'KF position, 389F
Antimony reflectivity and hardness, 41F
Antimony X-Ray data, 277, 279
Anti-scatter slit, 158
Anti-scatter slit choice, 157
Apatite, normative, 415, 416, 419, 420
  specific gravity, 379
  thermoluminescence, 472
Apatite accessory, 379
Apatite correction, 379, 381
Apatite diffraction, 232, 482
Apatite magnetic susceptibility, 119F
Apochromats, 66
Aqua regia, 337, 352
Aragonite, 217
  analysis, 218
  diffraction data, 232
  D.T.A. characteristics, 454
  D.T.A. peak temperature, 449
  fluorescence, 32
  interference figures, 81F
  staining, 27F, 28
  thin-section, 75
Araldite AY 105, 2, 4
Araldite MY 753, 13
Archeological pottery dating, 472
Archimedes' principle, 235, 236
Arfvedsonite diffraction, 231
Argentite D.T.A. characteristics, 455
Argentite D.T.A. peak temperature, 450
Argentite reflectivity and hardness, 41F
Argon, 341, 342
Argon:entrained air-hydrogen flame, 345
Argon:methane gas, 274
Argon X-Ray Data, 276
Armac 12D, 130

ARNOLD, R. G., 233, 473
Arsenic mass absorption coefficient, 289
Arsenic reflectivity and hardness, 41F
Arsenic spiking, 327
Arsenic X Ray data, 276
Arsenopyrite, 479
  diffraction data, 232
  D.T.A. characteristics, 455
  D.T.A. peak temperature, 450
  magnetic susceptibility, 119F
  reflectivity and hardness, 41F
ARRIENS, P. A., 474
ARVIDSON, R. E., 14, 473
Asama xenoliths, 222
Ascarite, 364
ATCHLEY, F. W., 72, 473
ATKINS, L., 295, 477
Atomic absorption amplifier, 340
Atomic absorption analysis, 473, 474
Atomic absorption analytical conditions, 349-
  352
Atomic absorption flame, 339
Atomic absorption lamp, 341F
Atomic absorption mode procedure, 346
Atomic absorption readout, 339
Atomic absorption spectrophotometer, 340F,
  488
Atomic absorption spectrophotometer adjust-
  ments, 339-345
Atomic absorption spectrophotometer compo-
  nents, 341F
Atomic absorption spectrophotometer procedure,
  345-348
Atomic absorption spectrophotometry, 333-353,
  473, 478, 479, 481
Atomic absorption spectrophotometry theory,
  333
Atomic Energy of Canada Ltd., 485
Attapulgite, 227
Attenuation, 297
  selection, 161, 163, 298, 299, 301
  sodium, 302
AUCOTT, J. W., 368, 473
Augite-ferroaugite series, 405
Augite, modal, 394
Augite series, 403F
Augite, subcalcic, 394
Automatic stops for goniometer, 160
AVO Ltd., 485
AZÁROFF, L. V., 132, 173, 473

Background correction, 327
  count, 182
  determination, 308-310, 309F, 316, 323, 326
  discrimination, 301
  line, 308
  position, 2θ, 308
BAILEY, E. H., 18, 473

BAIRD, A. K., 483
Bakelite, 295
  ring forms, 13
BAKER, F., 393, 473
Balance, adjustment, 462
  Berman density, 238, 239
  Cahn, calibration, 462
  Kraus-Jolly, 236
  Mettler K., 295
  T.G.A., 461F
  top-pan, 241
BAMBAUER, H. U. x, 197, 200, 201, 203, 473,
  482
Barite, interference figures, 81F
  magnetic susceptibility, 119F
  thin-sections, 75
Barium analyses in granite, 324
Barium chloride, 17, 18, 19, 20, 22
  diphenylamine sulfonate, 358
  standard preparation, 324
  X-Ray data, 277, 279
BARTH, T. F. W., 422, 425, 473
BARTON, P. B., 481
Basalt, A'KF position, 390F
  calculations, 411
  classification, 393-394
  crystallization index, 398F
  differentiation index, 399
  magmas, 476, 478
  magma origin, 484
  relative absorption, 328F
  solidification index, 395
  weathering index, 414
Basaltic andesite solidification index, 395
Basanite, 394
Base line, 458F
  determination, 467
  drift, 443, 444
  thermograms, 458
Base plate, 153
Basic rocks, ACF position, 390
Bausch and Lomb Inc., 485
  Spectrometer, 369
BAYLISS, P., 454, 473
B. D. H. Chemicals Ltd., 134, 236, 317, 324,
  485
Beaker, Berzelius, 358, 359, 360
Becke Line, 82, 98, 252, 254-256
  method, 254F, 255
Beckman Instruments Inc., 485
Beer's Law, 319
Beilby layer, 448
Bell jar, 3
  specific gravity, 235
Benioff Zone, 395
BENJAMIN, R. E. K., 121, 122, 473
BENSCH, J. J., 241, 473
Benzene, 121

Benzidine, 27F, 29
BEREK, M., 103, 473
Berek-Dodge method, 103, 104, 105F
BERGER, R. W., 215, 216, 482
BERGGREN, G., 456, 479
BERMAN, M., 275, 277, 279, 284, 473
Berman density balance, 235, 238, 239, 239F
  procedure, 239, 240
BERNAS, B., 337, 349, 351, 352, 473
BERRY, L. G., 235, 473, 479
Berthierite reflectivity and hardness, 41F
BERTIN, E. P., 264, 473
Bertrand lens, 80
Beryllium mass absorption coefficient, 288
Beta radiation source, 470
Betafite reflectivity and hardness, 41F
Bethlehem Instruments Co. Inc., 239, 485
Biaxial minerals, 248
Biaxial mineral orientation, 250
  refractive index determination, 249
  stereogram plotting, 100F
BILLINGS, C. K., 333, 334, 339, 352, 473
Binocular microscope, 113
Biotite, ACF position, 379, 381, 383F, 384
  AFM position, 391, 392F
  A'KF position, 388, 389F
  alkali content, 380
  analyses, 381, 477
  diffraction data, 233
  distinctive properties, 229
  D.T.A. characteristics, 456
  D.T.A. peak temperature, 449, 450, 451
  heated, 228
  magnetic susceptibility, 119F
  modal analysis, 379
  normative, 423, 424, 425
  specific gravity, 379
  X-Ray determination, 476
Birefringence, 252
  measurement, 251
  calculation, 252
BISCAYNE, P. E., 230, 474
Bisectrix figure, 112, 250
  plotting, 99, 102
BISKUPSKY, V. S., 335, 474
Bismuth detector conditions, 299F
  reflectivity and hardness, 41F
  X-Ray data, 277, 279
Bismuthinite reflectivity and hardness, 41F
Bixbyite reflectivity and hardness, 41F
Blackbody radiation, 467
Blank determination, 364, 367, 373
  disc, 303, 313
  preparation, 310
Blende (sphalerite) reflectivity and hardness, 41F
BLOSS, F. D., 79, 144, 146, 174, 175, 474, 475
Boehmite D.T.A. characteristics, 456
Bokaro coalfield, 222

Boling burner, 350, 351
BOLLIN, E. M., 454, 455, 474
Bomb, Teflon, decomposition, 335-337
Bond strength, silicates, 414
BOONE, G. M., 22, 23, 474
Borate fusion technique, 480
Borate glass, 292F
  glass background, 308
  glass blanks, 309, 313, 314
  glass corrections, 313
  glass discs, 295, 303
  glass disc comparison, 316
  glass disc preparation, 291, 293
  glass standards, 313, 314
Boric acid, 267, 268, 294, 295, 296, 322, 323,
    324, 334, 335, 337, 348, 349
  backing, 302
  solution, 357
BORG, I. Y., 183, 232, 474
BORLEY, G., 231, 474
Bornite D.T.A. characteristics, 454
  D.T.A. peak temperature, 450
  magnetic susceptibility, 119F
  reflectivity and hardness, 41F
Bostonite norm calculation, 418
BOUDETTE, E. L., 20, 476
Boulangerite reflectivity and hardness, 41F
Bournonite reflectivity and hardness, 41F
BOWEN, N. L., 73, 110, 111, 180, 482
BOWIE, S. H. U., 35, 40, 41, 474
Bracket adjusting device, 134, 135, 135F, 137
BRADSHAW, P. M. D., 64, 474
BRADLEY, F., 481
Bragg equation, 143, 161, 281, 284, 480
Brannerite reflectivity and hardness, 41F
Braunite reflectivity and hardness, 41F
Bravoite reflectivity and hardness, 41F
BREGER, I. A., 369, 474
Breithauptite reflectivity and hardness, 41F
BRINDLEY, G. W., 232, 474
British Chemical Standards, 348, 485
Broadening ratio, 187F, 188
Bromine, X-Ray data, 276
Bromoform, 120
Bronzite, 403F
Bronzite-hypersthene-inverted pigeonite series,
    405
Brookite magnetic susceptibility, 119F
BROWN, G., 225, 474
BROWN, G. M., 399, 403, 483
Brucite fluorescence, 32
  staining, 25-26, 476
BRUNTON, G., 147, 474
BRYNARD, H. J., 241, 473
BUCKLEY, D. E., 337, 349, 351, 474
Buehler Ltd. and Adolph I. Buehler, 1, 4, 7, 13,
    247, 485
  AB Whirlimet, 9

-Met AG, 1, 7, 485
BUERGER, M. J., 132, 173, 473
Bulb, heating, 367
BUNCH, T. E., 482
Burette, provision, 358
BURLEY, B. J., 233, 474
Burner, 341F
  assembly, 340
  cleaning, 341
  positioning, 340
  type, 341
  system, 339
Bushveld Complex, South Africa, 225
Bytownite diffraction data, 232

CABRI, L. J., 236, 474
Cadmium, X-Ray data, 277
  oxide standard, 168
CAFFYN, J. E., 31, 480
Cahn electrobalance, 459, 460, 461F, 485
  electrobalance calibration, 461, 462
  electrobalance operation, 464
Calc-alkali rock series, 395, 396F
Calc-alkaline rocks, 414
Calcareous rocks, ACF position, 390, 390F
Calciferous amphiboles, 406
  amphibole classification, 408F
Calcite, 217, 378
Calcite (104), 215
Calcite, ACF position, 383F, 384
  analysis, 218
  Antarctic, 471
  in carbonates, 376
  in fossil shells, 232
  glow curve, 468F, 469, 469F
  magnesian, 218, 476
  normative, 415, 416, 419, 420
  peak intensity, 214, 215
  staining, 25-28, 27F
Calcite:aragonite ratio, 475
Calcite corrections, 379, 381
  diffractogram, 214, 483
  :dolomite ratio, 216F, 476, 481
  D.T.A. characteristics, 453
  D.T.A. peak temperature, 451
  facies diagram, 377F
  fluorescence, 32
Calcium analysis, 335
  analysis settings, 350
  carbonate, 216
  carbonate disc, 296
  chloride, 362
  detection, 274, 281
Calcium detector settings, 298F, 299F
  interference, 332
  mass absorption coefficient, 286, 289
  matrix correction coefficient, 315
  orthosilicate, normative, 419, 422

Calcium oxide, 366, 367
  oxide, mass absorption coefficient, 327
  replacement by magnesium, 216-217
Calcium, sodium, potassium values in amphi-
      bole, 409
Calcium standard solution, 348
  X-Ray data, 276, 305
Calgon, 226
Calibration, eyepiece micrometer, 44-45
  field-of-view, 45-46
Calibration graph, 351, 352
Calomel fluorescence, 32
Camera alignment, 134-136, 136F
  mounting bracket, 135F
Canada balsam, 3, 9
Cancrinite, normative, 416
Capillaries, 132
Capillary cleaning, 340
  loading, 133, 133F
  plug, 366
Carbon determination, 361-364
Carbon dioxide absorption tube, 364
  dioxide determination, 361-364, 363F, 481
  dioxide evolution, 364
  dioxide mass absorption coefficient, 327
  dioxide procedure, 362
  D.T.A. characteristics, 457
  D.T.A. peak temperature, 449
  mass absorption coefficient, 288
  tetrachloride, 121, 236
  tetrachloride density, 238
Carbonates, 217-218
Carbonate (104), 217
Carbonate analysis by X-ray diffraction, 214-
      218
  detection, 476
  diffraction data, 232
Carbonate D.T.A. characteristics, 453, 454, 483
Carbonate rocks, composition plotting, 376-378
Carbonate staining, 24-30, 27F, 476, 483
  T.G.A. curve, 465F
  thermoluminescence, 472
  X-Ray diffraction, 479
Carborundum, 3
  reflectivity, 34
  SiC standard, 34
Cargille, R. P. Cargille Laboratories Inc., 244,
      247, 257, 259, 485
Carl Zeiss Inc., 488; see also Zeiss
CARMICHAEL, I. S. E., 401, 474
Carnegie Institution, 158
Carnegieite, 453
CARROLL, D., 225, 226, 228, 230, 231, 474
Carrollite, D.T.A. characteristics, 454
  D.T.A. peak temperature, 450
  reflectivity and hardness, 41F
Cascades trend, 400F
Cassiterite dissolution, 338

insolubility, 293
magnetic susceptibility, 119F
normative, 419, 420
reflectivity and hardness, 41F
Catanorm, **419-423**
Cation proportions, 420
Cattierite, 233
Celadonite distinctive properties, 229
heated, 228
Celsian, 186
Central Scientific Company, 485
Centrifugation, 123, 126, 352
Centrifuge tube, 352
tube, pointed, 353
special, 124, 125F
Cerium X-Ray data, 277, 279
Cerussite fluorescence, 32
staining, 27F, 28
CERVELLE, B., 232, 474
Cesium X-Ray data, 277, 279
Chabazite fluorescence, 32
Chalcocite magnetic susceptibility, 119F
D.T.A. characteristics, 454
D.T.A. peak temperature, 449
reflectivity and hardness, 41F
Chalcogenides, D.T.A., 474
Chalcophanite reflectivity and hardness, 41F
Chalcopyrite D.T.A. characteristics, 454
D.T.A. peak temperature, 449, 450
magnetic susceptibility, 119F
reflectivity and hardness, 41F
Chalcostibnite reflectivity and hardness, 41F
CHAMBERLAIN, J. A., 40, 479
Chamosite properties, 231
CHANDLER, J. C., 369, 474
CHAPPEL, B. W., 264, 266, 286, 292, 320, 322, 323, 474, 480
Characteristic temperature, 458
Chart paper, D.T.A., 446
CHATTERJEE, N. D., 483
CHAYES, F., 59, 474
Chemical analyses, 478, 479, 482
analysis, amphibole, 410
correction, 380
pyroxene, 401
wet, **354-373**
Chemical and Engineering Co., 2, 4, 486
Chemical standards, 287
supply companies, 488
Chemicals, film processing, 488
specialized, 488
Chisel, 113, 181
Chlorapatite fluorescence, 32; *see also* Apatite
Chlorine analysis, **331-332**
contamination, 332
detection, 274, 282
detection limits, 332
determination in silicates, 373

loss, 293
mass absorption coefficient, 289
spectrometer settings, 331
X-Ray data, 276
Chlorine X-Ray determination, 475
Chlorite, 227, 361, 477
Chlorite (001), 230
Chlorite (060), 232
Chlorite, ACF position, 383F, 384
AFM position, 392F
A'KF position, 389F
analyses, 477
analysis by X-Ray, 481
characteristics, 481
diffraction data, 232
dioctahedral, properties, 231
distinctive properties, 230
D.T.A. characteristics, 456, 457
D.T.A. peak temperature, 449, 450
heated, 228
identification, 474
magnetic susceptibility, 119F
swelling properties, 230
T.G.A. curve, 464, 465F
X-Ray diffraction, 480
Chloritoid, ACF position, 383F, 384
AFM position, 392F
A'KF position, 389F
CHODOS, A. A., 361, 473
Chromatographic cellulose, 294, 331, 332
Chrome spinel diffraction data, 234
Chrome steel, 127
Chromedia CF 11, 317
Chromel alumel thermocouple, 446
Chromite, 358, 477, 482
analyses, 330
composition, 233, 436
diffraction data, 233
hardness, 40-41, 41F
magnetic susceptibility, 119F
normative, 415, 416, 419, 420
reflectivity, 40-41, 41F, 437F
Chromium analysis settings, 351
detection, 274
detection conditions, 299F
diffraction lines, 274
Chromium mass absorption coefficient, 289
radiation, 283, 296
standard solution, 348
trioxide, 364
trioxide bubler, 364
tube, 274
tube voltage, 274
wavelength, 324
X-Ray data, 276
CHROMÝ, S., 259, 474
Ciba Ltd., 13, 486
C.I.P.W. norm, **414-418**

C.I.P.W. norm rules, 415
CLARK, G. L., 286, 474
CLARK, L. A., 232, 474, 479
Clay, modelling, 133F
Clays (001), 229
Clays, 479
  centrifugation on ceramic tile, 147
  constant humidity treatment, 227
  d-spacing, 227
  diffractograms, 228, 474, 480
  diffractogram procedure, 227-231
  disaggregation, 226
  D.T.A., 479
  D.T.A. characteristics, 456
  expansion of d-spacing, 227
  glycolation, 227
  heat treatment, 227
  heating results, 228
  mixed clay minerals, 227
  mixed layer clays, heated, 228
  mixed layer clays, properties, 231
  oriented mount preparation, 226
  powder press technique, 147
  smear on glass slide, 147
  specimens, 476
  specimens, oriented, 478
  specimen preparation, 147, 226
  suction on ceramic tile, 147
  unoriented mount preparation, 226
  X-Ray identification, 225-231, 474
Cleaning mineral separates, 115-116
Cleavage plane orientation, 104
Clerici's solution, 120, 121
CLEVELAND, J. H., 483
Clinopyroxenes, 483
  analyses, 478
  chemical composition, 405
  diffraction data, 233
  eclogite, 483
  X-Ray diffraction, 482
Coarse grained, definition, 46
Cobalt analysis settings, 351
  mass absorption coefficient, 289
  radiation, 140
  radiation for iron minerals, 227, 229
  standard solution, 348
  wavelength, 324
  X-Ray data, 276
Cobaltite D.T.A. characteristics, 455
  D.T.A. peak temperature, 450
  reflectivity and hardness, 41F
COCHRAN, M. C., 6, 7, 474
COCHRANE, R. H. A., 481
Coefficient of variation, 428, 429
Coffinite reflectivity and hardness, 41F
COLE, J. R., 174, 475
Colemanite fluorescence, 32
Collimator, 135, 136F, 137F

coarse, 274
change, 280
choice, 139, 303
for clays, 139
exit, 272
setting, 283
size, 137
Collodion, 134
Colmonoy, 127
Colorado Geological Industries, Inc., 1, 30, 486
Colorimeter, 369, 371
Colorimetric determinations, 372
COLUCCI, S. L., 72, 475
Columbia Scientific Industries, 439, 440, 459, 486
Columbite magnetic susceptibility, 119F
  reflectivity and hardness, 41F
Combustion tube, 362, 364
COMPSTON, W., 474
COMPTON, R. R., 475
Compton scattering, 63, 323, 481
Condensers, 370F
  conoscopic light, 91
Concentration calculation, 287
  mode, 339, 346
Conduction energy band, 465
Confidence coefficient, 433
  coefficient choice, 433
  interval, 430, 430F, 432
  interval, two-sided, 430
  interval, one-sided, 430, 433
  limits, 430, 433
Conoscopic light, 81
  method, 73
Contact metamorphism, 378
Conversion charts and tables, 144, 284, 490
Cook-Chas. W. Cook & Sons, Ltd., 114, 486
Cooling water, 442, 445
Coors porcelain tile, 226
COPELAND, D. A., 5, 475
Copper analysis settings, 351
  contamination, 293
  detection, 284
  mass absorption coefficient, 289
  metal, 272
  oxide, 272
  radiation, 140
  reflectivity and hardness, 41F
  standard solution, 349
  sulfate, 30
  tube, 362
  wavelength, 284
Copper X-Ray data, 276
  X-Ray spectrograms, 285F
Cordierite, 478
  ACF position, 383F, 384
  AFM position, 392F
  A'KF position, 389F

diffractograms, 222, 223F
distortion index, 222, 223, 476, 484
  perdistortional, 222
  in schist, 473
  staining, 19, 20-23, 474
  structural state, 221-224
  subdistortional, 222
  width index, 223
Cordierite-indialite series, 479
CORLETT, M., 473
Coronadite reflectivity and hardness, 41F
Corundum decomposition, 334
  normative, 396, 415, 419, 421, 424
Count-rate, suitable, 300
  through absorber, 312
Count-rate corrections, 303
Count-rate high, 310
Count-rate matrix corrections, 314
Count-rate reduction, 310
Counting error, high count rate, 310
Counting precision, 308
  procedure, 306, 313
  sequence, 306
  spurious, 306
  on standard, 306
  tabulation, 306
  time, 304
  time selection, 306
  value, erroneous, 311F
COURVILIE, S., 365, 366, 475
Covellite D.T.A. characteristics, 454
  D.T.A. peak temperature, 449
  magnetic susceptibility, 119F
  reflectivity and hardness, 41F
Cover glass, 9, 80
CRANSTON, R. E., 337, 349, 351, 474
CRAWLEY, W. D., 16, 475
Crescent Manufacturing Company, 318, 486
Cristobalite D.T.A. characteristics, 452
  D.T.A. peak temperature, 449
Crucible, gold-platinum alloy, 291
  graphite, 290, 293, 294, 295
  nickel, 338, 366
  palau, 290
  platinum, 183, 290, 333, 334, 339, 357, 359,
    368, 369, 372, 440
  washing, 360, 371
Crushing, 65
  minerals, 126
  rocks, 114-115, 126-128
Crystal adhesive, 80
  attachment, 80
  grinder, 250
  holder, 272
  holder sledge, 271
  holder adjustment, 271-273
  sphere, 250
Crystalline imperfections, 465

transitions, 459
Crystallinity index, kaolinite, 229
Crystallization index, 396-397, 398F, 399, 480,
    482
  index calculation, 397
  index versus solidification index, 397
Cubanite reflectivity and hardness, 41F
Cubic indexing, 173, 174-178, 474
  indexing quotients, 175-177
Cummingtonite, 412F, 483
  ACF position, 383F, 384
  A'KF position, 389F
  diffraction, 231
  -grunerite series, 478
  optical properties, 85, 86F
Cumulative frequency, 426
Cuprite reflectivity and hardness, 41F
Cuproscheelite fluorescence, 31
Cupsuptic pluton, Maine, 224
Cutrock Engineering Co., 1, 4, 7, 113, 486
Cutting rocks, 1-2
CUTTITTA, F., 479
Cyrtolite magnetic susceptibility, 119F

D-spacing quotients, 175-177
D spacing values, 170, 174
Dacite, 395
DANIELS, F., 484
Dark current, 467
Darkroom light, 138
Data display, 374-437
  sheet for X-ray spectrometry, 307F
Dating by thermoluminescence, 477
Davidite reflectivity and hardness, 41F
DAVIES, T. T., 232, 475
DAWSON, J. B., 452, 475
DAWSON, K. R., 16, 475
Dead time calculation, 312
  time correction, 312, 319, 320
  time determination, 310-312, 475
Debye-Scherrer powder camera, 133, 135,
    136F, 137F
Decomposition vessels, 336F
DEER, W. A., x, 32, 214, 233, 247, 253, 475
Degrees of freedom, 431, 432, 434
Dehydroxylation, 459, 464
  of amphibole, 453
Density, 235-246
  balance, 237F, 239F
  determination apparatus, 245F
  markers, 243
  separation, 120-126
  separation by centrifugation, 126
  separator, 123, 124F, 480
  standards, 489
  see also Specific gravity
DESBOROUGH, G. A., 225, 475
Desilification in norm calculation, 425

Desliming of grains, 116
Dessicator, 227, 294, 366, 368
  for pre-treatment, 443
Detector dead time, 310
  plateau, 297
  plateau determination, 161-162, 162F
  plateau width, 298
Detector voltage for operation, 161-163, 162F,
      283, 297, 301
  voltage adjustment, 166, 301
  voltage selection, 162, 296-299
  voltage selection for sodium, 302
Developing Tank, 141
Development of X-ray film, 141
DE VRIES, J. L., 264, 277, 279, 286, 477
DIAMOND, S., 147, 478
Diamond abrasive, 10
  blades, 2
  fluorescence, 32
  paste, 11, 12
  wear points, 6
Diamonds, 7
Diaspore D.T.A. characteristics, 456
  D.T.A. peak temperature, 449
Dicalcium silicate, normative, 415, 417
Dickite distinctive properties, 229
  heated, 228
DICKSON, J. A. D., 25, 475
Differential dispersion, 260
  thermal analyser, 440F, 489
  thermal analysis, 438-459, 476, 478, 479, 481
  thermal analysis of clays, 229
  thermal analysis conditions, 447
  thermal analysis operating procedure, 445-447
  thermal analysis of siderite and kaolinite, 473
  thermal analysis thermograms of mineral
      groups, 452
  thermal curve, 438, 448
  thermal peak measurement, 458F
  thermal peak temperature, 449
  thermal sample holders, 439, 441F
  thermocouple, 438, 440
  thermocouple amplification, 445
Differentiation, basaltic magma, 399
Differentiation index, 398-399, 482
  index plotting, 399
  trends, 397, 399
    of pyroxenes, 405
Diffraction angle, 142, 143, 272, 284
  angle, error, 143, 167
Diffraction camera samples, 480
  methods, 132-179
Diffractogram, intense lines, 170
  measurement, 169
  procedure, 158-161
  scan, 182
  of silicon, 160F
  start procedure, 159

Diffractometer alignment tools, 148F
  Philips, 151F
  specimen holder, 145F
  specimen preparation, 144-147
  techniques, 144-166
Digenite reflectivity and hardness, 41F
Digestion apparatus, 356
Diluents for heavy liquids, 120, 121, 479
Dilution, sample, 287, 448
  methods, special, 330
  standard solutions, 349
Dimethyl sulfoxide, 121
Diopside, 378, 406
  ACF position, 383F, 384
  allocation, 403
  carbonates, 376
  correction, 387
  diffraction data, 233
  end member, 401
  facies diagram position, 377F
  -hedenbergite series, 481
  Hedenbergite, Enstatite, Ferrosilite diagram,
      405
  mass absorption coefficient, 286
  mode, 385
  normative, 394, 415, 417, 419, 421, 422
  normative, magnesian, 396
  normative, provisional, 417
  specific gravity, 385
Diorite differentiation index, 399
Dioxan, 146
Disc, borate glass, 293
  cleaning with acetone, 293
  crystallization, 293
  graphite, 291
  preparation, 265-269
  preparation procedure, 265, 269
  storage, 293
Discs, standard specimens, 273
Disequilibrium, 385
Dishes, evaporation, 269
Dislocations, 466
Disorder Al/Si, 188, 222
Dispersing crystals, 271, 273, 282, 304, 477
  crystal choice, 281, 282, 283
  crystals d-spacings, 281, 282
  crystals wavelength tables, 284
Dispersing efficiency, 281
Dispersion curve, 259
  differential, 260F
Displacement rod, 366, 367
Distillation flask, 370F, 371
Divariant equilibrium, 385
Divergence slit, 156, 158
  slit assembly, 150, 152F
  slit choice, 156
Dodecylamine acetate, 130
DODGE, T. A., 103, 475

Dolomite, 217, 378
  ACF position, 383F, 384
Dolomite (104), 215
Dolomite:calcite ratio, 482
Dolomite calibration curve, 216F
  in carbonate rocks, **214-216**, 376
  correction, 387
  determination, 216
  diffractogram, 214
  D.T.A. characteristics, 454
  D.T.A. peak temperature, 450, 451
  facies diagram position, 377F
  mode, 385
  peak intensity, 215
  specific gravity, 385
Dolomite staining, **25-26**, 27F, 28, 30
  thermoluminescence, 472
DOMANSKA, E., 234, 475
Double, or sum peak, 310, 311F
Doublet resolution, 144
  X-Ray lines, 141
Dowel, wooden, 4, 9
Drainage by heat, 471F
  by ultraviolet, 471F
Drift, electronic, 301
Dunite, crystallization index, 398F
Du Pont, 269
Du Pont TFE fluorocarbon resin, 336
Durofix covering, 25
Dynamic gas, 445, 448
  gas flow, 440
Dynamothermal metamorphism, 378
Dynode resistors, 342
Dyscrasite reflectivity and hardness, 41F

Eberbach Co., 236, 486
EBERHARD, E., 473
Eclogite, pyroxene, 406
Edenite, 408F, 409, 411
  end member, 406
  normative, 423, 425
EDGAR, A. D., 232, 476
Electrical components, 488
Electrode, bimetallic, 356
Electromagnetic separation, 65
Electron microprobe, 361, 476, 478, 479, 482
Electrum reflectivity and hardness, 41F
Elektron Ltd., 486
Element concentration, 287
  concentration calculation, 321
EMBREY, P. G., 244, 475
EMERSON, D. O., 55, 56, 475
Emission mode, 339
EMMONS, R. C., 73, 99, 475
Emplectite reflectivity and hardness, 41F
Enargite magnetic susceptibility, 119F
  reflectivity and hardness, 41F
Encapsulator tool, 443

End point detection, 358
Endecotts Test Sieves Ltd., 115, 486
Endothermic peaks, 438, 451
Energy distribution curve, 164, 165F, 300, 301
Engis Ltd., 13, 486
Englehard Industries Inc., 290, 486
Enhancement of X-Ray fluorescence, 329
Enraf-Nonius, 143, 179, 250, 486
Enstatite, 401, 405
  diffractogram, 224
  normative, 419, 421
Epidote, 384
  ACF position, 383F, 384
  corrections, 386
  D.T.A. characteristics, 453
  D.T.A. peak temperature, 449, 451
  magnetic susceptibility, 119F
Epidote mode, 385
  specific gravity, 385
Epon 828 epoxy, 2, 3, 4, 14
Epoxide resin, 13
Epoxy, 2, 3, 4, 13
  curing, 2, 3, 5
  impregnation, 481
Equations, simultaneous, 437
ERGUN, S., 275, 277, 279, 284, 473
ERICKSON, K. P., 19, 480
Erichrome cyanine R, 369
Eringhaus compensator, 251
Error of estimate, 437
ESPOS, L. F., 475
Etching, hydrofluoric acid, 19
Ethylene glycol, 227
Eucryptite fluorescence, 32
EUGSTER, H. P., 232, 484
Euxenite magnetic susceptibility, 119F
  reflectivity and hardness, 41F
EVANS, E. H., 482
Excitation potential, 272, 275, 276, 278F, 300
  potential, K-Spectrum, 277
  potential, L-Spectrum, 279
Exit port, 136F, 137, 137F, 139
Exothermic peak, 438, 451
Exposure determination, 70
  meter, 69, 70
  meter adjustment, 70
  scale, 71
  time, powder camera, **139**
Extinction angles, 105, 107
  positions, 249
  position determination, 251
Eyepiece choice, **66-67**
  compensating flat-field, 66, 67
  field-of-view index, 66
  micrometer, 44, 45F
  micrometer calibration, 74-76

FABBI, B.P., 265, 267, 293, 294, 317, 331, 332, 475

Famatinite reflectivity and hardness, 41F
FANG, J. H., 144, 475
Farbwerke Hoechst, 14, 486
Fayalite diffractogram, 219
 normative, 419
Feigl's solution, 27F, 28
Feldspars, 385; see also Alkali feldspars; Plagioclase
Feldspar ($\overline{2}01$), 183
Feldspar, ACF position, 379
 alkali content, 380
 anomalous, 197
 diffraction data, **180-207**
 D.T.A. characteristics, 452
 flotation, 130, 131
 magnetic susceptibility, 119F
 separation, 118, 123
 staining, **16-20**, 473, 478
 structural state, 480
 thermoluminescence, 472
 unmixing, 181
Feldspathoids D.T.A. characteristics, 453
 staining, **24**, 481
Femic group, normative, 415, 419
Ferric iron determination, 354
 oxide, amphiboles, 407
Ferri-hornblende, 409
Ferro-augite optical parameters, 84F
Ferrofilter, 117
Ferro-hastingsites, 474
 diffraction, 231
Ferromagnetic separator, 116
Ferrosilite, 401, 405
 normative, 419, 421
Ferrous ammonium sulfate, 358, 359, 360
 ammonium sulfate standardization, 359
Ferrous:ferric conversion, 287
 :ferric ratios, 287, 361
Ferrous iron determination, **354-361**, 479
Ferrous iron, rocks, 354
Ferrous oxide determination, **359-361**
Ferrous solution standardization, 359
Fibres of powder specimens, 134
Field geology, 475, 478
Field-of-view number, 67
Field-of-view radius, 46
Filer's, 486
Film, 488
 labeling, 141
 loading, 138
 measurement, **141-144**, 142F
 measurement tabulation, 142
 punch and cutter, 138
 speed, 71
 X-Ray, 138
Filter candle, vacuum, 226
 light balancing, 71
 millipore, 126

millipore assembly, 125F
 paper pulp, 317, 323
 paper pulp discs, 319
 paper, Whatman, 372
Fine-grained, definition, 46
FINGER, L. W., 312, 475
Fink Index, 169
First order red plate, 105
First order X-Ray lines, 285
FISHER, G. W., 221, 235, 475
Fisher Scientific Co., 38, 486
Flame emission mode, 346, 347
 emission spectrophotometer, 347
 extinguishing, 343, 344, 345
 lighting instructions, 342, **343, 344,** 345
 type, 342, 350, 351
FLANAGAN, F. J., 287, 314, 329, 352,
 475, 476
Flash figure, 82, 103, 248, 249
FLEISCHER, M., 287, 322, 326, 476
FLINTER, B. H., 118, 119, 476
Flotation, **129-131**
 cell, 129, 129F, 480
Flotation, feldspar, 476
Flow counter, 271, 272, 273
 counter adjustment, 272
 counter operating voltage, 298
 counter pulse distribution, 301
 counter window life, 297
Fluorescence, 30
Fluorescent colors, 31
 screen disc, 135, 157
 yield, 274, 280, 286, 279F
 yield correction, 280
Fluoride distillation apparatus, 369, 370F
 standard solution, 371, 372
Fluorine detection, 274, 282, 283
 determination, **369-373**, 476, 477
 determination accuracy, 372
 mass absorption coefficient, 288
 normative, 416
 standards, 372
 topaz, 111, 234
 wavelength X-Ray data, 276
 X-Ray settings, 304
Fluorite diffraction lines, 178
Fluorite diffractogram, 170, 171
 fluorescence, 31
 magnetic susceptibility, 119F
 normative, 415, 416, 419, 420
 powder data card, 172F
 thermoluminescence, 472
Folded rocks, 483
Foliation surface, 47F
FONT-ALTABA, M., 40, 474
FORD, A. B., 20, 476
Formula weights, 415
Forsterite, 378

carbonates, 376
diffractogram, 219
facies diagram position, 377F
normative, 396, 419
FORTRAN computer programs, 422, 475, 477, 481
Forward slope, 117
Fractionation, 399
trends, 395, 400F
Franklin, New Jersey, 32
Frantz, S. G. Co. Ltd., 116, 117, 486
ferromagnetic separator, 116F
isodynamic magnetic separator, 117, 117F, 119
FRANZINI, M., 232, 476
FREEMAN, E. B., 474
Freibergite reflectivity and hardness, 41F
Frenkel defects, 466
FRENZEL, G., 474
Frequency, 426
distribution, **426-427**, 427F
FRIEDMAN, G. M., 26, 476
Fritsch Co., 114, 486
Friable rocks, 2, 3
FROESE, E., 483
FROST, M. T., 231, 474, 476
Funnel, long stemmed, 366
Furnace, 363F
choice in D.T.A., 442
cooling, 442
copper, low temperature, 362
heating, 463
high-temperature, 362
installation, 463
platform, D.T.A., 442
removal, 460, 464
Fusion disc tools, 292F
technique, 290

Gabbro differentiation index, 399
Gahnite magnetic susceptibility, 119F
Galena D.T.A. characteristics, 455
D.T.A. peak temperature, 451
magnetic susceptibility, 119F
reflectivity and hardness, 41F
Gallenkamp, 115, 128, 247, 486
Gallium X-Ray data, 276
Gamma radiation, 470
radiation dosage, 471F
Garnet (420), 207
Garnet (10 4 0), 207
Garnet, Adirondack, 478
AFM plotting, 393
chemical analyses, 214
composition, **207-214, 411-414**, 413F
decomposition, 334
determinative diagrams, 209, 210F, 211F, 212F
diffractogram, 207

end members, 208, 481
identification, 209-214
magnetic susceptibility, 119F
metamorphic grade, 413F
metamorphism, 480
mode, 385
properties, 483
refractive index, 209, 210F, 211F, 212F
specific gravity, 210F, 211F, 212F, 212, 213, 235, 385
skarn, 233
unit cell edge, 209, 210F, 211F, 212F, 413F, 414
unit cell edge determination, 207-208
unit cell edge versus R.I., 214
zone, 411, 413F
Gas density compensator, 274
control, 339
cylinder labeling, 342
density stabilizer, 272
flow, 272
flow in D.T.A., 439
flow proportional counter, 274, 297, 300
flow proportional counter dead time, 312
Gas flow proportional counter plateau, 298F
Gas purification, 362
Gaussian curve, 164, 427
Gedrite, ACF position, 383F, 384
A'KF position, 389F
Gel preparation, 478
Gemological Instruments Ltd., 121, 243, 486
Geochemical Standards, 475, 476
Geological sample, 428
specimens, 488
Geological Society of America, 224, 225, 476
Geoscience Instrument Corp., 11, 486
Gersdorffite D.T.A. characteristics, 455
D.T.A. peak temperature, 449, 450
reflectivity and hardness, 41F
Germanium data, 282
dispersing crystal, 332
X-ray data, 276
GHOSE, S., 231, 483
GIBBS, R. J., 147, 476
Gibbsite D.T.A., characteristics, 456
D.T.A. peak temperature, 449
GILBERT, C. M., x, 483
GILLERY, F. H., 232, 474
Glass beads, grinding, 295
borate, 291
disc, 290
disc cleaning, 295
disc, standard reference, 296
slide, size, 3
slide, smear, 144
specific gravity, 121
Vycor, 226
GLASSER, F. P., 452, 453, 476

Glaucodot reflectivity and hardness, 41F
Glauconite distinctive properties, 229
  heated, 228
Glaucophane, 407, 409
GLEASON, S., 31, 32, 476
Glen Creston, 114, 128, 486
Glide plane, 179
Gloves, asbestos, 291
Glow curve, 466
Glow curve apparatus, 467F
  curve, artificial, 470
    calcite, 468F
    interrupted, 469F, 470
    thermoluminescence, 467
  curve resolution, 469
Glue supplier, 489
Glycolation, 147, 222, 231, 474
Goethite D.T.A. characteristics, 455
  D.T.A. peak temperature, 449
  reflectivity and hardness, 41F
Gold decomposition, 334
  reflectivity and hardness, 41F
  X-Ray data, 277, 279
GOLDSMITH, J. R., 188, 217, 476
Goniometer, 144
  alignment and adjustment, 149, 153, **155-156**,
    269, 272, 273, 297, 480
  alignment tools, 148F, 150F
  angle, 272
  angular adjustments, 296
  angular calibration, **154-155**
  automatic stops, 160
  backlash, 168
  diffraction angle, 273
  scan, 160
  slit, divergence, **156**
  specimen mounting, 145F
  speed, 158
  vertical, 151F
  zero setting adjustment, 152, 154-155, 155F
Grade of ore, 434
GRAF, D. L., 476
GRAHAM, A. R., 233, 476
Grain size, 114
  size, powder preparation, 132
  size definition, 46
  size determination, **44-46**
Grain thin sections, **14**, 473
Granite, A'KF position, 390F
  differentiation index, 398
  mass absorption coefficient, 326
  matrix correction, 314, 316
  origin, 482
  weathering index, 414
Granitic rocks thermoluminescence, 472
Granodiorite, A'KF position, 390F
  differentiation index, 398
Granulite basic, pyroxene, 406

Graphical display, **374-425**
Graphite carrier, 295
  crucibles, 489
  disc, 291, 292F
  contamination, 290
  decomposition, 334
  D.T.A. characteristics, 457
  D.T.A. peak temperature, 450
  polycrystalline data, 282
  reflectivity and hardness, 41F
GREEN, D. H., 393, 394, 476
Greenschist facies, 378
Gregory, Bottley & Co., 486
Griffin and George Ltd., 13, 114, 115, 247,
    262, 486
GRIMSHAW, R. W., 452, 476
Grinding, **5-8**
  automatic, 7
  wheel, diamond, 7
  mill cleaning, 128
  mills, 127
Grossularite, 209
  ACF position, 383F, 384
  AFM position, 393
  in carbonates, 376
  correction, 386
  Facies diagrams position, 377F
  physical properties, 208
Groundmass estimation, 51
Grunerite diffraction, 231
Guinier-De Wolff camera, 179
Guinier-Lenne camera, 179
Guinier viewer, 143
GULBRANDSEN, R. A., 215, 216, 476
GUNATILAKA, H. A., 218, 476
Gypsum staining, 27F

HACH-ALI, P. F., 457, 483
Hackmanite fluorescence, 32
HAGNI, R. D., 3, 5, 476
HAINES, M., 26, 476
Halewood Chemicals Ltd., 348, 486
Halite normative, 415, 416, 419, 420
  thermoluminescence, 472
HALL, A., 373, 476
Halloysite, 227
  distinctive properties, 229
  D.T.A. characteristics, 457
  D.T.A. peak temperature, 449, 450, 451
  heated, 228
HAMILTON, D. L., 232, 349, 476
Hammer, 113
Hand extraction, 113-114
  press, 291
  specimens, cutting, **1-2**
    modal analysis, 49
    point counting, 51
  specimen size, 127

Hangdown tubes, removal, 460
  wire assembly, 463
Hardness calculation, 39
  determination, **37-40**
  load, 38
  microindentation, **35-40**
  on thin-sections, 36
  tester, 38F
HARKER, A., x, 476
Hartmann scale, 258F, 259
HARTSHORNE, N. H., 1, 73, 91, 93, 98, 476
Hartshorne's mounting apparatus, 80
  rotation apparatus, 79
  spindle stage, 79
HARVEY, R. D., 483
HARWOOD, D. S., 224, 476
Hastingsite, 411
HAUKKA, M. T., 296, 478
Hausmannite reflectivity and hardness, 41F
Hawaiian lavas, 479
Hawaiite, 395
Heating, controlled, 227
  controller, 466
  rate, 466, 467
  rate adjustment, 446
  rate choice, 444, 460
  stages, 257
Heavy absorber addition, 287
  element detection, 274, 275, 283
  liquids, 120, 244, 489
  liquid diluents, 244
  liquid recovery, **121-123**, 126, 473
  liquid recovery apparatus, 122F
  liquid separation, 65, 481, 483
  liquid specific gravity determination, 236, 243-
    246
  liquid storage, 121
  liquid traps, 122F
Hedenbergite, 406
  allocation, 403
  diffraction data, 233
  end member, 401
Heine condenser, 256
HEINRICH, E. W., x, 32, 247, 476
HEINRICH, K. F. J., xi, 286, 289, 476, 479
Hematite, 361
  accessory, 379
  D.T.A. characteristics, 455
  D.T.A. peak temperature, 450
  magnetic susceptibility, 119F
  mode, 385
  normative, 415, 417, 419, 421
  reflectivity and hardness, 41, 41F
  specific gravity, 379, 385
HENDERSON, C. M. B., 349, 476
Heraeus GmbH, 323, 486
Herasil silica glass, 323
HERBER, L. J., 130, 476

HERMES, O. D., 331, 477
Hessite reflectivity and hardness, 41F
Hexachlorobuta-1,3-diene, 236
  density, 238
Hexagonal indexing, 173
HEY, M. H., 232, 477
Heyden & Sons Ltd., 295, 486
HICKLING, N., 479
High alumina basalt series, 394
High-temperature Guinier-Lenne camera, 179
High-temperature peak, 469
HIGHAM, F., x
HIMMELBERG, G. R., 224, 477
Histograms, **426-427**, 427F
Hole, 465
Hollandite reflectivity and hardness, 41F
HOLMES, A., x, 1, 477
Homogenization, 183
HOOPER, P. R., 232, 295, 475, 477
Hooker Chemicals Co., 236, 486
Hopkin and Williams Ltd., 372, 486
HORMANN, P. K., 233
HORN, E. E., 234
Hornblende, 408F, 412F
  ACF position, 382, 383F, 384
  alkali content, 380
  correction, 386, 387
  D.T.A. characteristics, 453
  D.T.A. peak temperature, 450
  hornfels facies, 378
  magnetic susceptibility, 119F
  modal analysis, 379, 385
  normative, 424, 425
  separation, 118
  specific gravity, 379
Hornblendites, 411
HORMANN, P. K., 477
HORN, E. E., 477
HOSKING, K. F. G., 30, 477
Hot plate, 4, 9, 14, 292
HOWARTH, R. J., 425, 477
Hower, J., 325, 326, 327, 328, 329, 477
HOWIE, R. A., x, 214, 475
HUANG, P. M., 370, 372, 477
Humidity control, 227, 443
HUTCHISON, C. S., 41, 233, 330, 422, 423,
    425, 436, 437, 467, 468, 469, 470, 471,
    472, 477
HUTTON, J. T., 291, 293, 308, 313, 317, 480
Hydraulic press, 267, 267F, 268F, 269
Hydrochloric acid, 334, 335, 338, 339, 352,
    369, 372
  acid etching, 26
Hydrofluoric acid, 18, 23, 130, 131, 333, 337,
    349, 356, 358, 359
  acid deterioration, 21
  acid etching, 17, 21
  -sulfuric acid dissolution, 333

Hydrogen, 341, 342
  mass absorption coefficient, 288
  peroxide, 29, 227
Hydrogrossular, 209, 214
  physical properties, 200
Hydrophobic surface properties, 129
Hydroxides D.T.A. characteristics, 455
Hydrozincite fluorescence, 31
Hygroscopic sample, 368
Hypersthene, ACF position, 383F, 384
  A'KF position, 389F
  basalt, 393
  diffractogram, 224
  magnetic susceptibility, 119F
  normative, 393, 396, 415, 417, 419, 421
  normative, provisional, 417
Hypress diamond compound, 12, 13
Hyprocell paper lap, 12, 13
Hysol epoxi-patch, 14
HYTÖNEN, K., 481

ICI (Imperial Chemical Industries), 131, 487
Identification by powder data file, 170
Idocrase (see vesuvianite) diffraction data, 234
Idrialite fluorescence, 32
Igneous rocks, facies diagram positions, 390,
      390F
  rocks, intermediate, norms, 424
Ignition loss, 361
  loss determination, 317
Ilford Ltd., 487
Ilford Hypam rapid fixer, 141
  Industrial G X-Ray film, 138, 139, 141
  Phen-X developer, 141
  Phen-X replenisher, 141
Illite, 227
  ACF position, 390F
  distinctive properties, 229
  D.T.A. characteristics, 456
  D.T.A. peak temperature, 449, 450, 451
  heated, 228
Ilmenite, 361
  accessory, 379
  diffraction data, 232
  -geikielite series, 474
  magnetic susceptibility, 119F
  mode, 385
  normative, 415, 416, 419, 420
  reflectivity and hardness, 41F, 42
  separation, 118
  specific gravity, 379, 385
Image-splitting measuring eyepiece, 251
Immersion liquids, 80, 82, 247, 256
Immersion liquids, refractive index, 261-263
Immersion preparation, 247
Impregnation, 2-3
Impurity ions, 466
Incident light, 32

Indentation measurement, 39, 39F
  Vickers diamond, 37
Indexing powder diffractograms, 173-178
Indialite, 221
Indialite diffractograms, 222, 223F
Indicatrix orientation, 101
  symmetry plane, 99, 103
Indium X-Ray data, 277
Infrared lamp, 269
Infrared spectroscopy, 368
INGAMELLS, C. O., 293, 349, 477, 478
Ingram Laboratories, 7, 487
Initial rate stabilizer, 446
Inorganic analysis, 483
  index, 169, 171
Insertion device, 364
  device, water determination, 362
Intensity ratios, 320
  ratio, feldspar, 181-182, 187F
Interferences, 335
Interference equipment, 256
  figure, centered, 82
    oblique, 74
    uniaxial, 93F
  figure BXa, 74, 75
  figure measurement, 483
  figures, 81F
    biaxial, 98F, 101F
  filter, 33, 259, 263
    band-pass, 255
    continuous, 255
Interplanar spacings, 143
Interval estimation, 430
  midpoint, 426
  size choice, 426
Ionization radiation, 465
Ions in formula, pyroxene, 401
Iron, alkalis, magnesia diagram, 399, 400F
  ferrous, 287
    alloying, 293
    ferrous:ferric determination, 354-361
    ferrous:ferric ratios, 473
    fragments, removal, 117
    gravimetric factors, 354
    iron + magnesia:silica diagram, 401
    mass absorption coefficient, 289
    matrix correction coefficient, 315
    oxide calculation, 361
    oxidation, 354, 357, 361
    oxidation in amphibole, 453
    oxide in basalt, 394
    oxide dilution, 330
    oxide mass absorption coefficient, 327
    ratio, 399
    solidification, 445
    spectrometer settings, 297
    standard solution, 348, 357
    sulfate specimen disc, 297

sulfate heptahydrate, 356
total, 287
total determination, 354
X-Ray data, 276
X-Ray detection settings, 298F, 299F, 305
Iron analysis, 290, 293, 294, 295, 335
analysis settings, 350
coating, removal, 227
-constantin thermocouple, 446
detection, 274
enrichment, 398, 399
ferric, 287
Irradiated samples, 470
Irradiators, 488
Island arc, 395
Isogyre alignment, 98
centering, 81, 81F, 98F
'Isolite' standard capsule, 468, 468F
units, 469
Isostructural compounds, 168
Isothermal operation, 464
Isotropic minerals, 248
minerals, refractive index, 248
IZETT, G. A., 83, 85, 483

JACKSON, E. D., 224, 477
JACKSON, M. L., 353, 370, 372, 477, 478
Jacobsite reflectivity and hardness, 41F
Jadeite, 405, 406
allocation, 403
end member, 401
-Tschermaks molecule ratio, 406
JAFFE, H. W., 411, 412, 481
JAHANBAGLOO, I. C., 220, 221, 477
JAMBOR, J. L., 221, 477
Jamin Lebedeff interference equipment, 256, 263
Jamesonite reflectivity and hardness, 41F
Jaw crusher, 114
crusher, cleaning, 114
JEACOCKE, J. E., 422, 423, 425, 477
JENKINS, R., 264, 277, 279, 282, 286, 477
JENSEN, J. R., 6, 7, 474
JOENSUU, O. I., 476
JOHANNSEN, A., x, 1, 46, 417, 477
JOHNSON, G. G. jr., 284, 286, 483
JOHNSON, W., 231, 477
Johnson-Matthey and Co., 348, 487
Joint Committee on Powder Diffraction Stand-
ards (J.C.P.D.S.), 168, 172, 184, 199,
207, 215, 219, 224, 225, 227, 477, 487
JONES, F. T., 78, 79, 80, 477
JONES, J. B., 185, 186, 478
JORGENSEN, P. J., 40, 483

K and K Laboratories of California Inc., 130,
290, 487
K-Spectrum lines, 275
K-Spectrum wavelength, 276, 277

KAEBLE, E. F., 474
Kaliophilite, normative, 415, 417, 419
Kalsilite, normative, 398
Kaolinite, 227, 229
ACF position, 390F
-chlorite distinction, 474
D.T.A. characteristics, 457
crystallinity index, 229, 230F
diffractogram, 230F
disordered, distinctive properties, 229
D.T.A. curves, 478
D.T.A. peak temperature, 449, 451
heated, 228
ordered, distinctive properties, 229
particle size, 457
T.G.A. curve, 465F
KAP dispersing crystal, 282, 283, 302
Kaso-to andesite, 223
KATSURA, T., 399, 479
KATZ, A., 338, 349, 478
KEIL, K., 64, 478, 480, 482
KELLAM, J. E., 330, 479
KELLY, W. C., 456, 478
KELSEY, C. H., 417, 478
Kerosene, 9
KERR, I. S., 174
KERR, P. F., ix, 32, 247, 452, 454, 455, 457,
478
KIELY, P. V., 353, 478
KINTER, E. B., 147, 478
KISS, E., 355, 356, 478
KISTLER, G., 68, 481
KLEIN, C. jr., 231, 478
Knife edge, 154, 155, 155F
Knoop hardness, 36, 37, 481
indentation, 39
Kobellite reflectivity and hardness, 41F
KOCH, G. S. jr., 425, 426, 478
Kodak Ltd., 487
Kodak brown safelight filter, 138, 141
developing tank, 141
Photo-flo solution, 141
rapid fixer, 141
Kodachrome, 71
Kofler micro cold/hot stage, 257
KOPP, O. C., 454, 455, 478
Krantz - Dr F., 487
Kraus-Jolly density balance, 235, 236, 237F
density balance procedure, 237
KROCH, T. E., 336, 338, 478
KRÜGER, J. E., 454, 483
Krypton X-ray data, 276
KUELLMER, F. J., 180, 181, 186, 187, 188,
478
KULLERUD, G., 481
KUNO, H., 394, 395, 396, 478
Kupfferite, 409
KUSHIRO, I., 405, 478

KWIC guide, 169, 171
Kyanite, 384
  ACF position, 384
  A'KF position, 389F
  -almandine muscovite subfacies, 389F
  decomposition, 334
  diffraction, 231
  magnetic susceptibility, 119F
  zone, 413F
    garnet composition, 414
Kyoto Scientific Specimens Co., 1, 487

Laboratory suppliers, 489
Labradorite D.T.A. characteristics, 452
LADURON, D. M., 23, 478
LAHEE, F. H., ix, 478
Lakeside cement, 4, 14, 146
Lamp fitting, 346
  hollow-cathode, 339
  lighting, 346
  running, 345
  turret, 339
  type, 350, 351
LANGER, A. M., 452, 457, 478
LANGER, K., 223, 478
Lanthanum detector conditions, 299F
  interference, 293
  oxide, 290
  X-Ray data, 277, 279
Lap, 8, 10
  cast iron, 3
LARSON, R. R., 224, 476
Lattice type deductions, 178-179
Laurylamine acetate, 131
LAVES, F., 188, 476
Layered igneous rocks, 483
Layer silicates, removal, 352, 353
Lead analysis settings, 351
  oxide, 237, 266
  standard solution, 349
  X-Ray data, 277, 279
Leadhillite fluorescence, 32
LEAKE, B. E., 407, 408, 478
Least squares method, 435
LE COUTEUR, P. C., 481
LEE, D. E., 125, 126, 481
Leeds and Northrup Co., 466, 487
Leitz, Ernst, 39, 44, 251, 255, 256, 257, 487
  eyepieces, 67
  -Jelley refractometer, 261, 261F
  -Jelley refractometer procedure, 261, 262
Leitz Miniload-Pol hardness tester, 27, 28F
  objectives, 67
  Orthomat camera, 71
  photomicrographic camera, 69
  universal stage, 90
  universal stage objectives, 92
LE MAITRE, R. W., 296, 478

Lepidocrocite D.T.A. characteristics, 455
  D.T.A. peak temperature, 449
  reflectivity and hardness, 41F
Lepidolite D.T.A. characteristics, 456
Leucite D.T.A. characteristics, 453
  D.T.A. peak temperature, 450
  normative, 398, 415, 417, 419, 422
Level of significance, 431
LEVIN, S. B., 213, 478
LEWIS, C. L., 481
Light element detection, 274, 283
Lime content in amphiboles, 407
Limestone age determination, 472
  composition, 215
  dissolution, 338
Limit of cutoff, 434
  switches, 446
  switch operation, 460
Line equation, 435
Line equation, regression, 437F
Line intensities, 143
Lineation, 47F
Linear regression, 435-437
LINK, R. F., 425, 426, 478
Litharge, 366
Lithium analysis settings, 351
  carbonate, 290
  fluoride, 334
  fluoride crystal, 271
  fluoride data, 281, 282
  fluoride standard, 224, 225
  mass absorption coefficient, 288
  metaborate, 293, 477
  silicate melting, 445
Lithium standard solution, 349
  tetraborate, 290, 295, 296
  tetraborate glass discs, 478
Liquid trap, 340, 341F
Loading the camera, 137-139
Loellingite, composition, 474
  diffraction data, 232
  D.T.A. characteristics, 455
  D.T.A. peak temperature, 450
  reflectivity and hardness, 41F
Logarithm, natural, 320
Lommel's rule, 82
LONG, J. V. P., xi, 13, 478, 482
Lordon Ceramics, 147, 226, 487
Lower level adjustment, 166
  level potentiometer, 163, 164
  level setting, 165F, 299, 301
  level setting for strontium, 311F
  limit switch, 447
LOWITZSCH, K., 148, 155, 480
Lucite 44 acrylic resin, 269
LUTH, W. C., 349, 478
LYONS, P. C., 17, 478

McANDREW, J., 260, 478
McCrone Research Associates, 79, 487
MacDONALD, G. A., 399, 479
McDOUGALL, D. J., 466, 479, 472, 477
Mace de Lepinay half shadow wedge, 251
MacGREGOR, I. D., 234, 479
McINTYRE, D. B., 483
MACK, M., 144, 480
MACKENZIE, R. C., 439, 443, 444, 445, 448,
 456, 474, 475, 476, 479, 481, 483
McKINLEY, T. D., xi, 476, 479
McLAUGHLIN, R. J. W., 333, 439, 452, 453,
 458, 465, 479
McLEOD, C. R., 40, 479
Macroprobe sensitivities, 331
 spectrometer attachment, 331
MADLEM, K. W., 483
Maghemite reflectivity and hardness, 41F
Magma depth derivation, 395
Magnesia, iron, alkalis diagram, 399
Magnesian rocks on ACF, 390
Magnesite, 217
 carbonate rocks, 376
 D.T.A. characteristics, 454
 D.T.A. peak temperature, 450
 facies diagram position, 377F
 staining, 27F, 28, 29
Magnesium analysis, 290, 293, 335
 analysis settings, 350
 content in carbonate, 217
 detection, 274, 282, 283, 302
 detector conditions, 298F
Magnesium Elektron Ltd., 348
Magnesium mass absorption coefficient, 286, 288
 matrix correction coefficient, 315
 nitrate, 443
 oxide, 364
 oxide mass absorption coefficient, 327
 perchlorate, 362, 364
 replacement by Ca, 216
 spectrometer settings, 302
Magnesium standard solution, 348
 X-Ray data, 276
 X-ray detector settings, 304
Magneson, 28, 29
Magnet, hand, 117
Magnetic isodynamic separation, **116-120**
 isodynamic separator, 489
 isodynamic separator, Frantz, 116F, 117F
 separator, amperage, 118
 separator, isodynamic, 117
 separator slope and tilt, 119F
 susceptibilities of minerals, 116, 118, 119F,
  120, 476, 481
 stirrer, 358
Magnetite, 361
 accessory, 379
 D.T.A. characteristics, 455

D.T.A. peak temperature, 449, 450
 magnetic susceptibility, 119F
 modal analysis, 385
 normative, 415, 417, 419, 421, 424
 reflectivity and hardness, 41F, 42
 separation, 117, 118
 specific gravity, 379, 385
Major element analysis, **286-321**
 element analysis procedure, **303-308**
Malacon magnetic susceptibility, 119F
Malayaite, 32
 fluorescence, 31
Malaysia orogenic zones, 477
Mallard's constant, 75, 83, 88, 90
Manganapatite fluorescence, 32
Manganese analysis settings, 350
 background, 308
 detection, 274, 303
 mass absorption coefficient, 289
 matrix correction coefficient, 315
 oxide mass absorption coefficient, 327
 oxide magnetic susceptibility, 119F
 standard solution, 348
 wavelength, 324
 X-Ray data, 276
Manganese X-Ray detector settings, 305
 X-Ray peak, background, 309F
Manganite D.T.A. characteristics, 456
 D.T.A. peak temperature, 449, 451
 reflectivity and hardness, 41F
Manganous carbonate dilution, 330
Marcasite D.T.A. characteristics, 454, 478
 D.T.A. peak temperature, 449
 reflectivity and hardness, 41F
MAREL, H. W. van der, 448, 479
Margarite, ACF position, 379, 382, 383F, 384
 correction, 386
 mode, 385
 mode for ACF, 379
MARINENKO, J., 242, 243, 479
Marker, diamond tipped, 4
Marmatite magnetic susceptibility, 119F
MARSHALL, M., 368, 473
MASON, B., 235, 473, 479
Mass absorption coefficient, 61, 63, 226, 286,
 288-289, 321, 323, 325F, 326, 328,
 329, 479, 481
 absorption coefficient calculation, 319
 absorption coefficient of diluent, 320, 330
 absorption coefficient of diopside, 286
 absorption coefficient example, 319
 absorption coefficient measurement, 62-63,
  317-320
 absorption coefficient measurement procedure, 319
 absorption coefficient of oxides, 327
 absorption coefficient ratio, 324
 absorption coefficient rock/Al_2O_3, 325F

absorption coefficient tools, 318F
absorption coefficient values, 286
absorption effect, 286
Mathematical tables, 473, 490
Matheson Scientific, 487
  Scientific MX850, 294
Matrix absorption effects, 296
  calibration against Al_2O_3, **324-329**
  correction coefficients, **315**
  corrections, 477
    on borate discs, 313
Matrix corrections, granite example, 314
  corrections, simplified, 317
  effects X-Ray fluorescence, 479
Mats, asbestos, 292
Maucherite reflectivity and hardness, 41F
MAXWELL, J. A., 358, 367, 368, 373, 479
MAY, I., 242, 243, 479
Mean, 428, 429, 435
Means, difference, 435
Mean of sample, 428
  of sample, example, 428
Measurement of film, **141-144**, 142F
  units, conversions, 490
MEDARIS, L. G. jr., 221, 235, 475
Medium grained, definition, 46
Meker burner, 291, 335, 352, 371, 372
M.E.L. Equipment Co., 284, 487
Melatope, 75, 83, 88
  adjustment, 83F
  separation, 74, 75F
Merck, E., 38, 290, 487
Mercury, X-Ray data, 277, 279
MERTIE, J. B. jr., 76, 77, 479
Mesonorm, 480
  for granitic rocks, **423-425**
  for metamorphic rocks, 423
Metaborate glass, 294
Metahalloysite, heated, 228
Metal contamination, 127
Metals, hardness and reflectivity, 41F
Metamorphic facies, 382, 390, 391, 411
  grade, 406, 411-414
  paragenesis, **376-393**
  paragenesis, carbonates, **376-378**
  petrology, 482, 483
  textures, 482
Metamorphism, 476
Methanol, 26
Methyl cellulose, 267, 294, 295
  cellulose backing, 302
  orange, 372
Methylene blue solution, 24
  iodide, 120
MEYROWITZ, R., 121, 361, 479
Miargyrite reflectivity and hardness, 41F
Mica, 227
  ACF position, 380, 382

analysis, 382
  correction, 381
  cleavage, stereographic plotting, 112F
  diffraction data, 232
  dissolution, 478
  D.T.A. characteristics, 456
  heated, 228
  mode, 385
  orientation, 111
  removal from feldspar, 352, 353
  staining, **23-24**
  white, modal analysis, 379
  white, specific gravity, 379, 385
Micro, definition, 46
Microabsorption effects, 62
Microcline, 180, 196F
Microcline ($\bar{2}$01), 182
Microcline composition, 185
  composition graph, 183F, 185F
  diffractogram, 181F
  intermediate, 190-191, 195
  -low albite, 480
  maximum, 190-191
  optic axial angle, 111F
  powder photographs, 190-191
  X-Ray intensity, 64F
Microhardness determination, 483
Microindentation hardness equipment, 489
Microlite magnetic susceptibility, 119F
Micrometer eyepiece, 44, 45F, 75, 251
  eyepiece calibration, **74-76**
  stage, 44, 45F, 46
Microphotometer, 33, 33F
Microscopes and accessories, 489
Microscopy, 479
Microtec Development Laboratory, 7, 8, 487
Micro-Trim, 7, 8F
Middle-temperature peak, 469
Millerite D.T.A. characteristics, 454
  D.T.A. peak temperature, 450
  reflectivity and hardness, 41F
Millipore filter, 126
  filter assembly, 125F
MILLMAN, A. P., 40, 484
MILNER, H. B., x
Mineral crushing, 126
  decomposition, 334
  dissolution, **333-339**
  group powder data, **180-234**
  identification, 170
  identification by D.T.A., 448
  oil, 3
  optical properties, 482
  powder, 293
  powder smear, 144, 145F
  separates grain size, 115
  separation, **113-131**, 480
  separation by density, **120-126**

separation by flotation, **129-131**
separation, magnetic, **116-120**
specimen powder, 291
standard, 303
Minerals, ACF positions, **382-385**
A'KF positions, 388-390
Mineralight, 30
Mineralogy, 473, 479
  determinative, 484
  optical properties, 478, 483
Mining Chemicals Handbook, 131
Mining geology, 433
Minor element analysis, 335
  heavy element analysis, **321-330**
MITCHELL, B. D., 439, 443, 444, 445, 448, 479
MITCHELL, B. J., 330, 479
Mixing mills, 128, 295, 489
Mixtures, diffractograms, 171
MIYASHIRO, A., 221, 479
Modal analysis, **47-65**, 480, 481
  analysis for ACF, 379
    electron probe, 478
    fine-grained rocks, 61, 63
    of granite, **64**, 474
    of hand specimens, 49, 56
    by mineral separation, 65
    by point counting, **51-61**
    of porphyritic rocks, **53**, **59-60**
  analysis of rock slabs, **53-56**
    staining, 49, 53, 55, 56
    of thin sections, 49, **56-61**
    unit cell, 47F, **48-49**, 52, 60F
    by X-Ray diffraction, **61-64**, 480
  analysis calculation, 58, 60
  analysis decimal rounding, 474
  analysis procedure, 58
Mode calculation, 63, 65
  visual estimation, 50F, 51
Mohs' scale, 40
MOLDAN, B., 333, 352, 481
Molding die, 3
Molecular proportions, 376, 385, 415
  proportions for ACF, 381
  proportions for A'KF, 387
  proportions calculation, 380
  proportions CaO, SiO_2, MgO, 378
  proportions of pyroxenes, 401
  ratios, 391
  ratio SiO_2:CaO:MgO, 376
  weights, equivalent, 420
MÖLLRING, F. K., 69, 479
Molybdenite hardness, 36, 41F
  magnetic susceptibility, 119F
  reflectivity, 41F
Molybdenum X-Ray data, 277
Monazite magnetic susceptibility, 119F
Monochromatic light, 255, 256
  X-Ray beam, 63

Monochromator, 339
  interference filter, 33F, 255
  wavelength, 346, 348
Monoclinic indexing, 173
Monocular viewer, 149
Monsanto Chemicals, 348
Montmorillonite, ACF position, 390F
  distinctive properties, 230
  D.T.A. characteristics, 456
  D.T.A. peak temperature, 449, 450, 451
  expansion, 227
  heated, 228
  T.G.A. curve, 465F
MOORE, A. C., 320, 321, 479
MOOREHOUSE, W. W., x, 46, 479
MORELAND, G. C., 2, 4, 10, 11, 479
MORELLI, G. L., 232, 479
MORIMOTO, N., 232, 479
MORRIS, M. C., 482
MORSE, S. A., 253, 260, 479
Mortar and pestle, agata, 126, 132
Mortar and pestle, porcelein, 269
Mortar and pestle, steel, 114
Mortar, agate, 265, 358
  percussion, 115, 247
  steel, 294
MORTEANI, G., 233, 477
Muffle furnace, 227, 290, 334
Mugearite, 395
MUIR, I. D., 73, 479
MÜLLER, G., x, 480
MULLER, L. D., 118, 121, 480
MÜLLER, R. O., 264, 480
Mullite diffraction, 231
Muscovite, 384, 484
  ACF position, 379, 381, 383F, 384, 390F
  AFM point, 391
  A'KF position, 388, 389F, 390F
  alkali content, 380
Muscovite basal spacing, 232
  correction, 381
  distinctive properties, 229
  D.T.A. characteristics, 456
  D.T.A. peak temperature, 449, 450, 451
  exclusion from ACF, 381
  heated, 228
  magnetic susceptibility, 119F
  mode, 385
  modal analysis for ACF, 379
  normative, 425
  :paragonite ratio, 380, 484
  projection point, 391, 392F
  -sillimanite zone, 392F
  staining, 23

Nacrite distinctive properties, 229
  heated, 228
Nakamura half-shadow plate, 251

NANDI, K., 413, 414, 480
National Bureau of Standards, 487
Naumannite reflectivity and hardness, 41F
Nebulizer, 340, 341F
NEDOMA, J., 234, 475
Negatoscope, 143
Nepheline diffraction, 231, 482
  diffraction data, 233
  D.T.A. characteristics, 453
  D.T.A. peak temperature, 451
  fluorescence, 32
  normative, 394, 398, 415, 417, 419, 422, 425
  pyroxene norm, 405
  staining, 24
  thermoluminescence, 472
NESBITT, R. W., 53, 478, 480
NEUVONEN, K. J., 481
Niccolite D.T.A. characteristics, 455
  D.T.A. peak temperature, 449, 450
  reflectivity and hardness, 41F
Nickel analysis settings, 351
  detection, 284
  detector conditions, 298F, 299F
  mass absorption coefficient, 289
  sample holders, 440
  standard solution, 349
  wavelength, 284, 324
  X-Ray data, 276
  X-Ray spectrogram, 285F
Niggli's mg value, 407
Niggli molecular norm, 419-423
Niobium X-Ray data, 277
Nitric acid, 335
Nitrogen, 341, 342, 362
  bubbles, 129
  dynamic gas, 448
  -entrained air-hydrogen flame, 344
Nitrogen flow, 364
  saturation, 356
  supply, 354
Nitrous oxide, 341, 342
  oxide-acetylene flame, 352
  oxide-acetylene gas, 343
  oxide-propane flame, 344
NN-dimethyl formamide, 121
NOBLE, D. C., 87, 88, 89, 90, 105, 106, 107, 108, 109, 480
NOLD, J. L., 19, 480
Norm, 396
  calculations, 414-425, 483
  C.I.P.W., 393, 414-418, 478
  C.I.P.W. rules, 415
  computer printout, 423F
  computer programs, 425
  example of calculation, 397, 418
  for granitic rocks, 423-425
  for metamorphic rocks, 423-425, 473
  Niggli molecular, 419-423

weight percentage conversion, 417
Normal curve, 300, 300F, 427, 427F
  distribution, 432
NORMAN, M. B., 481
Normative minerals, 415
  minerals, formulae, 415, 419
  minerals, molecular weights, 415
NORRISH, K., 63, 140, 226, 264, 266, 288, 291, 292, 293, 308, 313, 317, 318, 320, 322, 480
Nujol, 92, 368
Numerical aperture, 74
Nylon cloth, 11, 12

Objective choice, 66-67
  field of view, 66
  flatfield, 66
  magnification, 66
  strain free, 66
Oblique illumination, 256
  illumination method, 254F
Ocular micrometer scale, 83
Oil immersion, 35
Oligoclase diffraction data, 232
  D.T.A. characteristics, 452
  X-Ray intensity, 64F
Olivine (130), 219
Olivine analysis, 482
  basalt, 393
  composition, 219-221
  composition graph, 220F
  diabase differentiation index, 399
  diffractograms, 219
  gabbro differentiation index, 399
  high angle reflection, 221
  intensity ratios, 221
  nephelinite, 394
  normative, 393, 415, 417, 419, 421, 422
  in pyroxene norm, 405
  specific gravity formula, 235
  synthetic, 221
  tholeiite, 393
  X-Ray data, 473
  X-Ray determination, 475, 477
  X-Ray diffraction, 484
Opal D.T.A. characteristics, 452
  D.T.A. peak temperature, 449
  fluorescence, 32
Opaque mineral identification, 32-43
  mineral tables, 474
  mineral staining, 15, 30
Operating mode buttons, 339
Optic axial angle, 247, 249, 475, 479, 480, 482, 483
  axial angle, air, 74
  axial angle chart, 75F, 77F, 87F, 105F
Optic axial angle estimation, 249
  axial angle, mineral, 74, 75, 76, 108, 110

plagioclase, 108F
  refractive index relationship, 76, 77F, 103,
    249
  topaz, 111
  universal stage, 480
axial angle histogram, 110
axial angle measurement, **82-90**, 83F, 87F, 102,
    103, 104, 105F
axial plane, 100F
axial angle on stereogram, 89F
Optic axis biaxial figure, 101F
  axis, horizontal setting, 96
    stereographic plotting, 94, 95F, 97F, 99,
      100F
  axis centering, 94
  axis figure, 99, 248, 249
  sign, 76, 77F, 95
  sign determination, 99, 101
Optical crystallography, 483
  indicatrix, 80
  indicatrix stereograms, 100F
  mineralogy, 483
  parameters, definition, 73
  properties, minerals, 476
  relief, 253
Ore body thermoluminescence association, 472
  microscopy tables, 481
  minerals, 481
  mineral analysis, 330
  minerals D.T.A. characteristics, 454
  minerals microhardness, 484
  minerals microscopy, 481
  mineral tables, 482
Organic Index, 169
Organic matter removal, 227
Orientation avoidance, **145-147**
  effects, X-Ray diffraction, 62
Orpiment reflectivity and hardness, 41F
Orthoclase, 180, 186, 195, 196F
  diffractogram, 181F
  normative, 398, 415, 419, 420, 422
Orthoclase normative, provisional, 416
  optic axial angle, 111F
  powder diffraction indexing, 192-193
  X-Ray intensity, 64F
Orthopyroxene (131), 224
  composition, **224-225**
  composition determination, 321
  X-Ray analysis, 475, 477
Orthorhombic indexing, 173
Orthoscopic method, 73
ORVILLE, P. M., 183, 185, 186, 187, 480
ØSTERGAARD, T. V., 123, 124, 480
Outcrop photograph, modal analysis, 56
  point counting, 51
OWEN, L. B., 134, 480
Oxide concentration, 287, 306
  concentration calculation, 313, 314, 316

concentration determination, 316
  D.T.A. characteristics, 455
  formula weights, 415
  hardness and reflectivity, 41F
  molecular weights, 401, 409, 410
  percentage calculation, 306
  proportions in amphibole, 410
  :silica plots, 401
Oxygen atomic proportions, 401
  care in use, 342
  mass absorption coefficient, 286, 288
Oxyhornblende, 409, 411

Paleoclimatological studies, 472
Palladium:ruthenium sample holders, 440
  X-Ray data, 277
Palygorskite, 227, 483
  D.T.A. characteristics, 457
  D.T.A. peak temperature, 449, 450
  heated, 228
  properties, 231
Panatomic-X film, 71
PAPIKE, J. J., 481
Paragonite, 384
  absence from A'KF, 390
  ACF position, 397, 381, 383F, 384
  alkali content, 380
  analyses, 387
  basal spacing, 232
  correction, 381, 393
  mode, 385
  mode for ACF, 379
  :muscovite staining, 478
  staining, 23-24
Parameters, 427
Pararammelsbergite reflectivity and hardness,
    41F
Pargasite, 408F
  end member, 406
PARKER, A., 414, 480
PARKER, A. B., 396, 397, 398, 399, 480
PARKER, R. L., 76, 480
PARL, B., 425, 436, 437, 480
Parr Instrument Co., 336, 336F, 487
PARRISH, W., 144, 148, 149, 155, 156, 480
PARSLOW, G. R., 425, 480
PARSONS, I., 480
Particle size effect, 286, 290, 296
PARTRIDGE, A. C., 129, 480
Pascall Engineering Co. Ltd., 128, 487
Paterson developing tank, 141
Path difference, 252
Peacock, 395
Peak angle, 458F
  area, 458
Peak:background ratio, 274
Peak, high temperature, 468F, 469F
  high temperature height, 471F

intensity, 287
intensity $\bar{2}01$ feldspar, 182
interfering, 309
low temperature, 470
middle temperature, 468F
position, 161
resolution, 156
scan, 308
temperature, 448
  characteristic, 449
  in D.T.A., 451
Peak height, 308, 458
Pearceite reflectivity and hardness, 41F
Pectolite fluorescence, 32
Pelitic rocks in ACF, 390
  rocks, on AFM, 391
    on facies diagrams, 390F
Penfield apparatus, 365, 365F
  tube, 366
Penta erythritol crystal, 271
  erythritol crystal data, 281
Pentlandite reflectivity and hardness, 41F
Perchloric acid, 369, 371
Peridotite crystallization index, 398F
  differentiation index, 399
Periplan eyepiece, 67
Peristerite, 200F, 202F, 203
Perkin-Elmer Corpr., 487
Perovskite magnetic susceptibility, 119F
  normative, 415, 417, 419, 422
Perspex plunger, 265, 266F
Perthite exsolution, 180
  lamellae, 181
  structural state, 188
  X-Ray determination, 478
PET dispersing crystal, 282, 331, 332
PETERSEN, L. E., 481
Petrofabrics, 111-112
Petrographic preparation materials, 489
  methods, 477
Petrography, 477, 479, 483
Petroleum fluorescence, 31
Petrology, 473
PETRUK, W., 232, 480
PHA-scope, Philips, 301
Phase contrast equipment, 256
Phenocryst extraction, 113
  :groundmass ratios, 50F
  visual estimation, 51
Philips alignment tools, 148F
  dispersing crystals, 282
  Electronics Instruments, 301, 487
  Gloeilampenfabrieken, 144, 269, 271, 274, 284, 487
  vertical goniometer, 147, 150F, 151F
  X-Ray equipment, 132
  X-Ray spectrometer, 264, 270F
  X-Ray tubes, 140

Phlogopite diffraction data, 233
  D.T.A. characteristics, 456
  D.T.A. peak temperature, 449, 451
Phosgenite fluorescence, 32
Phosphorescence, 30
Phosphoric acid, 24, 357, 358, 359, 360, 364
Phosphorus analysis, **332**
  detection, 274, 282
  mass absorption coefficient, 288
  matrix correction coefficient, 315
  pentoxide mass absorption coefficient, 327
  spectrometer settings, 332
  X-Ray data, 276
  X-Ray determination, 475
  X-Ray detection settings, 305
Photographic enlarger, 71
Photography of irradiated specimens, 72
Photometer, microscope, 257
Photomicrographic camera, 69
Photomicrograph scale, **68**
Photomicrography, **65-72**, 481
Photomicrography exposure time, 69
  low power, 67, **71-72**, 473
  procedure, 70
  system, 69
Photomultiplier, 341F, 467F
  adjustment, 346
  assembly, 341
  calibration, 468
  sensitivities, 341F, 342
  thermoluminescence, 466
  tube, 339
  tube types, 342
  voltage, 469
Picrite differentiation index, 399
Pigeonite inverted, 403F
Pigeonitic rock series, 394
Pipette, plastic, 359
Pitchblende reflectivity and hardness, 41F
Plagioclase, 384
  (131) peak, 199
  ($\bar{2}01$) peak, 232, 476
  ($\bar{2}41$) peak, 199
Plagioclase in basalt, 394
  cleavages, 253
  composition, 109, 200F, 201F, 202, 205F, 206F, 483
  composition determination, 73, **104-110, 204-206**, 321, 479
  composition graph, 106F, 107F, 109F
  correction, 386, 393
  delta θ_1, 199, 200F, 202F, 203
  delta θ_2, 199, 201F, 202F, 203
  diffraction data, 232
  diffractograms, 197, 198F, 482
  disordered, 199, 204
  D.T.A. characteristics, 452
  D.T.A. peak temperature, 450

extinction angles, 107F, 109F, 110
glass, 253
high, 199
intermediate structural state, 199, 203
low, 199
  composition, 202F
Plagioclase mode, 64, 385
normative, 419
optic axial angles, 108F
optical orientation, 202
ordered, 199, 204
orthoclase content, 204
peaks, 206F
powder patterns, 474
plutonic graphs, 106F
refractive index, 203, 253, 260
specific gravity, 235, 385
staining, **16-20**, 22, 481
structural state, **197-204**
structural state determination, **105-108**
twinning, 110
universal stage manipulation, 480, 482
volcanic graphs, 106F
X-Ray diagrams, 473
Planachromats, 66, 67
Planapochromats, 66
Plastic aerosol spray, 146
wrap, 146
Plateau, detector operating, 297
Platinel II thermocouple, 439, 446
Platinum crucible, 292F
laboratory ware, 489
-platinum 10% rhodium thermocouple, 440,
  446
reflectivity and hardness, 41F
10% rhodium sample holders, 440
X-Ray data, 277, 279
Plunger, aluminum, 291
Point estimate, 430
counter, hand specimen, 54F, 55F, 56
  Swift, 57
  thin-section, 57, 57F
counters, 482, 489
Point counting, **51-53**
counting grid, 50F
counting groundmass, 59
counting phenocrysts, 59
Polarizer vibration direction, 248
Polars rotation, 83
Polaroid film, 72
POLDERVAART, A., 396, 397, 398, 399, 480
Polished sections, 11
thick sections, **13-14**
thin sections, **9-13**, 331, 479, 483
Polishing, 9, 10, 12
Polybutyl metacrylate polymer, 269
Polyester film, 265
Polyethylene container, 337

Polypropylene window, 296
Polypropylene bottle, 349
Polystyrene vial, 337
Polytex Supreme lap cloth, 11
Polythene beaker, 338
bottle, 338
Polyvinyl acetate, 14
Pooled parameters, 435
sum of squares, 435
Population, 427
mean, 433
Porosity elimination, 16
Porous discs for D.T.A., 444
rocks, 2
Porphyritic rock modal analysis, 60F
Porphyroblast extraction, 113
Potassium, ACF position, 379
acid phthalate data, 281
analysis, 335
analysis settings, 350
bromate, 368
bromate 101 peak, 183
diffraction lines, 184
diffractogram, 184
standard, 166, 183, 197, 207
cation exchange, 206
chloride, 205, 338
chloride melting, 445
detection, 274, 281, 282
Potassium dichromate, 357, 358, 359, 360, 361
dichromate melting, 445
dichromate solution, 356, 357
ferricyanide, 25, 29
hydrogen sulfate, 348
hydroxide, 29, 354, 356
mass absorption coefficient, 289
matrix correction coefficient, 315
metasilicate in norm, 415, 416, 419, 420
oxide mass absorption coefficient, 327
permanganate, 358
plagioclase, 204
plagioclase diffractogram, 205
sodium, calcium diagram, 399
standard solution, 348
X-Ray data, 276
X-Ray detector settings, 305
Powder Data File, 185
Powder disc pressing, 267F
Powder diffraction, **132-179**
diffraction camera adjustment, **134-136**
diffraction camera capillaries, 132
diffraction camera details, 136F, 137F
diffraction camera diameter, 133, 143
diffraction camera exposure time, **139**
diffraction camera fibres, 134
diffraction camera loading, **137-139**
diffraction camera methods, **132-144**
diffraction camera, special, 179

diffraction data card, 169, 171, 172F, 489
Powder Diffraction Data File, **168-172**, 477
  Diffraction Data File, example of use, 170
Powder diffraction film measurement, 141, 143F
  diffraction lines indexing, 173-178, 179
  diffraction patterns, 168
  diffraction pattern cubic indexing, 174-178
  diffraction standards, 487
  pattern indexing, 483
  patterns, silicates, 474
  specimen preparation, 133F
Power supply, thermoluminescence, 466
POWERS, M., 284, 480
Preferred orientation, avoidance, **145-147**
Pressed powder sample discs, **265-269**, 294, 295,
    296, 299, 322, 323, 324, 327, 331
  powder specimens, background, 308
  powder disc preparation, 265, 268, 268F
  powder discs, tools, 266F, 268F, 475
Pressing, 3
Pressure metal bell, 445
Primary beam centering, **157-158**
Primary glare, 35
Propan-2-ol, 372
Propane gas, 341, 342
Proportional counter detector, 144, 161
Proustite D.T.A. characteristics, 455
  D.T.A. peak temperature, 449, 451
  reflectivity and hardness, 41F
PRZIBRAM, K., 31, 480
Psilomelane reflectivity and hardness, 41F
Pulse distribution curve, 164
  height analyser, 312, 332
    sodium, 302
  height analyser procedure, 300
  height analyser settings, 163-166, 165F, 283,
    299-302, 303
  height analysis, 322
  height curve, 300F
  height distribution, 311F
Pumice, 362
Pycnometer, 242F, 244, 245, 245F, 479
  design, 242
  method, 236, 242
  for specific gravity determination, **241-243**
Pyralspite, 209
Pyrargyrite reflectivity and hardness, 41F
Pyrite, 358
  accessory, 379
  cobaltiferous, 481
  decomposition, 334
  diffraction data, 233
  D.T.A. curve, 439F
Pyrite D.T.A. characteristics, 454, 478
  D.T.A. peak, 438
  D.T.A. peak temperature, 449
  hardness, 41F, 42
  magnetic susceptibility, 119F

mode, 385
  normative, 415, 416, 419, 420
  oxidation reaction, 438
  reflectivity, 34, 41F, 42
  specific gravity, 379, 385
Pyrochlore magnetic susceptibility, 119F
Pyrogallol solution, 354
Pyrolusite D.T.A. characteristics, 456
  D.T.A. peak temperature, 450, 451
  reflectivity and hardness, 41F
Pyrope, 209, 212
  physical properties, 208
Pyrophyllite, ACF position, 383F, 384
  A'KF position, 389F
  D.T.A. characteristics, 456
  D.T.A. peak temperature, 450
  T.G.A. curve, 465F
Pyroxene 060 peak, 225
  131 peak, 224
  analysis calculation, 401
  composition, **406**
  composition determination, **224-225**
  diffraction data, 233
  end member calculation, 404
  end member formula, 405
  end member rules, **405-406**
  formed from amphibole, 453
  formula, 401, 406
  formula calculation, 402
  diagram, 403F
  magnetic susceptibility, 119F
  normative, 405, 424
  regression equation, 225
  X position, 401
  Y position, 401
  Z position, 401
Pyrrhotite, 233
  D.T.A. characteristics, 454
  D.T.A. peak temperature, 450
  magnetic susceptibility, 119F
  reflectivity and hardness, 41F
  X-Ray diffraction, 476
  X-Ray method, 473

Qualitative analysis by D.T.A., **457-459**
  analysis by X-Ray fluorescence, **264-286**
Quantitative analysis by D.T.A., **457-459**
  analysis by X-Ray fluorescence, **286-321**
Quartz, 378, 385
  AFM position, 392F
  :calcite:dolomite ratio, 378
  in carbonates, 218, 376
  content estimation, 452
  D.T.A. characteristics, 452
  D.T.A. peak temperature, 450
  on facies diagrams, 377F
  fused, 470
  inversion α to β, 445, 452, 457

latite differentiation index, 398, 399
magnetic susceptibility, 119F
mode, 64
normative, 393, 398, 398F, 415, 417, 419,
    421, 425
normative, negative, 421
orientation, 111
refractive indices, 252
rod, 466, 467F
separation, 123
staining, **16-20**
tailings, 130
thermoluminescence, 472
tholeiite, 393
unstained, 17
X-Ray intensity, 64F
Quartzo-feldspathic rocks, ACF position, 390
-feldspathic rocks on facies diagrams, 390F

Radiation protection shielding, 154
Radioactive impurities, 466
Radiometric dating techniques, 323
RAGIAND, P. C., 331, 477
RAHDEN, H. V. R. von, 464, 481
RAHDEN, M. J. E. von, 464, 481
RAMDOHR, P., 43, 481
RAMÍREZ-MUÑOZ, J., 333, 352, 481
Rammelsbergite reflectivity and hardness, 41F
Random samples, 431
RAP dispersing crystal, 282, 283
Ratemeter setting, 283
    value, 159
Rawlplug Co. Ltd., 487
Readout, atomic absorption, 347
Realgar reflectivity and hardness, 41F
Receiving slit, 158
    slit assembly, 150
    slit choice, **156-157**
Recorder, 466
    adjustment, 446, 461
    calibration, 463
    deflection, 159
    speed choice, 158
    zero positioning, 446
Recovery of heavy liquids, **121-123**
REDFERN, J. P., 459, 461, 481
Reference azimuth, 83
    material, 445
    material choice, 443
References, **473-484**
Reflected microscopy, 474
Reflecting efficiency, 281
Reflectivity of chromite, 436
    determination, **33-35**
    of ore minerals, 479
    standards, 34
Refractive index, 2, 3, 4, 5
    index accuracy, 259

index change with temperature, 257
Refractive index determination, 80, 95, 103,
    247-263, 474, 478, 482
    index, isotropic, 248
        by temperature variation, 257
        uniaxial, 248
        wavelength dispersion, 259
    index determination, biaxial, 249
        uniaxial, 98
    index difference, 253
    index dispersion, 258F, 260F
    index of garnet, 208
    index immersion oils, 82, 83, 247, 256, **261-
        263**, 489
    index matching, 252, 253, 254, 254F, 255,
        256, 257, 259, 260, 263
    index measurement, 252
    index Nα, 76
    index Nβ, 74, 76
    index Nγ, 76
    index temperature coefficient, 257, 258F
Refractory silicate decomposition, 334
Refractometer, 261
    Abbe, 262F
    calibration, 263
    Leitz-Jelley, 261F
    procedure, 261, 263
    stage, 263
REICHEN, L. E., 233, 473
Reichert, C., 257, 487
REID, W. P., 15, 30, 481
Regression calculation, 436
    linear, **435-437**
Relative absorption, 326
    absorption, rocks, 329
    absorption rocks/Al_2O_3, 324, 326, 328F
    cumulative frequency, 426
    equivalent radiation dosage, 471
    frequency, 426, 427
Relief, **253-254**
    reduction, 254
Residue, mica free, 353
    study, 358
Retsch, K. G., 488
REYNOLDS, R. C., 63, 323, 481
Rhodium X-Ray data, 277
Rhodizonate reagent, 17, 18
Rhodizonic acid, 27F, 28
Rhodochrosite D.T.A. characteristics, 454
    D.T.A. peak temperature, 450
    staining, 27F, 29
Rhyolite, 395
    differentiation index, 398
    relative absorption, 328F
RIBBE, P. H., 111, 234, 481
Richterite, 408F
    D.T.A. characteristics, 453
    D.T.A. peak temperature, 451

end member, 406
Riebeckite, 411
  normative, 423, 424
RILEY, J. F., 233, 481
RILEY, J. P., 363, 364, 481
Ringsdorf-Werke GmbH, 290, 291, 488
RINGWOOD, A. E., 393, 394, 476
ROBERTS, A. L., 452, 476
ROBERTSON, F., 37, 40, 43, 481
ROBINSON, P., 411, 412, 481
ROBINSON, P. D., 474
Rock analysis, 337
  composition determination, 316
  cutting equipment, 488
  crushers, 489
  crushing, **114-115, 126-128,** 294
  decomposition, 334, 336
  disc, 268
  dissolution, **333-339**
  fabric, isometric, 47F
  forming minerals, 475
  mills, 489
  mineralogy from ACF, A'KF, 391
  plotting, on ACF, 382
    on facies diagrams, 377F
    on SiO$_2$:CaO:MgO diagram, 378
  powders, 293
Rock specimen powder, 291
  slab staining, 17, 18
  specific gravity, 235
  splitting machine, 113, 114
  standards, 295, 303, 475, 476
RODGERS, K. A., 425, 481
roller crusher, 114
ROSE, H. J. Jr., 225, 475
ROSENBERG, P. E., 111, 234, 481
ROSENBLUM, S., 118, 119, 481
Rotation angle correction, 96, 99
  angle correction formula, 94
  methods, **73-112**
ROY, N. N., 79, 481
ROYSE, C. F., 215, 216, 481
Rubber bag, 3
RUBÉSKA, I., 333, 352, 481
Rubidium acid phthalate data, 282
  chloride, 323
  determination, **323**
  determination by X.R.F., 474
  mass absorption coefficient, 289
  spectrometer settings, 323
  standard, 323
  X-Ray data, 276
RUPERTO, V. L., 19, 481
Rutile decomposition, 334
  reflectivity and hardness, 43
  magnetic susceptibility, 119F
  normative, 415, 417, 419, 421
RUTSTEIN, M. S., 233, 481

SAHAMA, Th. G., 220, 233, 481, 482, 484
Salic group, normative, 397, 415, 419
Sample, 427
  cup, 441F
    aluminum, 269
  dishes, 440, 441F
  divider, 128
  holders, 439, 448
  holder body material, 440
  holders for D.T.A., 441F
  holder loading, 443, 444
  measurements, 467
  packing, 444
  preparation, 182
    powder camera, **132-134**
    X-R-F, 483
  pre-treatment D.T.A., 442
  size, 430, 431
  splitters, 489
  splitting, **128,** 294
  weighing, 366
  weight, 443
  without preferred orientation, 474
Sampling, **126-127**
Sanidine, 180, 181, 185, 196F
  composition, 186
  composition graph, 187F
  determination, 186
  diffraction indexing, 192-193
  diffraction lines, 184
  diffractogram, 181F
  high, 195
  low, 195
  optic axial angle, 111F
  staining, 20
Saponite D.T.A. characteristics, 456
Saw, 1
  diamond, 8
Sawing, 1-2, **5-8**
  automatic, 7
Scanning rate, 158
Scandium spiking, 327
  wavelength, 324
  X-Ray data, 276
Scapolite, 474
  diffraction data, 233
  fluorescence, 31
Scheelite, 32
  fluorescence, 31
  magnetic susceptibility, 119F
  reflectivity and hardness, 41F
SCHENK, R., 68, 481
SCHIAFFINO, L., 232, 476
SCHMINCKE, H-U., x, 480
SCHOEN, R., 125, 126, 232, 481
SCHOUTEN, C., 43, 481
SCHREYER, W., 223, 478
SCHRYVER, K., 56, 481

SCHULZ, H., 234, 477
Scintillation counter detector, 144, 161, 271, 272, 275, 297, 300
  counter adjustment, 272
  counter dead time, 312
  counter operating voltage, 298
  counter plateau, 161, 162F, 299F
Screw axes, 179
Second-order transitions, 459
  -order X-Ray lines, 285
Secondary glare, 35
Sedimentary petrology, 480
  rocks on facies diagrams, 390, 390F
Selenite fluorescence, 32
Selenium spiking, 327
  X-Ray data, 276
Sensitivity, spectrometer, 304
Separation by density, 120-126
  by flotation, 129-131
  by hand, 113-114
  magnetic, 116-120
  of minerals, 113-131
Sepiolite, 227
Sepiolite D.T.A. characteristics, 457
  D.T.A. peak temperature, 449, 450
  heated, 228
  properties, 231
Serpentine diffraction data, 233
  D.T.A. characteristics, 457
  D.T.A. peak temperature, 450, 451
  fluorescence, 32
  minerals, 483
  T.G.A. curve, 465F
SHA dispersing crystal, 282
SHAND, S. J., 24, 481
SHAW, D. M., 474
Shell Chemical Co., 2, 4, 488
SHORT, M. A., 233, 481
SHORT, M. N., 481
Side tilt, 118
Siderite D.T.A. characteristics, 454
  D.T.A. peak temperature, 450
  magnetic susceptibility, 119F
  staining, 27F, 28, 29
Siegenite D.T.A. characteristics, 454
  D.T.A. peak temperature, 450
  reflectivity and hardness, 41F
Sieves, 489
  mesh size, 115
Sieving, 114, 247
  bank, 115
Silica:CaO:MgO diagram, 376, 377, 377F
  content amphiboles, 407
Silica deficiency, 379
  deficiency, normative, 417
  deficiency reduction, 417
  determination, 338
  glass, 323

mass absorption coefficient, 327
  minerals, D.T.A., 475
  minerals D.T.A. characteristics, 452
  saturation, 391
  standard, 368
Silica undersaturation, normative, 422
  compositions by gel, 476
Silicate rock dissolution, 474
Silicon analysis settings, 350
  detection, 274
  detector settings, 298F
  diffractogram, 160F, 167
  loss, 335
  mass absorption coefficient, 286, 288
  matrix correction coefficient, 315
  powder standard, 166, 185, 197, 217
  standard solution, 348
  value, amphibole, 409
  X-Ray data, 276
  X-Ray detector settings, 305
Sillimanite, ACF position, 383F, 384
  AFM position, 392F
  A'KF position, 389F
  decomposition, 334
  diffraction, 231
  magnetic susceptibility, 119F
  zone, 413F
  zone garnet composition, 414
Silt removal from clays, 226, 229
Silver hardness and reflectivity, 41F
  nitrate solution, 362
  standard, 168
  X-Ray data, 277
SINE, N. M., 287, 481
Sink-float density capsules, 243
SINKANKAS, J., 3, 481
Skaergaard, 398
  trend, 403F, 400F
SKINNER, B. J., 208, 209, 210, 211, 212, 233, 481
SKINNER, H. C. W., 232, 482
Skutterudite D.T.A. characteristics, 455
  D.T.A. peak temperature, 450
  reflectivity and hardness, 41F
SIADE, P. G., 478
SLEMMONS, D. B., 73, 104, 482
Slide holder, 9, 10F, 12
Slit shape, 134
  width, 346
  width control, 339
Slope ratio, 458, 458F
SMITH, C. H., 221, 234, 477, 479
SMITH, C. W., 129, 480
SMITH, D. K., 183, 232, 474
SMITH, J. R., 197, 482
SMITH, J. V., 221, 233, 482
SMITH, O. C., 30, 31, 482
SMITHSON, S. B., 54, 482

Smithsonite staining, 27F, 29
SNETSINGER, K. G., 330, 482
Sodalite fluorescence, 32
  staining, 24
Soda asbestos, 364
  lime, 362
  tremolite, 409, 411
Sodium, ACF position, 379
  analysis, 294, 295, 296, 335
  analysis procedure, 302
  analysis settings, 350
  carbonate, 334, 369, 371, 372
  carbonate fusion, 334
  carbonate, normative, 415, 416
  cobaltinitrite, 17, 18, 19, 20, 23
  cobaltinitrite preparation, 21
  cobaltinitrite stain, 181
  chloride standard, 234
  detection, 274, 282, 283
  detector conditions, 298F
  diphenylaminesulfonate, 358, 359
  dithionite, 227
  fluoride, 372
  fluoride solution, 369
  hydroxide, 28, 226, 352, 371
  hydroxide fusion, 338
  lamp, 262
  mass absorption coefficient, 288
  matrix correction, 315
Sodium metasilicate, normative, 415, 416, 419,
    421
  minerals, A'KF position, 390
  nitrate, 291
  nitrite, 356
  oxide mass absorption coefficient, 327
  pyrosulfate, 352
  spectrometer settings, **302**
  standard solution, 348
  sulfate, 352
  sulfate decahydrate, 445
  sulfate melting, 445
  value, amphibole, 409
  vapor lamp, 255
  X-Ray data, 276
  X-Ray detector settings, 304
  X-Ray peak, 302
  X-Ray peak background, 309F
Soiltest Inc., 128, 488
Solid phase reactions, 459
Solidification index, **395**
Solubilities in water, 120
Solution aspiration, 347
  dilution, 334
  storage, 337
Sorbitol hexa acetate data, 282
Southwestern Analytical Chemicals, 293, 488
Spatula, nickel, 338
  weighing, 444

Special X-Ray techniques, 330-332
Specific gravity accuracy, 246
  gravity, Archimedes' Principle, **236-241**
    garnets, 208
    heavy liquids, 120, 243-246
    heavy liquid recovery, **121-123**
    minerals on ACF, 379
    powder specimen, 241, 243
    pycnometer method, **241-243**
    pycnometer and heavy liquid method, 244-
      246
  gravity balance, 488
  gravity calculation, 238, 240, 241, 243, 246
  gravity determination, **235-246**, 473, 474,
    475, 482
  gravity errors, 235
  gravity glasses, 121
  gravity liquids, 120, 236, 238
  gravity liquid adjustment, 120, 244, 245
  gravity liquid diluent, 120, 121
  gravity markers, 243
  gravity procedure, 245
Specific heat of sample, 443
Specimen centering, 138
  centering device, 137F
  centering device bracket, 137
  discs, 319
  drying, 442
  holders, 5, 5F, 6F, 264, 265, 268
    acid resistant, 265
    aluminum, 265
    equivalence, 303
    Philips, 295
  labeling, 268
  mounting, **3-5**, 158
  mounting in camera, 137F
  preparation, 145F, 317, **333-339**
  preparation for diffractometer, **144-147**
  preparation pressing, 267F
  preparation tools, 266F
  preparation, X-Ray fluorescence, 286, 290
Specimen pre-treatment, 463
  rotation motor, 306
Spectrogram, 284
  X-Ray, 285F
Spectrometer adjustment, 303
  alignment, **269-271**
  alignment procedure, 271, 272
  angular limits, **281**, 283
  attachments, 318F, 319
  maximum 2θ limit, 280
  minimum 2θ limit, 280
  operation conditions, 283
  Philips PW 1540, 280, 283
  settings, 303, 304, 326
  specimen positions, 303
  X-Ray fluorescence, 264, 270F
Spectrometry standards, 287

X-Ray fluorescence, **264-332**
Spectrophotometers, 488
Spectroscopy scan procedure, **282-286**
Spessartite, ACF position, 383F, 384
  AFM position, 393
  A'KF position, 389F
  physical properties, 208
Sphalerite, 481
  composition determination, 321
  diffraction data, 233
  D.T.A., 478
  D.T.A. characteristics, 455
  D.T.A. peak temperature, 450
  iron content, 455
  magnetic susceptibility, 119F
  reflectivity and hardness, 41F
  unit cell spacing, 455
Sphene accessory, 379
  magnetic susceptibility, 119F
  normative, 415, 417, 421, 423, 424
  specific gravity, 379
Sphere diameter, 251
Spiking method, carbonates, 218
  with trace element, 324, 327
Spindle attachment, 80
  stage, 78-90, 477, 481, 483, 489
    2V determination, 480
  stage axis, 89
  stage construction, 78-79
  stage design, 78F, 79F
Spinel, 361
  composition, 234
  chrome, 479
  diffraction data, 233
  decomposition, 334
  hardness and reflectivity, 43
  magnesian, normative, 396
  magnetic susceptibility, 119F
  normative, 423, 425
  X-Ray method, 473
Spray chamber, 340
SPRY, A., x, 482
Stage micrometer, 44, 45F, 65, 68, 72
  refractometer, 263
Stain removal, 182
Staining, 113
  for carbonates, 475
  methods, 477
  of minerals, 481
  rack, 15, 16F
  techniques, **15-30**
Stainless steel sample holders, 440
Standard for 2θ, **166-168**
  for borate glass, 313
  deviation, 428, 429, 430F
  deviation example, 428
  deviation of sample, 428
  error of estimate, 437

fluoride solution, 372
  internal, 168
  iron solution, 357
  mineral, 303
  rocks, 303, 476
  solutions, 347, 348, 349
  solution making, 348
Standard solution, silicates, 349
  solution, whole rocks, 352
  X-Ray analysis, 287
Stannite reflectivity and hardness, 41F
Statistical data analysis, **425-437**, 478
  sample, 428
Statistics, 480
Staurolite, 384
  ACF position, 383F, 384
  AFM position, 392F
  A'KF position, 389F
  -almandine subfacies, 382, 383F
  decomposition, 334
  magnetic susceptibility, 119F
  water determination, 361
Steam generator, 370F, 371
STEGUN, I. A., 320, 473
Step-scanning control, 144
Stereographic net, 83
  net measurement, 84F, 86F, 89F
  net plotting, 84, 84F, 85, 86F, 87F, 89F, 94,
      97, 99, 102
    biaxial, 100F
    mica, 112F
    optic axes, 97F
  projection, optic axes, 95F
STEVENS, R. E., 18, 233, 322, 326, 473, 476,
    481, 482
STEWARD, E. G., 233, 481
STEWART, D. B., 189, 197, 484
Stibnite D.T.A. characteristics, 455
  D.T.A. peak temperature, 450
  reflectivity and hardness, 41F
Stillwater Complex, Montana, 225
Stilpnomelane, AFM position, 392F
  A'KF position, 388, 389F
Stirrer, magnetic, 338, 358, 360
Stirring rod, platinum, 359
Stock solution, 349
Stolzite fluorescence, 32
Stone D.T.A. system apparatus, 439, 440F
Stone sample holders, 441F
  thermogravimetric analysis apparatus, 459,
      459F, 461F
Stratigraphic correlation, 472
STRAUMANIS, M. E., 133, 245, 246, 482
Straumanis film arrangement, 141, 142F
Stromeyerite reflectivity and hardness, 41F
Strontianite staining, 27F, 28
Strontium analysis settings, 351
  carbonate, 323

carbonate disc, 296
chloride, 323
determination, 323
mass absorption coefficient, 289
radiation, 319
standard, 323
standard solution, 349
X-Ray data, 277
Structural formula, pyroxene, 401
state, alkali feldspar, **110-111, 186-197, 196F,** 484
cordierite, **221-224**
perthite, 188
plagioclase, **105-108, 197-204**
Structure disorder, 458
STUART, A., 1, 73, 91, 93, 98
Student's t, 431
Stuffing box, 362
Subcalciferous amphiboles, 406
Subduction zone, 395
Sulfo-salts hardness and reflectivity, 41F
Sulfides hardness and reflectivity, 41F
Sulfide rock standards, 481
Sulfur detection, 274, 282
detector settings, 298F
loss, 293
mass absorption coefficient, 288
matrix correction coefficient, 315
X-Ray data, 276
X-Ray detector settings, 305
Sulfuric acid, 333, 335, 339, 356, 357, 358, 359, 362
Sum of squares, 429
SUNDERMAN, H. C., 252, 482
Supply companies, **485-488**
Surface diffusion, 296
SWAINE, D. J., 457, 482
SWANSON, H. E., 168, 482
SWEATMAN, T. R., xi, 318, 322, 330, 482
Swift, James, 56, 57, 488
SWINEFORD, A., 478
Sylvanite reflectivity and hardness, 41F
System international units, 490
Systematic absences, 178

T-distribution, 431, 432
T-values, 431
TABORSZKY, F., x, 482
Take-off angle, 153
Talc, 378
ACF position, 383F, 384
A'KF position, 389F
carbonates, 376
facies diagrams, 377F
D.T.A. characteristics, 456
D.T.A. peak temperature, 451
T.G.A. curve, 465F
Tantalite reflectivity and hardness, 41F

Tantalum detector conditions, 299F
X-Ray data, 277, 279
TAYLOR, R. M., 63, 140, 226, 480
TAYLOR, W. Ö., 481
Teallite D.T.A. characteristics, 455
D.T.A. peak temperature, 450
Techtron-Varian, 488
Teflon beaker, 339
decomposition bomb, 336F, 348, 349, 350, 351, 352, 354, 355F, 489
decomposition bomb design, 336
decomposition bomb leakage, 336
decomposition bomb procedure, **335-337**
electrostatic charge, 356
pouring spout, 336
Tellurbismuth reflectivity and hardness, 41F
Tellurium X-Ray data, 277, 279
Tema, N. V., 488
laboratory disc mill, 127, 265
Temperature calibration, 445
characteristic, 458F
for clays, 227
coefficient, 257
scale, 448
stability, 281
TENNANT, C. B., 215, 216, 482
Tennantite, 119F
reflectivity and hardness, 41F
Tenorite reflectivity and hardness, 41F
Teric 402, 131
Tetrabromethane, 120
Tetrachlorethylene, 243
Tetraethyl orthosilicate, 348
Tetragonal indexing, 173
Tetrahedrite magnetic susceptibility, 119F
reflectivity and hardness, 41F
Tetrahedron Al_2O_3-FeO-MgO-K_2O, 391, 392F
amphibole, 412F
Texme cloth, 12
Thallium X-Ray data, 277, 279
Thallous formate, 121
malonate, 121
Theoretical frequency distribution, 427
Thermal analysis methods, **438-471**
peak area, 458F
Thermocouples, 439, 441F
adjustment, 444
reference, 463
selection, 448, 462
type switch, 446
Thermogram drift, 443
report of conditions, 447
Thermogravimetric analysis, **458-465**
analysis apparatus, 459, 459F
analysis procedure, 460
balance, 489
curves, 465F
curves interpretation, 464

Thermoluminescence, 72, **465-472**, 479, 484
  apparatus, 466, 467F
  applications, 472
  calibration, 477
  calibration curve, 471, 471F
  dating, 477
  drainage, 470
  glow curve, 468F
  glow curve method, 467
  index, 472
Thermoluminescence, natural, 466
  photography, 475
  residual, 470
Thernardite fluorescence, 32
  normative, 415, 416
Thin film sections, 484
Thin section apparatus, 475
  section covering, **9**
  section grinding holder, 474
  section holder, 10F, 21
  section point counter, 57F
  section point counting, 51
  sections, polished, **9-13**
  section preparation, **1-14**, 476
  section removal, 14
  section staining, 15, 18, 19
  section staining rack, 16F
Thin sectioning equipment, 8F, 488
Thingmuli trend, 400F
  volcano, Ireland, 401, 474
Tholeiite, 393
Theoliitic series, 394, 394F, 395, 396F
THOMPSON, J. B. jr., 391, 392, 411, 482
Thorianite magnetic susceptibility, 119F
  reflectivity and hardness, 41F
Thorite magnetic susceptibility, 119F
Thorium X-Ray data, 277, 279
THORNTON, C. P., 397, 482
Tie lines, 378, 382, 403F, 405, 411
  lines, crossing, 385
TILL, R., 218, 476
TILLEY, C. E., 393, 394, 402, 403, 405, 406,
      409, 410, 411, 484
TILLING, R., 233, 483
Time constant, 158
Timer activation, 297
Tin analysis, 482
  analysis in ores, 322, 330
  detector conditions, 299F
  radiation, 310
  solidification, 445
Tin spectrometer settings, 301
  X-Ray data, 277, 279
Titanhornblende, 409
Titania in amphiboles, 407
Titanite, normative, 419, 421; *see also* Sphene
Titanium analysis, 335
  analysis settings, 351

counter conditions, 299F
detection, 274
detector conditions, 298F
dioxide mass absorption coefficient, 327
mass absorption coefficient, 289
matrix correction coefficient, 315
peak and background, 309F
standard solution, 348
X-Ray data, 276
X-Ray detection settings, 305
Titration, 358, 359, 360
  cell, 354
  end point, 356
  monitor, 355F
Titrator, automatic, 358
TOBI, A. C., 74, 75, 88, 482
Toluene, 236, 243
  density, 238
Tonalite, A'KF position, 390F
Tongs, crucible, 291
  crucible, Blair type, 359
  platinum-tipped, 335, 357, 360
Tools for determining mass absorption coeffi-
      cient, 319
  for disc preparation, 265, 291
Tool steel cylinder, 265
TOONG, K. S., 482
Top-pan balance, 241
Topaz diffraction data, 234
  fluorine content, 481
  magnetic susceptibility, 119F
  optic axial angle, 111
Tourmaline analysis by X-Ray diffraction, 477
Tourmaline decomposition, 334
  diffraction data, 234
  D.T.A. characteristics, 453
  D.T.A. peak temperature, 451
  magnetic susceptibility, 119F
Trace element analysis, **321-330**
  element analysis, no matrix correction, 329
  element analysis procedure, 326
  element analysis of similar rocks, 329
  element calculation, 322, 326
  elements in carbonates, 328
  element concentration procedure, 327
  element variation diagram, 400F
  elements versus major element diagram, 401
Trachyte, 395
Trachyandesite, 395
Transmission mode, 339, 346
Tray, cooling, 366, 367
Tremolite, 378, 408F, 409, 411
  ACF position, 383F, 384
  -actinolite corrections, 387
  in carbonate rocks, 376
  end members, 406
  D.T.A. characteristics, 452, 453
  D.T.A. peak temperature, 451

on facies diagrams, 377F
formula calculation, 377
mode, 385
Triangular diagrams, plotting, 374, 375F
diagrams, reading, 375, 375F
variation diagrams, 374-376, 375F
Triclinic indexing, 173
Triclinicity, alkali feldspars, **188**
index, **188**
Tridymite D.T.A. peak temperature, 449
Triethylene tetramine, 5
Trigonal indexing, 173
TROCHIM, H. D., x, 482
TRÖGER, W. E., x, 94, 247, 253, 482
Trona fluorescence, 32
Tropaeolin OO, 27F, 29
Trypan blue solution, 22, 26
Tschermakite, 408F, 409
end member, 406
Tschermak's molecule, 401, 405, 406
molecule allocation, 403
Tube weighing, 367
cap, 366
Tungsten analysis, 274
anode X-Ray tube, 274
carbide, 128
filament, 140
K-spectrum, 275
radiation, 283, 296
standard, 168
tube lines, 285F
wavelength, 284
X-Ray data, 277, 279
X-Ray lines, 274
TURNER, F. J., x, 379, 381, 382, 386, 390, 391, 482, 483
TUTTLE, O. F., 73, 110, 111, 180, 397, 482
Tweezers, platinum, 360
Twin planes, 466

Ugrandite series, 209, 214
Ullmannite reflectivity and hardness, 41F
Ultrabasic rocks on ACF, 390F
Ultra Carbon Co., 293, 294, 295, 488
Ultrasonic cleaning, 12
Ultraviolet colors, 32
fluorescence, 476, 480
lamps, 489
light, **30-32**
Ultraviolet Products Inc., 30, 488
Ultraviolet radiation, 470
Uniaxial inclined interference figures, 93F
minerals, 82, 248
mineral orientation, 250
mineral refractive indices, 248
mineral, universal stage, **93-98**
Union Carbide Corporation, 5, 488
Uni-Seal Decomposition Vessels, 336, 336F, 488

Unit cell edge, cubic, 178
cell, modal analysis, **48-49**
cell edge, garnet, 208
cell face, 48
cell selection, 48
United Kingdom Atomic Energy Authority, 488
Units of measure, 490
Univariant equilibrium, 385
Universal stage, **90-112**, 250, 473, 475, 489
stage, biaxial minerals, **98**, **103**
conoscopic method, 91, **93-103**
plagioclase determination, **104-110**
uniaxial minerals, **93-98**
stage accessories, 91
stage accessory objective, 92
stage axes adjustment, 91
stage axes nomenclature, 90, 91F
stage glass hemispheres, 92
stage height adjustment, 92
stage rotation angle correction, 94, 96, 99
stage setting-up, **91-92**
Uraninite magnetic susceptibility, 119F
reflectivity and hardness, 41F
Uranium minerals fluorescence, 31
X-Ray data, 277, 279
Uranoan monazite magnetic susceptibility, 119F
U.S. Geological Survey, 158, 287, 352
Geological Survey rock standards, 331, 475, 476
U.S. National Bureau of Standards, 358
Uvarovite physical properties, 208
UYTENBOGAART, W., 43, 482

Vacant lattice sites, 466
Vacuum, 3
path, 274
pump, 5
stability, 281
Valence electron, 465
Vanadium analysis, 330-331, 337, 482
analysis settings, 351
detection, 274
standard solution, 349
trace analysis, 331
wavelength, 324
X-Ray data, 276
Varian Techtron Inc., 333, 339, 488
Techtron atomic absorption spectrophoto-meter, 340F
Variance, 428, 429
of sample, 428
of sample, example, 428
Variation diagrams, 399-401, 400F
diagrams, triangular, **374-376**
Vegard's Law, 214, 221
Vermiculite, 227

distinctive properties, 229
D.T.A. characteristics, 456
D.T.A. peak temperature, 449, 450
heated, 228
VERNON, M. J., 474
Vesuvianite, 475
  ACF position, 383F, 384
  in carbonates, 376
  correction, 386, 387
  diffraction data, 234
  on facies diagrams, 377F
  mode, 385
  specific gravity, 385
Vickers hardness, 13, **35-37**
  indentation, 39F
Vickers Instruments, 33, 251, 255, 488
  microindentation hardness tester, 37
Vickers reflecting microscope, 33F
VILLIGER, H., 174, 475
VISWANATHAN, K., 174, 204, 205, 206, 231,
    233, 473, 482, 483
VITALIANO, C. J., 425, 483
VIVALDI, J. L., 457, 483
VOGEL, A. I., 356, 483
VOGEL, T. A., 9, 12, 483
Volatile loss, 296, 337, 459, 464
  matrix correction, 315
Volcanic rocks, Japan, 478
  series index, 395
  series diagram, 394-395
Volumetric flask, 334, 335, 337, 338, 349, 357,
    370F, 372
Vycor glass slides, 226

WADELL, J. S., 481
WAGER, L. R., 399, 403, 483
WAHLSTROM, E. E., ix, 73, 483
WAITE, J. M., 217, 483
WALDBAUM, D. R., 231, 478
WALSH, J. N., 373, 476
WARD, A. M., x
Ward's of California, 488
Ward's Natural Science Establishment, 1, 4, 7, 14,
    113, 236, 488
WARNE, S. St. J., 27, 30, 454, 473, 483
Washing, 65
  vessel, 122, 122F
Water absorption tube, 364
  in amphiboles, 407
  analysis by infrared, 473, 474
  boiling point, 445
  determination, **361-369**, 481
  determination apparatus, 363F, 365F
  determination procedure, 364, 366
  mass absorption coefficient, 327
  total, determination, **365-367**
  total, determination apparatus, 475
  tube, 365F

Wavelength, 139
  choice, 350, 351
  dispersion, 259
  K-spectrum, 276, 277
  L-spectrum, 279
  selector, 339
  X-ray tables, 284
Weathered rocks, 2
Weathering index, 414, 480
WEBB, T. L., 454, 483
WEBBER, G. R., 481
Weight loss, 368, 464
WELDAY, E. E., 295, 483
WESTBROOK, 40, 483
Whatman CF 11, 294, 331
  ashless cellulose powder, 317
WHEELER, E. P., 22, 23, 474
WHITE, A. J. R., 406, 483
WHITE, E. W., 284, 286, 483
White radiation, 140
WHITTAKER, E. J. W., 231, 233, 483
WHITTEN, E. H. T., ix, 483
Widia, 127
Wig-L-Bug mixer mill, 318, 324, 328, 330, 331
WILBURN, F. W., 452, 475
WILCOX, R. E., 78, 80, 81, 82, 83, 85, 483
WINCHELL, A. N., x, 1, 32, 247, 483
WINCHELL, H., x, 32, 75, 76, 88, 209, 211,
    233, 247, 375, 483
WILKINSON, J. F. G., 231, 483
Willemite fluorescence, 31
WILLIAMS, H., x, 46, 483
Window calculation, 300
  determination, 299
  potentiometer, 164
  setting, 164, 165F, 300F
    for strontium, 311F
WINKLER, H. G. F., 379, 382, 388, 390, 391,
    483
Witherite fluorescence, 32
  staining, 27F, 28
WITTRY, D. B., xi, 476, 479
Wolframite magnetic susceptibility, 119F
  reflectivity and hardness, 41F
Wollastonite, 401, 405
  in carbonates, 376
  in facies diagrams, 377F
  fluorescence, 32
  normative, 415, 419, 421, 422
    provisional, 417
  in skarn, 233
WONG, Y. C., 482
WOO, C. C., 120, 121, 483
WOODBURY, J. L., 9, 12, 483
WOODS, 481
WRAY, J. L., 484
WRIGHT, J. B., 414, 483
WRIGHT, T. L., 186, 189, 191, 195, 196, 197, 484

Wright eyepiece, 251
  combination quartz wedge, 251, 252
Wurtzite diffraction data, 233

Xenotime magnetic susceptibility, 119F
  refractive indices, 259, 260F
X-Ray absorption, 476
  angular tables, 473
  conversion charts, 480, 489
  detector operating conditions, 161-163
  diffraction, 482, 484
  diffraction equipment, 489
  diffraction patterns, 482
  diffraction of silicates, 481
  diffraction tables, 475
  diffractometer, alignment, 147-156
  diffractometer techniques, 144-166
  film development, 141
  film measurement, 142F
  filter, 140
  fluorescence, 373
  fluorescence analysis, 475, 477
  fluorescence analysis data sheet, 307F
  fluorescence special techniques, 330-332
  fluorescence specimen preparation, 286, 290
  fluorescence spectrometer, 72, 264, 270F, 489
  fluorescence spectrometer alignment, 271
  fluorescence spectrometer settings, 304
  fluorescence spectrometer specimen tools,
      266F
  fluorescence spectrometry, 264-332, 473, 477,
      480
  fluorescence spectrometry conversion tables,
      480
  fluorescence spectrometry data tabulation, 306
  fluorescence spectrometry scan procedure, 282-
      286
  fluorescence spectrometry theory, 264
X-Ray generator voltage, 140
  intensity, 134, 286
  lines, 143
    L-spectrum, 275
    M-spectrum, 277
  macraprobe, 477
  peak intensities, 322
  peak ratios, 361
  powder cameras, 132-144
  powder camera adjustment, 136
  powder camera loading, 137-139
  powder diffraction, 64, 132-179, 451, 473
    modal analysis, 61-64
  powder diffraction peak intensity, 62, 64
  powder diffractogram procedure, 158-161
  powder patterns, 476
  powder sample preparation, 132-134
  properties, 474
  reflection $\bar{2}01$, 181
  tube adjustment, 273

  tube choice, 274-275
  tube current, 274
  tube operating conditions, 139 141, 274
  tube rating, 140
  tube voltage, 275
  wavelengths, 139, 274, 277
    L-spectrum, 279
  wavelength conversion tables, 284
  wavelength shift, 361
  wavelength tables, 483
Xylol, 9

YODER, H. S., 197, 220, 232, 233, 393, 394,
      402, 403, 405, 406, 409, 410, 411, 482,
      484
YOUNG, B. B., 40, 484
Yttrium mass absorption coefficient, 289
  X-Ray data, 277
YUND, R. A., 233, 481

ZECK, H. P., 224, 484
ZEIDLER, W., 14, 484
Zeiss, Carl, 34, 44, 251, 255, 257, 262, 263,
      488
  eyepieces, 67
  objectives, 66
  photomicrographic camera, 69
  photomicroscope, 71
  universal stage, 90
ZELLER, E. J., 471, 484
ZEN, E - AN, 232, 484
Zeolites D.T.A. characteristics, 453
Zernike condenser, 256
Zero adjustment, 346
  calibration, 349
Zinc analysis settings, 351
  detection, 284
  mass absorption coefficient, 289
  oxide, 369
  standard solution, 349
  sulfides, 481
  wavelength, 284
  X-Ray data, 276
  X-Ray spectrogram, 285F
Zincite reflectivity and hardness, 41F
Zinkenite reflectivity and hardness, 41F
Zircon decomposition, 334, 478
  dissolution, 338
  D.T.A. characteristics, 453
  D.T.A. peak temperature, 451
  fluorescence, 32
Zircon magnetic susceptibility, 119F
  normative, 415, 416, 419, 420
  thermoluminescence, 472
Zirconium mass absorption coefficient, 289
  X-Ray data, 277
Zirconyl chloride octahydrate, 369
  fluoride, 372

Zoisite, ACF position, 383F, 384
  correction, 386
  D.T.A. characteristics, 453
  D.T.A. peak temperature, 451

mode, 385
ZUSSMAN, J., x, xi, 140, 146, 179, 214,
  233, 475, 478, 479, 480, 483,
  484